Windows Server® 2022 & PowerShell®

ALL-IN-ONE

by Sara Perrott

Windows Server® 2022 & PowerShell® All-in-One For Dummies®

Published by: **John Wiley & Sons, Inc.**, 111 River Street, Hoboken, NJ 07030-5774, www.wiley.com

For general information on our other products and services, please contact our Customer Care Department within the U.S. at 877-762-2974, outside the U.S. at 317-572-3993, or fax 317-572-4002. For technical support, please visit www.wiley.com/techsupport.

Wiley publishes in a variety of print and electronic formats and by print-on-demand. Some material included with standard print versions of this book may not be included in e-books or in print-on-demand. If this book refers to media such as a CD or DVD that is not included in the version you purchased, you may download this material at http://booksupport.wiley.com. For more information about Wiley products, visit www.wiley.com.

Library of Congress Control Number: 2022930649

ISBN 978-1-119-86782-1 (pbk); ISBN 978-1-119-86784-5 (ebk); ISBN 978-1-119-86783-8 (ebk)

SKY10033272_021122

Table of Contents

Introduction

Microsoft continues to improve on its server operating system with its release of Windows Server 2022. Although Windows Server 2022 doesn't introduce any huge sweeping changes, it offers some great new features, including advancements to security, services, and system administration.

Security features have been improved upon to offer better encryption support for Server Message Block (SMB) protocol. A new offering with Windows Server 2022 is a secured core server, which essentially allows system administrators to harden their system at the hardware/firmware level.

As more and more customers have moved to the Azure cloud, Microsoft has added new features to better support Windows Server 2022 administration in the Azure cloud, as well as on-premises for hybrid environments. One of the new features is the ability to patch without reboots (yes, it's a thing!).

About This Book

Windows Server 2022 & PowerShell All-in-One For Dummies provides something for everyone — from the junior system administrator just getting their start, to the seasoned system administrator looking to improve their skills.

I try to cover as many of the everyday topics that you would need to know as a system administrator and explain things that are outside of your daily work. My goal with this book is to help you understand not just the what and the how, but also the why.

This isn't the kind of book that you pick up and read from start to finish, and it's probably not the kind of book you'll read on the beach. Instead, this book is a reference — the kind of book you can pick up, turn to just about any page, and start reading. It's divided into eight minibooks, each covering a specific aspect of working with Windows Server 2022 or PowerShell.

You don't have to memorize anything in this book. Pick it up when you need to know something. After you find what you're looking for, put it down and get on with your life.

Within this book, you may note that some web addresses break across two lines of text. If you're reading this book in print and want to visit one of these web pages, simply key in the web address exactly as it's noted in the text, pretending as though the line break doesn't exist. If you're reading this as an e-book, you've got it easy — just click the web address to be taken directly to the web page.

Foolish Assumptions

I had to make some assumptions about you as I wrote this book:

- I assume that you want to know more about Windows Server 2022 and PowerShell and you've worked with some version of Windows Server in the past.
- I assume that you're a system administrator, and that you have the permissions to do the things mentioned in this book. Some of the procedures require you to have administrator access.

Icons Used in This Book

As you read through the book, you'll see icons in the margin. I use those icons to grab your attention. Here's what each of these icons mean:

TIP

Anything marked with the Tip icon will save you time or frustration or just generally make your life easier — at least your system administrator life (I can't do anything about your relationship with your parents).

WARNING

If you see a Warning icon, take heed! Anything marked with this icon could be destructive or at the very least give you a major headache.

TECHNICAL STUFF

When you see the Technical Stuff icon, this is usually where I go full nerd and add some more in-depth technical information. If you want to let your inner geek flag fly, read these with gusto! But if you're in a hurry and just want to get the information you absolutely need, you can pass these by.

If something is really important — important enough for you to commit it to memory — I mark it with the Remember icon.

Beyond the Book

In addition to what you're reading right now, this product also comes with a free access-anywhere Cheat Sheet that includes helpful tips and tricks to navigate and administer Windows Server 2022. To get this Cheat Sheet, simply go to `www.dummies.com` and type **Windows Server 2022 & PowerShell All-in-One For Dummies Cheat Sheet** in the Search box.

Where to Go from Here

I'm a traditionalist, so I recommend starting with Book 1, Chapter 1. This is where you find out about the new things that await you in Windows Server 2022. From there, it's entirely up to you! You can read the book in order, or skip around, letting your curiosity be your guide.

One last note: I highly recommend that you create a test environment as you go through this book and experiment with different components of the Windows Server operating system. I try to call attention to potentially destructive procedures, but it's your responsibility to ensure that you're practicing in a safe environment, ideally not your production environment.

1

Installing and Setting Up Windows Server 2022

Contents at a Glance

IN THIS CHAPTER

» Getting an overview of the features new to Windows Server 2022

» Making sense of the Windows Server 2022 editions

» Looking at the different Windows Server 2022 user experiences

» Recognizing the benefits of Server Manager

» Working with the Windows Admin Center

» Making your data center bigger and better

Chapter 1
An Overview of Windows Server 2022

Windows Server 2022 is the latest version of Microsoft's flagship server operating system. This chapter has something for everyone. If you're already familiar with Windows Server, I discuss the new features that Windows Server 2022 brings to the table. If you haven't worked with Microsoft Server operating systems much before, you'll appreciate the information on the editions and user experiences that you can use, depending on your needs.

Extra! Extra! Read All About It! Seeing What's New in Windows Server 2022

With each new version of Windows Server, Microsoft introduces new and innovative technologies to improve administration, add needed functionality, and improve security. Here are some of the new features in Windows Server 2022:

- **Secured-core server:** These systems have special hardware that enables them to use advanced security features. Trusted Platform Module (TPM) 2.0 is a standard feature, for example, which can be used for a variety of things, including secure key storage and improved boot time protection from BitLocker. Microsoft also provides increased protections against firmware-level attacks and virtualization-based security (VBS).
- **Improvements to Domain Name System (DNS) security:** DNS requests can now be made via Hypertext Transfer Protocol Secure (HTTPS), using an encrypted channel. This new feature is referred to as *DNS-over-HTTPS,* which is shortened to DoH. If you're like me, you probably pictured Homer Simpson yelling "D'oh!"
- **Transport Layer Security (TLS):** In Windows Server 2022, both HTTPS and TLS 1.3 are enabled by default to better protect your network and Internet activity. For more information on this feature, check out Book 5, Chapter 3.
- **Server Message Block (SMB) security:** Windows Server 2022 brings several improvements for SMB security. SMB can now be used over the QUIC protocol instead of Transmission Control Protocol (TCP). This allows you to take advantage of TLS 1.3 encryption. SMB Direct now supports encryption with little to no performance impact. Previously, if you enabled SMB encryption, direct data placement was disabled due to the impact to performance. Plus, traffic between storage clusters in Storage Spaces Direct can now be encrypted.
- **Azure hybrid capabilities:** In Windows Server 2022, you have Azure Arc, which provides centralized management of servers, and Azure Automanage: Hotpatch, which allows for rebootless updates. I discuss these features a bit more toward the end of this chapter.
- **Windows Admin Center:** Improvements have been made to Windows Admin Center to add support for the new secured-core server features.
- **Improved support for Windows Containers:** Container image sizes have been greatly reduced, and increased support for Kubernetes has been added. I discuss these subjects in greater detail in Book 8.
- **Network performance:** TCP and User Datagram Protocol (UDP) performance have both been improved in Windows Server 2022.

- **Microsoft Edge browser:** Microsoft Edge is now the browser of choice on Microsoft products. It has replaced Internet Explorer as the default browser in Windows Server 2022.
- **Storage improvements:** Migrating data has been made simpler with improvements to the Storage Migration Service. Storage Spaces Direct has also gotten some love with two big improvements. The first new feature, user adjustable storage repair speed, allows users to specify how many resources should be allocated to repairing data or servicing active storage needs. The second new feature is the storage bus cache being available on non-clustered systems, which allows you to create tiered storage on a stand-alone server.

Deciding Which Windows Server 2022 Edition Is Right for You

Windows Server 2022 comes in three editions: Essentials, Standard, and Datacenter. In the following sections, I walk you through each edition so you can determine which one is right for you.

Essentials

The Essentials edition is ideal for small organizations (usually no more than 25 to 50 users). It provides enough basic functionality to do most jobs and is a cost-effective solution for small organizations. Features of the Essentials edition include the following:

- Supports up to two CPU cores
- Supports a maximum of 64GB of random access memory (RAM)

Standard

The Standard edition is ideal for environments with little to no virtualization or when used as a guest operating system. Features of the Standard edition include the following:

- Up to two Hyper-V containers and unlimited Windows containers
- HGS and Nano Server support
- Storage Replica (with some limitations)

Datacenter

The Datacenter edition has the same features as the Standard edition and some additional features that make it the ideal edition for organizations with a lot of virtualization needs, the desire to do software-defined networking, or that need advanced storage options. Some of these features include the following:

- » Unlimited Hyper-V containers in addition to unlimited Windows containers
- » Unlimited Hyper-V virtual machines and support for shielded virtual machines
- » Storage Replica (unlimited) and Storage Spaces Direct
- » Software-defined networking
- » Network controller
- » Host Guardian Hyper-V support

Note: There is a specialized version of Windows Server Datacenter referred to as the Azure Edition. Windows Server 2022 Datacenter: Azure Edition provides greater integration with the Microsoft Azure cloud. You can only get this version through Microsoft Azure by installing it as a virtual machine in Azure. You can't install it on your own on-premises systems or run it on your own hypervisors. New features include the following:

- » Azure Extended Network
- » Hotpatching
- » SMB over QUIC
- » Shielded VM support

Note: You won't see Datacenter: Azure Edition called out in this book specifically. However, many of the topics I cover in this book can be applied to Datacenter: Azure Edition.

Walking the Walk: Windows Server 2022 User Experiences

Windows Server 2022 has two user experiences to choose from. What you use will depend on the workload you're wanting to support, as well as organizational requirements. In this section, I explain the Desktop Experience and the Server Core experience, as well as some pros and cons of each.

Desktop Experience

Desktop Experience is what you would consider to be the standard graphical user interface (GUI) that you may have used in previous versions of the Windows Server operating systems. It allows you to interact with the system with buttons and menus rather than through the command line. Server with Desktop Experience can be managed through Group Policy if attached to an Active Directory domain, and workgroup (non-domain) servers can be managed via local Group Policy.

TIP

Desktop Experience tends to be the easier form of server installation and administration for beginning system administrators, but I highly recommend that you don't rely on the GUI (shown in Figure 1-1). Become a PowerShell ninja instead! PowerShell is a very versatile language and can be used on a variety of systems, including some of the newer versions of Linux.

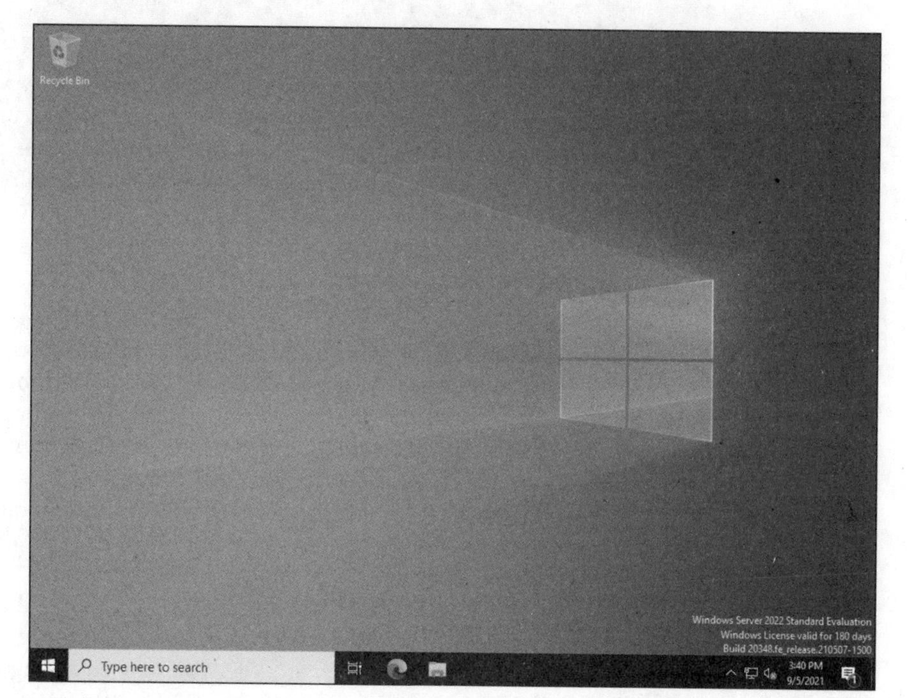

FIGURE 1-1: Server with Desktop Experience.

Server Core

Server Core (shown in Figure 1-2) provides a much simpler interface if you connect to the console. You're greeted by a somewhat familiar-looking command window that prompts you for your username and password. After you've logged in, by default you're presented with the sconfig window. When you choose to exit to command line from sconfig, you're given a PowerShell window to interact with. Initial configuration is done with the sconfig utility, though it could be done through a PowerShell script or PowerShell Desired State Configuration (DSC). This experience can be managed through Group Policy if attached to an Active Directory domain or through local Group Policy if they're workstation servers.

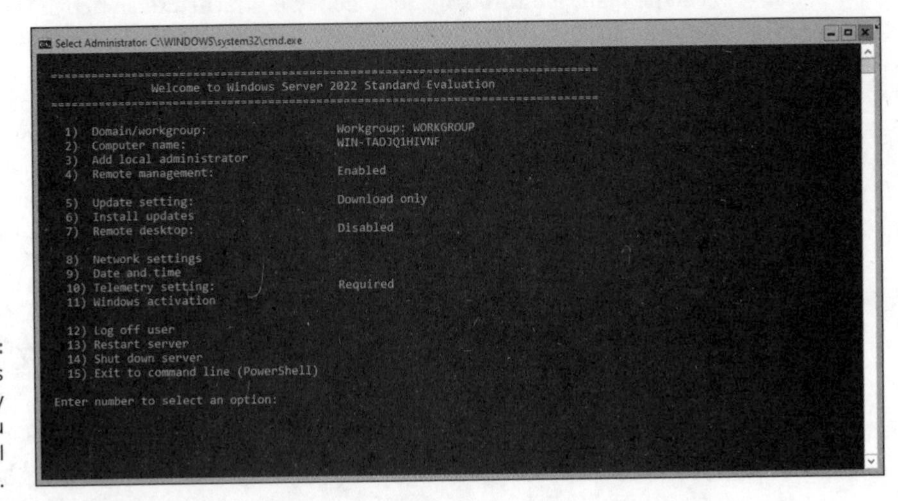

FIGURE 1-2: Server Core's sconfig utility is where you perform initial configuration.

Nano

Nano provides an even simpler interface and a much more limited console, which is referred to as the Recovery Console. It isn't available through the regular installer on the disc; instead, you have to download the container image from Microsoft. Nano has a much smaller footprint, both in terms of disk and compute needs, than Desktop Experience or Server Core. Because it has a smaller overall footprint, the attack surface is also reduced. Windows Server Nano 2022 is available only as a container base operating system image and can only be run as a container on a container host.

Nano can't be managed through Group Policy. You need to use PowerShell DSC instead if you want to manage Nano at scale. You may be asking why you would even use Nano when it's such a limited version of the operating system. If you need to run container workloads that use .NET, Nano is an excellent candidate because it has been optimized to run .NET Core applications.

If you want to check it out, you can download the Nano server images from Microsoft's container registry on DockerHub with this command:

```
docker pull mcr.microsoft.com/windows/nanoserver
```

You can also go to https://hub.docker.com/_/microsoft-windows-nanoserver to see more information about the current Nano server container image. This includes a description of the image, reviews, and additional resources.

If you want to find out more about containers and using Docker commands, check out Book 8.

Note: You won't really see Nano discussed in depth anywhere in this book because you're far more likely to encounter the Desktop Experience or Server Core installations of Windows Server 2022.

Seeing What Server Manager Has to Offer

When you first install Windows Server 2022 and you log in, the first screen that you're greeted with is Server Manager (see Figure 1-3). This screen gives you a central area to do all the configuration tasks you need to do on your server. It presents a handy menu to manage all the roles and features installed on your server as well.

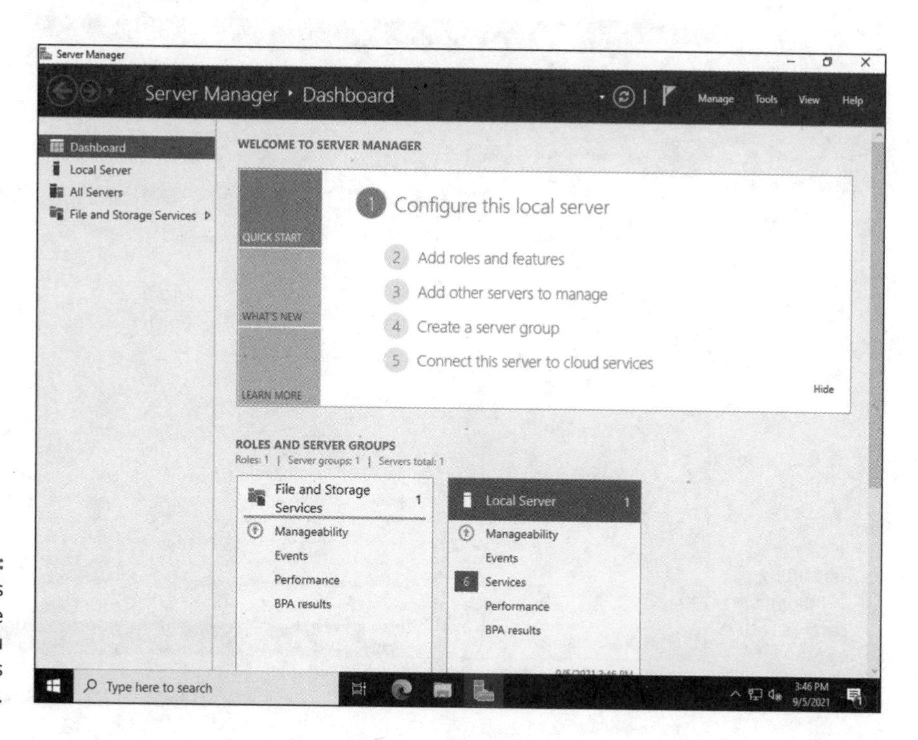

FIGURE 1-3: Server Manager is the landing page you get when you log in to Windows Server 2022.

Server Manager will allow you to manage remote servers, not just the local server. The remote servers need to be added to Server Manager before they can be managed, and some firewall ports may need to be opened to allow full functionality. After remote servers are added, you can run PowerShell against them and perform basic management tasks like shutting down, connecting via Remote Desktop Protocol (RDP), and so on. You can manage up to 100 remote servers with Server Manager. This number may be lower depending on what you're running on the manage servers. If you're running large workloads, then you may not be able to manage as many.

REMEMBER

Server Manager can be used to manage the same operating system it's installed on, as well as operating systems that are older than what is installed. It can't manage the operating system on a server that's running a newer version of the operating system. For example, a server running Server Manager on Server 2019 can't manage a server running Windows Server 2022.

Figure 1-4 shows some of the options available through the Server Manager menu. You may notice that Remote Desktop Connection is grayed out. This is because I was logged on the server that is in the window.

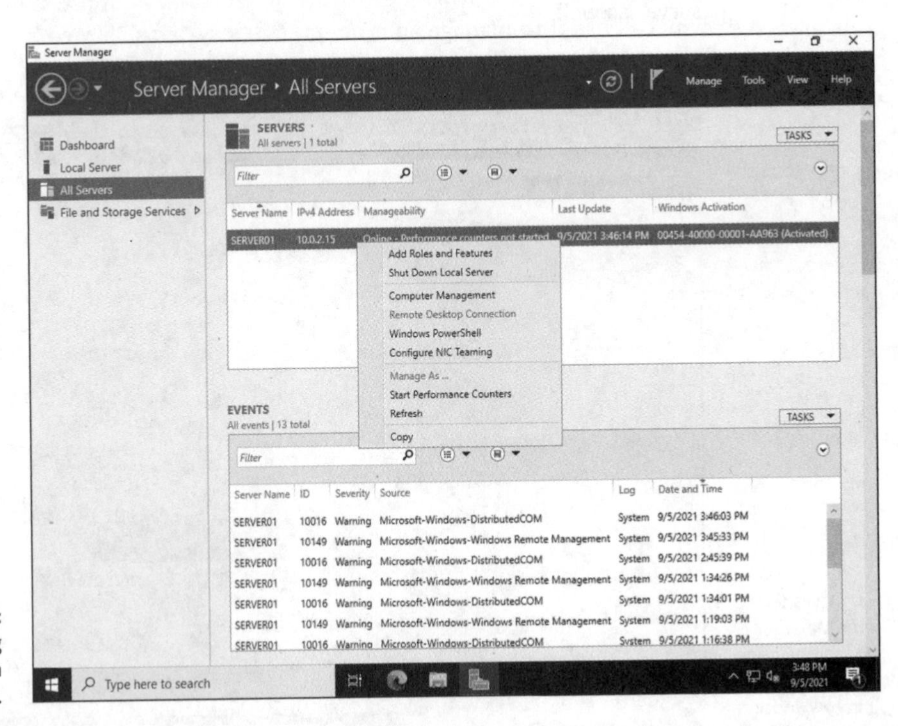

FIGURE 1-4: Managing servers with Server Manager.

Here's a list of some of the more commonly used features of Server Manager:

- » Managing local and remote servers
- » Managing roles and features on servers (To install or remove roles and features, the target system must be running at least Server 2012)
- » Starting management tools like Windows PowerShell and MMC snap-ins
- » Reviewing events, performance data, and results from the Best Practices Analyzer

Windows Admin Center: Your New Best Friend

Windows Admin Center is a newer server management tool from Microsoft. Microsoft has been investing heavily in Windows Admin Center, and it shows. You can use it to manage your on-premises systems, as well as your systems in Azure. Windows Admin Center is accessible through your browser and allows you to perform nearly all your administrative tasks through the same interface. Best of all, it's free! You just need to pay for the license of the operating system it's running on.

Admin Center can be used to administer Windows Server 2022, 2019, 2016, 2012R2, and 2012 with full support for all functionality.

By default, Windows Admin Center uses TCP port 6516, so you need to allow this through your server firewalls depending on how your network is architected. To access the Windows Admin Center Dashboard, you need the hostname of the system that Admin Center is installed on. In Figure 1-5, notice that the address is localhost:6516. That's because I've installed it on a Windows 10 client in Desktop mode. Desktop mode is typically used by a single system administrator, as opposed to Gateway mode, which is available for a larger number of staff.

The first screen (refer to Figure 1-5) shows your connected devices.

If you click one of the devices in the list, you get a management view specific to that device. For Figure 1-6, I clicked on server2022-dc. You see an overview of the system as well as some management options. On the left side of the screen, there are many more options you can work from.

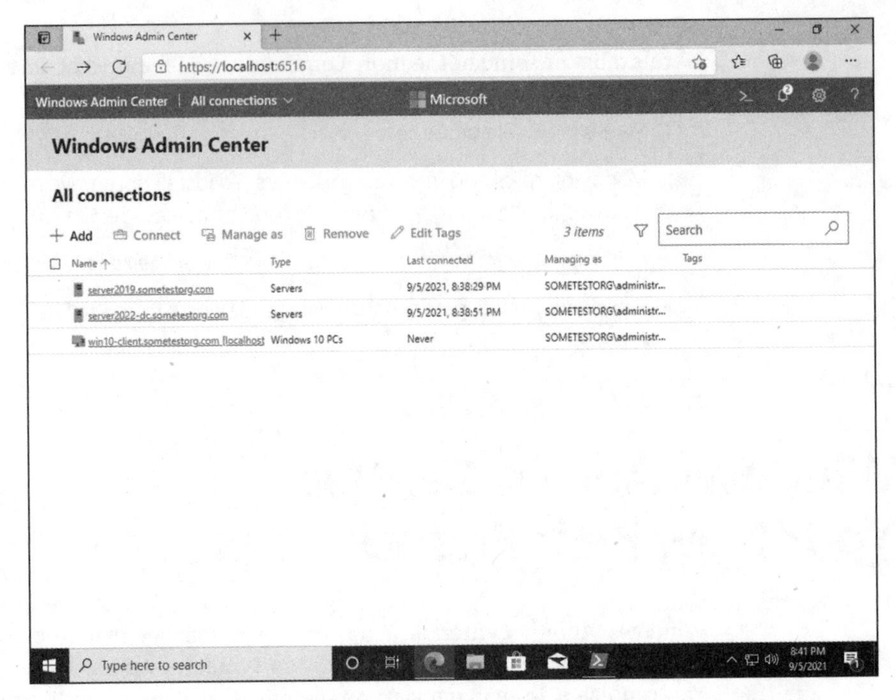

FIGURE 1-5: You can see all your connected devices on the All Connections page.

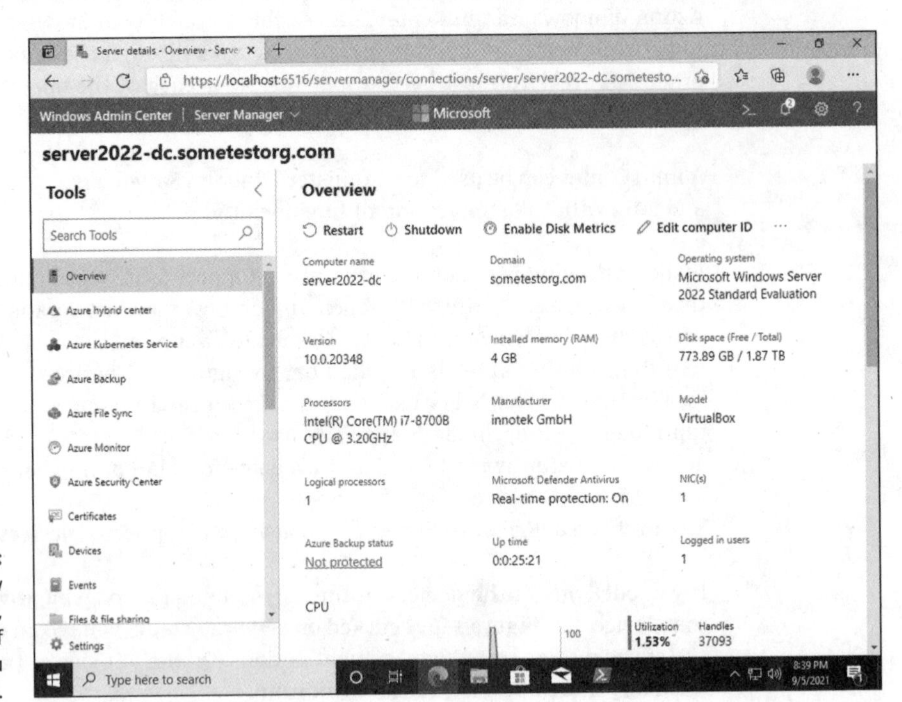

FIGURE 1-6: The Overview page shows, well, an overview of the device you clicked.

Installation of Windows Admin Center is simple. You download the Microsoft Installer (MSI) package from the Microsoft Windows Admin Center website (`www.microsoft.com/en-us/cloud-platform/windows-admin-center`). Before you install it you need to decide if you're simply going to install it on your desktop client or if you want to install it on a server. My recommendation would be to use your desktop if you're just trying it out or if you manage only a few servers. If you're going to use Windows Admin Center in all its glory, install it on a server so that all your administrators can get to it. You'll be their hero!

You can install Windows Admin Center on a supported version of Windows 10 or on Windows Server 2016 and newer. To manage older servers — including 2012 and 2012 R2 — you need to install Windows Management Framework 5.1 on each of those servers.

When you install Windows Admin Center on Windows 10, it's installed in Desktop mode, which means that you access it using `https://localhost:6516`. When Windows Admin Center is installed on a server, it installs in gateway mode, which can be accessed with the server name in the URL (for example, `https://servername`). No port number necessary!

TIP

Localhost refers to the local loopback address on a system which can also be accessed at the IP address 127.0.0.1.

TECHNICAL STUFF

Installing Windows Admin Center onto a domain controller is not supported. As you might imagine, this would be a terrible idea! Because Windows Admin Center exposes its services via a web page, it provides a point of attack that would not normally be there.

Some of the coolest features of Windows Admin Center include the following:

- Centralized server management
- Integration with Azure so you can manage on-premises and cloud resources from the same console
- Cluster management tools built into Windows Admin Center
- Showscript, which allows you to see the PowerShell scripts that are being run to do your administrative work

REMEMBER

The only browsers currently supported are Microsoft Edge and Google Chrome. Firefox and Internet Explorer have not been tested and are not officially supported.

Extending and Improving Your Datacenter

Windows Server 2022 allows you to take advantage of some very powerful features in the Azure cloud. Some of the functionality requires that you run Server 2022 Datacenter: Azure Edition; others are not so strict. I won't go into a ton of depth on these features because they're a better topic for an Azure book, but you should know what they are and what they're capable of.

Azure Arc

Azure Arc is a newer service that allows you to manage both Azure and on-premises assets with the Azure tool set. Windows Server 2022 is one of several operating systems that can be managed by Azure Arc. You need only install the Azure Connected Machine agent.

Azure Automanage: Hotpatch

Azure Automanage: Hotpatch is in preview at the time of this writing; you can preview it on Windows Server 2022 Datacenter: Azure Edition. It works by establishing a baseline with the latest cumulative update (CU) that was published. From there, each month hot patches are released that can be installed and require no reboot. Let me repeat that: NO REBOOT! It's pretty much the dream of a system administrator to not have to reboot for patching. When the baseline is updated with a new CU, which happens approximately every three months, *then* you'll have to reboot.

» Using a special boot mode to bypass server issues

» Performing a memory test on your server

» Using the Command Prompt to troubleshoot

» Troubleshooting with third-party boot utilities

Chapter 2
Using Boot Diagnostics

As a system administrator, you'll get the inevitable call one day about a server that just won't start. Maybe the server is in a continuous boot loop. Maybe the server just hangs. Your mission, should you choose to accept it, is to figure out why the system is having issues starting and then fix the issue.

This chapter discusses basic tools and techniques to troubleshoot issues that are causing your system to not be able to boot properly.

Accessing Boot Diagnostics

The first step to figuring out what's going wrong with your system is to access the boot diagnostic utilities that ship with Windows Server operating systems.

From the DVD

If the server that is having boot issues is a physical server, you can use a DVD or a USB flash drive to access the boot diagnostics menu. It's very rare to have physical media on hand anymore, so, chances are, you'll need to download the ISO file for Windows Server 2022 from the Microsoft website and burn the image to the DVD or USB flash drive.

An ISO file is a duplicate of what's on a physical disc.

After you have the disc ready to go, you need to insert the disc or the USB flash drive into the server and boot from it. You may need to change the boot order on the server so that the boot order will start with the DVD drive or the USB flash drive before the hard drive. You can make this change by accessing the basic input/output system (BIOS). On server systems, this option is available when the system is booting. The key you have to press to access the BIOS will depend on the firmware manufacturer that created the BIOS or Unified Extensible Firmware Interface (UEFI). Some systems simply offer you a boot menu when you press F12, which allows you to select the DVD drive or USB flash drive for a one-time boot.

When you've figured out how to boot from the DVD or USB flash drive, follow these steps:

1. **Boot from the DVD or USB flash drive.**
2. **When you see the message** `Press any key to boot from CD or DVD,` **press any key.**

   The installation wizard for Windows Server 2022 runs.
3. **On the first screen, click Next.**

   This screen is just asking for language, time and currency format, and keyboard or input method. You can safely accept the defaults.
4. **On the next screen, you see the big Install Now button. Don't click that! Instead, look in the lower-left corner for the Repair Your Computer link (see Figure 2-1), and click that.**
5. **On the next screen, click Troubleshoot.**

   This gives you your available options (see Figure 2-2):

   - **Command Prompt:** Allows you to do advanced troubleshooting and is especially helpful if you need to repair boot files. You can use the diskpart utility to work with the drive, and the bootrec command to rebuild or repair the boot files.
   - **System Image Recovery:** Allows you to restore your system from an image created by a backup utility. You'll be asked to choose a target operating system to restore, and then you'll be shown available backups you can use.

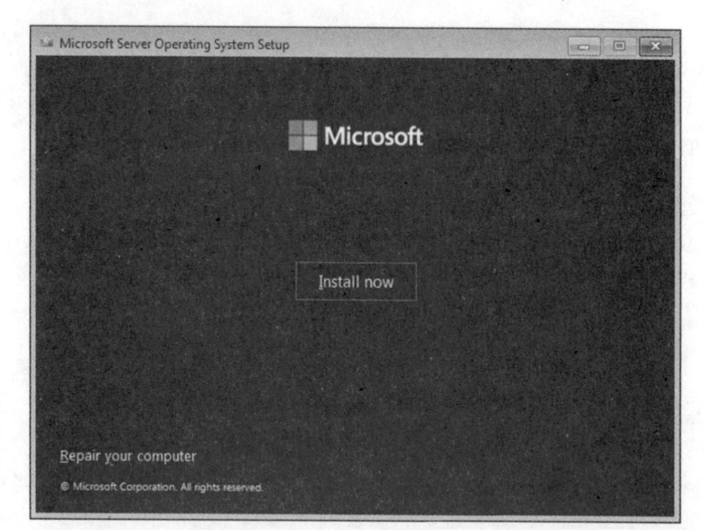

FIGURE 2-1: Look for the Repair Your Computer link in the lower-left corner.

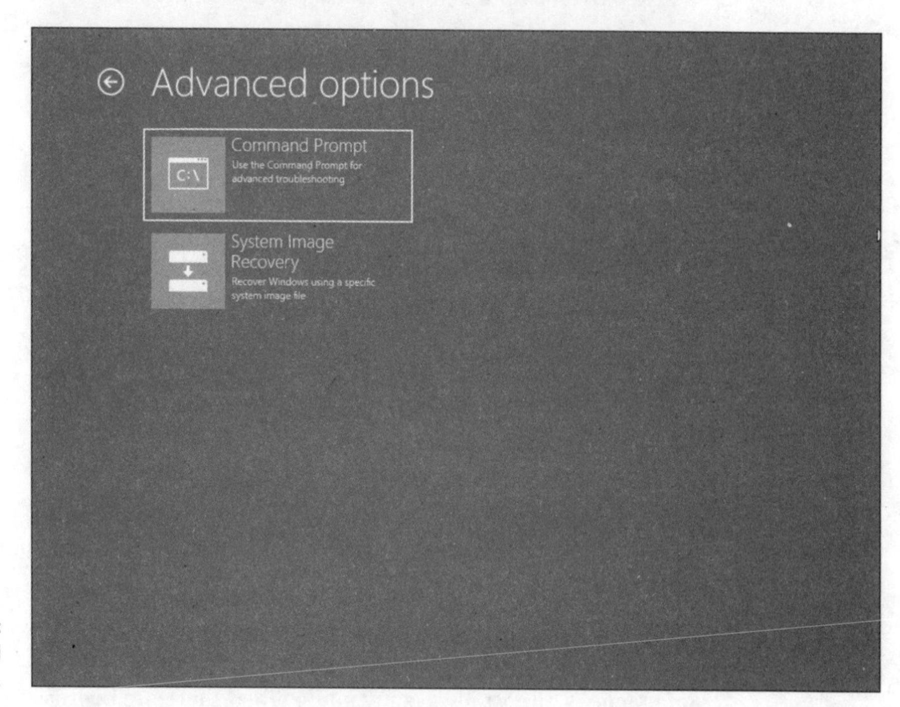

FIGURE 2-2: The Advanced Options screen.

Using Boot Diagnostics

Using Advanced Boot Options

The Advanced Boot Options menu gives you, the system administrator, a number of utilities to troubleshoot various system issues.

Advanced Boot Options is a menu that has been around in Windows operating systems for a very long time. There are two ways to get to it:

- The first option is the nightmare of every system administrator, in which the system has an issue, reboots, and then enters into the Advanced Boot Options menu, indicating that there was a problem.
- The second and less scary option is when a system administrator chooses to boot into Advanced Boot Options menu. This may be done for a number of reasons. I've done it to troubleshoot issues with drivers and to investigate and remove malware from a potentially infected machine.

To enter into the Advanced Boot Options menu, follow these steps:

1. **Click the Start menu and then click the Settings icon.**
2. **Click Update & Security and then click Recovery.**
3. **Under Advanced startup, click the Restart Now button, shown in Figure 2-3.**
4. **Click the Continue button.**
5. **Click the Troubleshoot button.**
6. **Choose Startup Settings.**
7. **Click Restart.**

When the Advanced Boot Options menu is up, you're presented with a number of options, shown in Figure 2-4. I describe these options in the following sections.

Safe Mode

Safe Mode is almost always my go-to when there are boot issues with a system. Whenever new hardware or software has been installed, or if I suspect that a system may be having issues because of a malware infection, I turn to Safe Mode.

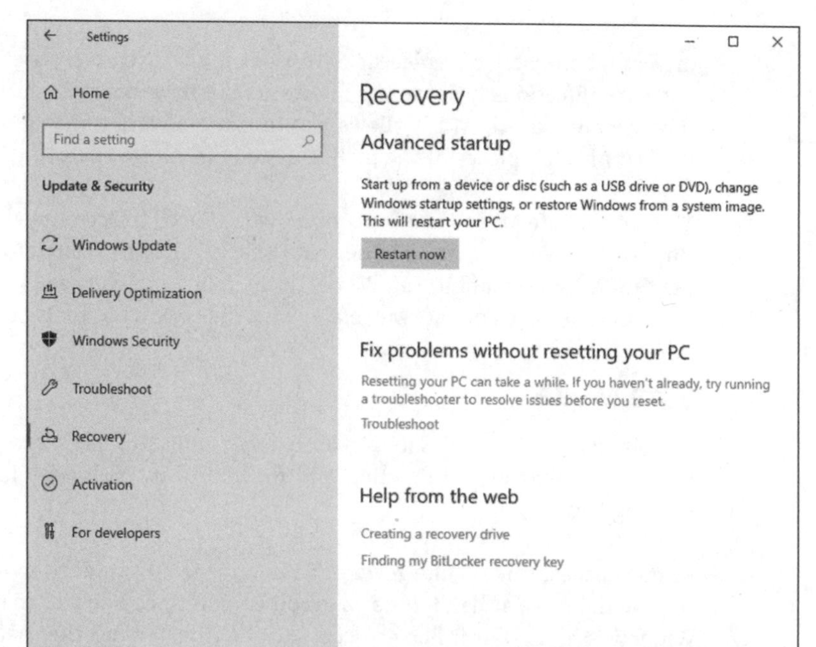

FIGURE 2-3: Your journey into the Advanced Boot Options menu starts with the Restart Now button.

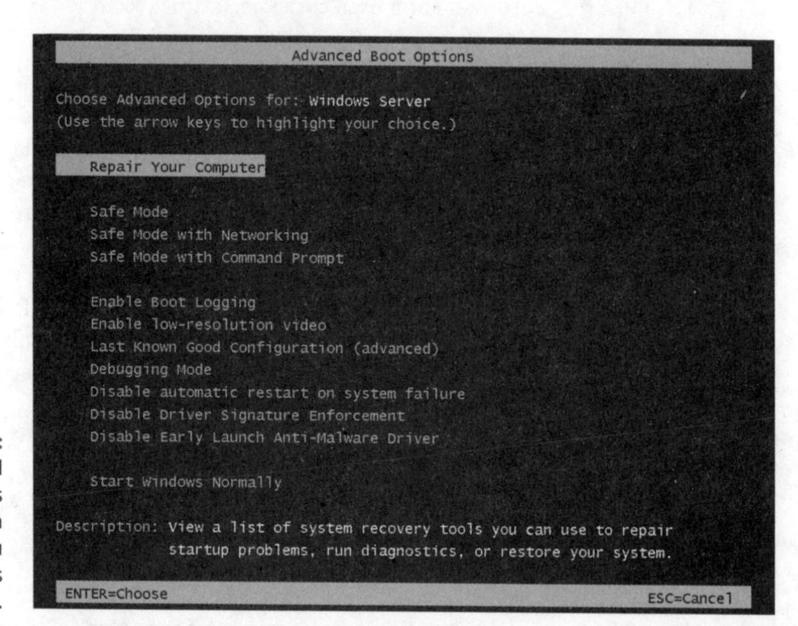

FIGURE 2-4: In the Advanced Boot Options menu, you can choose what you want Windows Server to do.

You may be asking, "What is Safe Mode, and why is it such a big deal?" Safe Mode starts Windows with the bare-minimum services and drivers it needs in order to run. Safe Mode is crucial for troubleshooting issues where a bad driver is causing a boot loop. By going into Safe Mode, you can troubleshoot what's wrong with the driver and uninstall or replace it. Safe Mode is also extremely useful with potential malware infections because the malware may have dependencies it needs to run that are not loaded, which allows you to run malware removal tools and destroy the last bits and pieces of the malicious code from the operating system.

The type of Safe Mode I use depends on what I need to accomplish. For instance, if I'm just troubleshooting an issue that I suspect may be related to drivers, most of the time I use regular old Safe Mode. In the following sections, I walk you through the different forms of Safe Mode and why you may want to use each of them.

Safe Mode

This is just regular old Safe Mode. It loads only the basic services and drivers needed for Windows to function and for you to interact with it. Nothing more, nothing less.

In most cases, this regular form of Safe Mode is all you need to troubleshoot and resolve the issue at hand. It has a graphical interface like you're used to seeing in Windows Server, but it has no access to the Internet or other network resources. In essence, it's a stand-alone machine.

Safe Mode with Networking

Safe Mode with Networking is similar to regular Safe Mode, except the system will also load the drivers needed for the network interface card (NIC) to function properly. This is useful if you need to download software from the Internet (for example, drivers or diagnostic software) or over a network share.

Safe Mode with Networking is most useful when you're trying to resolve a software or driver issue. It allows you to download replacement software or replacement drivers while still in Safe Mode. Then you can replace the misbehaving driver or incompatible software with a known good version and then boot successfully.

Safe Mode with Command Prompt

In Safe Mode with Command Prompt, you bypass the Explorer desktop environment. This can be especially useful if the desktop is not displaying properly for whatever reason.

If you like Server Core, you'll like this version of Safe Mode. If you aren't as comfortable with the command window as you would like to be, having a cheat sheet available may help you.

I recommend Safe Mode with Command Prompt when the issue that needs to be fixed has something to do with graphics. The problem may be due to a driver, graphics rendering, or removing a malware infection that relied on graphical components like wallpapers and screensavers.

Enable Boot Logging

If you need to see which drivers were installed as the system started up, you should choose Enable Boot Logging. This will create a file called `ntbtlog.txt`, which lists all the drivers that were installed when the operating system started. The file is stored in your Windows system directory; typically, this will be `C:\WINDOWS`. Incidentally, this is the same list you see flash by on the screen when you boot into Safe Mode.

Enable Low-Resolution Video

This setting is very useful if you're having display issues, most commonly after changing display settings to something your monitor doesn't support. It uses the currently installed video driver but starts with lower resolution (typically 640 x 480) and refresh settings.

Last Known Good Configuration

Last Known Good Configuration is helpful in fixing issues with booting that occur because the Windows Registry has been damaged. Most commonly, this occurs due to user misconfiguration or from updates or patches. When you choose Last Known Good Configuration, the Registry is reverted so that it matches the settings it had the last time the system booted successfully.

Any time you use something that modifies the Registry in any way, be extra cautious. There's no way to undo using Last Known Good Configuration. If it doesn't fix the issue, or it makes matters worse, you'll need to restore from a backup.

Directory Services Restore Mode

This option only appears on a server that is a domain controller (and, therefore, it isn't shown in Figure 2-4). Directory Services Restore Mode (DSRM) is a special form of Safe Mode made for domain controllers that allows you to repair or recover an Active Directory database.

TIP

To use this utility you need to know the DSRM password that was set when the domain controller was initially created. If you don't know the password, you can use the ntdsutil tool to change the password. You need to have access to the Command Prompt on the system in question to run it.

If all of this is Greek to you, don't worry! I cover Active Directory in depth in Book 2, Chapter 5. For now, think of Active Directory like a special database that stores information on users, computers, sites, and other objects in your network. This database can be crucial to your organization, so knowing how to restore it if it becomes damaged is a very useful skill.

Debugging Mode

If you're a hard-core system administrator and you want to get your feet wet using a kernel debugger, this option is for you!

The *kernel* is a program that is one of the first to run when your server boots (the kernel loads right after the bootloader); it has total control over everything on your system.

Debugging Mode turns on kernel debugging, which allows you to work with the kernel debugger to examine states and processes that are running at the kernel level. This can be very useful for troubleshooting issues with device drivers that cause the infamous blue screen of death and issues with the central processing unit (CPU). You can look at the kernel memory dump on the system that is having the issue, or you can view the kernel memory dump remotely on another system via a serial connection. The information from the Debugging Mode is typically made available over the COM1 port (assuming you have a serial port and it's assigned to COM1). On newer systems that don't have a serial port, you can also access this information over USB.

TECHNICAL STUFF

Kernel debugging is not for the faint of heart. For more information on how to set up your system for kernel debugging with either serial or USB connectivity, check out the following articles:

- **Serial connection:** `https://docs.microsoft.com/en-us/windows-hardware/drivers/debugger/setting-up-a-null-modem-cable-connection`
- **USB connection:** `https://docs.microsoft.com/en-us/windows-hardware/drivers/debugger/setting-up-a-usb-3-0-debug-cable-connection`

COM ports were typically presented as serial ports with RS-232 connectors on older systems. On newer systems, these have been replaced with USB ports. USB stands for *Universal Serial Bus* — it's still a serial connection.

Disable Automatic Restart on System Failure

Eventually, every system administrator has a system that will continuously try to start, fail, reboot, and then try to start, fail, reboot, and so on. This situation is known as a *boot loop.* If you're experiencing a boot loop on one of your systems, you can get the system to stop automatically restarting by choosing Disable Automatic Restart on System Failure from the Advanced Boot Options menu.

Disabling automatic restart can be very helpful if the system is getting the blue screen of death and you need to get the information being displayed. When the system halts on its next blue screen, you'll have all the time you need to copy down the information.

Disable Driver Signature Enforcement

By choosing the Disable Driver Signature Enforcement option, you're basically telling the system that it's okay to load drivers that aren't digitally signed. Microsoft requires drivers to be digitally signed by default, and it will prevent unsigned drivers from running. Microsoft does this because, when a driver is digitally signed, it's seen as being authentic because you can verify from the digital signature that it came from the vendor it claims to be from. Digital signatures also guarantee that the driver hasn't been altered in any way since it was released by the vendor.

Digital signatures use a code-signing certificate to encrypt the hash of a file. (Hashes are unique thumbprints — any change to the file will change the hash.) That encrypted hash is then bundled with the certificate and the executable for the driver. When the end user installs the driver, the hash of the file is decrypted with the public key in the certificate. The file gets hashed again on the end user's system, and the new hash is compared to the decrypted hash. If they match, the driver hasn't been tampered with.

If you choose to disable driver signature enforcement, you'll be able to load unsigned drivers. Choose this option at your own risk: You could end up installing malware that presents itself as an unsigned driver.

Disable Early Launch Anti-Malware Driver

Malware that installs after Windows has booted will most likely be seen by the antivirus software that is installed on the system. But the problem is, virus writers began writing malware called *rootkits*. These rootkits can be very difficult to get rid of because they install and execute *before* the operating system has booted. Many of the more sophisticated rootkits began installing drivers that start *really* early in the boot process of the system. This can make them extremely difficult to find and remove.

Microsoft does its best to evolve and respond to threats and prevent them whenever possible. In this case, it came up with the early launch anti-malware (ELAM) driver. Certified antivirus vendors whose products support early launch can get their products' drivers to launch before the Windows boot drivers, which allows them to scan for malicious processes on boot. Pretty cool, right?

But what happens if a legitimate boot driver for Windows gets flagged as malicious? Your server won't boot. So, Microsoft gives you the ability to turn off this feature, by choosing Disable Early Launch Anti-Malware Driver, to allow the boot driver to launch like normal.

WARNING

This feature is a great one to have on. Only disable it if you absolutely have to, and then only until the issue is resolved.

Performing a Memory Test

What happens if your server is crashing unexpectedly or throwing blue screens when you least expect it? That can be a difficult question to answer. These symptoms could occur because of corrupted software or because of hardware failure. Memory is a great place to start with your troubleshooting efforts, and Windows Server 2022 includes a built-in memory diagnostic utility called the Windows Memory Diagnostics Tool.

You can run the Windows Memory Diagnostics Tool by pressing the Windows Key+R, typing **mdsched.exe**, and clicking OK. If you do nothing, the Windows Memory Diagnostics Tool will run in Standard mode. You can interrupt it at any time by pressing F1 to enter the Options screen and change the settings. Your options are as follows (see Figure 2-5):

- **Test Mix:** The test mix is the set of tests you want the tool to run:
  - **Basic:** Runs three tests on your memory and is the fastest option.
  - **Standard:** Runs the same tests on your memory as Basic, and adds five additional tests. It takes longer to complete than Basic.

- **Extended:** Runs the same tests as Standard and adds nine additional tests. This test is the most detailed and takes the longest to complete.

TIP

If you don't know what each of these tests is looking for, Standard is a good starting point for your tests. Extended takes longer, so if you don't need the extra tests, you may not get any worthwhile information from running them. That said, it won't hurt your server to run any of the three tests.

» **Cache:** Cache sets the cache setting (cache is used to improve the speed of memory access for things that are frequently accessed by the CPU) for each test you're going to run. The cache should be disabled if you're running tests that require direct access to the memory. Your options are as follows:

  - **Default:** In most cases, Default is the appropriate setting. It selects the correct cache setting for the test that's being run.
  - **On:** Forces the cache on for the tests.
  - **Off:** Forces the cache off for the tests.

» **Pass Count (0–15):** Pass count controls how many times the whole test mix you selected will run. If it's set to 5, the selected test mix will run through its tests five times. The default for this setting is to make two passes.

After you've made your selections, press F10 to apply the settings, and the scan will restart.

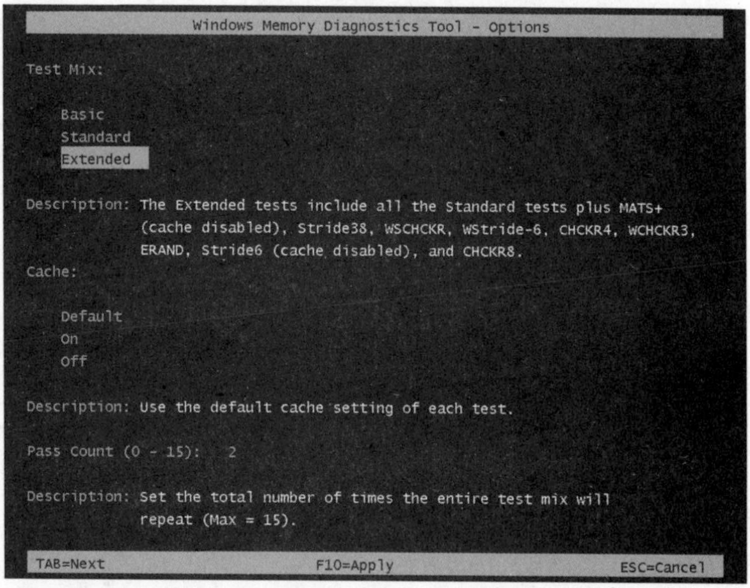

FIGURE 2-5: Windows Memory Diagnostics Tool options.

Using the Command Prompt

When all else fails, the Command Prompt is always there. I've had to troubleshoot many issues over the years where I was saved because the Command Prompt was available. Corrupted system files? Open the Command Prompt and run `sfc /scannow`. Damaged hard drive? Open the Command Prompt and type **chkdsk /f /r**.

In Table 2-1, I list some of the most helpful tools that I've used over the years. The majority of these commands need the Command Prompt to be running with administrator credentials. To run the Command Prompt as an administrator, choose Start➪Windows System, right-click Command Prompt, click More, and then select Run as Administrator, or if you can bring up Task Manager, you can choose File➪Run New Task and type **cmd.exe**.

TABLE 2-1 Troubleshooting with the Command Prompt

Name	Command	Description
System File Checker	`sfc /scannow`	This utility checks system files to see if they match what's expected by comparing the signature of the system file on the server with the signature of a cached copy of the same file. The cached files are stored in a compressed folder located at `C:\Windows\System32\dllcache`. If a corrupt system file is found, it's replaced.
Check Disk	`chkdsk /f /r`	This utility repairs file system errors and marks bad sectors so the operating system doesn't use them anymore. The `/f` will tell the utility to fix any issues it finds, and the `/r` will locate the bad areas (sectors) on the disk. This can take a while. Kick it off, and grab a cup of coffee.
Driverquery	`driverquery`	This utility queries the system for all the hardware drivers that are installed on Windows. This can be very helpful if you're running into issues with systems that have similar hardware and you want to know if they have a driver in common.
BCDEdit	`bcdedit`	This utility is covered in depth in Book 1, Chapter 4. For now, just know that it allows you to edit the boot configuration on your Windows server.

Working with Third-Party Boot Utilities

This chapter wouldn't be complete without a brief look at third-party utilities that are designed to help diagnose and resolve boot issues, or to at least assist with recovery. Table 2-2 lists two of my favorites, along with their cost and a brief description.

TABLE 2-2 **Third-Party Boot Utilities**

Name	Cost	Description
Ultimate Boot CD	Free	This is one of my all-time favorite utilities. It includes multiple diagnostic and recovery tools. To use it, you boot to the disc. It's that easy! Go to `www.ultimatebootcd.com` for more information.
Trinity Rescue Kit	Free	Trinity Rescue Kit is full of great features, this is also a very useful utility. Go to `https://trinityhome.org` for more information.

IN THIS CHAPTER

» Installing the prerequisites to support a successful installation

» Doing a clean install of Windows Server 2022

» Upgrading from a previous version of Windows Server

» Performing a network install with Windows Deployment Services (WDS)

Chapter 3
Performing the Basic Installation

You've made the decision: You want to install Windows Server 2022. Great! You may be wondering what's next. One of the most important things you can do to ensure a successful installation is make sure that you're meeting all the prerequisites for Windows Server 2022. By ensuring that you have the appropriate hardware to meet the needs of the operating system, you can definitely save yourself some headaches later.

When you've got everything necessary to install Windows Server 2022, you're ready to go. In this chapter, I walk you through how to perform a clean install as well as an upgrade install. I also explain how to do a network install with Windows Deployment Services.

TIP

You should know that you can't change between Server Core and Server with Desktop Experience anymore. This capability was removed in Windows Server 2016, in order to support the newer Windows 10 desktop experience on the server, rather than the older legacy desktop experience you had with Windows Server 2012 R2. If you install Server Core, and then change your mind and decide you actually want Server with Desktop Experience, you need to reinstall it. If you try to use the Windows Server installation media to move between Core and Desktop experience, you won't be given the option to keep anything.

Making Sure You Have What It Takes

Microsoft publishes the prerequisites for each of its operating systems. Some of the hardware requirements are independent of which edition of Windows Server you're planning to install; other hardware requirements vary based on whether you're installing Server with Desktop Experience or Server Core.

Windows Server 2022 is available only as a 64-bit operating system; there is no 32-bit version available. When you run the installer, you're presented with options for the Standard edition or Datacenter edition. At the same time, you choose whether you want to install Server Core or Server with Desktop Experience.

WARNING

Where I discuss minimum requirements in this section, it's important to understand that these are the *bare minimums* to successfully install Windows Server 2022. You should *not* expect your server to perform well if you give it the specs listed here. For any real workload, your server should have faster processors, more processor cores, and more memory.

So, what are the absolute bare minimums that you have to meet in order to install Windows Server 2022? Read on.

VERSION AND EDITION REQUIREMENTS

For the most part, there aren't many differences between the minimum requirements of Server Core and those of Server with Desktop Experience. The one very important exception to that is the amount of random access memory (RAM). The minimum requirement to install Server Core is 512MB of RAM; Server with Desktop Experience needs a minimum of 2GB of RAM.

If you're installing Windows Server 2022 Standard, you should base the hardware specifications on the requirements of the workload you're intending to run. If you choose to run Windows Server 2022 Datacenter, you may want to look at installing better hardware. A great example of a use case for the Datacenter edition is as a Hyper-V host. The Datacenter edition doesn't limit you in terms of how many virtual machines (VMs) you can run. Your hardware will really be the limiting factor. In this use case, you would want multiple cores and a lot of memory.

Central processing unit

The central processing unit (CPU) is the brains of the outfit. It processes instructions made by the program and/or applications. The CPU requirements for Windows Server 2022 are pretty easily met by most modern processors:

- **1.4 GHz 64-bit processor:** Considering that the operating system is an x64 system, it makes sense that the processor must also be an x64 processor. Even a cheap server with a lower-end processor should be able to meet the 1.4 GHz requirement with flying colors.
- **Supports No Execute (NX):** When the NX bit is enabled on certain areas of memory, the processor will not execute anything in that memory space, which can provide protection against malware. Areas protected by the NX bit usually contain things like processor instructions or data storage.

  TECHNICAL STUFF

  Intel may refer to this technology as XD (short for Execute Disable), while AMD processors refer to it as Enhanced Virus Protection (EVP).
- **Supports Data Execution Prevention (DEP):** DEP provides additional protection against malware that may target memory locations.
- **Supports CMPXCHG16b, LAHF/SAHF, and PrefetchW:** These settings are specific to the processor, and there are multiple whitepapers published on the specifics. CMPXCHG16b is an instruction set supported by most modern x86_64 processors. Load AH from Flags (LAHF)/Store AH into Flags (SAHF) is needed to support virtualization. PrefetchW provides improvements to performance when using AMD processors. You don't need to memorize these things — just know that these processor features can speed up execution of tasks and add some additional security features as well.
- **Supports Second Level Address Translation (Extended Page Table [EPT] or Nested Page Table [NPT]):** This feature is especially important if you're planning on running Hyper-V. It improves the performance of the VMs on the system and takes some of the pressure off the hypervisor, which can, in turn, improve hypervisor performance.

TIP

You may be curious how you can tell if your CPU supports these requirements. Microsoft offers a tool that is part of the Sysinternals suite named Coreinfo; this tool tells you what your processor is capable of supporting. You can download Coreinfo for free from the Microsoft website (`https://docs.microsoft.com/en-us/sysinternals/downloads/coreinfo`). The file you download is a compressed zip file, so you need to extract it first. Then launch a command prompt to run the utility. To run Coreinfo, simply type **coreinfo** into the command window and you get a report of all available and unavailable features. Available features are marked with an asterisk (*), and unavailable features are marked with a hyphen (-), as shown in Figure 3-1.

FIGURE 3-1: Running the Coreinfo utility on a Windows system.

Random access memory

Random access memory (RAM) is used by the server to store things that you need to access right now and things that you may need to access in the near future. RAM is much faster than persistent storage, so a server that has lots of RAM will perform far better than a system with very little RAM. As I mention in the "Version and edition requirements" sidebar, earlier in this chapter, Server Core requires a minimum of 512MB of RAM, while Server with Desktop Experience requires a minimum of 2GB of RAM. The RAM must also be Error Correcting Code (ECC)-type memory. ECC-type memory is able to correct single-bit errors (for example, if electrical interference flips a bit in error, using the parity bit can ensure that the data in memory is corrected).

Storage

There's no fancy formula or calculation here. If you want to install Windows Server 2022, you need a minimum of 32GB of hard drive space. Remember that this is the absolute bare minimum to install the operating system. If all you have is 32GB, you won't have room to install anything else. If you're limited on storage space, according to Microsoft, Windows Server Core is approximately 4GB smaller than Windows Server with Desktop Experience.

Network adapter

A server does you no good if you can't access it. The network adapter, also referred to as the network interface card (NIC), gives your server a way to talk on your network. For Windows Server 2022, your network adapter will have to support at least gigabit ethernet. Your network adapters may be *onboard*, meaning that they're a part of the motherboard, or they may be on a NIC, which plugs into a PCI Express slot.

Your network adapter should support the Pre-boot Execution Environment (PXE). This is what the majority of organizations use today to image systems from a central imaging server like Windows Deployment Services or System Center Configuration Manager.

DVD drive

Not all servers come with DVD drives anymore. There are so many more options for installing operating systems like booting from flash drives or booting from the network that many system administrators don't bother with DVDs. That said, if you want to install from a DVD, you need to ensure that you have a DVD drive. The drive can be internal or external.

UEFI-based firmware

Unified Extensible Firmware Interface (UEFI) has replaced the traditional legacy Basic Input/Output System (BIOS) at this point. I highly recommend that you choose UEFI rather than BIOS. It'll be required if you want to use some of the advanced features like secure boot.

Trusted Platform Module

The majority of motherboards come with a Trusted Platform Module (TPM) chip nowadays. If you plan on doing disk encryption with BitLocker, this is a must-have item.

Monitor

It goes without saying that you need to be able to see what's going on with your server when you're installing your operating system. Windows Server 2022 requires a Super Video Graphics Array (SVGA) connection with a minimum of 1024 x 768 screen resolution. You can accomplish this by attaching a physical monitor to the server or by viewing the video stream through a KVM.

TECHNICAL STUFF

KVMs allow you to use one keyboard, monitor (video, in the acronym), and mouse to administer multiple servers. The older KVMs required you to be physically on site to use the keyboard, monitor, and mouse. Modern KVMs allow you to administer your servers remotely through a web service, and they provide similar functionality to what you would get if you physically plugged in a keyboard, monitor, and mouse to your server.

Keyboard and mouse

You can connect a keyboard and mouse directly to the server during imaging or you can present them to the system via a KVM. Either way, you need a keyboard and a mouse of some kind to interact with the system.

Performing a Clean Install

Clean installs are my preferred way to go. By performing a clean install, you're far less likely to run into issues caused by bad drivers, corrupted system files, or misconfigurations. In this section, I walk you through how to do a clean install of Windows Server 2022.

In this section, I assume that you've already booted to whatever media you're going to use for the installation (DVD, flash drive, and so on), and you're on the starting installation screen for Windows Server 2022. If you've done this, you should see a screen that looks like Figure 3-2. From this screen, follow these steps:

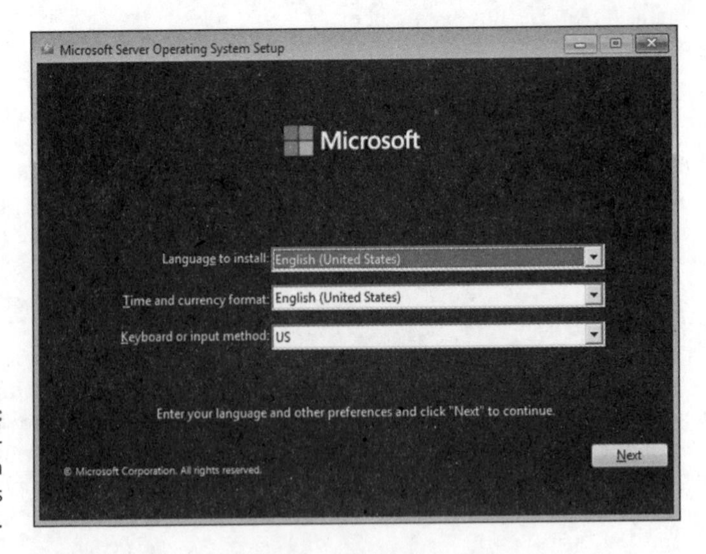

FIGURE 3-2: The first installation screen for Windows Server 2022.

1. **Select the appropriate settings for your locality and click Next.**

   In my example, I've chosen the following:

   - **Language to Install:** English (United States)
   - **Time and Currency Format:** English (United States)
   - **Keyboard or Input Method:** US

   After you click Next, the screen shown in Figure 3-3 appears.

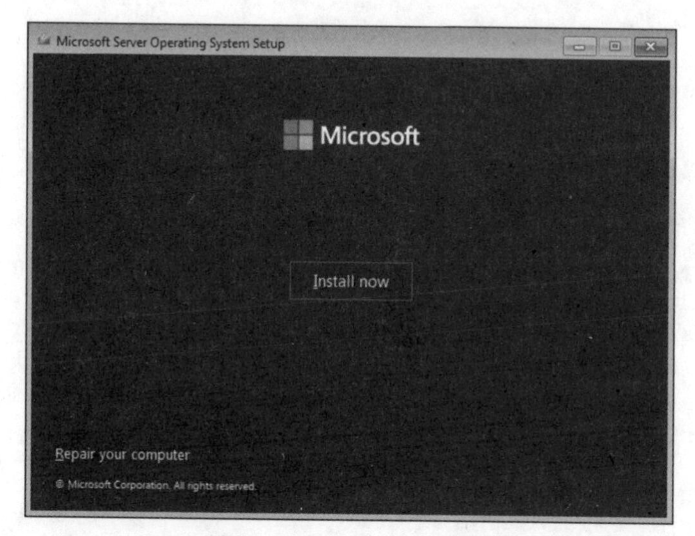

FIGURE 3-3: The Windows Server 2022 Install Now button.

2. **Click Install Now.**
3. **On the next screen, choose which version of the operating system you want to install and click Next.**

   The default selection is for Windows Server 2022 Standard (shown in Figure 3-4). If you prefer, you can select Windows Server 2022 Standard (Desktop Experience), Windows Server 2022 Datacenter, or Windows Server 2022 Datacenter (Desktop Experience).

4. **On the next screen, check the I Accept the License Terms box and click Next.**
5. **On the next screen, choose Custom.**

   The other option is for upgrade installations.

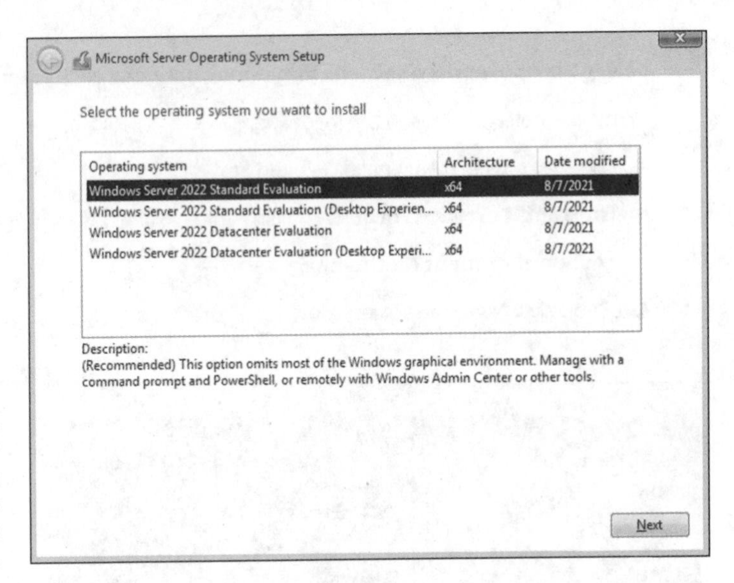

FIGURE 3-4: Choosing your desired edition and experience of Windows Server 2022.

6. **On the next screen, select the partition on which you want to install Windows and click Next.**

   In Figure 3-5, you can see that this is Drive 0.

   Windows Server 2022 begins installation and restarts after it's finished. That's when the real fun begins!

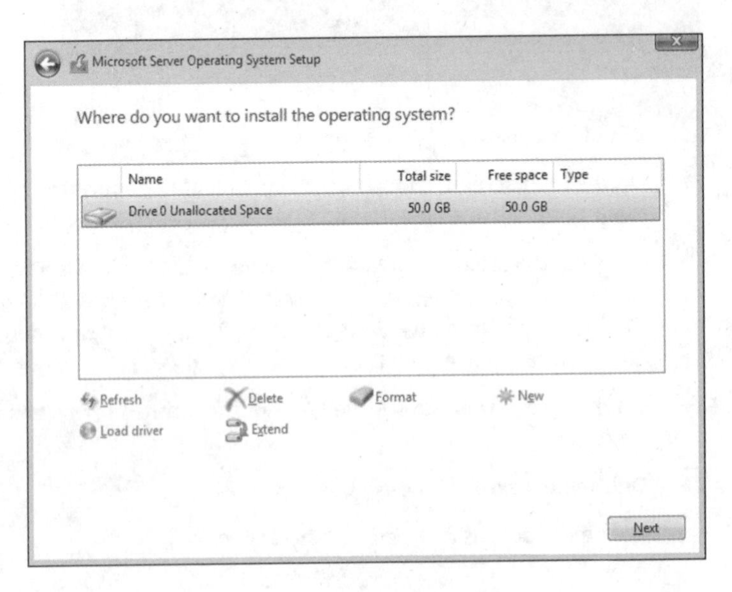

FIGURE 3-5: Choose where to install Windows.

Upgrading Windows

When considering an upgrade install, you need to ensure that the version of the operating system you're starting with is able to be upgraded to Windows Server 2022. Table 3-1 tells you which operating systems you can upgrade from and which edition of Windows Server 2022 you can upgrade to.

TIP

As a general rule, you can update to the newest Windows Server operating system directly as long as you're within the last two major releases.

TABLE 3-1 **Windows Server 2022 Upgrade Compatibility Matrix**

If you're running this edition . . .	You can upgrade to these editions . . .
Windows Server 2016 Standard	Windows Server 2022 Standard or Datacenter
Windows Server 2016 Datacenter	Windows Server 2022 Datacenter
Windows Server 2019 Standard	Windows Server 2022 Standard or Datacenter
Windows Server 2019 Datacenter	Windows Server 2022 Datacenter

You also need to check with your application vendors to ensure that the applications on the server are compatible with Windows Server 2022. If they aren't, then you may need to upgrade your applications before you upgrade the server operating system.

TIP

There is no direct upgrade path from Windows Server operating systems that are older than Windows Server 2016. If you're migrating from an older server, start with a clean installation. If you can't use a clean installation, you'll need to upgrade to either Windows Server 2016 or Windows Server 2019 to be able to upgrade to Windows Server 2022.

After you've verified that you're on a compatible version, you can begin the upgrade install. For this example, I'll start with a Windows Server 2019 Standard installation and upgrade it to Windows Server 2022 Standard. Follow these steps:

1. **Log in as the administrator on the system that you want to upgrade.**
2. **Insert the disc or other installation media into the system that you're wanting to upgrade, and run** `setup.exe`**.**

   The next screen asks if you want to download updates and drivers ahead of time (see Figure 3-6).

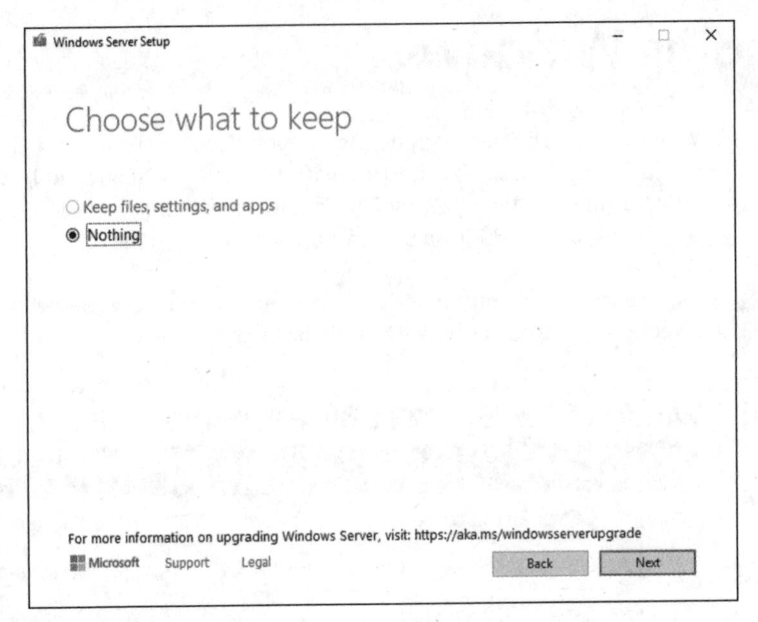

FIGURE 3-6: Choose to either keep your files and settings or start fresh.

3. **On the Install Windows Server screen, click Next.**

   Any relevant updates will be downloaded. You may be required to reboot before you're able to continue.

WARNING

   On the Install Windows Server screen, you'll see a Change How Setup Downloads Updates link. Clicking this link allows you to not patch the server at the time of installation. I don't recommend doing this unless the server doesn't have an Internet connection, because it will be potentially vulnerable to attack until it's patched.

4. **Select Windows Server 2022 Standard with Desktop Experience (or whichever version you want) and click Next.**
5. **Read through the license terms if you have time on your hands, and then click Accept.**

   On the next screen, you choose whether to keep your personal files and apps or keep nothing. If you're sticking with the same experience (Core or Desktop), you'll see both options. If you're changing the experience, the only option you'll have will be to keep nothing.

6. **If you have the option, select the Keep Personal Files and Apps radio button, and then click Next.**

   If you want to start clean, you can select Nothing. Just be aware that you will lose data if you choose this option. In my case, because I'm using evaluation media, I can only choose Nothing. This screen is shown in Figure 3-6.

If the Keep Files, Settings, and Apps option is grayed out, it may be because you're trying to change the user experience (for example, from Core to Desktop), or you may be using evaluation media rather than retail media. In either case, you have to choose Nothing.

The installer fetches any applicable updates and presents you with a summary screen.

7. **If everything looks correct, click Install.**

   The installer begins the upgrade install to Windows Server 2022. It may restart several times during this process.

   That's it — you're done!

Performing a Network Install with Windows Deployment Services

Windows Deployment Services (WDS) is a role that can be installed on a Windows Server operating system. It serves as a combination of a Preboot Execution Environment (PXE) server and a Trivial File Transfer Protocol (TFTP) server and enables you to install Windows over a network connection by choosing the network interface card as the boot device.

Installing WDS is fairly straightforward. You can choose to install it as a standalone server or integrate it with Active Directory. You tell it what the boot file is that you want to use. The easiest one to start with is the `boot.wim` file on the Windows Server installation media, which contains the Windows Preinstallation Environment (WinPE). This is typically located under the Sources directory on the installation media.

From there, you need to create the installation files. The simplest way to get started with this is to copy `install.wim` from the Windows Server 2022 installation media (again in the Sources directory) to the system that will serve as your WDS server. You'll have the same edition and experience options that you would've gotten from the installation wizard on disc. After WDS is fully configured, it serves images over the network. All you need to do is tell your new server to boot from the network.

If you're doing a network install, and the server isn't in the same subnet as the WDS server, you need to set Dynamic Host Configuration Protocol (DHCP) options 66 and 67. Option 66 specifies the hostname or IP address of the WDS server, and Option 67 is the bootfile name. You may also need to create a firewall rule to allow UDP ports 67 and 68 if there is a firewall between the two networks.

If you use the default `boot.wim` option from the installation media, you'll receive a deprecation notice, but you can proceed in configuring with the `boot.wim`. Future versions of Windows Server won't support this option. You can, however, use custom `boot.wim` files. Microsoft recommends moving to either Microsoft Endpoint Configuration Manager or Microsoft Deployment Toolkit products for more granular customization and deployment of images.

» Getting an overview of the configuration process

» Providing the information your server needs to be set up properly

» Updating Windows Server 2022 with the latest patches, hotfixes, and everything in between

» Customizing Windows Server 2022 to your preferences

» Configuring your server startup options with BCDEdit

Chapter 4
Performing Initial Configuration Tasks

Now that you've installed Windows Server 2022, it's time for the fun to begin! As an administrator, your next task after installing the server operating system is to configure it to do what you want it to do.

Microsoft introduced the Server Manager feature in Server 2008, and it was updated heavily in Windows Server 2012 to support Remote Management, as well as multi-server management. Server Manager is your starting location for the majority of the configuration tasks that you need to accomplish on your server if you're working on a server that has Desktop Experience.

If you're working on a Server Core system, you won't use Server Manager on the console. Instead, you'll use the sconfig utility to do your initial configuration, assuming that you aren't deploying Server Core images that are already configured for your environment. Of course, you can use Server Manager to administer your Server Core systems remotely, with a little setup initially to get things going. I cover that subject in my overview of the configuration process.

Understanding Default Settings

When Windows Server 2022 is first installed, there are some settings that are created or set by default. Typically, these are things that you'll want to change, such as setting the server's name, setting an IP address, joining the server to a domain, and so on. Table 4-1 covers these default settings and discusses what they're set to out of the box to give you a better idea of what you're starting with.

TABLE 4-1 Windows Server 2022 Default Settings

Setting	Default Value	Description
Computer Name	WIN-<*randomstring*>	This will be a randomly generated name starting with WIN-. You should change the name based on your organization's naming standards. When you change the name, you'll be required to restart the system.
IP Address	Assigned by DHCP	By default, your brand-new server is using DHCP to automatically receive an IP address. If your organization uses DHCP to manage IP addresses, you're good to go. If not, you may need to set a static IP address.
Domain or Workgroup	Workgroup named WORKGROUP	Windows Server 2022 begins life joined to a workgroup named WORKGROUP. If it's going to be a standalone server, then that setting may work well for you. Servers in workgroups are not domain joined. If your server needs to be joined to a domain, you'll want to change this setting. Doing so will require a reboot.
Windows Update	Automatic update download	Updates are downloaded automatically, but they aren't installed until you allow them to be.

Setting	Default Value	Description
Microsoft Defender Firewall	Public and private profiles: On Core OS functionality: Allowed	In its default state Microsoft Defender Firewall has a public and a private profile. Core functionality needed for the operating system to function is allowed automatically. The domain profile will appear if the server becomes domain joined.
Microsoft Defender Antivirus	Real-time protection: On	Provides real-time virus/malware scanning. It prevents malware from installing and/or running on your server. Automatic sample submission is also enabled by default. This sends sample files to Microsoft for analysis.
Roles and Features	Some roles/features are installed	Some roles and features are enabled out of the box to allow the server basic functionality. It's important to note that just because a role or feature is selected, that doesn't mean that the role as a whole is installed.
Remote Management	Enabled	Allows the server to be managed by PowerShell remotely. Also allows applications or commands that require Windows Management Instrumentation (WMI) to manage the server.
Remote Desktop	Disabled	Allows users to connect to the desktop of the server remotely. Allowed users can be configured individually or by security groups.

Getting an Overview of the Configuration Process

When you start with a freshly installed server, it isn't configured to do much of anything. You'll need to take some basic configuration steps. Some of these steps are the basics like setting the day and time; others are tasks that will allow you to manage your systems remotely.

Here's the basic process:

- Activate Windows Server 2022.
- Set the date, time, and time zone.

- Change the computer name.
- Add to the domain (if there is one to join).
- Configure the networking.
- Configure the server to receive Windows updates.
- Add roles and features.
- Setup the Windows Server OS for remote administration.
- Configure the Windows Server firewall.

You can find the specifics on how to do each of these tasks in the following section.

Providing Computer Information

When you're deploying new servers, you have to perform certain tasks, such as activating the operating system with a valid Microsoft product key, setting the time zone, changing the name, and adding the server to the domain. In this section, I explain how to provide information for the server on both Windows Server 2022 with Desktop Experience and Server 2022 Core.

Windows Server 2022 with Desktop Experience

Many system administrators got their start with the graphical user interface (GUI) of a Windows Server operating system. Windows Server 2022 continues the tradition of the GUI with the Desktop Experience installation. Let's take a look at what is involved with configuring Windows Server 2022 with Desktop Experience.

Activation

One of the first things that you do after installing the Windows Server operating system is activate it with a valid product key. You can do this through the desktop interface or through PowerShell.

In this section, I cover activating through the desktop interface. I cover activation through PowerShell in the later section on activation for Server Core.

1. **Log into the server.**

   Server Manager opens automatically.

2. **In Server Manager, click Local Server in the navigation pane.**
3. **To start the activation process, click the Not Activated hyperlink next to Product ID.**

   A dialog box launches automatically asking for the product key.
4. **Enter your product key and click Next.**

   You're prompted to activate Windows.
5. **Click Activate.**

   You get a confirmation that Windows has been activated.
6. **Click Close.**

   You're left on the Activation screen shown in Figure 4-1, where you see that your version of Windows is now activated.

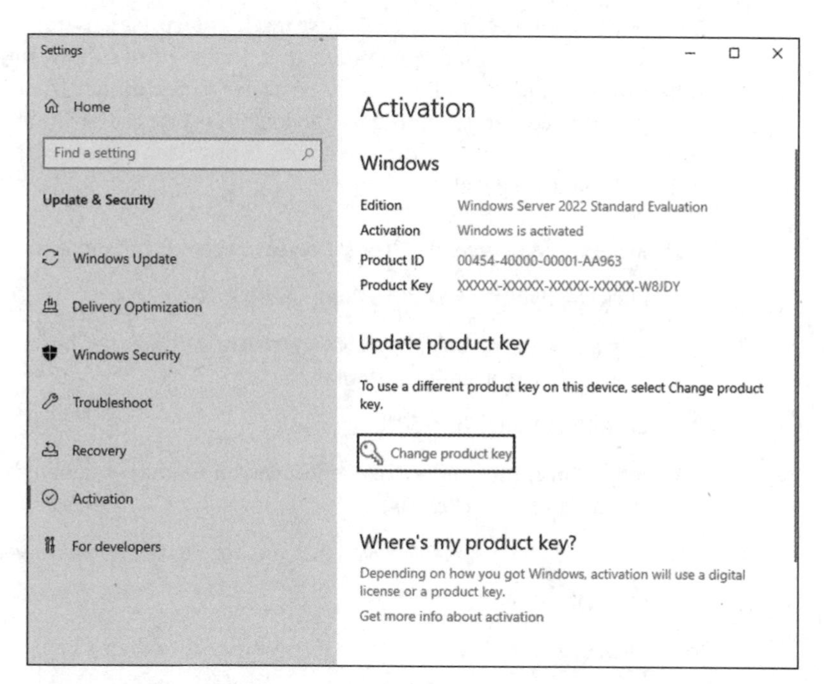

FIGURE 4-1: The Activation screen showing that Windows Server 2022 is activated.

Time zone

Setting the time zone is a common task in the server provisioning process. You may want to set the server to the time zone that you are in, or to the same time zone as a corporate office located elsewhere. This is common if your servers are in a co-location and you want them to be on the same time zone as your local systems.

1. **In Server Manager, click Local Server in the left-hand menu.**
2. **Click the hyperlink next to Time Zone.**

   This may already be set to the correct time zone for your area.
3. **Click the Change Time Zone button.**
4. **Select your time zone from the drop-down list.**
5. **If you're in an area that uses Daylight Saving Time, click the check box next to Automatically Adjust Clock for Daylight Saving Time. If you do not use Daylight Saving Time, leave the box unchecked.**
6. **Click OK to exit the Time Zone Settings dialog box, and then click OK again to exit the Date and Time dialog box.**

Computer name and domain

Setting the computer name is a must in an enterprise environment. Most organizations have a naming convention that you need to follow, but the names the organization requires will certainly be easier to remember than the default randomly generated name. Joining to the domain is one of the simpler steps, but also one of the most important steps to enable centralized authentication management and configuration capabilities.

1. **In Server Manager, click Local Server in the left-hand menu.**
2. **Click the hyperlink next to Computer Name.**

   This will be the default name that starts with WIN- and will be followed by a random string of letters and numbers.
3. **Click the Change button.**
4. **In the Computer Name field, enter the name that you want for your server, and then click OK.**

   A dialog box appears telling you that you need to restart the server.
5. **Click OK.**
6. **Click the Close button in the System Properties dialog box.**

   You're prompted to either Restart Now or Restart Later.
7. **Click Restart Now if you want to reboot the server immediately. Click Restart Later if you want to finish other administrative tasks you may have first.**

   If you click Restart Later, you'll need to manually reboot the server when you're ready.

8. **To join a domain, perform Steps 1 through 3.**
9. **In the Computer Name/Domain Changes dialog box, click the Domain radio button, and enter the name of the domain you want to join.**
10. **Click OK.**

    A dialog box appears telling you that you need to restart the server.
11. **Click OK.**
12. **Click the Close button in the System Properties dialog box.**
13. **Click Restart Now or Restart Later.**

    After the restart, the server will be joined to the domain.

Configure networking

Your server will use a dynamically assigned IP address by default. If this is not desirable, you'll want to set a static IP address so that the server will continue to use the same address.

1. **In Server Manager, click Local Server in the left-hand menu.**
2. **Next to Ethernet, click the hyperlink that says IPv4 Address Assigned by DHCP, IPv6 Enabled.**
3. **Right-click your network adapter (it should be called Ethernet), and click Properties.**
4. **Click Internet Protocol Version 4, and then click the Properties button.**

   By default, the server is set to obtain an IP address automatically and obtain DNS server addresses automatically. If this is what is desired, then no changes are necessary.
5. **If you need to make changes, select Use the Following IP Address.**
6. **Fill in the IP address, subnet mask, and default gateway.**
7. **Manually enter the addresses for the preferred DNS servers.**

   See Figure 4-2 for an example.
8. **Click OK to close the dialog box.**
9. **Click OK one more time to exit out of Ethernet Properties.**

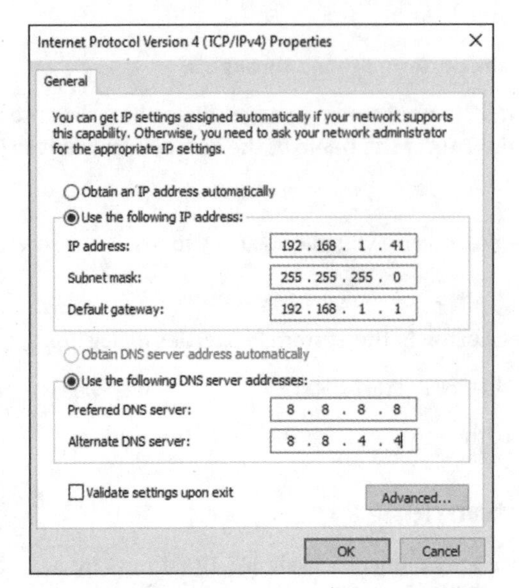

FIGURE 4-2: The Internet Protocol Version 4 Properties dialog box.

Windows Server 2022 Core

Many system administrators have configured a Windows Server with a GUI, but not many have used Windows Server Core. As you see in this section, Windows Server Core has a simple interface, and when you learn how to navigate it, you may find it simpler to work with than Windows Server with Desktop Experience.

Activation

Windows Server Core gives you a few different options for activating your copy of Windows Server 2022. In this section, I cover activating via sconfig, as well as activating via PowerShell.

ACTIVATING WITH SCONFIG

Sconfig is the built-in configuration utility in Windows Server Core. It's a text-based menu that allows you to do the majority of your initial configuration tasks all from one central location. By default, sconfig launches automatically after you've logged in.

1. **From the sconfig utility, type** 11 **for Windows Activation and press Enter.**
2. **Type** 3 **to install your product key.**
3. **Enter your 25-character product key in the dialog box that pops up, and then click OK.**

After the key is installed, you see a message saying the key was installed successfully.

4. **Close the window by clicking the red X, or by pressing Enter twice.**
5. **When you're back on the sconfig screen, type** 2 **to Activate Windows, and then press Enter.**

   A Command Prompt window launches again with the slmgr.vbs script to perform the activation. Assuming there are no errors, this will complete with no message.

6. **Close the window by clicking the red X or by pressing Enter twice.**

ACTIVATING FROM POWERSHELL

After you've logged into Windows Server Core, you're presented with the sconfig utility. From there, you can activate your copy of Windows. To set the license and do the activation from the command line, you'll need to select menu option 15, "Exit to command line (PowerShell)". To activate, you have to set the key. You do this with the Windows Server License Manager script, slmgr.vbs.

TECHNICAL STUFF

The slmgr.vbs script allows you to work with your Windows Server product keys in different ways depending on the parameter that you use along with it. In the example in this book, I use both `-ipk` and `-ato`. The `-ipk` parameter is used when installing product keys, and the `-ato` parameter is used to specify online activation.

To install the product key that will be needed for your version of Windows Server 2022, use the following command with the parameter `-ipk`. Just replace *<productkey>* with your 25-character license key, including the dashes.

```
slmgr.vbs -ipk <productkey>
```

You get a dialog box that tells you the product key installed successfully. Click OK.

After the license key is installed, you use the same script with the `-ato` parameter to do an online activation of your copy of Windows. You do that with the following command:

```
slmgr.vbs -ato
```

If the activation was successful, you get a dialog box that says the product was activated successfully (see Figure 4-3).

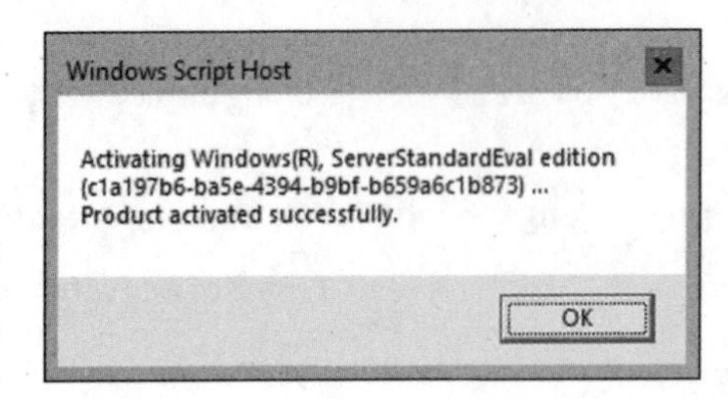

FIGURE 4-3: Using slmgr. vbs to activate Windows Server.

Time zone

Much like activation in Windows Server Core, you can set the time zone via sconfig or PowerShell. In this section, I cover both methods. The great thing about PowerShell version is that it will work on Windows Server with Desktop Experience as well.

SETTING THE TIME ZONE WITH SCONFIG

Sconfig is the built-in configuration utility in Windows Server Core. Because it's a simple text-based menu, it provides a simple way for administrators to configure the time zone without needing scripting knowledge to do so.

1. **From the sconfig utility, type** 9 **to go into the settings for Date and Time.**

   The Date and Time dialog box appears.

2. **Click the Change Time Zone button.**
3. **Select your time zone from the drop-down list.**
4. **If you're in an area that uses Daylight Saving Time, click the check box next to Automatically Adjust Clock for Daylight Saving Time. If you do not use Daylight Saving Time, leave the box unchecked.**
5. **Click OK to exit out of the Time Zone Settings dialog box, and click OK once more to exit out of the Date and Time dialog box.**

SETTING THE TIME ZONE FROM POWERSHELL

If you prefer to work in PowerShell, you can also set the time zone from there. This utilizes the control command to call the Control Panel's Date and Time screen.

In PowerShell, type the following:

```
Set-TimeZone -Id <Time Zone Id>
```

TIP

If you aren't sure what your time zone ID is, you can run `Get-TimeZone -ListAvailable` to see all the time zones you can choose from.

Computer name and domain

Setting the name and adding a server to a Windows domain are some of the most common activities that system administrators do with new servers. With Windows Server Core, there are two methods that you should know to complete this task: sconfig (the configuration utility in Windows Server Core) and PowerShell.

SETTING THE COMPUTER NAME WITH SCONFIG

The sconfig utility in Windows Server Core makes it simple to change the name of your server with its text-driven menus. Follow these steps:

1. **In the sconfig utility, type** 2 **to change the computer name.**

   You're prompted to enter a new name.

2. **Enter the new name, and press Enter.**

   You need to restart your computer to apply the change.

3. **Type** yes **to reboot now or** no **to reboot later.**

ADDING TO A DOMAIN WITH SCONFIG

When the server has the correct name, you may want to add it to a Windows domain. You can do this with the sconfig utility as well.

1. **In the sconfig utility, type** 1 **to change the domain.**
2. **Type** D **to join a domain and press Enter.**
3. **Give it the name of the domain you want to join and then press Enter.**
4. **Enter the name of an authorized user and press Enter.**
5. **Enter the password of the user and press Enter.**

   You need to restart your computer to apply the change.

6. **Click** yes **to reboot now or** no **to reboot later.**

SETTING THE COMPUTER NAME FROM POWERSHELL

Although sconfig is a nice utility, you may want to be able to script the changes that you want to make. Whenever this is the case, PowerShell can be very helpful. From running batch scripts in the Command Prompt, to running PowerShell scripts in PowerShell, both methods work regardless of whether you're on Windows Server Core or Windows Server with Desktop Experience.

1. **From the sconfig utility, type** 15 **to exit to command line (PowerShell).**

   The PowerShell window opens on your Server Core box.

2. **Use the** `Rename-Computer` **command to change the name of your server:**

   ```
   Rename-Computer -NewName <new-name>
   ```

3. You get a message stating that the NetBIOS name will be truncated if your name is longer than 15 characters.

4. **If you receive this message, type** Y **and then press Enter to accept.**

ADDING TO A DOMAIN FROM POWERSHELL

The ability to script the joining of the domain is a useful skill if you're going to be deploying any quantity of servers. Not only does adding a domain via PowerShell make it simpler to do, but it also helps to ensure that there are no mistakes in the process of joining the domain.

1. **From the sconfig utility, type** 15 **to exit to command line (PowerShell).**

   The PowerShell window opens on your Server Core box.

2. **Use the Add-Computer command to add the server to the domain.**

   Here's an example:

   ```
   Add-Computer -DomainName "your_domain_name" -Restart
   ```

   A dialog box appears asking for a username and password.

3. **Enter a username that is authorized to add systems to your Active Directory domain and enter the corresponding password.**

4. **Click OK.**

   The server restarts.

Configure networking

Before you can set the IP address for the adapter with PowerShell, you need to find out what the index of your interface is. You can do this by typing the following:

```
Get-NetAdapter
```

The output lists all network adapters. In this case, you want the one that says Ethernet. After you have the index number, you can set the IP address and the DNS servers. On my server, the index is 4.

Use the following command to set the static IP address. `InterfaceIndex` is the index number for my network card, `IPAddress` is the IP address I want to assign, `PrefixLength` is the subnet mask that I want to use, and `DefaultGateway` is the gateway address for the local network (see Figure 4-4).

```
New-NetIPAddress -InterfaceIndex 4 -IPAddress 192.168.1.50
   -PrefixLength 24 -DefaultGateway 192.168.1.1
```

TECHNICAL STUFF

I haven't discussed PowerShell much at this point, and this is a more complex bit of PowerShell. The `New-NetIPAddress` is a cmdlet that allows you to work with IP addresses on Windows Server systems. The parameters that come afterward, like `-InterfaceIndex`, help to identify the object you want to work with (the network adapter, in this case) or to make changes to the settings, like the `-IPAddress` parameter where you specify the IP address you want to set on the network adapter.

```
Administrator: C:\Windows\system32\cmd.exe
PS C:\Users\Administrator> Get-NetAdapter

Name                      InterfaceDescription                    ifIndex Status       Mac
                                                                                       Add
                                                                                       res
                                                                                       s
----                      --------------------                    ------- ------       ---
Ethernet                  Intel(R) PRO/1000 MT Desktop Adapter          4 Up           08-


PS C:\Users\Administrator> New-NetIPAddress -InterfaceIndex 4 -IPAddress 192.168.1.50 -Pref
ixLength 24 -DefaultGateway 192.168.1.1


IPAddress         : 192.168.1.50
InterfaceIndex    : 4
InterfaceAlias    : Ethernet
AddressFamily     : IPv4
Type              : Unicast
PrefixLength      : 24
PrefixOrigin      : Manual
SuffixOrigin      : Manual
AddressState      : Tentative
ValidLifetime     : Infinite ([TimeSpan]::MaxValue)
PreferredLifetime : Infinite ([TimeSpan]::MaxValue)
SkipAsSource      : False
PolicyStore       : ActiveStore

IPAddress         : 192.168.1.50
InterfaceIndex    : 4
InterfaceAlias    : Ethernet
```

FIGURE 4-4: Setting the IP address with PowerShell.

To set the DNS Server after that, the command uses the same index number for my network card. `ServerAddresses` is used to identify the DNS servers that the system should use (see Figure 4-5). If you have more than one, you can separate them with a comma.

```
Set-DNSClientServerAddress -InterfaceIndex 4 -ServerAddresses
  8.8.8.8, 8.8.4.4
```

FIGURE 4-5: Setting the DNS servers with PowerShell.

Updating Windows Server 2022

After you have installed your brand-new Windows Server, and maybe even done some of the basic configuration work like changing the name and joining the domain, you'll want to update the server. Updates contain fixes for security vulnerabilities and new features, and should always be installed before turning a server over to the team that requested it.

Windows Server 2022 with Desktop Experience

Considering how important it is to stay up to date on Windows Server updates, most organizations are going to set up automatic updates. You may have a server that can't be set to receive updates automatically, or there may be an emergency patch that was issued and you want to apply it right away. In this section, I explain how to do automatic updates and manual updates.

Automatic updates

Most organization use automatic updates. The following directions walk you through setting up your server to reach out to Microsoft's update servers (the default behavior).

TECHNICAL STUFF

Many organizations have patching solutions that handle the scheduling of updates, and could still be considered an automatic update because the tool will schedule the deployment of approved patches.

1. **Click the Start menu and type** gpedit.msc.
2. **Navigate to the Windows Update section by clicking on Computer Configuration, then Administrative Templates, then Windows Components, and finally Windows Update.**
3. **Double-click Configure Automatic Updates.**
4. **Select Enabled.**

   You're given configuration options.

   Under Configure Automatic Updating, you can see that it's set to Auto Download and Notify to Install. This is the default setting.
5. **Click the drop-down box and select the setting that works best for your environment.**

   In my case, I've chosen Auto Download and Schedule the Install. See Figure 4-6 for an example.
6. **Click OK to save the change.**

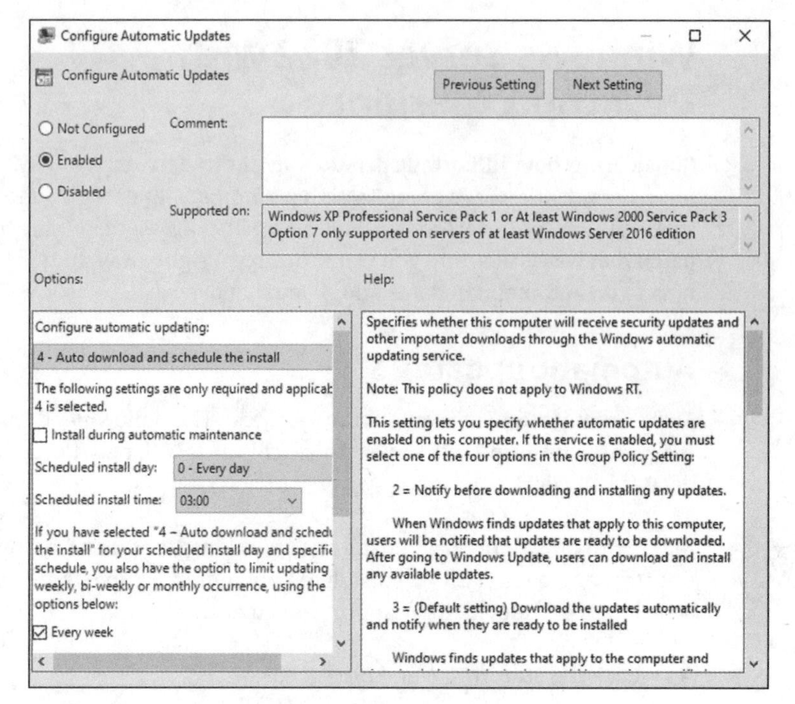

FIGURE 4-6: Using the Local Group Policy Editor to change the automatic updates setting in Windows Server 2022.

Downloading and installing updates

You hear about the next big security vulnerability on the news media, and vendors release patches to the vulnerability very quickly after that. When a security vulnerability impacts your Windows Server systems, you may want to start a manual update — that way, your systems are protected outside of your normal patching windows. If your organization uses a patching solution, the patch may be pushed from that system, but there are always a few systems that don't take the patch for whatever reason. You may have to manually update when that occurs.

1. **With Server Manager open, click Local Server in the left-hand menu.**
2. **Click the hyperlink next to Last Checked for Updates.**

   This may say Never if it hasn't been run yet.
3. **Click the Check for Updates button.**

   The server will check to see if there are any updates available.

Windows Server 2022 Core

Windows Server Core has the same needs when it comes to receiving updates from Microsoft that Windows Server with Desktop Experience does. In this section, I show you how to set up automatic updates and how to perform manual updates from PowerShell.

Automatic updates

There are two ways you can enable automatic updates on Server Core: using the sconfig utility and using PowerShell.

SETTING UPDATES TO AUTOMATIC VIA SCONFIG

The text-driven menu provided by the sconfig utility makes enabling automatic updates very simple. You can set up automatic updates in just four quick steps:

1. **From the sconfig menu, type** 5 **to configure Windows Update settings, and then press Enter.**

   You're given the choice of selecting A for automatic download and install, D for download only (which is the default), or M for manual updates.

2. **Type A for automatic download and installation of Windows updates.**

   You get a text confirmation that the change was successful.

3. **Press Enter to exit the updates section.**

SETTING UPDATES TO AUTOMATIC VIA PowerShell

To set updates to automatic via PowerShell, you need to navigate to `C:\Windows\system32` and stop the Windows Update service. It may already be stopped. Then you can use the script program to execute scregedit.wsf. Adding the switch `/AU 4` enables automatic updates, `/AU 1` would disable automatic updates. The following example enables Windows updates:

1. **Stop the Windows Update Server service.**

   ```
   net stop wuauserv
   ```

2. **Set automatic updates to 4 which is enabled.**

   ```
   cscript scregedit.wsf /AU 4
   ```

3. **Start the Windows Updates Server service.**

   ```
   net start wuauserv
   ```

If you would like to see an example of what this looks like and what the responses should be, please see Figure 4-7.

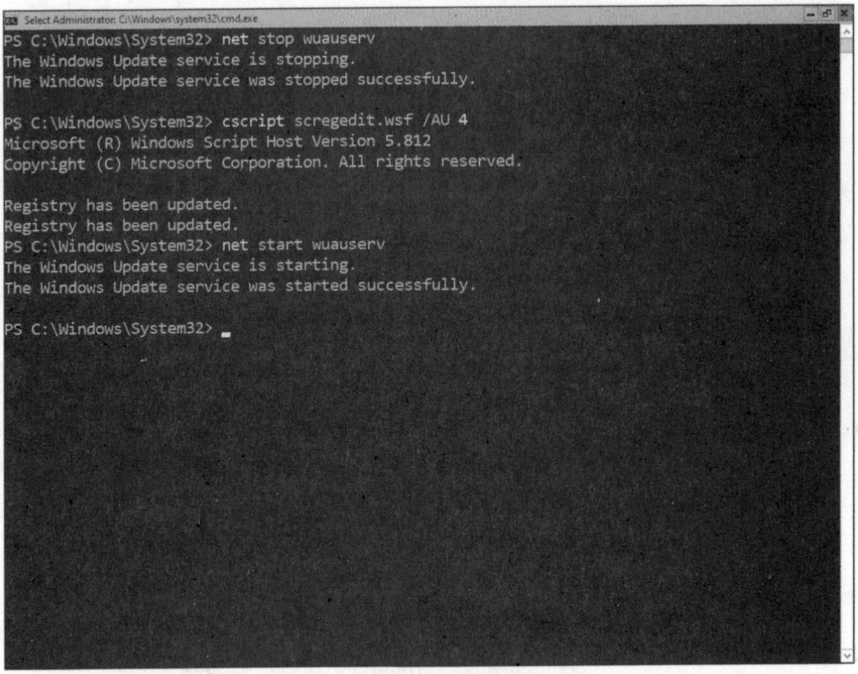

```
Select Administrator: C:\Windows\system32\cmd.exe
PS C:\Windows\System32> net stop wuauserv
The Windows Update service is stopping.
The Windows Update service was stopped successfully.

PS C:\Windows\System32> cscript scregedit.wsf /AU 4
Microsoft (R) Windows Script Host Version 5.812
Copyright (C) Microsoft Corporation. All rights reserved.

Registry has been updated.
Registry has been updated.
PS C:\Windows\System32> net start wuauserv
The Windows Update service is starting.
The Windows Update service was started successfully.

PS C:\Windows\System32> _
```

FIGURE 4-7: Setting automatic updates in PowerShell.

Downloading and installing updates

To force Server Core to then detect and install any available updates, simply type the following command and press Enter.

```
wuauclt /detectnow
```

Customizing Windows Server 2022

After your Windows Server operating system is installed, the next step is to customize it and make it your own! This involves things like installing roles and features, setting up remote administration, and configuring the firewall.

Windows Server 2022 with Desktop Experience

I'll start the customization discussion with the Desktop Experience. When you log into a server with Desktop Experience enabled, by default Server Manager will launch. A lot of the configuration and customization tasks you may have can be accomplished from Server Manager.

Adding roles and features

Roles and features are added in Windows Server 2022 with Desktop Experience through Server Manager.

1. **Open Server Manager.**
2. **Choose Manage⇨Add Roles and Features.**
3. **On the Before You Begin page, click Next.**
4. **On the Select Installation Type page, click Next.**
5. **On the Select Destination Server page, click Next.**
6. **Check the check box next to the role that you want to install and click Next.**

   For this demonstration, I've chosen File Server under File and Storage Services (see Figure 4-8).

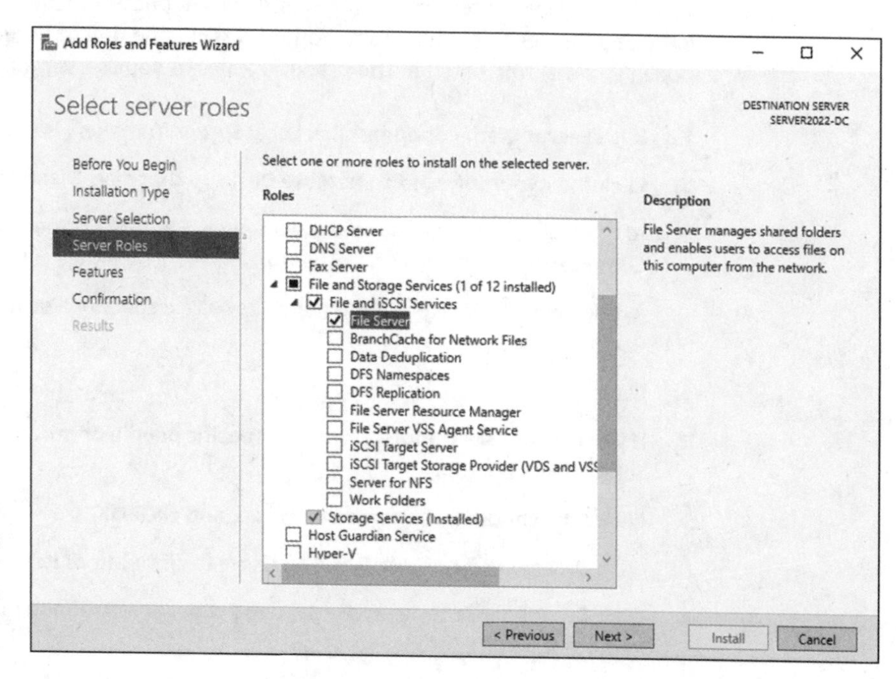

FIGURE 4-8: The select Server Roles Screen with File Server selected.

Performing Initial Configuration Tasks

7. **On the next screen, select any features you may want to install and then click Next.**
8. **If you want the server to restart automatically if needed for the role you installed, you can select the Restart the Destination Server Automatically if Required check box. If a restart is not needed, or you don't want it to restart, leave the check box unchecked.**
9. **Click Install to install the roles and/or features you selected.**

Enabling remote administration

REMEMBER

Remote Management is enabled by default and allows for remote administration through PowerShell. Remote Desktop is a separate setting that allows you to connect to the server and work with it directly.

When a server has Desktop Experience, administrators often prefer to work with the server over Remote Desktop. This is disabled by default; you enable it to use it. If the firewall on the server is enabled and does not have Remote Desktop enabled, you won't be able to connect to it. You need to enable the Remote Desktop – User Mode (TCP-In) rule listed in the Inbound Rules of your server's firewall.

1. **With Server Manager open, click Local Server in the left-hand menu.**
2. **Click the hyperlink next to Remote Desktop that says Disabled.**
3. **In the dialog box that appears, select Allow Remote Connections to This Computer.**

   A dialog box appears telling you that a firewall exception will be made for Remote Desktop.
4. **Click OK.**
5. **If you want to set remote access for specific people or groups, click the Select Users button.**
6. **Click Add, choose your person or group, and click OK.**
7. **Click OK again on Remote Desktop Users to close out of it.**
8. **Click OK one more time on the System Properties screen to enable Remote Desktop.**

Configure Windows Firewall

Assuming that you're going to use the Windows Firewall on your server, you need to know how to enable applications through the firewall. By allowing inbound traffic, you enable the server to do the job you plan on using it for.

1. **From Server Manager, select Local Server on the left-hand side.**
2. **Click the Private: On link next to Microsoft Defender Firewall.**

   The Firewall & Network Protection app opens.
3. **Click Allow an App through Firewall.**
4. **Select File and Print Sharing and enable it for the Private profile by selecting the check box under Private (see Figure 4-9).**
5. **Click OK to save your changes.**

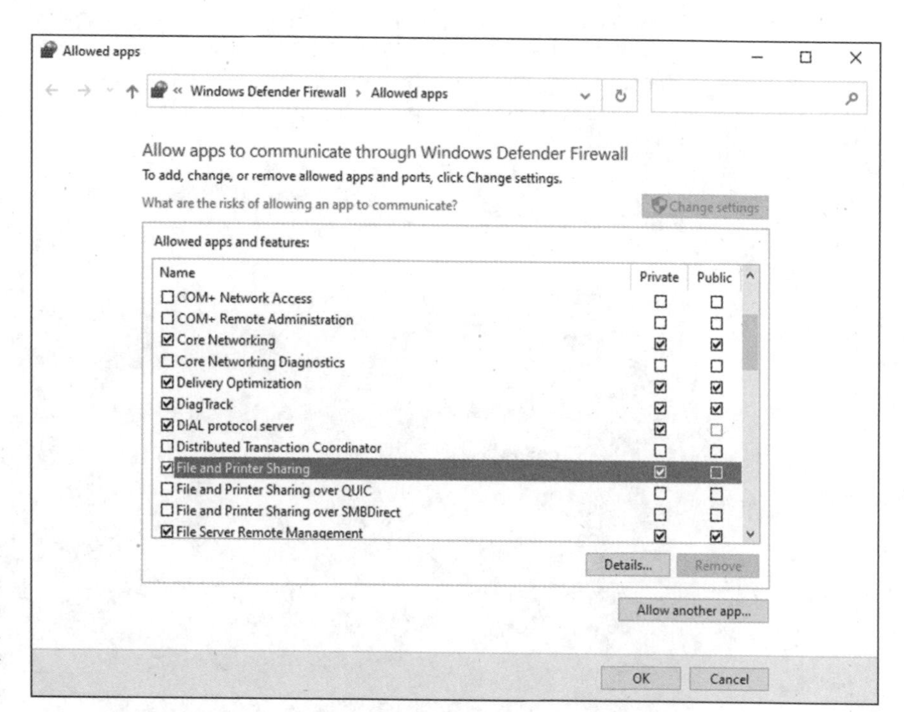

FIGURE 4-9: Allowing an app through Microsoft Defender Firewall.

Windows Server 2022 Core

Whether you're running PowerShell commands against your Windows Server Core system while connected to the console or through remote PowerShell, you can do much of your configuration work with just a few PowerShell commands.

Adding roles and features

To get really good working with Server Core, half of the battle you face is learning how to find the things you want. In Server with Desktop Experience, you have the GUI to guide you. Not so with Server Core.

Let's look at the example I used with the Desktop Experience server. You want to install the File Server role. Before you can install the role, you need to find out what to call it. By using `Get-WindowsFeature`, you can find the names of the roles and features you're interested in. If you have an idea of what the name is, you can do a wildcard search. In the following example, I've used `*file*` to indicate that I want the `Get-WindowsFeature` cmdlet to return results that have the word *file* in them.

```
Get-WindowsFeature *file*
```

When you type the preceding command, you get three results of items that have *file* in their names. You can see File Server under Display Name. For the installation command, you need the name under the Name column. In this case, it's FS-FileServer. Now you're ready to install it! Use the following command to install the File Server (see Figure 4-10):

```
Install-WindowsFeature FS-FileServer
```

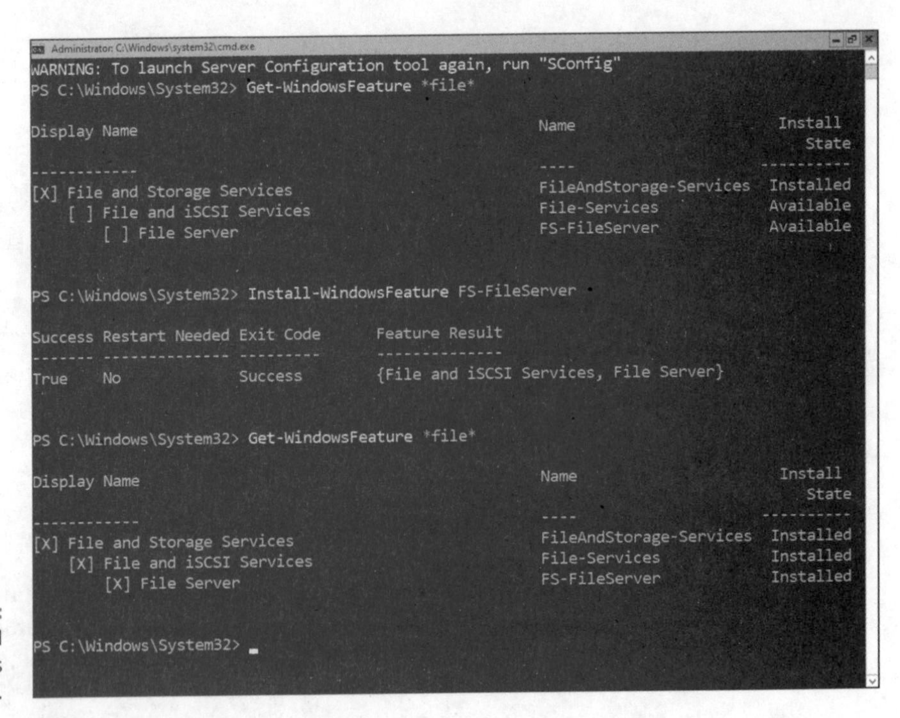

FIGURE 4-10: Using PowerShell to install roles and features.

You see a progress bar as the feature is installed. After it's installed, if you run the first command again, you see that all three results are now installed. File and iSCSI Services was installed because File Server relies on it.

Enabling remote administration

Remote Management is enabled by default in Windows Server 2022. If it was disabled in your environment, you can enable it by running the `Configure-SMRemoting` command. This allows you remotely administer your server with Server Manager.

```
Configure-SMRemoting -Enable
```

To be able to administer the server remotely with PowerShell, you need two additional commands. `Enable-PSRemoting` configures PowerShell to receive remote commands that are sent to your system. `Winrm quickconfig` will analyze and automatically configure the WinRM service for you. This is very helpful when you just want it to work and don't need to customize it. The command starts the WinRM service if it isn't already started, and ensures that WinRM is set to automatically start. It also configures listeners for HTTP and HTTPS, and ensures that the Windows firewall is allowing HTTP and HTTPS traffic inbound.

The `Enable-PSRemoting` command will not give you any output if it succeeds. You'll simply be presented with the PowerShell prompt again.

```
Enable-PSRemoting -force
```

Running `winrm quickconfig` is a little different. After it runs its analysis, it tells you what needs to be changed and asks for a yes or no as to whether it can make the necessary changes. Select Y and press Enter. If everything looked good during the analysis, you'll be told that WinRM is already running and is already set up for Remote Management instead of the yes/no question.

```
winrm quickconfig
```

Configure Windows Firewall

Working with the Microsoft Defender Firewall on Server Core is pretty simple. You need to find the name of the rule you want to work with first. You can do that with the `Get-NetFirewallRule` command (see Figure 4-11). Using the `Format-table` command at the end makes the output more easily readable. Try the command without it — you'll see what I mean!

```
Get-NetFirewallRule *remote* | Format-table
```

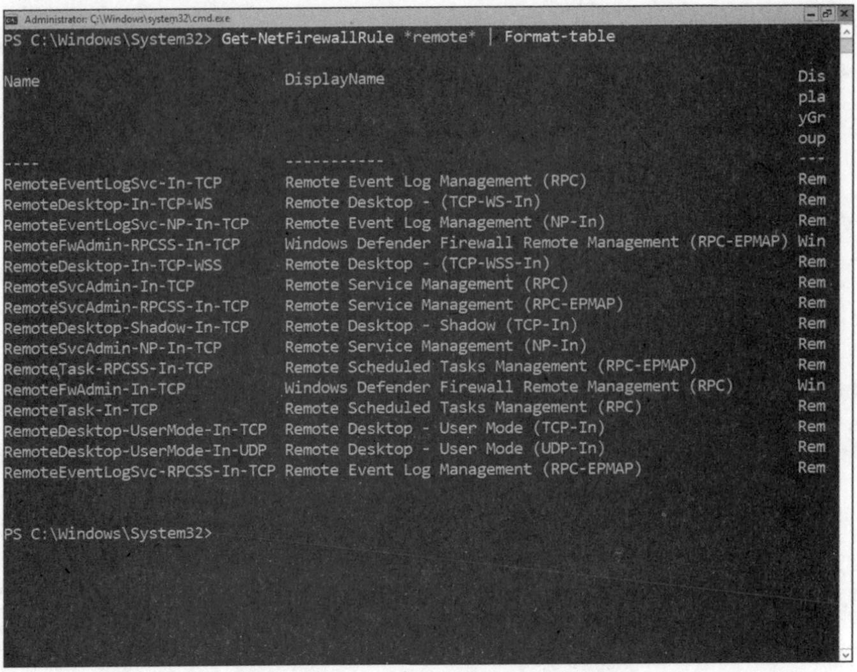

```
Administrator: C:\Windows\system32\cmd.exe
PS C:\Windows\System32> Get-NetFirewallRule *remote* | Format-table

Name                            DisplayName                                                 Dis
                                                                                            pla
                                                                                            yGr
                                                                                            oup
----                            -----------                                                 ---
RemoteEventLogSvc-In-TCP        Remote Event Log Management (RPC)                           Rem
RemoteDesktop-In-TCP-WS         Remote Desktop - (TCP-WS-In)                                Rem
RemoteEventLogSvc-NP-In-TCP     Remote Event Log Management (NP-In)                         Rem
RemoteFwAdmin-RPCSS-In-TCP      Windows Defender Firewall Remote Management (RPC-EPMAP) Win
RemoteDesktop-In-TCP-WSS        Remote Desktop - (TCP-WSS-In)                               Rem
RemoteSvcAdmin-In-TCP           Remote Service Management (RPC)                             Rem
RemoteSvcAdmin-RPCSS-In-TCP     Remote Service Management (RPC-EPMAP)                       Rem
RemoteDesktop-Shadow-In-TCP     Remote Desktop - Shadow (TCP-In)                            Rem
RemoteSvcAdmin-NP-In-TCP        Remote Service Management (NP-In)                           Rem
RemoteTask-RPCSS-In-TCP         Remote Scheduled Tasks Management (RPC-EPMAP)               Rem
RemoteFwAdmin-In-TCP            Windows Defender Firewall Remote Management (RPC)           Win
RemoteTask-In-TCP               Remote Scheduled Tasks Management (RPC)                     Rem
RemoteDesktop-UserMode-In-TCP   Remote Desktop - User Mode (TCP-In)                         Rem
RemoteDesktop-UserMode-In-UDP   Remote Desktop - User Mode (UDP-In)                         Rem
RemoteEventLogSvc-RPCSS-In-TCP Remote Event Log Management (RPC-EPMAP)                      Rem


PS C:\Windows\System32>
```

FIGURE 4-11: Using `Get-NetFirewallRule` to find rules.

The preceding command looks for any rules that have remote in the name. You can see each rule and whether it's enabled.

Let's enable the Remote Firewall Management rules. These would allow you to administer this server's firewall from another system. The rules you're interested in are `RemoteFwAdmin-In-TCP` and `RemoteFWAdmin-RPCSS-In-TCP`.

Here are the commands you'll use to enable these (see Figure 4-12):

```
Set-NetFirewallRule -Name "RemoteFwAdmin-In-TCP" -Enabled True
Set-NetFirewallRule -Name "RemoteFwAdmin-RPCSS-In-TCP" -
  Enabled True
```

If the commands complete successfully, you'll get no response. You'll be returned to the PowerShell prompt. If you run your search again, you'll see that these rules are now enabled.

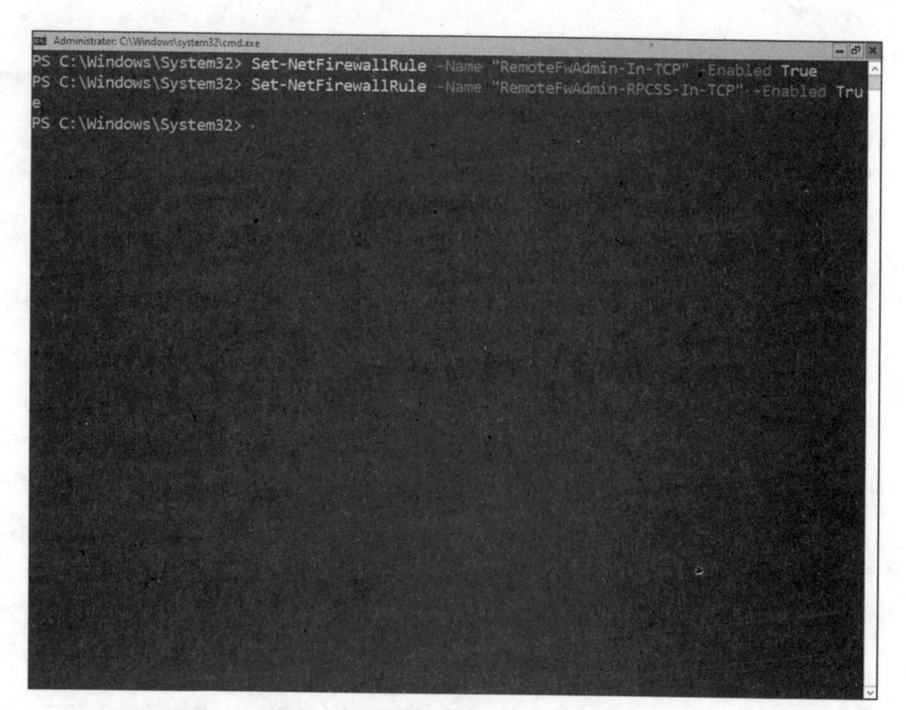

FIGURE 4-12: Using PowerShell to set firewall rules and validate that they're enabled.

Configuring Startup Options with BCDEdit

With Windows Server 2008, Microsoft introduced a utility called BCDEdit, which allows you to manipulate the Windows boot configuration data (BCD) store. The BCD is used to tell the operating system how it should boot; it contains all the boot configuration parameters needed to support that function. This replaced the older `bootcfg.exe` utility that was used to edit the `boot.ini` file pre–Windows Vista. You must be a member of the local Administrator's group on a system to use BCDEdit. This is an advanced utility that is useful in troubleshooting issues that are preventing a server from booting properly.

REMEMBER

You may need to disable or suspend both BitLocker and Secure Boot on a system before you can use BCDEdit.

WARNING

Mistakes made using the BCDEdit utility could leave your system unable to boot at all. Always make sure that you either have a good backup of the system, or if you don't have a good backup, export the current settings from BCDEdit so that you can restore them if needed. You can export the current boot configuration database by typing **BCDEdit /Export** ***<export_path>***. If you need to restore from that export, the command is very similar. You need only type **BCDEdit /Import** ***<path_to_export>***.

Table 4-2 lists some of the more common options available for BCDEdit.

TABLE 4-2 BCDEdit Common Options

Option	Description
/bootdebug	Enables or disables boot debugging.
/dbgsettings	Configures the type of debugging connection.
/debug	Enables or disables kernel debugging.
/delete	Deletes boot entries from the datastore — use with caution!
/deletevalue	Deletes or removes a boot entry option — use with caution!
/displayorder	Sets the order used by the boot manager when displaying the multiboot menu.
/enum	Lists all the entries in the boot configuration datastore.
/export	Exports the contents of the BCD; can be used as a backup to restore the BCD.
/import	Imports the contents of an exported file; can be used as a restore option if needed.
/set	Sets a value in a boot option.

Most often, you'll use `bcdedit /set` to make changes to your boot configuration datastore. Before you make any changes, you need to know what your BCD looks like currently. You can use the `/enum` option to do that. In Figure 4-13, you can see the current settings for the Windows Boot Manager and the Windows Boot Loader.

You may notice that the description in the Windows Boot Loader just says Windows Server. Maybe you want it to be more descriptive than that. You can change it with `bcdedit /set`. You need the ID of the object that you're wanting to work on. In this case, you're wanting to edit the Windows Boot Loader; the identifier that you can see in Figure 4-13 is `{current}`. The full command you type will look something like this:

```
bcdedit /set {current} description "Windows Server 2022
  Standard"
```

TIP

This command will work perfectly in a command prompt, but if you try to run it in PowerShell, you'll need to put quotes around the identifier. For instance, if I were to run the command in PowerShell I would type it like this:

```
bcdedit /set "{current}" description "Windows Server 2022
  Standard"
```

When you get the message `The operation completed successfully`, use `bcdedit /enum` again. You'll see your new description. See Figure 4-14 for my example.

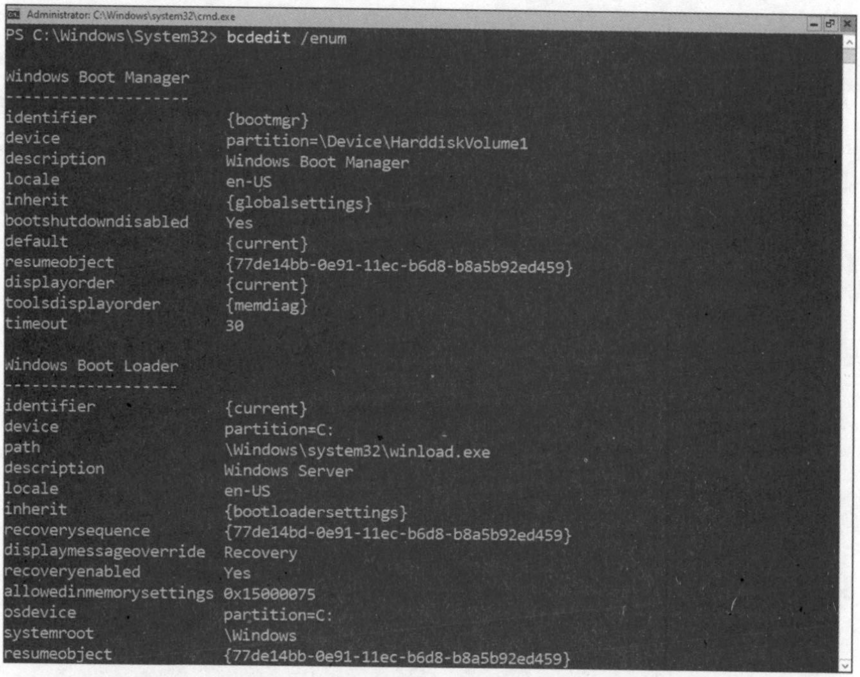

```
Administrator: C:\Windows\system32\cmd.exe
PS C:\Windows\System32> bcdedit /enum

Windows Boot Manager
--------------------
identifier              {bootmgr}
device                  partition=\Device\HarddiskVolume1
description             Windows Boot Manager
locale                  en-US
inherit                 {globalsettings}
bootshutdowndisabled    Yes
default                 {current}
resumeobject            {77de14bb-0e91-11ec-b6d8-b8a5b92ed459}
displayorder            {current}
toolsdisplayorder       {memdiag}
timeout                 30

Windows Boot Loader
-------------------
identifier              {current}
device                  partition=C:
path                    \Windows\system32\winload.exe
description             Windows Server
locale                  en-US
inherit                 {bootloadersettings}
recoverysequence        {77de14bd-0e91-11ec-b6d8-b8a5b92ed459}
displaymessageoverride  Recovery
recoveryenabled         Yes
allowedinmemorysettings 0x15000075
osdevice                partition=C:
systemroot              \Windows
resumeobject            {77de14bb-0e91-11ec-b6d8-b8a5b92ed459}
```

FIGURE 4-13: Using `bcdedit /enum` to see the current settings of the boot configuration datastore.

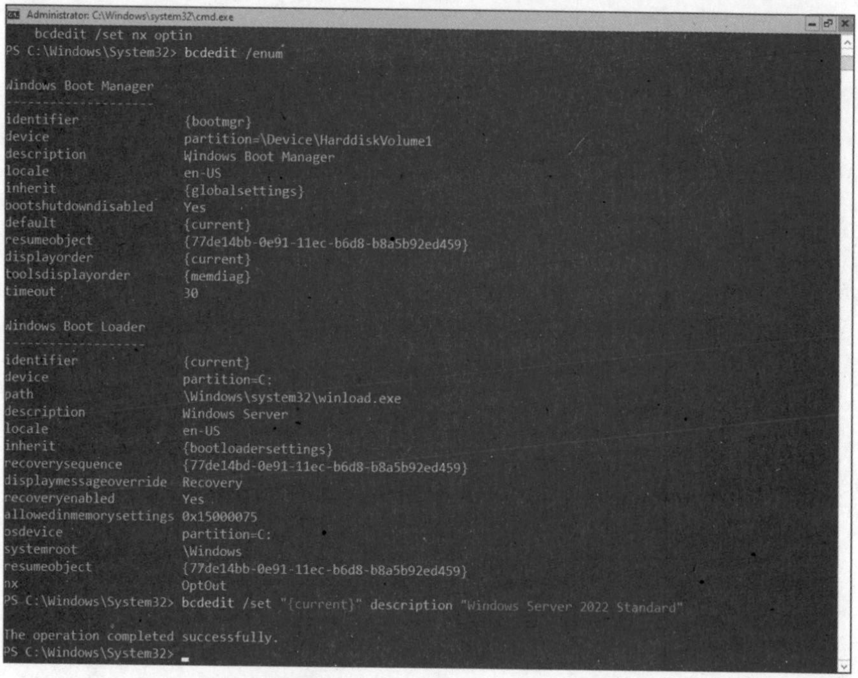

```
Administrator: C:\Windows\system32\cmd.exe
    bcdedit /set nx optin
PS C:\Windows\System32> bcdedit /enum

Windows Boot Manager
--------------------
identifier              {bootmgr}
device                  partition=\Device\HarddiskVolume1
description             Windows Boot Manager
locale                  en-US
inherit                 {globalsettings}
bootshutdowndisabled    Yes
default                 {current}
resumeobject            {77de14bb-0e91-11ec-b6d8-b8a5b92ed459}
displayorder            {current}
toolsdisplayorder       {memdiag}
timeout                 30

Windows Boot Loader
-------------------
identifier              {current}
device                  partition=C:
path                    \Windows\system32\winload.exe
description             Windows Server
locale                  en-US
inherit                 {bootloadersettings}
recoverysequence        {77de14bd-0e91-11ec-b6d8-b8a5b92ed459}
displaymessageoverride  Recovery
recoveryenabled         Yes
allowedinmemorysettings 0x15000075
osdevice                partition=C:
systemroot              \Windows
resumeobject            {77de14bb-0e91-11ec-b6d8-b8a5b92ed459}
nx                      OptOut
PS C:\Windows\System32> bcdedit /set "{current}" description "Windows Server 2022 Standard"

The operation completed successfully.
PS C:\Windows\System32> _
```

FIGURE 4-14: Using `bcdedit /set` to alter the description of the Windows Boot Loader entry.

Performing Initial Configuration Tasks

Why would you want to change the name on the Windows Boot Loader? Consider the example of a multiple boot system that has the same operating system on both disks. The disks are used for very different purposes, so you want to ensure that you remember which is which. Being able to change the descriptions will simplify choosing the appropriate disk in the boot menu. BCDEdit can also be used to change the order of the boot menu. This is useful if you want to set one of your disks to be first in the list and the default disk to boot to after a certain amount of time.

2

Configuring Windows Server 2022

Contents at a Glance

IN THIS CHAPTER

» Using Server Manager to administer your server

» Understanding the server roles available in Windows Server 2022

» Understanding the server features available in Windows Server 2022

Chapter 1 Configuring Server Roles and Features

Being familiar with Server Manager, and how to find the tools that you need will make your life as a system administrator much simpler. From knowing how to access the basic information about your server (like its hostname, IP address, and activation status) to installing new roles and features, Server Manager offers you a central administration point to start from.

This chapter starts with an introduction to Server Manager and discusses where the tools are located that will enable you to be able to perform your job. Then I explain the roles and features that are available in Windows Server 2022.

Using Server Manager

Server Manager is where you'll spend a great deal of time with a brand-new server. It launches right after you log in and is a central management area for the server you're logged into.

When you first log in, Server Manager will launch with the Dashboard selected. The Dashboard has a large tile at the top, called a Quick Start tile, that has typical initial server configuration tasks available as hyperlinks, shown in Figure 1-1. If you want this large tile to go away, you can click Hide in the lower-right corner.

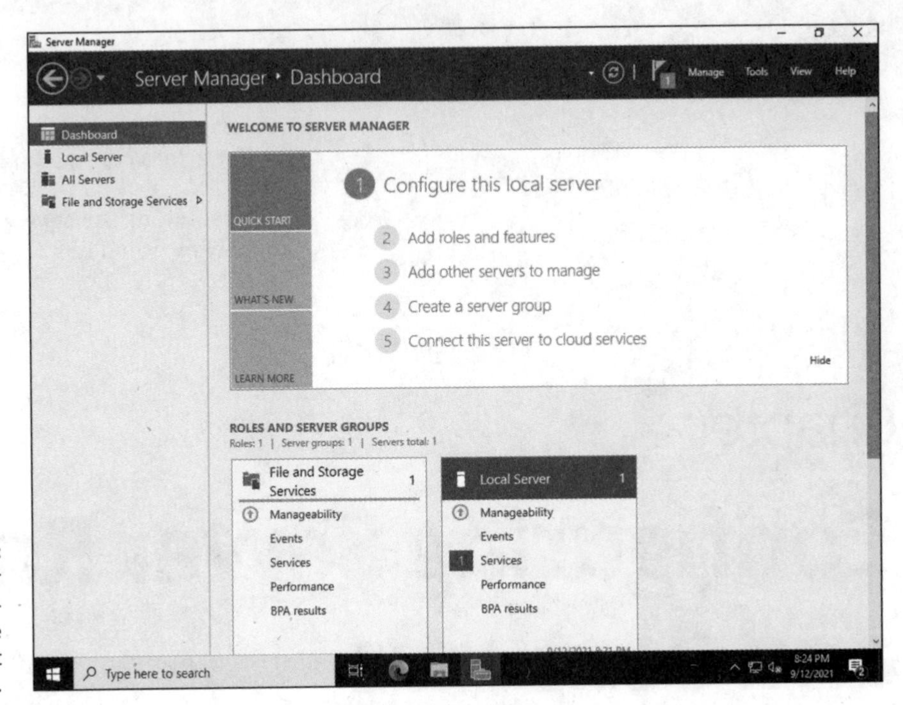

FIGURE 1-1: The Server Manager Dashboard with the Quick Start tile at the top.

Below the Quick Start tile are tiles for all the roles that are installed on the server. With a fresh installation of Windows Server 2022 with no customizations made, you'll have tiles for File and Storage Services and Local Server. These tiles are very useful because they can tell you very quickly if your server is healthy. If the tile is green and has a little up arrow, that means that the service is up and running properly. If the tile is red and it has a little down arrow, that means that the service either ran into problems or isn't running. You can click the individual tiles to get more information on the individual roles. For instance, if I click the File and Storage Services role tile, I can see events related to the services running that are supporting this role, shown in Figure 1-2. On the left side are the configuration options for the local server's storage, including volumes, disks, and pools.

Roles and features

Roles and features allow you to add functionality to your server. A *role* is something you want to use your server to do. For instance, you may install the Active Directory Domain Services role so that you can make this server a domain controller. A *feature* is typically used to support a role. In this case, you would also install the management tools for Active Directory Domain Services (you'll be prompted to install this feature when you select the role). The management tools are a feature.

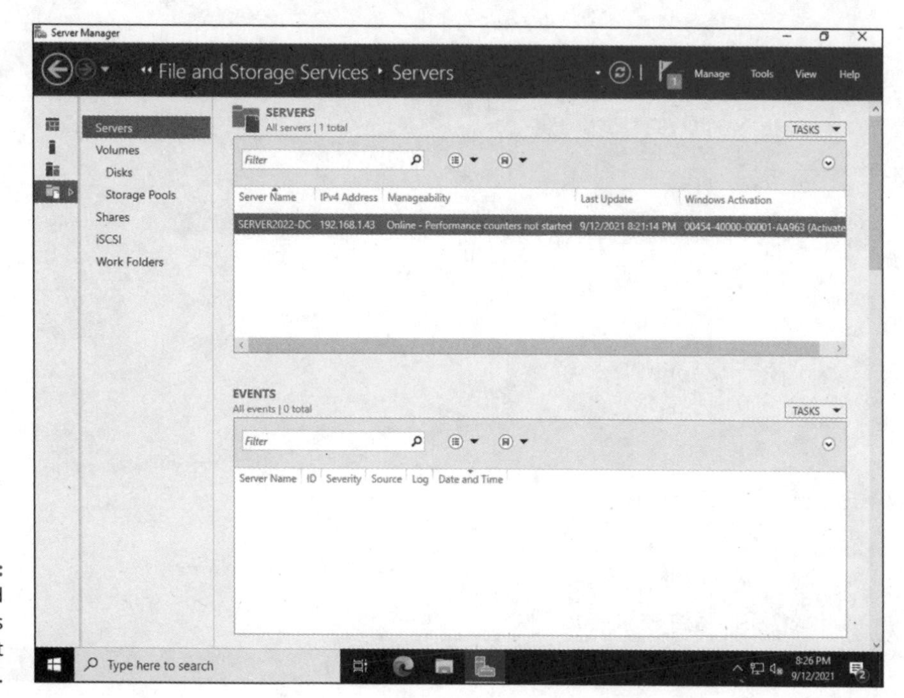

FIGURE 1-2: The File and Storage Services management window.

Diagnostics

Server Manager gives you quick and easy access to many of the diagnostic tools that you may need over your career. To access them, simply click Tools in the top menu, shown in Figure 1-3.

Here are a few of the items in the Tools menu that have helped me over the years:

- **Event Viewer:** When I'm troubleshooting an issue, my first stop is almost always the Event Viewer. I start with the System, Application, and Security logs, and then I get into role- and/or component-specific logs if needed.
- **Performance Monitor:** Performance Monitor is a very useful tool when you need to be able to measure very specific metrics in relation to performance like central processing unit (CPU) idle time, interrupt time, user time, and so on. It provides hundreds of counters out of the box and can be set to start when a certain criteria or threshold is met. This tool is most useful for troubleshooting issues related to performance like slowness and/or freezing.
- **Resource Monitor:** If you just want a quick look at how your system is doing, Resource Monitor is great for that. It has a simple summary screen that gives you information on your CPU, disk, network, and memory usage. This tool focuses on resource utilization and could help troubleshoot issues like low disk space, an overutilized disk, or insufficient network bandwidth.

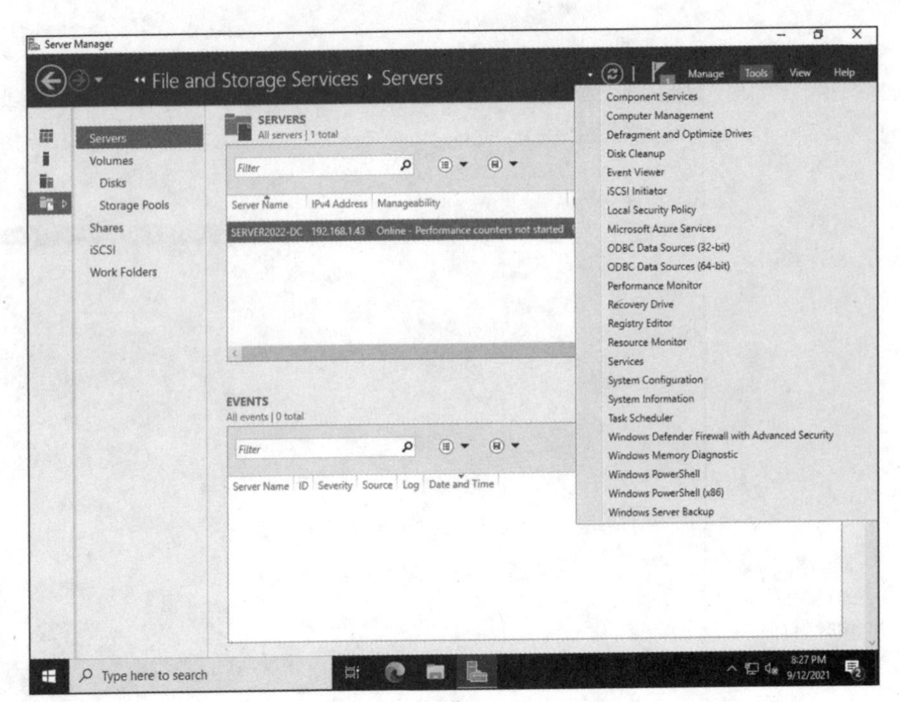

FIGURE 1-3: The Tools menu in Server Manager.

- **System Configuration:** I use System Configuration when I'm having boot issues. The Boot tab gives me several helpful options, including whether I want to use safe boot, if I want to create a boot log, and if I want to use a very basic video setting.
- **System Information:** When you need to know what your hardware specs are or what some of your settings are, this should be your go-to utility. It can even tell you what version of BIOS you're running and what mode it's running in.
- **Windows Memory Diagnostic:** This is the same memory diagnostic that I cover in Book 1, Chapter 2. You're simply launching it from the menu rather than typing in the name of the program.
- **Windows PowerShell:** PowerShell is always handy. You can query settings and export to a text file if need be.

Configuration tasks

The majority of your configuration tasks will be done in the Local Server section in Server Manager. When you click Local Server, you're presented with a Properties page that displays current server information, shown in Figure 1-4. The great thing about this page is that every setting is a clickable hyperlink. If you click it, you're taken to where you can configure that individual setting. Pretty cool, right? When you have a freshly built server, it simplifies the configuration process because you can change the hostname and IP address from here, as well as update the server, add it to a domain, even activate the operating system.

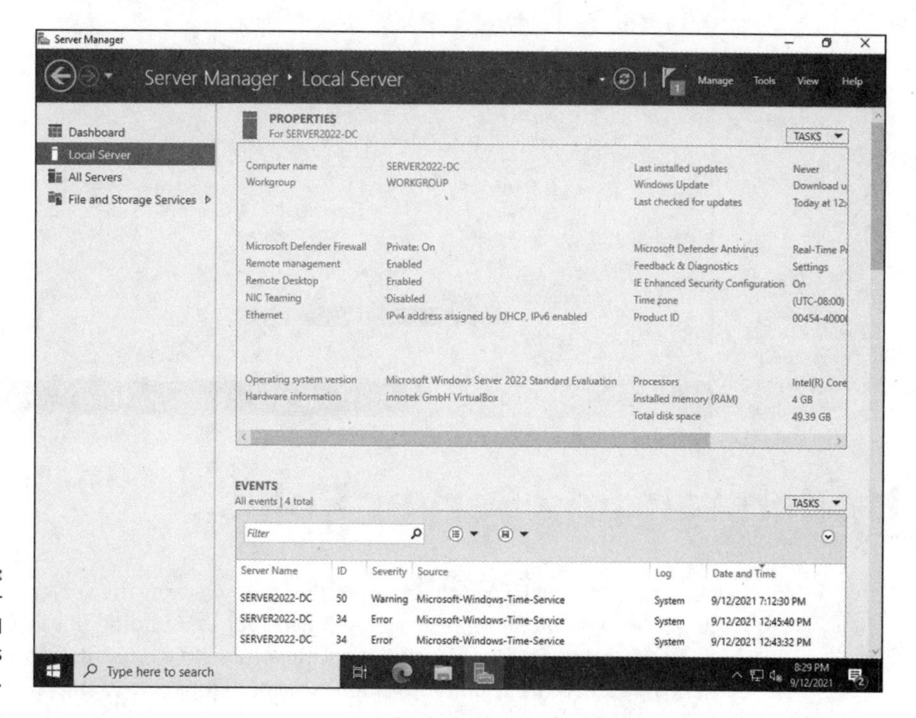

FIGURE 1-4: The Server Manager's Local Server properties screen.

Configure and Manage Storage

I mention earlier that, by default, all Windows Server 2022 systems have the File and Storage Services role installed. This gives you an easy-to-access menu to work with your server's storage. When you click File and Storage Services in the navigation menu, you're presented with several options; select Disks. This is where you can bring new disks online, initialize the disks, and create volumes, shown in Figure 1-5.

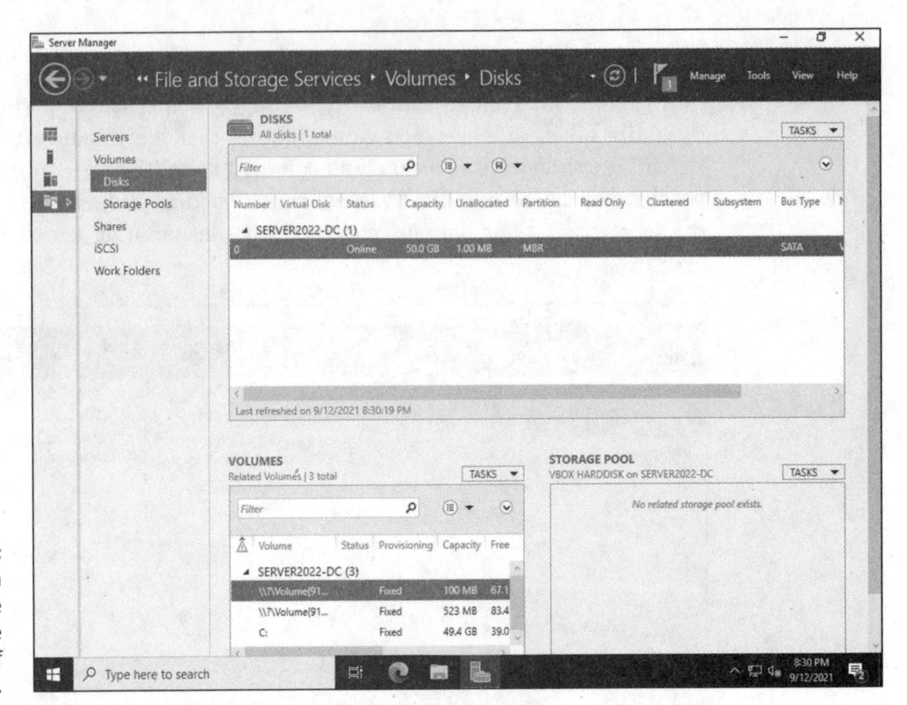

FIGURE 1-5: Working with disks in the File and Storage Services area of Server Manager.

Understanding Server Roles

As I mention earlier, a role is something that we want the server to do. By installing roles, you make servers useful. Maybe you're building out an Active Directory infrastructure, or maybe you're creating a robust virtualization platform. Regardless of what you're trying to accomplish, you'll most likely start by installing a role.

Let's take a look at the roles that are part of Windows Server 2022.

Active Directory Certificate Services

Active Directory Certificate Services (AD CS) is a role that allows you to create a public key infrastructure (PKI) in your organization that will allow you to issue your own internal certificates. This may include certificates for your domain controllers so they can support Lightweight Directory Access Protocol (LDAP) over Secure Sockets Layer (SSL), or certificates for internal web servers, or even code-signing certificates for scripts that will run on your organization's systems. You can install certificate authorities (CAs) and provide additional services like Online Certificate Status Protocol (OCSP), which provides lookups for certificate revocation information, and Network Device Enrollment Service (NDES), which allows network devices to enroll for certificates without domain credentials.

AD CS has a lot of moving pieces. If you're interested in learning more about AD CS, including how to install and configure it, check out Book 5, Chapter 6.

Active Directory Domain Services

Active Directory Domain Services (AD DS) gives you the ability to store information about users and other network objects in a directory service. You can organize these objects in a hierarchical structure with forests, domains, and organizational units (OUs).

Active Directory contains a global catalog, which contains information about every single object in the directory, and is required for successful logon to the domain. With Active Directory, it's simple to search for and locate specific objects if you know a little information about them.

If you're interested in AD DS, you can learn more about installing and configuring AD DS in Book 2, Chapter 5.

Active Directory Federation Services

Active Directory Federation Services (AD FS) can provide single sign-on capabilities to organizations that are utilizing AD DS. It allows those with an Active Directory account to use that account on applications that are outside the boundaries of their Active Directory (for example, a web application hosted by a business partner), or applications that don't rely on Active Directory accounts for authentication at all. By creating a *federation* (the sharing of identity information), the user can be authenticated via his company's Active Directory and can then be authenticated to the business partner's web application with a claim. The business partner simply has to configure their web application to trust the incoming claims.

Active Directory Lightweight Directory Services

Active Directory Lightweight Directory Services (AD LDS) is a Lightweight Directory Access Protocol (LDAP)–based directory service similar to AD DS. It's designed to be used with directory-enabled applications, and it's especially handy for an organization that may want to establish a directory of customer accounts, but keep that directory separate from the organization's AD DS infrastructure.

It can be used as an identity provider with AD FS for both authentication and the generation of claims to web applications that are configured to understand federation.

Active Directory Rights Management Services

Active Directory Rights Management Services (AD RMS) allows businesses to create and enforce policies to protect their data. The rules are created on the AD RMS server but continue to protect documents even if they leave the premises. For example, you can set the policy to allow documents to only be accessible for a brief amount of time, after which the recipient can no longer open them. You can take away the ability to print the document or copy text out of it with copy/paste.

AD RMS is not perfect. It won't prevent someone from taking a screenshot of the data in a sensitive document (there aren't many rights management products that can prevent this activity). Plus, the applications on the client side must support RMS. The functionality exists in the Microsoft Office suite of applications, SharePoint, and Exchange Server. You can also make Internet Explorer compatible with an add-on.

Device Health Attestation

The Device Health Attestation role was added in Windows Server 2016. It gives administrators a way to verify that a device is healthy as it boots. It can measure several different settings and is configured with whichever settings the system administrator or network administrator wants to track. This role is often used for systems to validate that they're safe before they're allowed to connect through remote access services like DirectAccess or other virtual private network (VPN) services.

The settings Device Health Attestation can validate include the following:

- Is BitLocker enabled?
- Is Early Launch Anti-Malware (ELAM) enabled?
- Is Secure Boot enabled?
- Is Code Integrity enabled?

Dynamic Host Configuration Protocol

Dynamic Host Configuration Protocol (DHCP) is a system administrator's best friend for sure. Without DHCP, you had to manually assign an IP address and track which IP addresses were assigned. DHCP automates that process. It can automatically assign IP addresses out to systems on a lease-based system. When the lease has gotten to 50 percent of the configured lease duration time, the client will request that the IP address be renewed. If a system needs to keep the same IP address, you can set a reservation for that IP address. For as long as the system in question has the same network interface card, it will get the same IP address. As an additional bonus, you can set DHCP options for each scope that is defined. These options may tell the systems in the scope where they can find their gateway server, their DNS servers, where an imaging server might reside, and so on.

If you're interested in finding out more about DHCP, check out Book 2, Chapter 5, where I cover installing DNS and DHCP. Be sure to also check out Book 2, Chapter 6.

Domain Name System

Domain Name System (DNS) is a very useful service that helps map hostnames to IP addresses. It's because of DNS that you can type `www.dummies.com` in your web browser, which is really easy to remember, instead of having to remember an IP address like 13.32.254.23. Let's face it, the human brain remembers words and phrases better than numbers.

DNS can resolve hostnames to IP addresses and also can do reverse lookups, which map IP addresses to hostnames. When dealing with network devices that deal only with IP addresses, this can be extremely useful.

If you're interested in finding out more about DNS, check out Book 2, Chapter 5, where I cover installing DNS and DHCP. Be sure to also check out Book 2, Chapter 6. In addition, there is a whole section on securing your DNS infrastructure in Book 5, Chapter 7.

Fax Server

The Fax Server role can give a server the ability to act as a fax machine. The server enables users on the network to send and receive fax messages. The server is handling the actual message transmission and requires a fax modem with a connection to a telephone line, as well as a network connection so that it can communicate with your users on the network.

This type of setup is far more efficient than having multiple physical fax machines hanging around the office. The coolest thing about this role is that it can be configured to send faxes to your users by email, and they can send an email or Word document to the server and have it faxed out.

File and Storage Services

The File and Storage Services role has quite a few components that you can install. By default, on a fresh install of Windows Server 2022, the Storage Services component is installed. None of the following components under File and iSCSI Services is installed:

- **File Server:** Manages folder shares and lets users access those shares from the network.
- **BranchCache for Network Files:** A bandwidth optimization technology that caches the contents of servers at your main site with servers at branch sites.
- **Data Deduplication:** Saves disk space by eliminating duplicate data on drives; a single copy is left intact and links are put in place of the file in the other locations.
- **DFS Namespaces:** Allows you to use a logical namespace to access groups of shared folders on different servers, but it appears to be a single folder with multiple subfolders to end users.
- **DFS Replication:** Synchronizes folders across multiple servers.
- **File Server Resource Manager:** Allows you to manage and classify data on your file servers.
- **File Server VSS Agent Service:** Allows you to enable volume shadow copies on your system, which will take backup copies (snapshots) of your files and/or volumes even if something is using them.
- **iSCSI Target Server:** Services and management tools for iSCSI targets. iSCSI allows you to send SCSI commands for storage over regular TCP/IP networks and enables organizations to have a storage area network (SAN) that is not cost prohibitive.
- **iSCSI Target Storage Provider:** Allows applications connected to an iSCSI target to make volume shadow copies of the data on virtual iSCSI disks.
- **Server for NFS:** Allows the server to serve files to Unix and Linux systems that use the NFS protocol.
- **Work Folders:** Synchronizes files across multiple computers.

Host Guardian Service

This role was introduced for the first time in Windows Server 2016. It manages and releases keys for Hyper-V hosts that are considered trusted (known as *guarded hosts*). This allows the guarded hosts to power on shielded virtual machines (VMs) and perform live migrations. It uses two services to do its work:

- **Attestation Service:** Validates the identity of the hosts that are communicating with it as well as their configuration
- **Key Protection Service:** Gives access to the encrypted transport keys that allows the guarded hosts to work with the shielded VMs

If you want to learn more about shielded VMs, check out Book 7, Chapter 2.

Hyper-V

Installing the Hyper-V role installs a hypervisor on to the Windows Server operating system. On Server Standard edition, you're limited to two VMs; you can run an unlimited number of VMs on Server Datacenter edition. Datacenter edition also includes the ability to work with shielded VMs.

I cover Hyper-V in great detail in Book 7.

Network Controller

Network Controller is a newer role that was introduced in Windows Server 2016. It's only available in the Datacenter edition, not the Standard edition. Network Controller allows you to configure, monitor, program, and troubleshoot your physical and virtual network infrastructure. To do this work, it can leverage Windows PowerShell or the Representational State Transfer (REST) application programming interface (API) to communicate with the devices. If your organization wants to begin exploring Software-Defined Networking (SDN), this is a great way to start. Being able to use PowerShell to work with the Network Controller could be very powerful, but the REST API will allow you to build integrations with other products, including those that would not understand PowerShell. The communication is done through HTTP/HTTPS, so you don't have to worry about opening any uncommon network ports to support REST APIs either.

Network Policy and Access Services

Network Policy and Access Services installs the Network Policy Server (NPS). This provides services like RADIUS and offers authentication, authorization, and accounting (AAA). NPS is very commonly used for authentication of network devices and VPN clients.

Note that you can only install this role on Server with Desktop Experience.

If this sparked your curiosity, check out Book 4, Chapter 3, where I cover the installation and configuration of NPS as a RADIUS server.

Print and Document Services

By installing the Print and Document Services role, you can turn your server into a network print server. This centralizes the management of printing, from working with queues to setting your desired default configurations for network printers. These are commonly things like printing in black and white or printing double-sided.

Remote Access

The Remote Access role allows you to do a few different things. It can provide connectivity to your network with DirectAccess and VPNs, and also offers a web application proxy. At its core, Remote Access is designed to be a VPN solution. Routing and Remote Access Service provides a traditional VPN service to support connectivity to your internal network, while DirectAccess offers end users a more seamless experience with VPN-like functionality. Your users will not have to stop or start their VPN connections; with DirectAccess, they're connected to your organization when they have a good Internet connection. If you install the web application proxy, you can publish HTTP- and HTTPS-based web applications to devices on and off your network. The Routing functionality provides very similar functionality to a traditional router, including network address translation (NAT) and other methods needed to perform routing on an IP network.

Remote Desktop Services

Previously known as Terminal Services, Remote Desktop Services lets users access virtual desktops to run software just as they would if they were on their own desktops. This can be very helpful when you have limited licenses for applications, and the application can be used in this way. It can be especially helpful for client/server-style applications where upgrades can be an overwhelming effort due to configuration changes that need to occur after an upgrade. You can make the changes on each RDS server once, instead of having to do it on hundreds of desktops.

Volume Activation Services

This role creates a Key Management Service (KMS) server, which can manage all the keys for your Windows products and take care of automatic keying and activation for domain-joined systems, servers and clients alike. You can even set requirements like requiring systems to check in with the KMS server every 15 days or the key will no longer be valid. This can help to ensure that laptops find their way back on premises for patches and other things at least every 15 days as well.

Web Services

Web Server installs the Windows-based web server known as Internet Information Services (IIS). IIS can be used to host multiple websites and supports many of the server-side languages you know and love, like PHP and ASP. It also provides support for FTP services. With the Microsoft Web Platform installer, setting up applications like ASP.NET, Microsoft SQL Server, and non-Microsoft applications like WordPress or Joomla is very simple.

Windows Deployment Services

Windows Deployment Services (WDS) makes managing images for servers and desktops very simple. WDS is part Preboot Execution Environment (PXE) server and part Trivial File Transfer Protocol (TFTP) server with a nice, user-friendly graphical user interface (GUI) console to manage it. If you aren't familiar with PXE, it allows a server with no operating system to boot from the network so that a system administrator can configure it and choose an operating system image for it. TFTP is used to transfer the image over the network. Images are saved as `.wim` files and can be kept up to date with tools already available on the system. Systems that are imaged by WDS are booted from their network interface card (NIC) and are able to get the settings for the WDS server from DHCP options 66 and 67.

Windows Server Update Services

Windows Server Update Services (WSUS) is exactly what the name implies: a server role that installs software, which allows you to centrally manage security patches and other updates for all your Microsoft products. It scales well and can be deployed as a single server that does it all, or as an upstream server that downloads updates from Microsoft and then makes those updates available to other downstream WSUS servers.

Understanding Server Features

Roles get a lot of attention, but features provide the necessary support for roles and other applications to perform their functions. Features can provide everything from frameworks to support applications to management tools and encryption functionality.

TIP

If you aren't seeing a feature that you're wanting to use, check whether you're using Standard or Datacenter. Several features are only available in the Datacenter edition of Windows Server 2022.

.NET 3.5

Provides support for .NET 3.5 and legacy support for .NET 2.0 and .NET 3.0 APIs. APIs allow applications to interact with the operating system or services. This may be required for the application you're are trying to install; vendor documentation will usually be very explicit in telling you what needs to be installed as a prerequisite for the application.

.NET 4.8

Windows Server 2022 ships with the newer .NET 4.8, which is installed by default. This feature can also add support for ASP.NET 4.8 and adds support for WCF Services. Many newer applications are taking advantage of these features. Check with your application vendor to see if it supports .NET 4.8.

Background Intelligent Transfer Service

Background Intelligent Transfer Service (BITS) is used to transfer files between servers and clients and will provide progress information on the status of those transfers. It's very commonly used by the Windows operating system to download updates.

The cool thing about BITS is that if the connection is for some reason lost, BITS will suspend the transfer. When the connection is back up, BITS will resume the transfer as if something happened.

There is a service installed by default with the Windows Server 2022 OS, so you don't need to install this feature unless an application requires the feature to serve its purpose.

BitLocker Drive Encryption

BitLocker Drive Encryption is responsible for encrypting the entire hard drive and its contents on systems where it has been enabled. On modern systems with a TPM 1.2 or later chip, BitLocker ensures that the system has not been tampered with while the system was offline. Assuming the hardware checks out okay, it will boot.

Systems with older TPM chips can still use BitLocker, but it is not as user-friendly. Older TPM chips (pre-1.2) do not check for system integrity like the newer TPM chips do either.

TPM stands for Trusted Platform Module. It's a chip on your computer's motherboard, and it's what generates the keys that BitLocker uses to provide the full disk encryption. It keeps half of the key, and the other half of the key is stored on disk. This prevents a thief from stealing a BitLocker-encrypted hard drive and booting it in another system.

BitLocker can lock the startup process until the user enters a PIN. This ensures that the user is the authorized user and will prevent data loss from an unencrypted drive if the system is stolen while offline.

BitLocker Network Unlock

The BitLocker Network Unlock feature was introduced in Windows Server 2012. It gives systems the ability to automatically unlock BitLocker if the system is on the corporate network. This can make patch management simpler if a company is using a Wake on LAN technology to wake systems up for patching or software installations.

BitLocker Network Unlock does have some dependencies on DHCP, so make sure that you're running DHCP in your environment if you want to use it.

BranchCache

You may recall from the File and Storage Service role that BranchCache is a bandwidth optimization technology that copies files from main office file servers and caches the content locally at remote (branch) locations. For users at the remote locations, this means that they can access files with decreased latency. This also means that their network traffic is not crossing the network to the main office to retrieve files, which can improve bandwidth utilization significantly.

Turning on the BranchCache feature on a server enables you to turn your server into a hosted cache server or a BranchCache-enabled content server.

Client for NFS

If you have Unix- or Linux-based file servers in your environment that are using the NFS protocol, installing the Client for NFS feature will allow your server to access the NFS shares, so long as the shares allow anonymous access.

Containers

If you want to run Hyper-V Containers or Windows Server Containers, you need to enable this feature. If you want a higher degree of isolation and want to go with Hyper-V containers, you need to enable the Hyper-V role in addition to the Containers feature.

When the containers feature is installed, you have more steps to getting to a working container host like installing Docker and pulling base images. In the Standard edition, you can have unlimited Windows containers, but you can only have two Hyper-V containers. In the Datacenter edition, both Windows containers and Hyper-V containers are unlimited.

Containers are an exciting new technology that was first introduced in Windows Server 2016. For more on containers, check out Book 8.

Data Center Bridging

If your server is going to be used for clustering or for storage, you should consider enabling Data Center Bridging (DCB). DCB allows you to prioritize certain kinds of traffic over others (think of it like a traffic cop). It allows you to utilize your hardware for better bandwidth allocation as well.

Direct Play

Direct Play is a part of the DirectX API and has been deprecated. You must have Desktop Experience enabled to be able to enable the Direct Play feature. You may still run into applications that require the Direct Play API, though, this is unlikely to be found on a server because it was traditionally used for gaming.

Enhanced Storage

Enhanced Storage enables support for additional functions that are available when you use Enhanced Storage–compatible devices. These devices have built-in safety features that can require you to authenticate before you can access the data on the drive. This is very commonly used in USB flash drives.

Failover Clustering

Failover Clustering is a feature used to provide high availability to server roles. It's often used for file servers, Hyper-V hosts, and database applications like Microsoft SQL Server. If a server in a failover cluster fails, services can be moved almost seamlessly to another server in the cluster. Systems in a failover cluster are referred to as *nodes*. Failover clusters take advantage of shared storage so that all nodes have access to the same data. If a failover event occurs, the transition from node to node can be as seamless as possible, because each node has access to the same storage.

For more on high availability with the Failover Clustering feature, check out Book 7, Chapter 5. There, I discuss Failover Clustering in relation to Hyper-V, but the way it works is pretty similar regardless of which application is using it.

Group Policy Management

Group Policy Management is a Microsoft Management Console (MMC) for managing group policies across your environment. It allows you to create, edit, delete, and assign group policies all the way down to an OU level. It can also be used to enforce a Group Policy Object.

Host Guardian Hyper-V Support

If you want to provision shielded VMs on your Hyper-V hosts, you need to install this feature. It's available in the Datacenter edition only. This allows the Hyper-V server to communicate with the Host Guardian Service.

I/O Quality of Service

Enabling this feature will allow you to set quality of service settings for your applications, including maximum I/O and bandwidth limitations.

IIS Hostable Web Core

The IIS Hostable Web Core feature allows you to write your own custom applications that can host core IIS functionality on their own. Your application will be able to serve HTTP requests and use its own configuration files (`applicationHost.config` and `web.config`) instead of the configuration files used by the traditional full Web Server (IIS) role installation. After the IIS Hostable Web Core is installed, you can open a browser and type **http://localhost**. This will load the traditional IIS splash screen even though the Web Server role is not installed.

Internet Printing Client

Internet Printing Client allows you to connect to and print to printers on the network or Internet using the Internet Printing Protocol (IPP). It does require that Desktop Experience be installed and, as such, is not available in Server Core.

IP Address Management Server

IP Address Management (IPAM) was a breath of fresh air to network administrators and system administrators who had to manage multiple DNS and DHCP servers. It provides a centralized management pane for both DNS and DHCP and is able to help you locate available IP addresses, available subnets, and so on. Best of all, it supports multiple Active Directory forests, so it really can be a single pane of glass for your organization.

You can read more about IPAM in Book 2, Chapter 6.

LPR Port Monitor

Line Printer Remote (LPR) Port Monitor enables your server to print to a printer that is shared using Line Printer Daemon (LPD). This will typically be a Unix or Linux server being used as a print server.

Management OData IIS Extension

This feature gives you the ability to expose PowerShell cmdlets through an OData-based web service that runs on IIS. OData is a data access protocol that allows you to query and update data. To use this feature, you need to install the Web Server role.

Media Foundation

Media Foundation allows you to work with media files. You can transcode. You can analyze media files. You can even generate thumbnail images for media files. It also offers DirectX Video Acceleration and an enhanced video renderer (EVR). Media Foundation supports many of the codecs, sources, and sinks that you would expect, including AVI, DV, H.264, MP3, and MP4.

Message Queueing

Message Queueing is often used by applications to deliver messages to other applications. It guarantees message delivery and provides routing, security, and messaging based on priority between applications. Applications are able to send and receive messages from the queues. This is very useful when you need the guarantee that the message will get to its end destination, or when an application may not be able to get the message right away because it's busy or offline. If your application requires Message Queueing, your vendor will list it as one of its requirements.

Microsoft Defender Antivirus

This feature is installed by default and was previously known as Windows Defender. It's a built-in next-generation antivirus solution that's able to look at files and process behaviors for things that appear malicious. It gets regular updates from signatures and through machine learning and threat research.

Multipath I/O

What happens if your server is connected to its storage through a network switch, and that switch goes down? Your server can't communicate with its storage any longer. With Multipath I/O, you can allow your server to use multiple paths to your SAN — you could have connections through two separate switches, for instance. If you have connections to two separate switches, your server will still be up and able to access its storage even if one of the switches goes offline. This allows you to build a truly fault-tolerant storage network.

Multipoint Connector

If you're using Multipoint Services, the Multipoint Connector allows the system to be managed by the Multipoint Manager and the Multipoint Dashboard.

Network Load Balancing

Network Load Balancing (NLB) allows you to spread traffic across multiple servers, which can improve response times because the traffic is evenly distributed. It's popular with web servers, especially with a stateless application, where the user's request can be served by a server in the NLB cluster. As load increases, you can simply add more servers to the cluster, and when you need to do maintenance on a particular server, you do so while keeping the other servers in the NLB cluster up. An NLB cluster focuses on reliability and performance, not high availability or fault tolerance.

Network Virtualization

Network Virtualization allows you to create virtual network overlays on the same physical network. If you want to start working with software-defined networking (SDN), this is an important feature to examine. By taking advantage of network virtualization, you can automate the provisioning of networking resources, in addition to other server automation projects you may be working on.

Peer Name Resolution Protocol

Peer Name Resolution Protocol allows applications to register and resolve names on your computer so that other computers on the network can also communicate with these applications. This is especially helpful for systems that are in workgroups, rather than being domain-joined.

Quality Windows Audio Video Experience

Quality Windows Audio Video Experience, also known as qWave, is a networking platform for A/V streaming applications on home IP networks. When it's installed on a Windows Server OS, it only provides rate of flow and prioritization services.

RAS Connection Manager Administration Kit

The RAS Connection Manager Administration Kit (CMAK) feature serves a very simple purpose: to create profiles for connecting to remote servers and remote networks. This feature is only available in the Desktop Experience.

Remote Assistance

The Remote Assistance feature allow you, as the support person, to offer remote assistance to your end users. You can view and control the user's desktop from the server. Remote Assistance requires Desktop Experience.

Remote Differential Compression

Remote Differential Compression can help to optimize bandwidth. It's able to look at a source and destination object and will only transfer the differences between the objects, rather than transfer the object as a whole.

Remote Server Administration Tools

Remote Server Administration Tools (RSAT) is your best friend as a system administrator. Installing the RSAT feature will give you all the snap-ins and command line management tools to manage roles and features. Typically, when you install a role or a feature that has a management tool in RSAT, you'll be prompted to install the management tool at the same time. Installing RSAT is traditionally reserved for client-side devices to aid in remote administration tasks, so unless you're going to administer roles and/or features on the server, you shouldn't install the tools there.

To install RSAT on your client device, you can download the RSAT installation package from the Microsoft website. For Windows 10 client systems that are not on the October 2018 update, you can download RSAT at `www.microsoft.com/en-us/download/details.aspx?id=45520`. If you're using Windows 10 and have the October 2018 update, you don't need to download the RSAT installer. It's included as a Feature on Demand, which you can install from the operating system directly.

REMEMBER

If you install the role or feature through PowerShell, you have to specify the management tools to get RSAT to install — for example, `Install-WindowsFeature -Name Web-Server -IncludeManagementTools`.

RPC over HTTP Proxy

This feature is typically used to support VPN clients that need to communicate over HTTP. It relays RPC traffic over HTTP, as the name suggests.

Setup and Boot Event Collection

This feature was first introduced in Windows Server 2016. With this feature enabled, you can set up your server as a collector, which can be used to gather lots of different types of events from other systems as they boot up or as they go through the setup process. You can view the events once they're collected with Event Viewer as you're used to.

Simple TCP/IP Services

This feature is provided for backwards compatibility and should not be installed unless it's required. It's a collection of utilities used on the command line. The utilities respond to telnet requests on specific ports. Quote of the Day, for example, will give you a random quote when you telnet to port 17.

SMB 1.0/CIFS File Sharing Support

If you enable this feature, you're enabling support for Common Internet File System (CIFS) clients and/or CIFS servers to connect over SMB v1.0, which is an insecure protocol at this point. Only use this feature if absolutely necessary. At this point, it's a huge security vulnerability and should never be enabled. If you have devices or applications that require SMB v1.0, you need to work with your organization to get those devices and/or applications replaced.

SMB Bandwidth Limit

The SMB Bandwidth Limit feature allows you to categorize your SMB traffic and limit the amount of traffic you want to allow by category. This is especially helpful when you're doing live migrations on Hyper-V hosts and you want to limit the amount of bandwidth that the live migrations are able to use so that you don't negatively impact your other VMs or your end users.

SMTP Server

The SMTP Server in Windows Server 2022 is a basic email server. It can be used as an organization's main email server so long as the organization is small. Just keep in mind that it's nowhere near as robust as Exchange Server.

Simple Network Management Protocol Service

Simple Network Management Protocol (SNMP) is used by many organizations to monitor devices for events and status. Enabling the SNMP Service gives you the ability to accept events from other servers and devices.

To configure SNMP after installation, launch the Service Control Manager MMC console (`services.msc`) and locate the SNMP Service. In the Properties for the service, you can set the community string and which hosts you want to receive SNMP packets from.

Software Load Balancer

This feature provides outbound network address translation (NAT), provides inbound NAT, can load-balance between multiple instances of applications, and can check to make sure that an instance of the application is healthy before sending traffic to it. This is excellent for SDN because you can configure it though PowerShell. It operates at Layer 4 of the OSI model, the Transport Layer.

Storage Migration Service

Storage Migration Service is new to Windows Server 2022. It allows you to painlessly inventory your data and settings on a server and then transfer that data and the configuration settings to a newer server. The new server can then take over the identity of the old server. Applications and users don't need to change anything on their end.

This is a great new feature! Think of that old Windows Server 2008 system that's still hanging around because everyone is afraid to touch it. Storage Migration Service is your answer to that old server. You can migrate data from systems as old as Windows Server 2003, but the destination server has to be Windows Server 2012 R2 or newer.

TIP

Using a Windows Server 2022 system as the destination is recommended because you can install the Storage Migration Service Proxy, which can double the transfer performance over older versions of Windows Server.

Storage Migration Service Proxy

When this feature is installed on the destination server, the transfer performance of the Storage Migration Service is almost doubled. This is only available on Windows Server 2022.

Storage Replica

Storage Replica adds the capability to replicate synchronously or asynchronously across servers or clusters. This is great for disaster recovery!

When Storage Replica was first introduced, it was only supported in the Datacenter edition. With Windows Server 2022, Storage Replica was made available in the Standard edition as well, though it does have some limitations when it's installed on Standard: It can only replicate a single volume, volumes can have only one partner, and volumes can only be a max size of 2TB.

System Data Archiver

This feature is installed by default on Windows Server 2022 and is also new to Windows Server 2022. Its job is pretty simple: It's responsible for collecting and archiving system data from the server.

System Insights

One of the challenges of being a system administrator is forecasting what your compute and storage needs are. New to Windows Server 2022 is System Insights. This cool new feature includes analytics and machine learning to predict based on usage what your needs may be. It's a very useful tool when doing capacity forecasting, and can cover compute, storage, and networking needs. No more guesswork!

Telnet Client

The Telnet Client allows you to connect to a Telnet Server using the Telnet protocol.

Use this carefully! It can be very useful for troubleshooting, but because it sends information in plain text, you don't want to send usernames or passwords through it.

TFTP Client

The TFTP Client feature allows you to interact with a TFTP server. With this feature installed, you can read from and write to a remote TFTP server. This may be beneficial for network administrators in particular to pull/push images on network equipment.

VM Shielding Tools for Fabric Management

Fabric in this context is referring to a guarded fabric, which provides a more secure infrastructure for shielded VMs to run on.

This feature should be installed on the Fabric Management Server. It includes utilities that can be used by solutions that manage the fabric.

WebDAV Redirector

The WebDAV Redirector allows you to connect to WebDAV sites and access files on the sites through a mapped drive. This is great from a compatibility standpoint because some applications don't support WebDav, but they can absolutely understand how to access files on a mapped drive.

Windows Biometric Framework

The Windows Biometric Framework allows you to use fingerprint devices or facial recognition to authenticate to Windows. This includes the Windows Biometric Service, which supports the Windows Biometric Framework API. Client applications are able to leverage the API to take advantage of biometric authentication.

Windows Identity Foundation 3.5

The Windows Identity Foundation 3.5 feature provides a .NET 3.5 framework for building claims-aware applications. You should only use this if for some reason you need to code against .NET 3.5 or .NET 4.0. Windows Identity Foundation is included in .NET 4.5 and is no longer a separate feature that needs to be installed.

Windows Internal Database

The Windows Internal Database is a relational database intended to support Windows roles and features such as AD RMS, WSUS, and Windows System Resource Manager. This is not designed to replace SQL Server. It's really only intended to support roles and features in the Windows Server operating system.

Windows PowerShell

Windows Server 2022 includes both PowerShell 5.1 and the PowerShell ISE installed by default. In most cases, this is all you need to work with the server with PowerShell. You can also install the older PowerShell 2.0 Engine, PowerShell Web Access, and PowerShell Desired State Configuration (DSC).

PowerShell is such a broad topic that Book 6 is devoted to it.

Windows Process Activation Service

When you install the Windows Process Activation Service, you can provide features you would normally get with IIS and HTTP applications to non-HTTP applications using Windows Communication Foundation (WCF) services. Additionally, IIS 10.0 takes advantage of Windows Process Activation Service to do message-based activations over HTTP.

Windows Search Service

Windows Search Service (WSS) can analyze a set of documents and extract useful information, typically metadata, which can then be queried later on. The processing of indexing can be pretty CPU intensive, but the service will throttle itself or even pause indexing if the user experience might be impacted by indexing. By leveraging the indexing, your users will notice a performance improvement when they do a search on your file server.

Windows Server Backup

Windows Server Backup is a built-in backup utility. It can be used to back up and restore data, and can perform full backups, system state backups, volume backups, and specific folder backups. You can even do a bare metal backup, which will allow you to completely restore your system should it need to be rebuilt.

Backups can be saved on a local drive or on a remote server share, and they can be run once or scheduled to run as often as needed.

Windows Server Migration Tools

Microsoft tried to make migrating to a newer version of Windows Server as painless as possible with Windows Server Migration Tools. This feature, when installed, can migrate roles, features, OS settings, and shares. I think that the greatest value of this tool is being able to move roles from several versions of the operating system back. This can be a very scary proposition for some system administrators, but Windows Server Migration Tools really simplifies the process and reduces the risk of things going wrong.

Windows Standards-Based Storage Management

Do you need to discover, manage, and monitor your storage devices? Are the management interfaces using the SMI-S standard? If you answered yes to both questions, then this is the feature for you. Installing this feature will add several Windows Management Instrumentation (WMI) classes and Windows PowerShell cmdlets to the server and will allow you discover, manage, and monitor compatible devices.

Windows Subsystem for Linux

The Windows Subsystem for Linux (WSL) originally gave you the ability to run a form of an Ubuntu-based bash shell on Windows. It has been enhanced since that early start and now allows you to install a full version of Linux from the Windows Store. It should be noted that WSL does not give you the full graphical Linux experience. It gives you terminal access and was primarily designed with developers in mind.

Windows TIFF IFilter

When the Windows TIFF IFilter is enabled, you can search TIFF files for text using optical character recognition (OCR). The IFilter can be used by the Windows search utility and will allow you to do full text searches of the TIFF files on your systems. It's worth noting that the text needs to be clear. If the TIFF image is the result of a scan and the scan quality is poor, or includes images, the IFilter may not be able to read the TIFF file as well.

WinRM IIS Extension

Enabling WinRM IIS Extension allows you to manage the server remotely from a client that is using WS-Management (WS-Man), like PowerShell remoting for example.

WINS Server

WINS Server maps out NetBIOS names to IP addresses. A Windows system could register itself with WINS, and then other systems in the workgroup could query the WINS server for that system's IP address. This has, for the most part, been replaced by DNS in most organizations.

Wireless LAN Service

If your server needs to connect to a wireless connection, you'll need to install this feature. Wireless LAN Service allows the server to find wireless network adapters and manages both the wireless connections and wireless profiles.

WoW64 Support

WoW64 Support is installed by default on Windows Server 2022 and allows you to run 32-bit applications on a 64-bit system.

XPS Viewer

The XPS Viewer is installed by default on Windows Server 2022 with Desktop Experience. It allows you to read XPS documents and assign permissions or digitally sign XPS documents as well.

IN THIS CHAPTER

» Working with Device Manager

» Adding new hardware with the Add Hardware Wizard

» Configuring hard drives and performing other storage-related tasks

» Completing printer-related setup and configuration tasks

» Performing other configuration tasks

Chapter 2

Configuring Server Hardware

Whenever you're working with a brand-new server, and sometimes when you're working with an older one, you'll have configuration tasks you need to perform. Some of these tasks are related to the operating system and the applications you want to install; others are related to the server hardware.

When you first install a new server, one of the first things that you'll want to do is verify that all the hardware is functional. This means checking to ensure that there is a driver present for each piece of hardware and that none of the devices is having issues.

With an older server that is having issues, you need to be able troubleshoot whether the hardware is the issue. This is especially important if the software has not been changed, but the server has suddenly started freezing, crashing, or performing slowly.

In this chapter, I walk you through some different methods of working with server hardware and explain your options.

Working with Device Manager

Device Manager has been front and center for working with hardware in Windows operating systems for a very long time. In fact, it was introduced for the first time to the Windows Server operating system with Windows Server 2000. Its strength comes from its simple interface that makes it very easy to spot hardware devices that are having difficulties. Icons next to each section and driver tell you at a glance if there is an issue with a device in that section and which device is having the issue. Here are the icons you may run into:

- **An orange triangle with an exclamation point in the middle:** Indicates that the device is having a problem. It may still be functioning but should be checked out.
- **A white circle with a black arrow pointing down:** The device is disabled. You can enable the device by right-clicking the device and choosing Enable Device.
- **A white circle with a lowercase *i* in blue:** This does not mean there is a problem of any kind. Instead, it indicates that the device was manually configured and is not using Use Automatic Settings.
- **A white circle with a blue question mark:** This means that a driver has been installed that should be compatible with the device, but it is not the device's intended driver, because that driver could not be located.

Opening Device Manager

There are multiple methods to open Device Manager. In this section, I present one of the shortest methods to open it. Microsoft simplified getting to many of the administrative tools that are commonly used on the Windows Server operating system, and this is one of those tools.

1. **Right-click the Start menu.**
2. **Select Device Manager from the menu.**

   Device Manager looks similar to Figure 2-1.

Configuring how Device Manager displays

The View menu in Device Manager gives you different ways to visualize the hardware in your system. You can logically group your hardware in different ways that make sense to you, which can make it easier to troubleshoot.

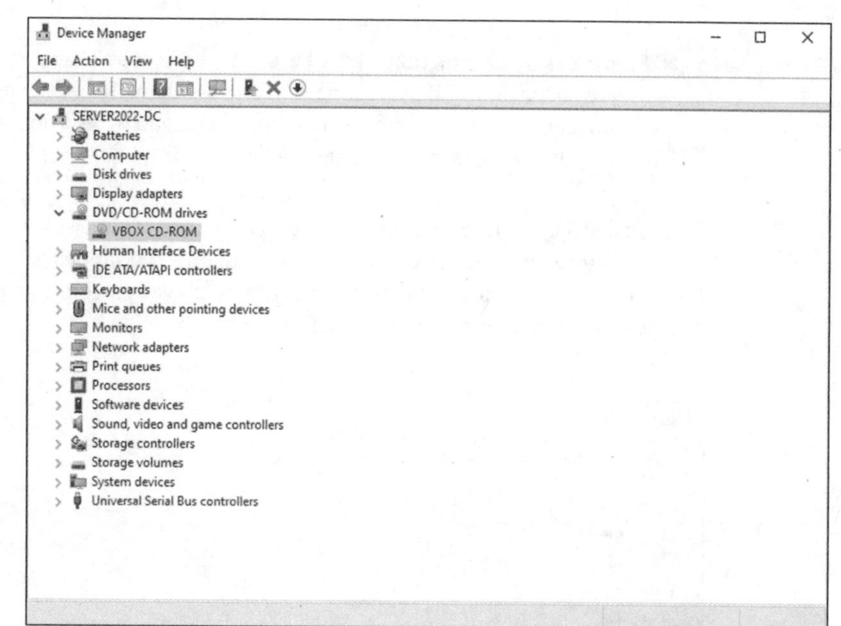

FIGURE 2-1: Device Manager on Windows Server 2022 will look familiar to those who have worked with Windows in the past.

When you choose the View menu, you have the following options:

- **Devices by Type:** This groups the hardware logically by the type of hardware. For instance, if your server has multiple network interface cards, they will all be grouped under Network Adapters. If you have multiple hard drives, they'll be grouped under Disk Drives. This is the default view that Device Manager opens up to.
- **Devices by Connection:** The devices in your server are able to do their jobs because they're all connected to each other somehow. This view sorts them by how they're connected. Many of your storage devices and network adapters will show up under the PCI Bus because that's typically how they're connected.
- **Devices by Container:** This view groups devices by container IDs. These container IDs may be assigned to the device, or the device may inherit a container ID from a parent object. For instance, one of the parent containers is your server. Some of the items grouped under that container may be communications ports, ATA channels, processor cores, and so on.
- **Resources by Type:** This view will sort all your server's devices into resource types. The types are direct memory access, input/output, interrupt request, and memory. I discuss these in more detail in this chapter, in the "Understanding resources" section.

» **Resources by Connection:** This view sorts your system's resources by how they're connecting and interacting with one another. This view was very helpful back when it was common to have to research interrupt request (IRQ) conflicts but may not be as widely used now.

You can also customize what information you're shown in Device Manager. Choose View⇨Customize and the Customize View dialog box appears (see Figure 2-2). From here, you can change which items are displayed to you. I've never really had a good reason to change from the default view.

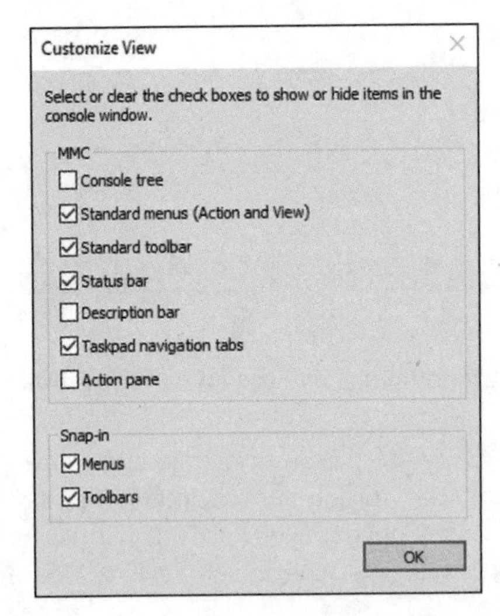

FIGURE 2-2: Customizing your Device Manager view.

Viewing devices that are not working properly

Typically, if a device is broken, it will show up in one of two ways:

» If the device is having issues but is still somewhat functional, it will have an orange triangle on it with an exclamation point in the middle.

» If the device is not working at all and has a white circle with a black downward-facing arrow, that means the device is disabled.

When you have a device that's having an issue, it will automatically be expanded in Device Manager when you open it. That makes it much simpler for you as the system administrator to find the device that's having problems. In Figure 2-3, you can see I have a device with an issue. The DVD/CD-ROM drive is disabled, which is evident because you can see the downward-facing arrow next to its name.

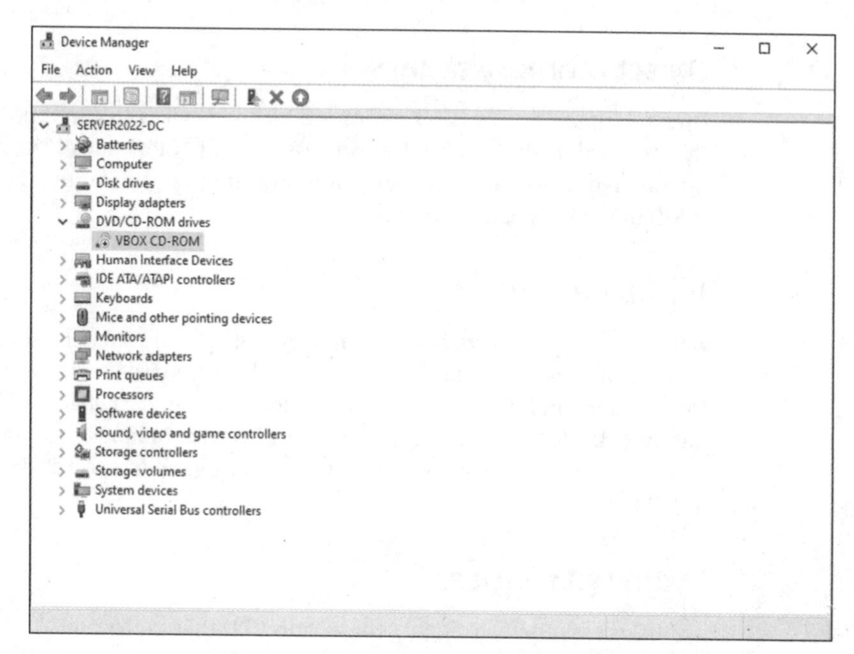

FIGURE 2-3: Device Manager makes it simple to identify devices with problems.

The DVD/CD-ROM can be re-enabled to see if it's having issues. I can simply right-click it and choose Enable Device. In my case, it will function properly because I disabled it to take the screenshot.

When I view the Properties screen for a broken device, I can find out more about what the issue is. In the case of a missing driver, I can have Windows automatically search to see if it's able to find a compatible device. Most often this will fix the problem. If not, then your best bet is to go to the vendor's website to download the drivers manually.

Understanding resources

Understanding the resources that are represented in Device Manager can be very helpful when troubleshooting technical issues on the system. In some cases, like input/output (IO) and IRQ, having duplicate addresses can cause issues with devices. The devices may stop working or may cause your system to freeze or crash.

If you've gone through the content for the A+ certification, this section will probably be old news and you can feel free to skip it. If you haven't worked on your A+ certification yet, or you're new to servers and computing, read on!

Direct memory access

Direct memory access (DMA) allows attached devices to communicate directly to the memory, taking the central processing unit (CPU) out of the equation as the middleman. This can speed up communication significantly. Typically, around 16MB of memory can be used for DMA.

Input/output

I/O, at its simplest, is defined as any device that can transfer data to or from your server. This could include input devices like keyboards and mice, output devices like printers and monitors, and other devices that are capable of both input and output (like disk drives). In the context of Device Manager, on the Resources tab you can see which I/O addresses have been assigned to specific devices by looking at I/O Range.

Interrupt request

An IRQ is essentially a signal sent to your CPU that "interrupts" what it's doing so that the device or software that sent the interrupt can get attention. This is used to support multitasking as the processor is told when it's needed, instead of having the CPU ask (poll) while it's idle. Interrupts are assigned to specific devices and are not shared. If two devices have the same IRQ, you'll get an IRQ conflict. This used to be an issue that system administrators needed to be aware of, but because most devices today are plug and play, it's not really an issue most system administrators have to work on.

Memory

Random access memory (RAM) is great for loading things quickly because it's a solid state technology. In Device Manager, you can view the memory ranges that are assigned to the individual devices under the Resources tab for each device.

Viewing hidden devices

Sometimes devices aren't visible in Device Manager when you first open it. Devices that are not currently connected and devices that are not plug and play are examples of devices that won't show up in Device Manager by default.

For devices that are not plug and play, choose View⇨Show Hidden Devices. This causes the non-plug-and-play devices to show up.

To view devices that were once installed but are no longer attached to the server, you need to open a Command Prompt or PowerShell window as an administrator and enter the following command:

```
SET DEVMGR_SHOW_NONPRESENT_DEVICES=1
```

Then in the same window, launch Device Manager by typing **devmgmt.msc**. When Device Manager opens, choose View⇨Show Hidden Devices. You see "ghost" entries for the devices that once existed. This is helpful if old device drivers are causing issues with new devices or applications because you can remove them when you can see them.

Scanning for new devices

If you've installed a new device but it isn't showing up in Device Manager, you can start a scan for new hardware by choosing Action⇨Scan for Hardware Changes. This will scan for new hardware, and if it locates it, attempt to locate and install a driver for the hardware.

Working with older devices

As a system administrator, you may be asked to install a device that is older and not recognized by the computer when it's plugged in. For these non-plug-and-play devices, your best bet is to use the Add Legacy Hardware option in Device Manager. You can get to it by choosing Action⇨Add Legacy Hardware from inside the Device Manager window.

Viewing individual device settings

While in Device Manager, viewing the settings of individual devices is simple. You can double-click the device, or you can right-click the device and choose Properties. On most devices, you have a minimum of four tabs:

- **General:** The General tab (shown in Figure 2-4) contains basic information regarding the device, such as the name of the device, the type of device, who manufactured it, and the device's location on the system.

TIP

One of my favorite parts of this tab is the Device Status box at the bottom. If you're troubleshooting an issue, the Device Status box can be very helpful. When your device is working properly, it will look like what you see in Figure 2-4.

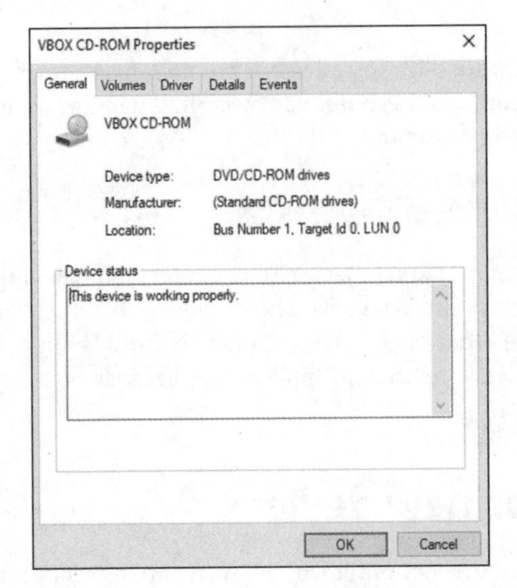

FIGURE 2-4: The General tab for a VirtualBox CD-ROM drive.

» **Driver:** The Driver tab gives you information regarding the installed driver and allows you to perform several driver-related management tasks. It starts out with the name of the driver and then lists the provider of the driver, the date the driver was made available, the version of the driver, and if the driver is signed it includes the identity of the organization that signed it.

The following buttons allow you to manage the drivers on your system from this tab:

- **Driver Details:** Provides more details regarding the driver, including where the driver files live and what their names are.
- **Update Driver:** You can click this button to update your drivers. You're given two options. You can search automatically for updated drivers, which searches your system and the Internet for updates, or you can browse to the driver software if you know where the updated drivers are located on your system.
- **Roll Back Driver:** Say you updated you driver, and now your device isn't working properly. You can click Roll Back Driver to go back to the previously installed driver.
- **Disable Device:** Clicking Disable Device disables the device in Device Manager. When you disable a device, the icon next to the device name will have a downward-facing arrow indicating that it is currently disabled.

- **Uninstall Device:** If you need to uninstall the device altogether, you can press this button. This option is useful for removing stale drivers that weren't removed when hardware was uninstalled, for removing drivers that have become corrupted, or for manually removing device drivers that are causing stability issues on your system.

» **Details:** The Details tab doesn't look like much when you first click it, but it actually contains a wealth of information regarding your device. By default, the drop-down box will have Device Description selected, but you can choose from many other properties to get more information on the specifics of your device.

» **Events:** The Events tab contains events that have occurred to the specific device. On a new system, there won't be much to look at — you have the initial Device Started and Device Configured entries. If the server has been around for a while and you're trying to troubleshoot an issue, click View All Events, and the Event Viewer appears with a custom filter so that you can look at events specific to the device.

Some devices have additional tabs that add further management capabilities, such as Power Management and Resources. Power Management gives you the ability to allow the computer to shut off a device to conserve power, or allow the device to wake the computer if it's in Sleep mode. The Resources tab, when it's available, tells you which memory ranges a device is using, which IRQ it's using, and if there are any conflicting devices.

Updating drivers

Updating drivers from Device Manager is pretty simple. You have two options: You can right-click the device and choose Update Driver, or you can right-click the Device, choose Properties, and click the Drivers tab.

After you select Update Driver through either method, you see the dialog box shown in Figure 2-5. Here, you're presented with the choice to search automatically for the driver or browse the computer for the software. Searching automatically is typically the easiest method; it searches the local computer and then searches the Internet for updated drivers. If for some reason it isn't able to locate the drivers and you know where they're located, choosing to browse your computer for the driver software may be the best option.

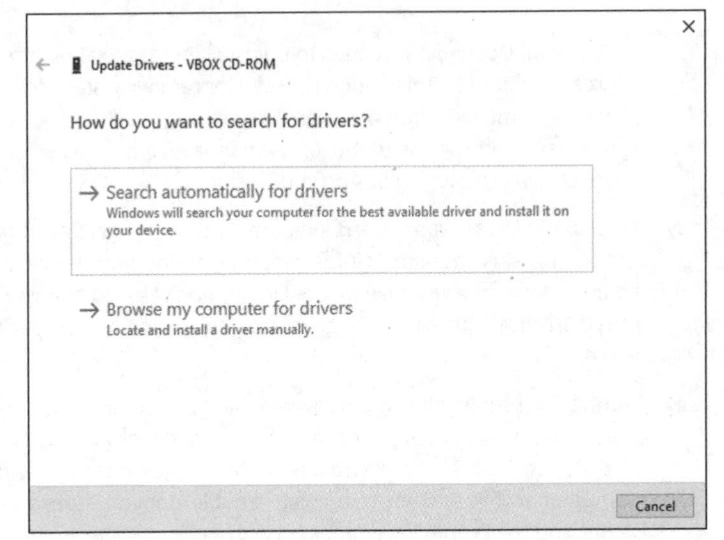

FIGURE 2-5: You have two options when updating drivers in Windows Server 2022 through Device Manager.

Configuring power management

Some of the devices in Device Manager have a Power Management tab. This tab lets you configure the behavior of the device in regards to — wait for it — power management. You have two or three options, depending on what the device is:

- **Allow the Computer to Turn Off This Device to Save Power:** This option is most useful on devices that will run off battery power like laptops. You most likely don't want components of your server to turn off, though, so you may want to uncheck this box.
- **Allow This Device to Wake the Computer:** A device with this check box selected is able to wake the server if it has gone to sleep. This feature is used by network cards that need to be able to wake a server for patching, and by input devices like mice and keyboards so that they can wake up the attached system.
- **Only Allow a Magic Packet to Wake the Computer:** This setting is available on network adapters and is only available if Allow This Device to Wake the Computer is enabled. The magic packet is also known as a Wake-on-LAN packet and is a special broadcast packet designed to wake the system.

Using the Add Hardware Wizard

With most devices being plug and play, you shouldn't ever need to use the Add Hardware Wizard. In fact, it was removed from the Control Panel back in Windows 7, because driver installation is handled automatically. If for some reason the device driver is not being installed, maybe because the device is too old or because it's too new, you can manually launch the Add Hardware Wizard. Download the drivers from the manufacturer's site first, and then follow these instructions:

1. **Right-click the Start menu and choose Windows PowerShell (Admin).**
2. **Type** hdwwiz.exe **and press Enter.**

   The Add Hardware Wizard launches (see Figure 2-6).

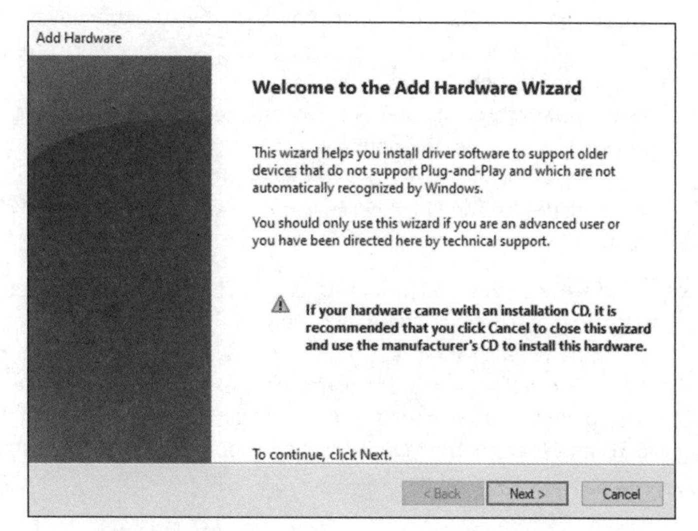

FIGURE 2-6: The Add Hardware Wizard.

3. **Click Next.**
4. **Select the radio button next to Search for and Install the Hardware Automatically (Recommended) and click Next.**
5. **If the wizard finds new hardware, it will tell you what it found and you can install the driver.**
6. **If the wizard did not find new hardware, it will ask you to choose a hardware category. Choose one and click Next.**
7. **Choose the manufacturer and the model of your device and click Next.**

   The wizard installs the driver.

It is worth noting that there are multiple ways to get to the Add Hardware Wizard that are not documented in steps that I've just described. Find the way that works best for you to access these items — don't be afraid to experiment.

Performing Hard-Drive-Related Tasks

Central to most of the work you'll do on your servers is the hard drive. Hard drives store the operating system and the data that your organization needs to do what it does best. There are two main types of storage in use today.

- **Hard disk drives (HDDs):** For quite a few years, magnetic storage (known as hard disk drives) were the only option. These drives were available at varying speeds and capacities; higher-speed and higher-capacity drives were more expensive. Because they were mechanical drives, they were prone to wearing out, and as technology improved, magnetic storage was no longer able to keep up with some of the more storage-intensive workloads because of their physical constraints. Magnetic storage hasn't gone away — it's still around. Due to their lower cost, HDDs are still an excellent choice for data storage when high speed and high performance aren't as important as keeping cost down.
- **Solid state drives (SSD):** Solid state drives have no moving parts; they store data on non-volatile flash memory chips. As technology improves, their lifespans increase, their capacities continue to grow, and their prices come down. For most system administrators, SSD drives are the staple of operating system drives because they're so fast. If you have an application that needs constant steady performance and high input/output operations per second (IOPs), SSDs will meet that need.

TIP

In many scenarios, system administrators and storage administrators take a best-of-both-worlds approach: They create tiered storage. Higher-tier storage is composed of the faster and more expensive SSD drives, while lower-tier storage, which is used for infrequently accessed data, is made up of the less expensive HDDs.

Choosing basic or dynamic disks

When you first initialize a disk, it's a basic disk. For most users, this works just fine. Basic disks are a simple solution and, by default, how new disks are created. However, if your organization wants to use more advanced disk features, you need to upgrade to a dynamic disk. Dynamic disks support

- **Simple:** A simple dynamic disk is very similar to a basic disk. When you first convert a basic disk to a dynamic disk it will be a simple volume.
- **Striping:** Data is written across multiple drives (or *striped*). This can improve write performance, but it's not a fault-tolerant solution. You need a minimum of two disks to take advantage of this functionality, and if you lose one disk you have lost the entire striped set. Striping is sometimes referred to as RAID 0.
- **Spanned:** A spanned volume is a logical volume created from combining multiple physical hard drives. This can be really cool if you have a couple of smaller hard drives, but you need one big hard drive. The downside to this approach is that, like striping, there is no fault tolerance if one of the drives ceases to function. You lose all the data on the drive.
- **Mirroring:** Data is written identically to two or more drives. This provides fault tolerance because there is no data loss if one of the drives becomes inoperable. Mirroring is sometimes referred to as RAID 1.
- **RAID 5:** A RAID 5 volume is also referred to as a *parity volume*. It requires a minimum of three disks, one of which is used for parity. That parity bit guarantees that if one of the drives goes down, there will be no data loss. Of course, if you lose more than one drive, you'll suffer data loss.

Using multipath I/O

Picture yourself in the datacenter working on installing a new server. You're working on getting it cabled in, grumbling about the inevitable rat's nest of cables

TALKING THE DISK TALK

Before I continue into the exciting world of storage, here's a quick rundown on some terms I use in this chapter:

- **Disk:** The physical or virtual storage device that is presented to the server to allow for the storage of data.
- **Partition:** A disk can be broken into multiple partitions. These partitions are smaller chunks of the disk that may or may not contain a file system.
- **Volume:** Volumes have names and are what you present to your users. They have a file system and can store data of all kinds. You can have multiple volumes on disks and in partitions.

in the way, when you realize you accidentally unplugged the power from the rack switch that is used for connections to the Internet Small Computer Systems Interface (iSCSI) storage. You brace yourself for the inevitable slew of angry phone calls, but nothing happens. Then you realize that every system has a second network cable going to the iSCSI switch in the rack next door. You breathe a sigh of relief — you have multipath I/O (MPIO).

MPIO is a technology that allows a server to recognize more than one path to its storage area network (SAN) storage. This is commonly set up for Fibre Channel and iSCSI storage networks to provide fault tolerance for the storage network. In addition to that, depending on how it's configured, it can help increase performance because you can use both connections at the same time.

Installing MPIO is pretty straightforward, but the configuration can be a little more complex. MPIO is offered as a feature in Windows Server 2022. Here's how to install MPIO:

1. **With Server Manager open, choose Manage⇨Add Roles and Features (as shown in Figure 2-7).**

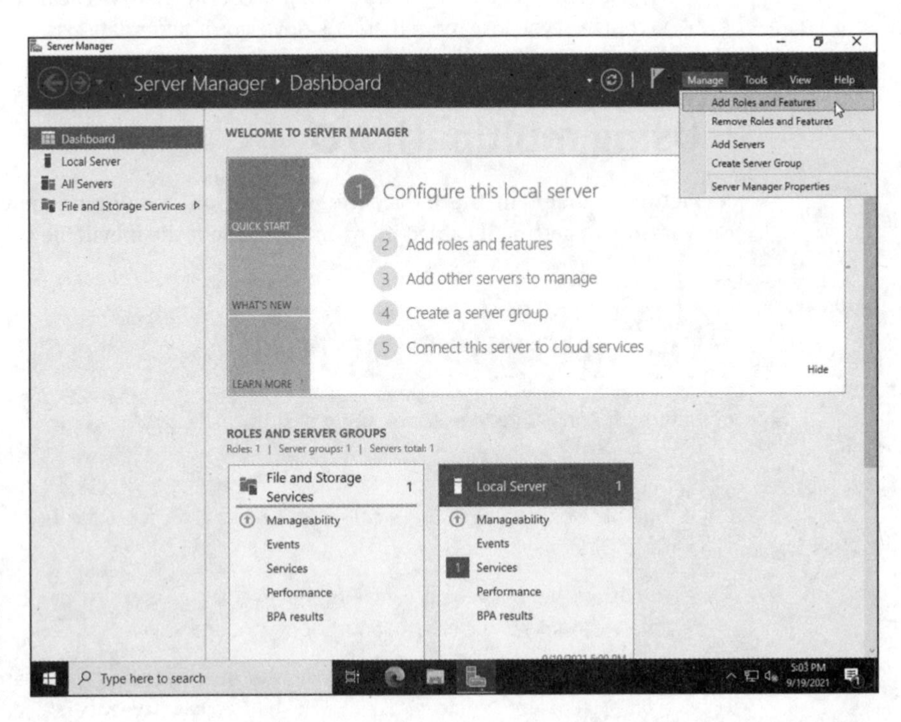

FIGURE 2-7: Installing the MPIO feature in Windows Server 2022.

2. **On the Before You Begin screen, click Next.**
3. **On the Select Installation Type screen, select Role-Based or Feature-Based Installation and then click Next.**
4. **On the Select Destination Server screen, your server will be highlighted; click Next.**
5. **On the Select Server Roles screen, click Next.**
6. **On the Select Features screen, scroll down and select Multipath I/O (as shown in Figure 2-8) and click Next.**

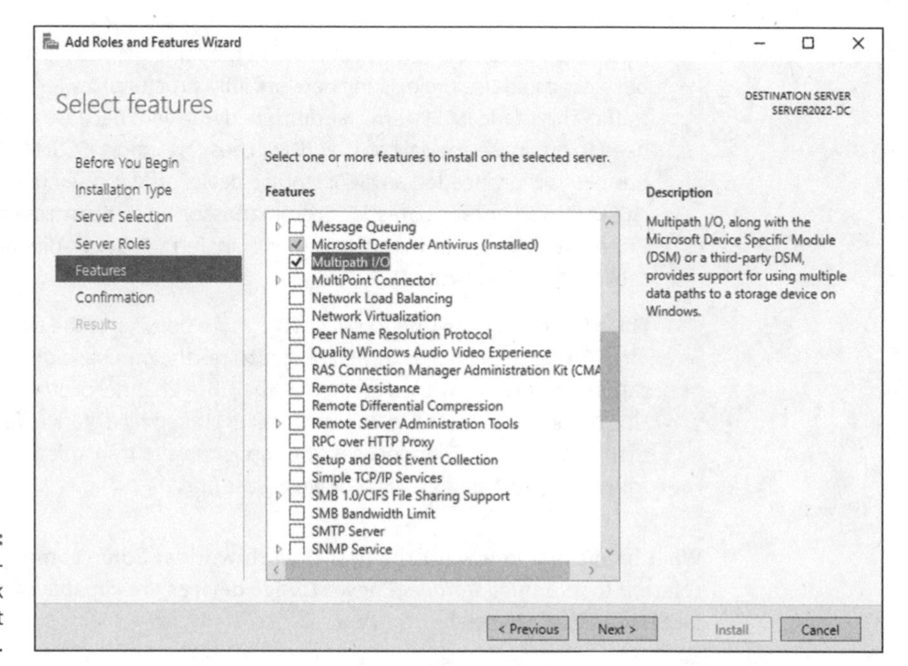

FIGURE 2-8: Select the Multipath I/O check box on the Select Features screen.

7. **On the Confirm Installation Selections screen, click Install.**
8. **When the installation is finished, click Close.**

With MPIO installed, you can configure it to work with your iSCSI storage or even Serial Attached SCSI (SAS) storage.

Working with storage area networks

For years, the local storage on the server was all you had to work with. Being limited to the storage in the server had several problems. It wasn't fault tolerant (unless you used RAID, with the exception of RAID 0), it wasn't easily scalable, and the capacity was limited to however many disks would fit in the server.

Somebody came up with the idea of creating drive arrays to alleviate those issues. The drive arrays could have expandable trays of disks. Those trays could be used to mirror other trays or to use RAID if fault tolerance was the main goal. These large drive arrays needed a network that could support the growing amount of data being pushed into them. And so the SAN was born.

There are many different protocols in use in SANs, but here are the two most frequently used in enterprise environments today:

- **iSCSI:** iSCSI became a favorite of small to medium businesses because the business could use the existing network infrastructure to support the storage traffic. This made iSCSI a smaller upfront investment because there was no need to purchase special cards and switches to support iSCSI traffic. The business simply needed an iSCSI storage device, and it could use iSCSI. The iSCSI protocol encapsulates SCSI commands inside a Transmission Control Protocol (TCP) packet. That data is then transferred over traditional Ethernet cabling to its destination.
- **Fibre Channel:** Fibre Channel is still the more popular of the two protocols, mainly due to speed and reliability. It requires the purchase of special network cards and special switches to support the Fibre Channel protocol. Although Fibre Channel offers fast data transfers, it's limited by distance, far more so than iSCSI is. Fibre Channel uses fiber-optic cable to transmit data. Data is transmitted as flashes of light down a glass fiber.

What happens if you want the best of both worlds? Some companies sell what is referred to as *unified storage.* These storage devices are capable of supporting both iSCSI and Fibre Channel.

Understanding Storage Spaces Direct

Storage Spaces Direct is a feature that was introduced in Windows Server 2016 Datacenter edition. It was a great way for organizations that couldn't afford a SAN to be able to achieve the dream of highly available and highly scalable storage. It took advantage of the local storage on the servers to create a pool of storage that the entire cluster of servers, called *nodes,* could take advantage of.

To use Storage Spaces Direct on multiple servers, you need to install the Failover Clustering and Hyper-V roles on each server. You need a minimum of 10 Gb Ethernet between the clustered systems, and remote direct memory access (RDMA) is recommended.

Setting up the storage is fairly straightforward. You create a storage pool from the physical disks, and from there you create logical disks and then volumes. I'll show you how to do this on a standalone server, but the process is very similar with clustered servers. First, you'll create a storage pool. Then you'll create logical disks and volumes.

Creating the storage pool

The storage pool is the starting point when building out the storage for Storage Spaces Direct. Here's how to create the storage pool:

1. **From Server Manager, click File and Storage Services in the left-hand menu.**
2. **Click Storage Pools.**

   If you have available disk drives, they show up under Physical Disks on the lower-right side. Notice under Storage Pools that there is an entry called Primordial. This is created by default with the available disks (see Figure 2-9).

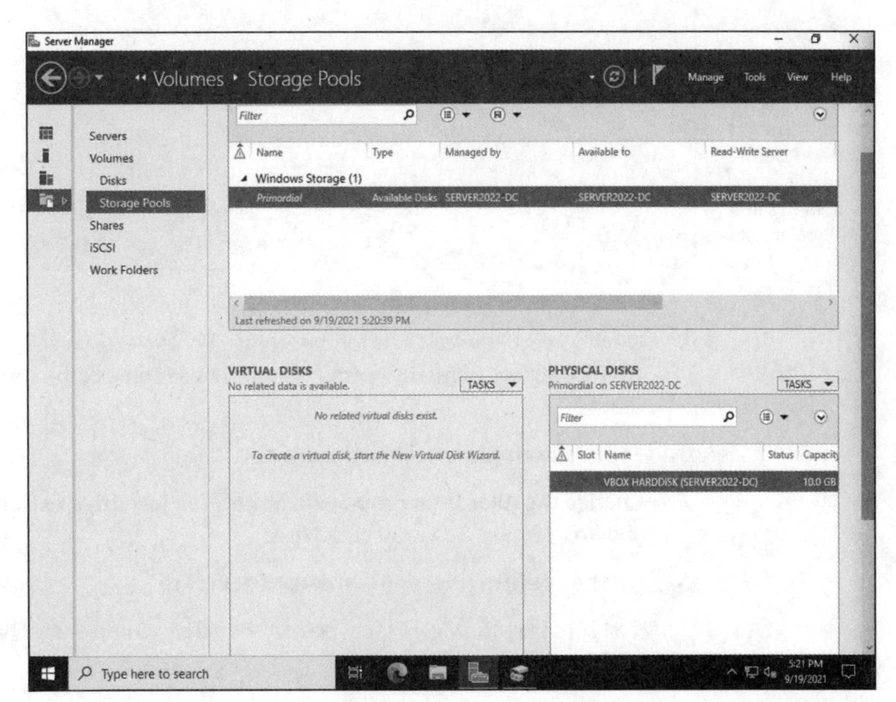

FIGURE 2-9: The Storage Pools screen showing available physical disks.

3. **Right-click where it says Primordial and choose New Storage Pool.**
4. **On the Before You Begin screen, click Next.**
5. **On the Specify a Storage Pool Name and Subsystem Screen, enter a name for the pool and click Next.**

   I'm naming our pool "Pool1." Notice in Figure 2-10 that the primordial pool is selected by default.

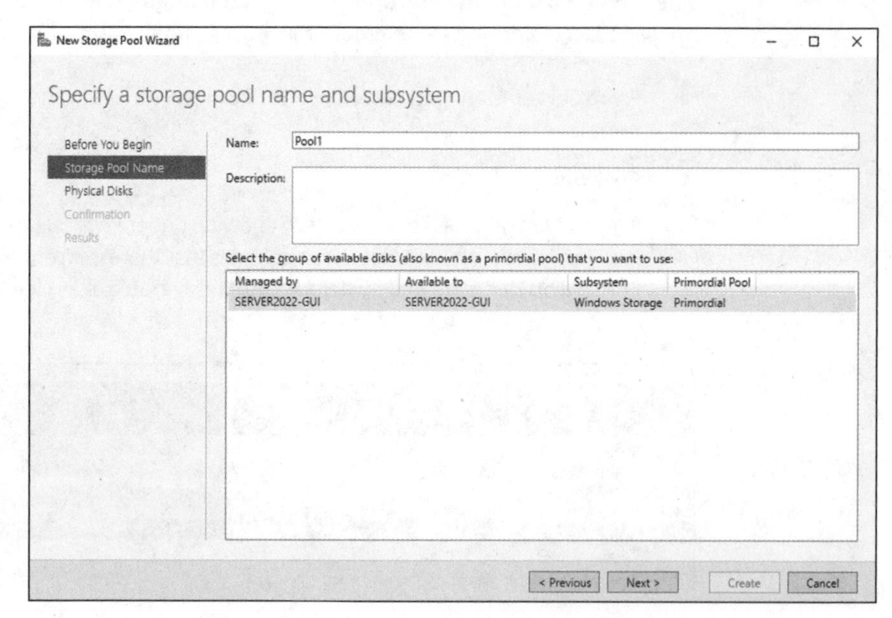

FIGURE 2-10: Choose a name for your storage pool and choose the primordial pool to select physical disks from.

On the Select Physical Disks for the Storage Pool screen, you see all the disks that were in the primordial pool. You can choose some of the disks or all of the disks.

6. **For this example, select all the disks.**
7. **Change the Allocation drop-down box on the last drive to Hot Spare, as shown in Figure 2-11, and click Next.**
8. **On the Confirm Selections screen, click Create.**

   If all goes well, the View Results screen should say Completed.

9. **Click Close.**

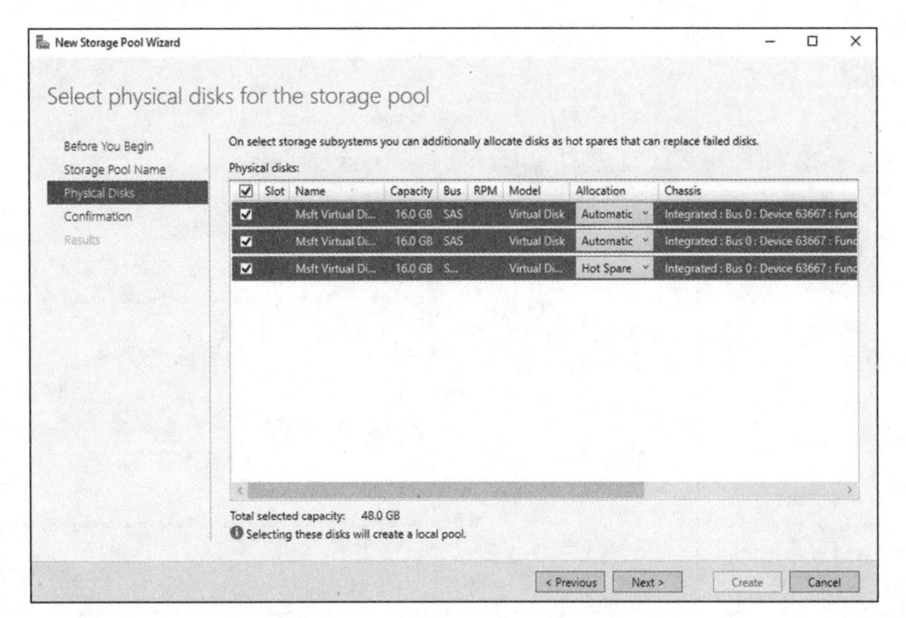

FIGURE 2-11: Configuration of the physical disks in the storage pool.

Creating a logical disk

Now that the pool is created, you can create a logical disk. Follow these steps:

1. **Right-click Pool1 and choose New Virtual Disk, or click the To Create a Virtual Disk, Start the New Virtual Disk Wizard hyperlink in the Virtual Disks box.**

   You will be asked to select the storage pool you want to work with. You should only see the one pool that we created so far, Pool1.

2. **Select Pool1 and click OK.**
3. **On the Before You Begin screen, click Next.**
4. **On the Specify the Virtual Disk Name screen, name the disk Disk1, as shown in Figure 2-12.**
5. **Click Next.**
6. **On the Specify Enclosure Resiliency screen, leave the Enable Enclosure Awareness check box unchecked and click Next.**

   The next screen is the Select the Storage Layout screen. You have three options: Simple, Mirror, and Parity. These are very similar to the RAID levels as far as what they do. Simple is a non-RAID disk, Mirror is similar to RAID 1, and Parity is similar to RAID 5.

7. **Select Simple, as shown in Figure 2-13, and click Next.**

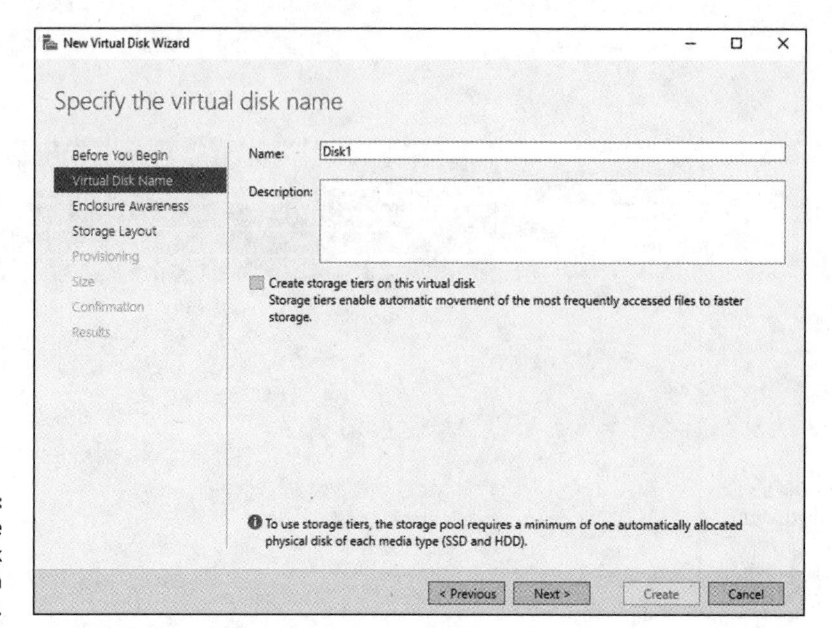

FIGURE 2-12: You have to name your virtual disk before you can use it.

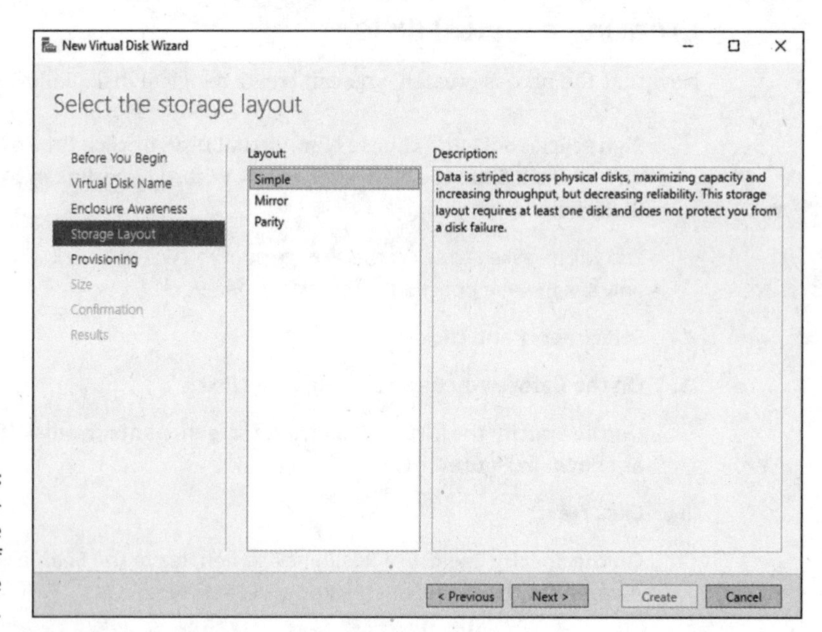

FIGURE 2-13: Selecting the storage layout can be very important if you need it to be fault tolerant.

8. **On the Specify the Provisioning Type screen, Select Thin for the Provisioning Type, and click Next.**
9. **On the Specify the Size of the Virtual Disk screen, tell it how big you want the disk to be and click Next.**

   In my example, I've chosen to make the virtual disk 5GB.

10. **On the Confirm Selections page, click Create.**
11. **If everything succeeded, click Close.**

Creating a volume

Now that you've created a storage pool and a virtual disk, you're ready to create a volume. Follow these steps:

1. **Right-click Disk1 and choose New Volume, as shown in Figure 2-14.**

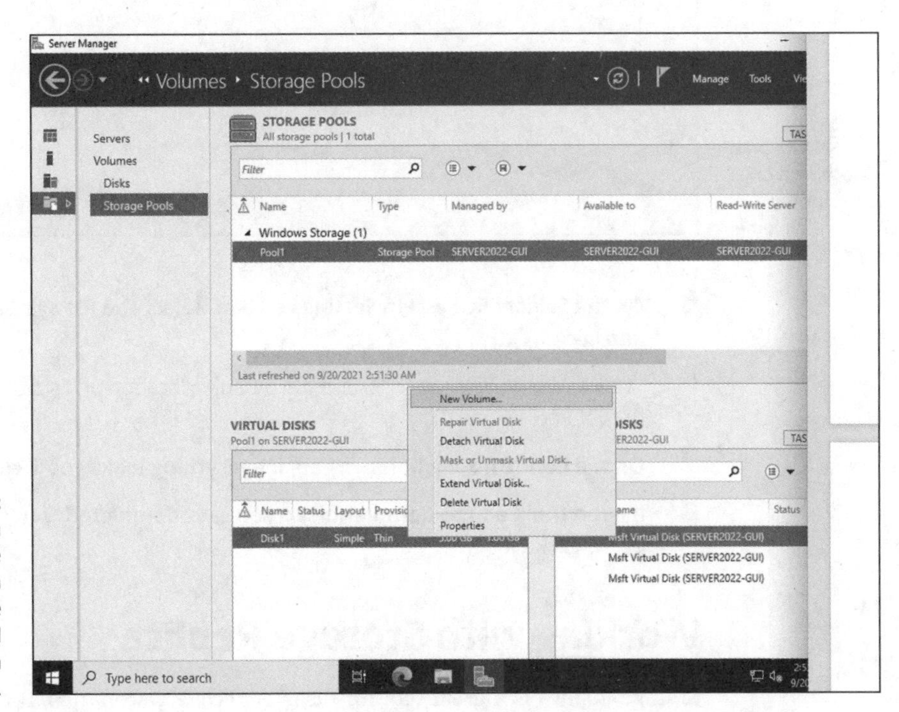

FIGURE 2-14: Creating a volume that the operating system can use is simple after the virtual disk has been created.

2. **On the Before You Begin screen, click Next.**
3. **On the Select the Server and Disk screen, you should only have one server and disk at this point, so you can simply click Next.**
4. **On the Specify the Size of the Volume screen, enter a size and click Next.**

   I've entered 3GB.

5. **On the Assign to a Driver Letter or Folder screen, select a drive letter or specify a folder and click Next.**

   I've kept it simple and chosen a drive letter. I was automatically given the letter F, as you can see in Figure 2-15.

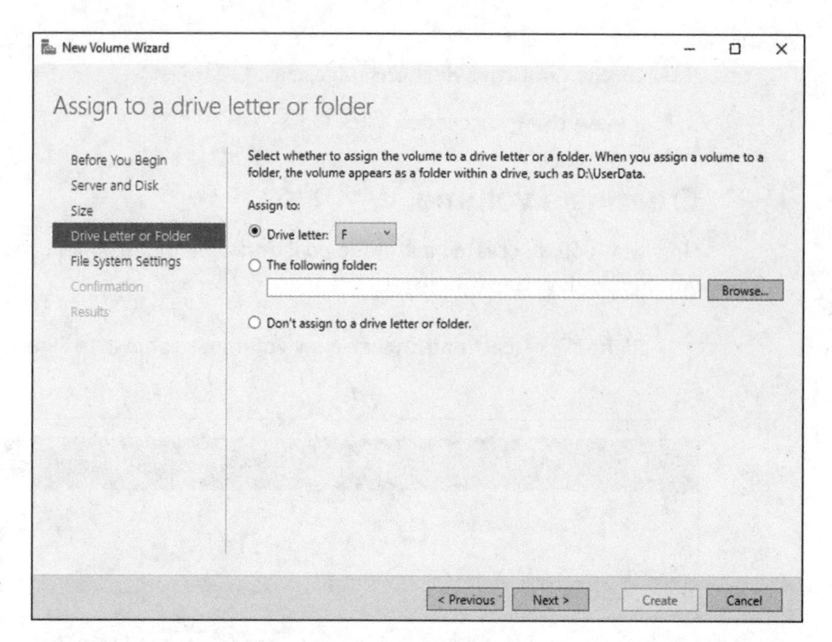

FIGURE 2-15: Choosing a drive letter for my new volume.

6. **On the Select File System Settings screen, select the file system settings and click Next.**

   I'll stick with the default NTFS and the default allocation unit size, and I'll name the volume, Volume1.

7. **On the Confirm Selections screen, if everything looks good, click Create.**
8. **If everything on the Completion screen says Completed, you're good to go, so click Close.**

Working with Storage Replica

Storage Replica is a great tool for disaster recovery scenarios. It can do synchronous or asynchronous block-level replication of volumes between servers or clusters. By implementing stretch clusters, you could even replicate across distant sites.

Storage Replica was initially released in Windows Server 2016 Datacenter edition. As of Windows Server 2019, you could also take advantage of Storage Replica if you were using the Standard edition. This is still true in Windows Server 2022, although it does have a few limitations:

- Storage Replica is only able to replicate one volume, not an unlimited number of volumes.

- A volume can have only one replication partnership, unlike Datacenter where a volume can have an unlimited number of partners.
- Storage Replica will only support sizes up to 2TB, while Datacenter has no limit on size.

Using Storage Quality of Service

Storage Quality of Service (QoS) allows you to centrally monitor the storage performance of virtual machines (VMs) when you're using Hyper-V and the Scale-Out File server roles. Note that you can use cluster shared volumes instead of Scale-Out file server if desired. Storage QoS accomplishes a few things:

- It ensures that one VM can't use up all the storage bandwidth.
- It monitors storage performance.
- It uses policies to determine minimum and maximum I/O for VMs.

One of the simplest ways to experiment with Storage QoS is to set up a failover cluster and create a cluster shared volume. If you do this, then Storage QoS is set up automatically. You can view it under the Cluster Core Resources.

Encrypting with BitLocker

Securing data has never been more critical than it is today. With BitLocker, you can protect your data from would-be thieves by encrypting your entire data drive. BitLocker can encrypt both fixed drives and removable drives. Fixed drives will in general use a Trusted Platform Module (TPM) chip to save the cryptographic key, while BitLocker To Go utilizes either a password or a smart card to unlock it.

In the following sections, I explain t what a TPM is, and walk you through installing BitLocker as well as configuring BitLocker To Go and BitLocker.

Understanding Trusted Platform Modules

A Trusted Platform Module is a special chip on the motherboard designed to store passwords, certificates, and cryptographic keys. For the most seamless BitLocker experience, you'll want a TPM chip that is version 1.2 or newer. If your system doesn't have a TPM, you can still use BitLocker, but you'll have to use Group Policy to override the TPM requirement. You'll have to enter a password to unlock the system. I cover how to setup BitLocker without a TPM later in this chapter in the section called, "Knowing what to do if there's no TPM module."

Installing BitLocker

To install BitLocker on Windows Server 2022, you need to install the BitLocker feature. Follow these steps:

1. **With Server Manager open, choose Manage⇨Add Roles and Features.**
2. **On the Before You Begin screen, click Next.**
3. **On the Select Installation Type screen, click Role-Based or Feature-Based Installation and click Next.**
4. **On the Select Destination Server screen, click Next.**
5. **On the Select Server Roles screen, click Next.**
6. **On the Select Features screen, click select the BitLocker Drive Encryption check and click Next.**

   You'll be asked if you want to install additional features that are required for BitLocker Drive Encryption, as shown in Figure 2-16.

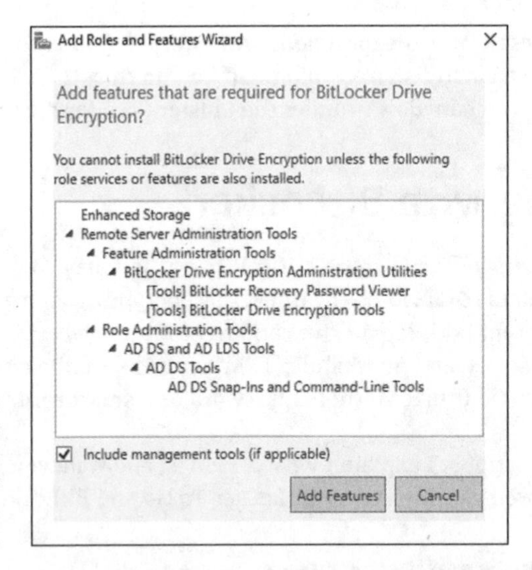

FIGURE 2-16: Adding features that are needed for BitLocker Drive Encryption to work its magic.

7. **Click Add Features and then click Next.**
8. **On the Confirm Installation Selections screen, select the Restart the Destination Server Automatically if Required check box.**

   The BitLocker Drive Encryption feature requires a reboot.

9. **Click Install.**

After the server reboots, you can move on to configuring BitLocker.

WARNING

Anytime you look at encrypting your data, be sure that you have a good backup to recover from in case something goes wrong. If you don't have a good backup, and the encryption process runs into an issue, you will need to restore from your backup.

Configuring BitLocker To Go

To configure BitLocker To Go, you need to navigate to where the management utilities are hidden and then you can start to play. Follow these steps:

1. **Click Start, scroll down to Windows System, and choose Control Panel.**
2. **Click System and Security.**
3. **Under BitLocker Drive Encryption, click Manage BitLocker (see Figure 2-17).**

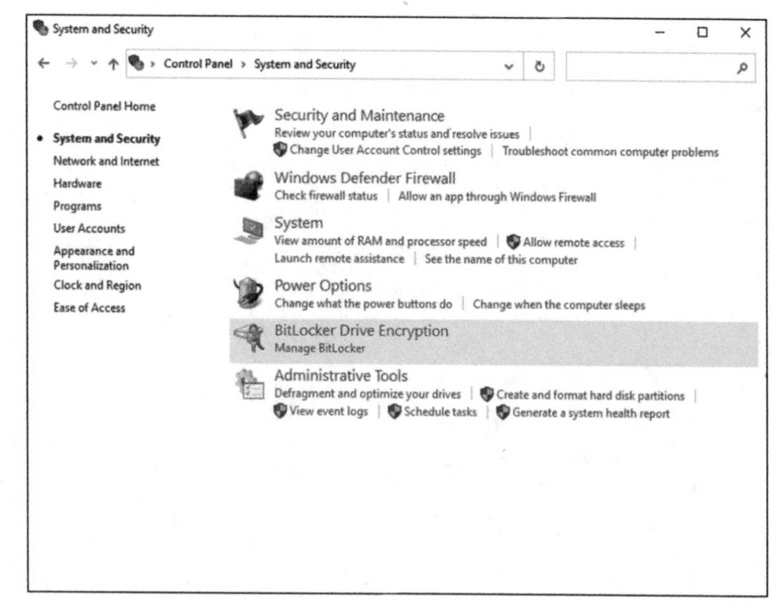

FIGURE 2-17: Getting to the BitLocker Management screen from the Control Panel.

4. **On my system, BitLocker is currently off, so I'll click Turn on BitLocker for Volume1 (see Figure 2-18).**

   Because Volume1 is considered a removable disk, the TPM is not used, and you have to set a password or tell it to use a smart card. This is to support BitLocker To Go.

5. **Set a password, as shown in Figure 2-19.**
6. **Click Next.**

   You're asked how you want to back up your recovery key. You can save to a file or print the recovery key.

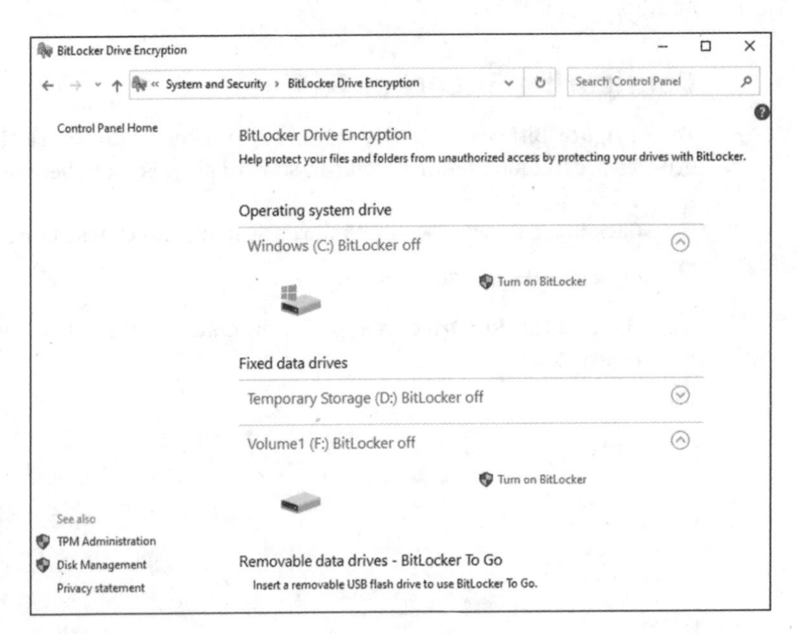

FIGURE 2-18: Turning on BitLocker for Volume1.

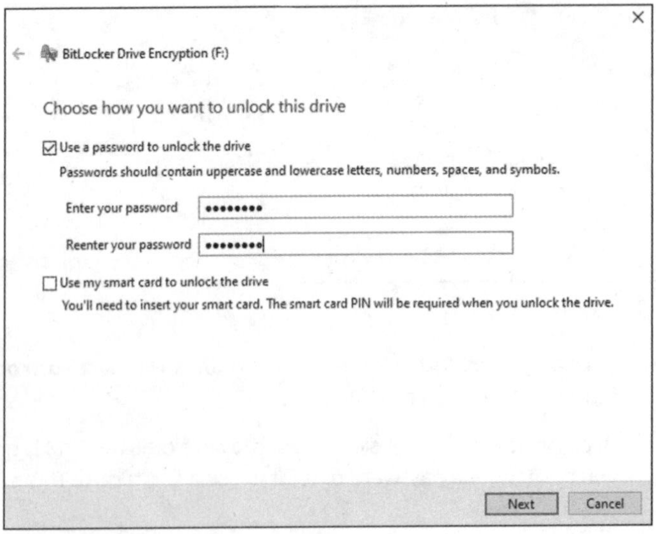

FIGURE 2-19: Setting a password for BitLocker To Go.

7. **Save it, and then click Next.**

   You're asked how much of your drive you want to encrypt.

8. **Choose Encrypt Used Disk Space Only and click Next.**
9. **On the Choose Which Encryption Mode to Use screen, select Compatible Mode and click Next.**

   It's generally safest to go with Compatible Mode for removable media.

   You're asked if you're ready to encrypt the drive.

10. **Click Start Encrypting.**

    After the encryption is complete, you see the status of BitLocker on the BitLocker To Go drive. It should say BitLocker On.

Configuring BitLocker

To configure BitLocker, you need to navigate to where the management utilities are. These steps are the same as Steps 1 through 3 in "Configuring BitLocker To Go," so I'll start the instructions after Step 3. If you need a refresher on how to get to the BitLocker Management screen, review steps 1 through 3 in "Configuring BitLocker To Go" before proceeding.

1. **On my system, BitLocker is currently off on the C drive, so I'm going to click Turn on BitLocker for the C drive.**

   On the How Do You Want to Back Up Your Recovery Key screen, you can choose to save to a file or print the recovery key.

2. **Save the file and click Next.**

   You're prompted to choose how you want to encrypt your drive. You can encrypt the entire drive, or just the used space only. You should choose whatever your organization requires.

3. **I'll choose Encrypt Used Disk Space Only.**
4. **Click Next.**
5. **On the Choose Which Encryption Mode to Use screen, select New Encryption Mode and click Next.**
6. **On the final screen, select the Run BitLocker System Check check box.**

   This ensures that BitLocker can read the keys before it encrypts the drive.

7. **Click Continue.**

   You're dropped back out to the BitLocker screen and asked to restart your system.

8. **Click Restart Now.**

   After the system comes back up, BitLocker will begin the drive encryption. After the encryption is done, it will say BitLocker On.

Knowing what to do if there's no TPM module

You may run into a server that doesn't have a TPM module. When you try to turn on BitLocker, you get an error saying that a compatible TPM could not be found. This doesn't mean that you can't use BitLocker; it just means that there will be more work involved to get it to encrypt the data drive. Follow these steps:

1. **Click the Start Menu and choose Run.**
2. **Type** gpedit.msc.
3. **Click Local Security Policy, click Computer Configuration, click Administrative Templates, click Windows Components, click BitLocker Drive Encryption, and click Fixed Data Drives, as shown in Figure 2-20.**

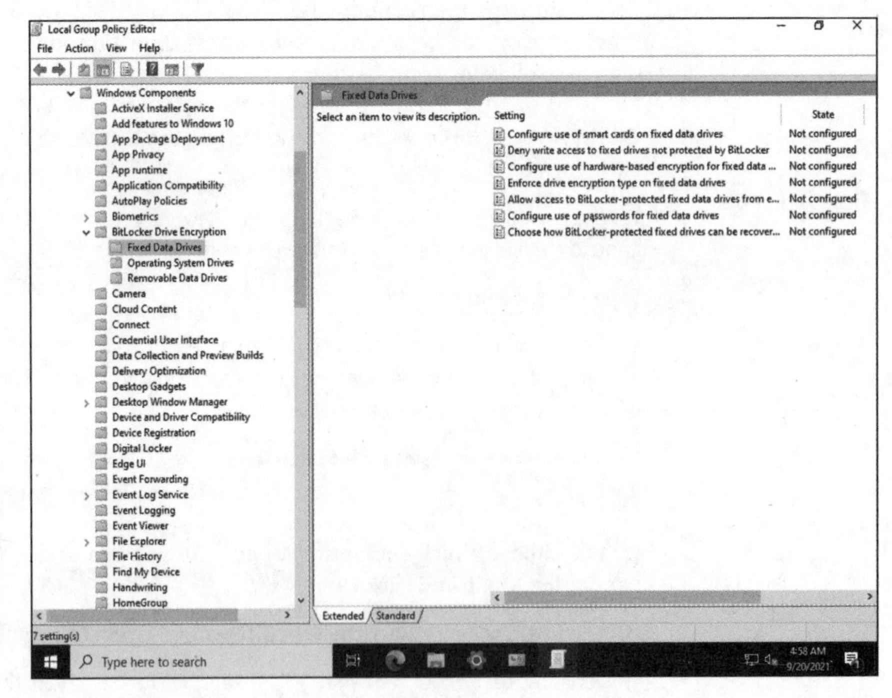

FIGURE 2-20: Configuring BitLocker to work without a TPM module involves editing the local security policy.

4. **Double-click Configure Use of Hardware-Based Encryption for Fixed Data Drives.**
5. **Select Enabled and ensure that the Use BitLocker Software-Based Encryption When Hardware Encryption Is Not Available check box is selected (see Figure 2-21).**

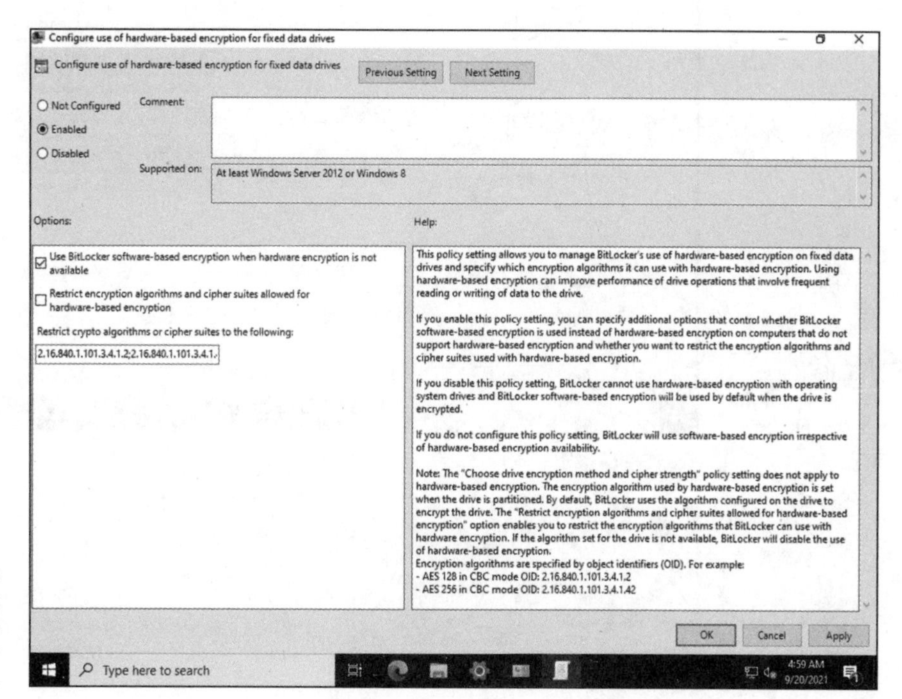

FIGURE 2-21: Enabling software-based encryption for BitLocker.

6. **Click OK.**
7. **Click Local Security Policy, click Computer Configuration, click Administrative Templates, click Windows Components, click BitLocker Drive Encryption, and click Operating System Drives, as shown in Figure 2-22.**
8. **Double-click Require Additional Authentication at Startup.**
9. **Choose Enabled.**
10. **Make sure the Allow BitLocker without a Compatible TPM check box is selected, as shown in Figure 2-23.**

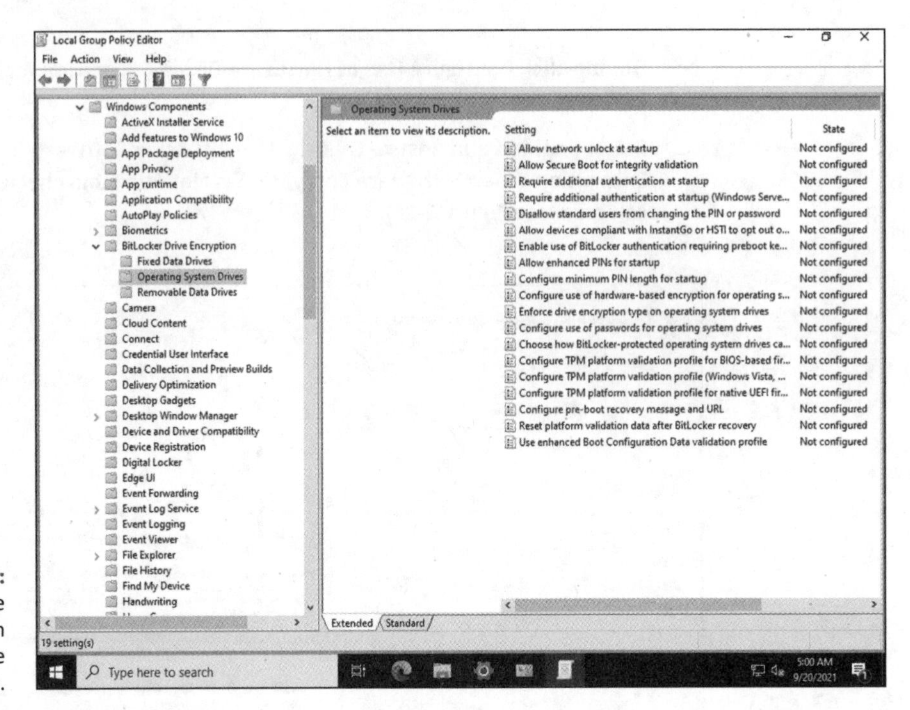

FIGURE 2-22: Allowing alternate authentication methods for the system drive.

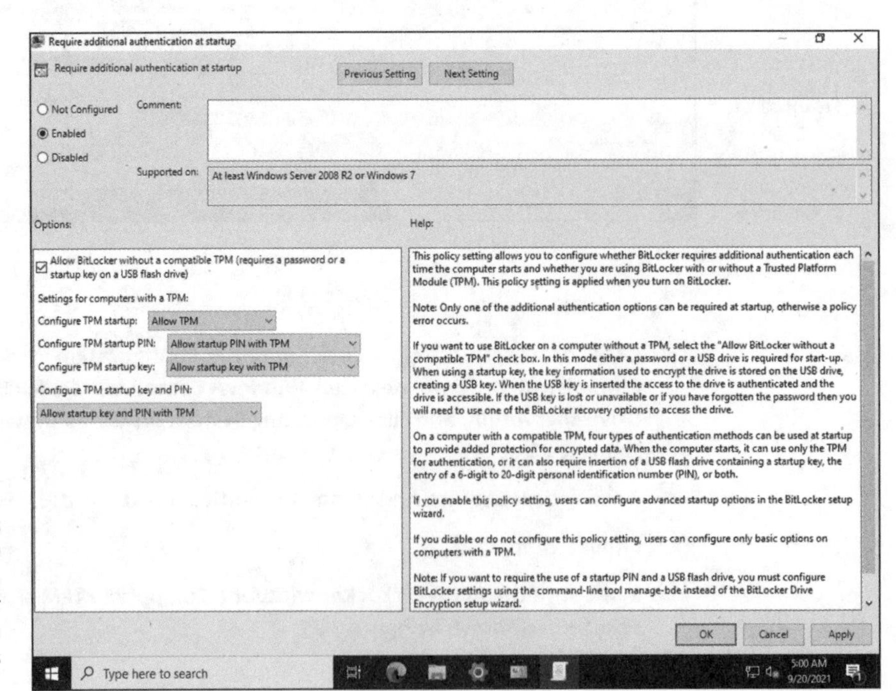

FIGURE 2-23: Letting the operating system drive encrypt or decrypt without a TPM requires this setting.

11. **Click OK.**

12. **Close out of Local Group Policy Editor.**

13. **Now go ahead and go back in to the BitLocker screen.**

    You should have no issue getting it to work now.

Performing Printer-Related Tasks

Printers are a pretty important resource. If your print server goes down, you'll start getting calls very quickly. Conversely, if users are trying to print, they'll get very cranky if they can't.

There are a few different ways you might install a server. A locally attached printer, for example, will most likely be connected through a USB cable. That printer may be used by one user or, in a small office setting, it may be shared from that workstation so that multiple users can print to it. Network-attached printing is very common as well, from homes with a wireless printer to large organizations that have printers on their local area networks (LANs). Some organizations may use a print server to manage their print queues centrally. You print to the server's print queue, and it sends the job to the printer through the network.

Using the Printer Install Wizard

The Printer Install Wizard walks you through the installation of your printer. You can launch it by going to Settings (the gear in the Start menu) and selecting Devices and then Printers & Scanners. From there, you simply click Add a Printer or Scanner and the wizard launches.

If your printer is found, the wizard is a nice easy type of install. If your printer is not found, you have a little more work to do. Here's how to add a printer that just doesn't want to cooperate:

1. **Click The Printer That I Want Isn't Listed.**

2. **In my case, my printer is on my network so I choose Add a Printer Using a IP Address or Hostname, as shown in Figure 2-24.**

3. **Click Next.**

4. **Fill in the IP address of your printer (as shown in Figure 2-25).**

   Your IP address will be different from mine, so you'll need to check your printer to see what IP address it's using. Make sure you choose TCP/IP Device from the Device Type drop-down list.

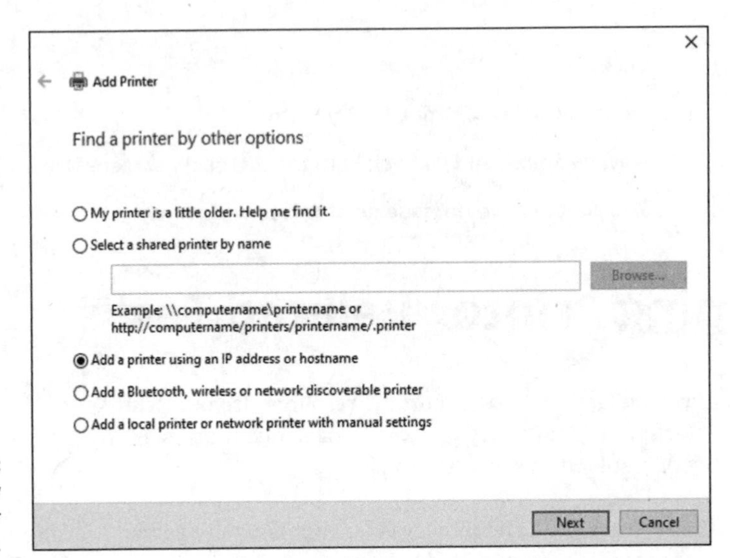

FIGURE 2-24: Selecting how you'll find your printer.

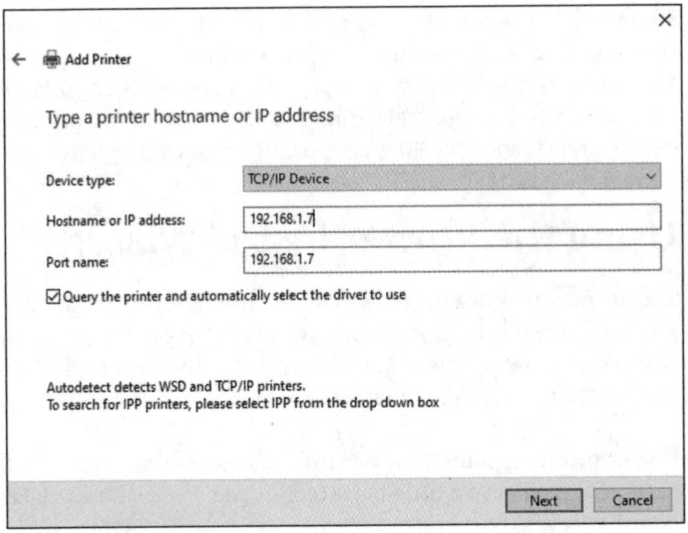

Add Printer

Type a printer hostname or IP address

Device type: TCP/IP Device

Hostname or IP address: 192.168.1.7

Port name: 192.168.1.7

Query the printer and automatically select the driver to use

Autodetect detects WSD and TCP/IP printers.
To search for IPP printers, please select IPP from the drop down box

Next Cancel

FIGURE 2-25: Configuring the IP address of your printer.

5. **Click Next.**

   If all goes well, the system will be able to contact the printer and figure out which driver it needs.

6. **Click Next.**

   On the Printer Sharing screen, you can decide if you want to share the computer with another system.

7. **I'm going to select Do Not Share This Printer.**
8. **Click Next.**

   The last page gives you the option to set the new printer as the default printer.
9. **Leave that check box selected and click Finish.**

TIP

If you need to use the more advanced print features that are available from your printer, but the Windows drivers aren't giving you the options you need, you may need to download the drivers from the manufacturer's website. The drivers are typically packaged in an executable file to aid in installation.

Configuring Server Hardware

Configuring print options

To get to the printer management screen, start in the Printers & Scanners menu (see the preceding section). In the Printers & Scanners section, you see the printer you just added. Select it by clicking it, and then click the Manage button. When configuring individual printers, there are two options to look at:

» **Printing Preferences:** Printing Preferences allows you to set what the default print settings will be for you printer. From here, you can set the orientation of your printing. If you click the Advanced button, more settings are available — those settings depend on the driver that is installed for your printer. My printer settings are shown in Figure 2-26.

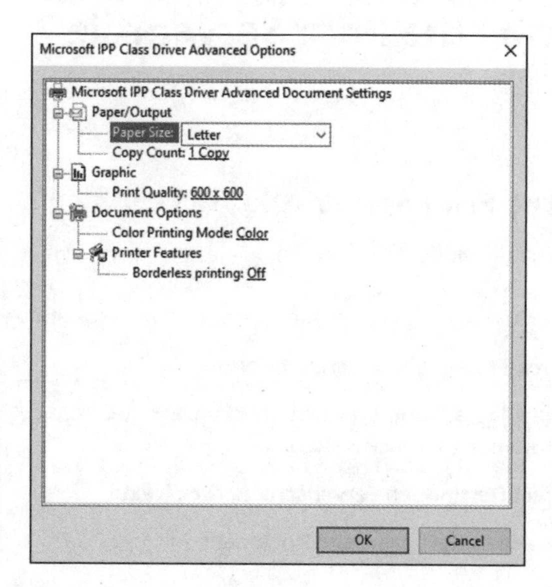

FIGURE 2-26: Advanced Options gives you more settings to work with.

- **Printer Properties:** The Printer Properties option gives you configuration items for the printer. For instance, if you choose after installation to share the printer or change the driver, or if you need to change the IP address, you can do that with Printer Properties. You get a screen similar to Figure 2-27.

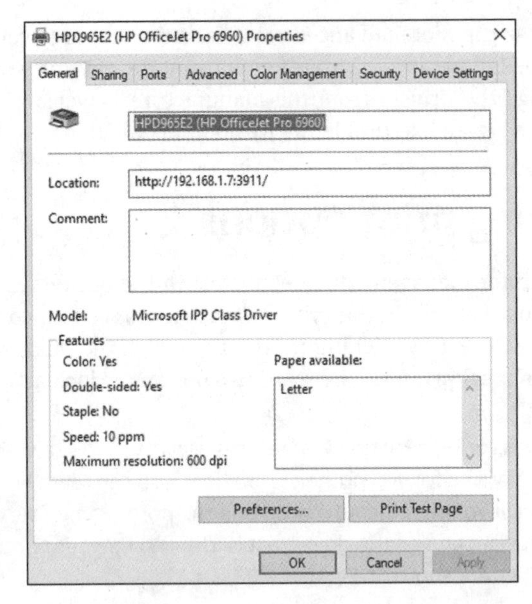

FIGURE 2-27: The Printer Properties screen allows you to change configuration items.

Configuring the Print Server role

In Windows Server 2022, Print Server is a role that can be installed and then configured.

Installing the Print Server role

Here's how to install the Print Server role:

1. **With Server Manager open, choose Manage⇨Add Roles and Features.**
2. **On the Before You Begin screen, click Next.**
3. **On the Select Installation Type screen, choose Role-Based or Feature-Based Installation and click Next.**
4. **On the Select Destination Server screen, click Next.**
5. **Scroll down and select Print and Document Services.**

6. **Click Add Features when you're prompted to do so.**

   Your screen should look like Figure 2-28.

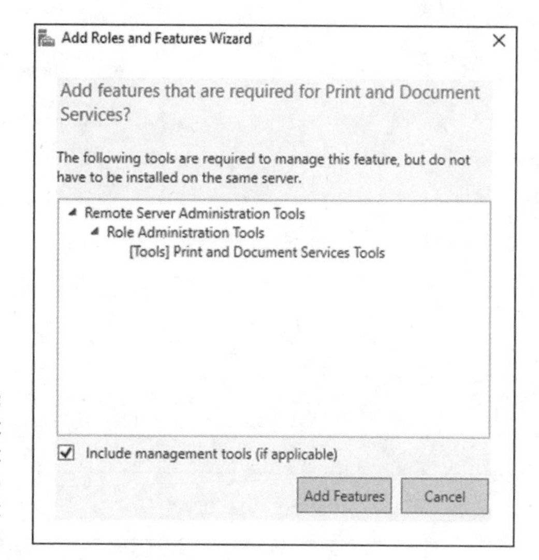

FIGURE 2-28: Install the Print and Document Services role to set up your print server.

7. **Click Next.**
8. **On the Select Features screen, click Next.**
9. **On the Print and Document Services screen, click Next.**
10. **On the Role Services screen, select Print Server, as shown in Figure 2-29, and then click Next.**
11. **On the Confirm Installation Selections screen, select Install.**
12. **On the Installation Progress screen, click Close after the installation is complete.**

Configuring the Print Server role

Now that you've got the Print Server role installed, you can configure it. Here's how to add your first printer:

1. **From Server Manager, choose Tools⇨Print Management.**
2. **Expand Print Servers, right-click your print server, and choose Add Printer, as shown in Figure 2-30.**

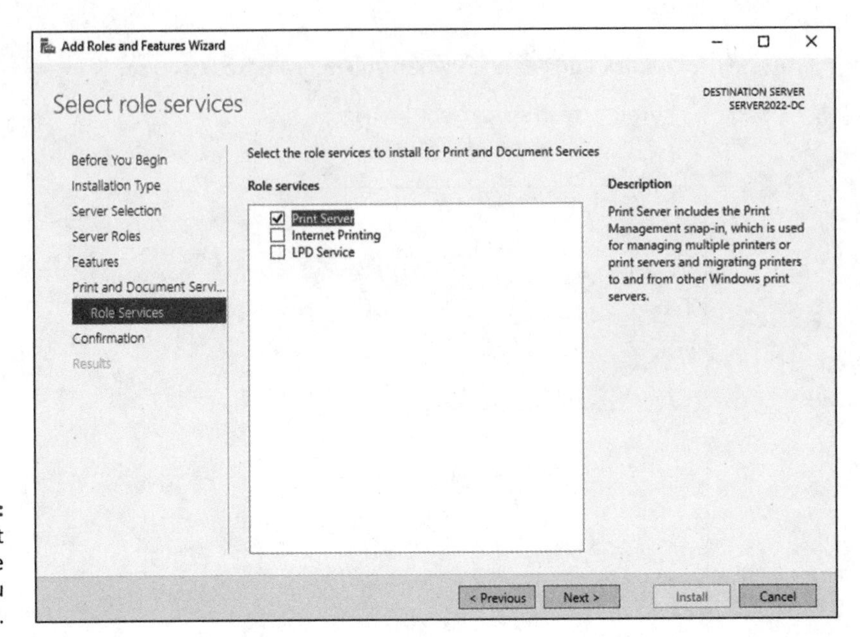

FIGURE 2-29: Select Print Server for the role that you want to install.

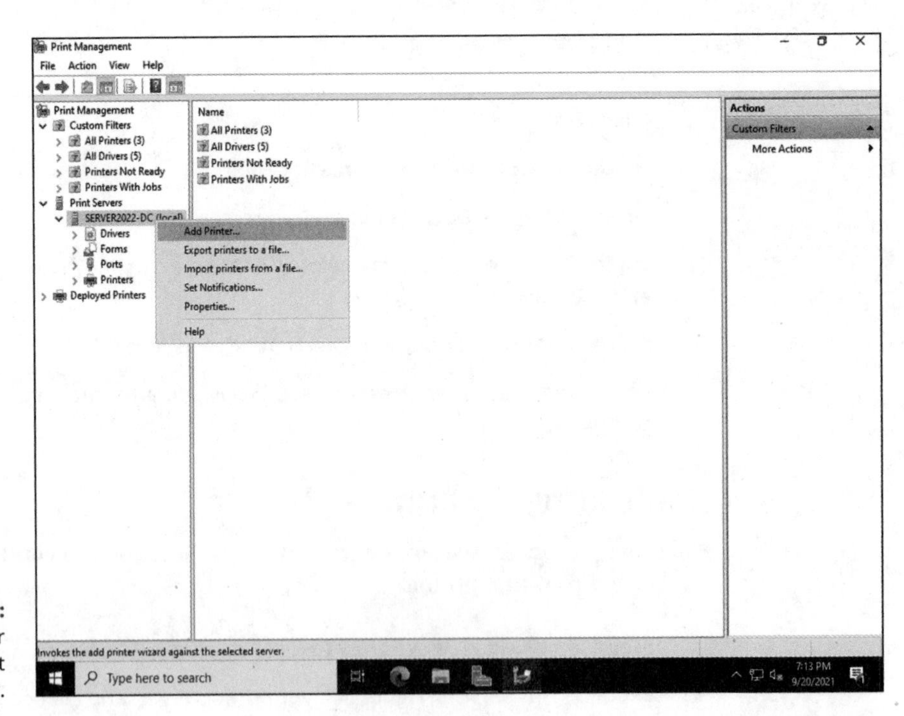

FIGURE 2-30: Adding a printer to the Print Server.

3. **Select Add an IPP, TCP/IP or Web Services Printer by IP address or hostname and click Next.**

4. **Enter the IP address and click Next.**

   On the next screen you can name your printer. In an enterprise, I recommend doing this.

5. **I'll call my printer Main Office Printer #1, as shown in Figure 2-31.**

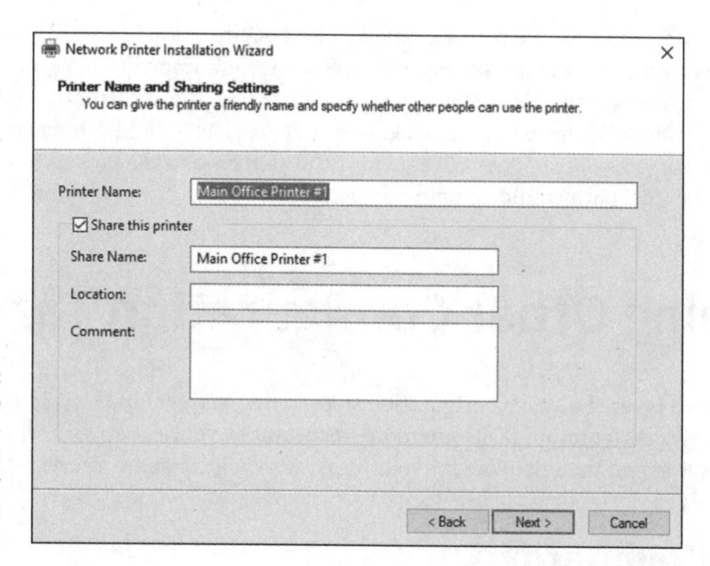

FIGURE 2-31: Naming the printer and adding a location makes it easier for your users to find the printer.

6. **Click Next.**

   If the wizard finds the printer, the next screen tell you that the printer was found and shows you a summary of the printer information.

7. **Click Next.**

   The server installs the drivers for the printer.

8. **After the drivers are installed, click Finish to exit the wizard.**

Connecting to a Printer on a Print Server

Connecting to a printer on a print server is pretty simple. You can add the printer with a Universal Naming Convention (UNC) path like `\\servername\printername`. For this to work, the server's name must resolve in DNS. If you don't have the printer name in DNS, you can do the UNC with the IP address instead of the server name. In enterprise environments, it's common to use Group Policy to manage printers for users and/or computers.

Using the UNC path to an object, like a share on a file server or a printer, is a convenient way to access resources without knowing where they're located. In the example of a file server, for instance, the local path to a file might be `E:\SomeFolder\SomeShare\SomeFile.txt`. The UNC path would be `\\ServerName\SomeShare\SomeFile.txt`. If the share was ever moved to a different disk, the UNC path would remain the same.

Performing Other Configuration Tasks

I've covered a lot of configuration steps in this chapter, and I'm almost done. The last few configuration tasks are more important to your experience when working with the server then they are for your users, who will consume a service from the server.

Keyboard

From the Settings menu, go to Devices and then Typing. Right-click the keyboard you want to manage and select Keyboard Settings. The Typing screen (shown in Figure 2-32) appears. From here, you can configure your keyboard to match what you need it to do.

Mouse

From the Settings menu, go to Devices and then Typing. The Mouse screen (shown in Figure 2-33) appears. From here, you can adjust button configuration, cursors, and the behavior of the scroll wheel.

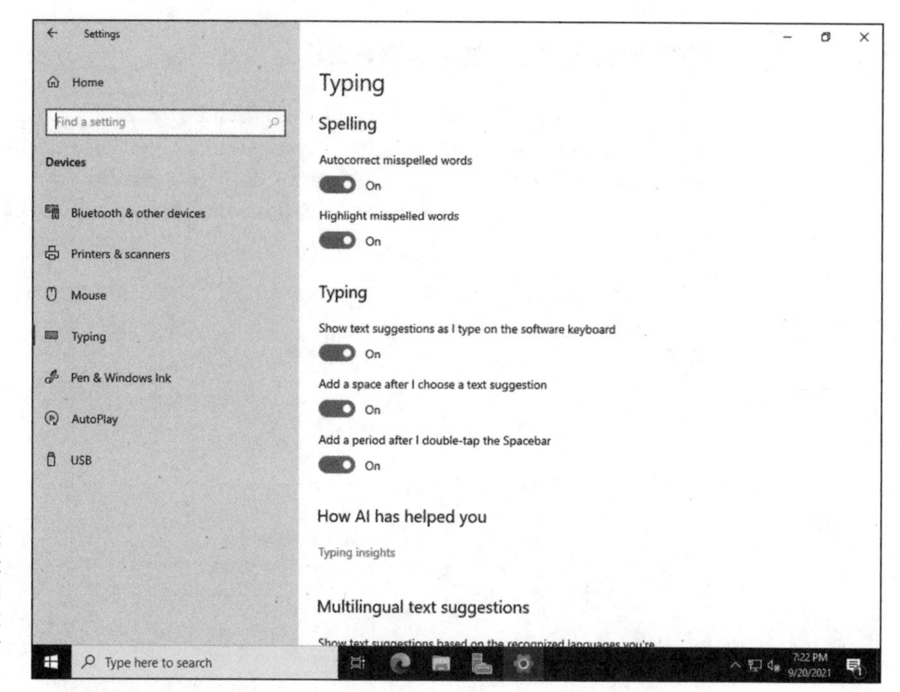

FIGURE 2-32: You can adjust the keyboard settings to work better with your typing speed.

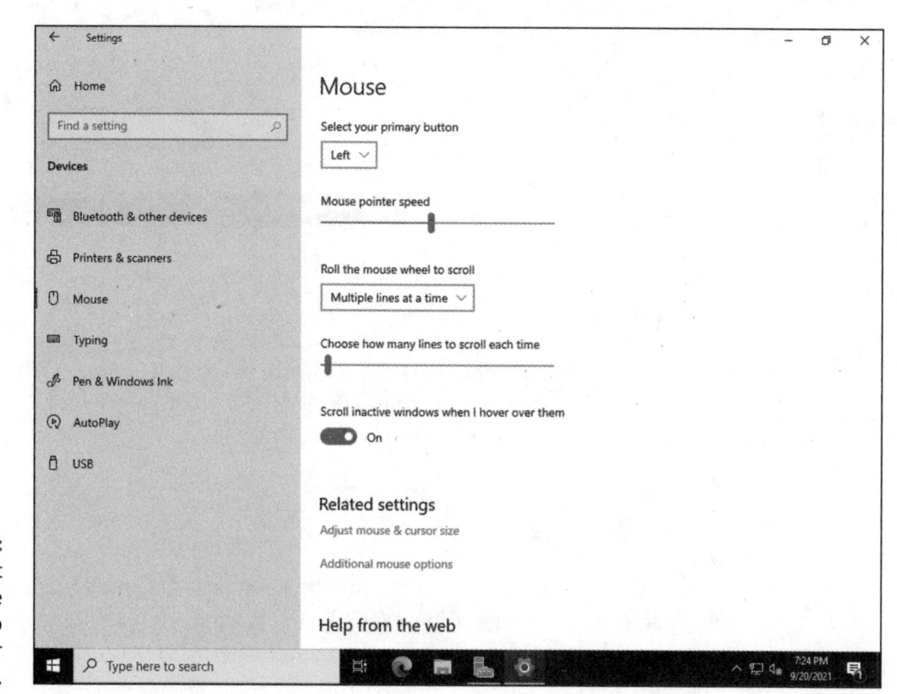

FIGURE 2-33: You can adjust the mouse settings to customize your experience.

Power management

From the Settings menu, go to System and then Power & Sleep. You'll be presented with some very basic settings. However, you can get very granular by clicking Additional Power Settings (see Figure 2-34). This will allow you to choose power plans to customize how your system will balance power and performance.

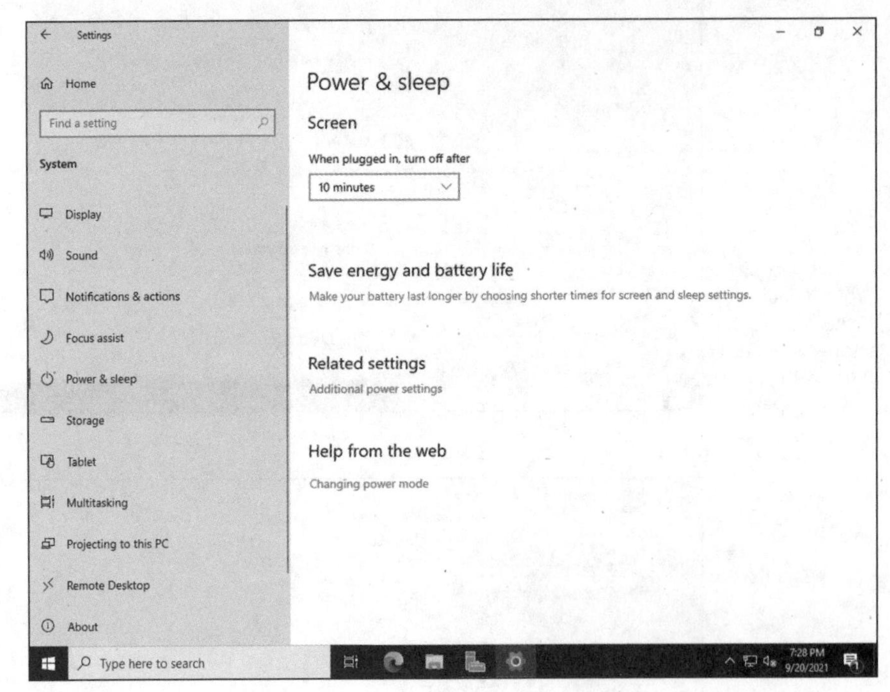

FIGURE 2-34: Advanced Power Settings lets you create a granular power management scheme.

Sound

From the Settings menu, click System, and then click Sound. With this screen, you can customize your default sound devices and recording devices. You can troubleshoot your sound devices from here as well.

Language

From the Settings menu, click Time & Language and then click Language. From this screen, you can add new languages, set new default languages, and change keyboard layouts if desired. The Language screen on your server should look similar to Figure 2-35.

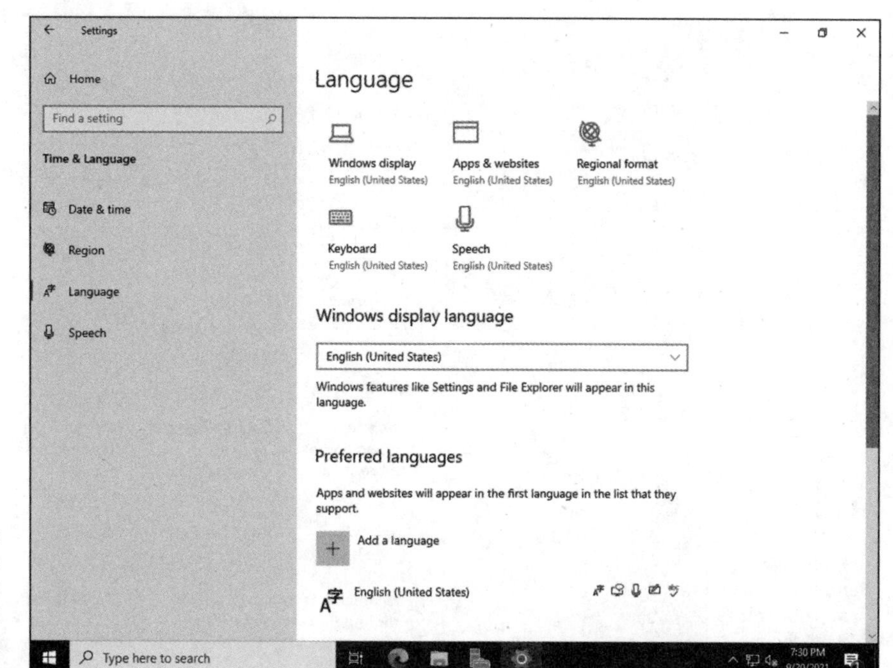

FIGURE 2-35: The Language screen allows you to adjust settings for language and keyboard layouts.

Fonts

You may need to install new fonts on the server. This may happen because the company standard requires a different font than normal. I've even had to install barcode fonts before. To work with Fonts, open the Settings menu. Click Personalization and then click Fonts. What I love most about the new interface is that it gives you visual samples of each of the fonts and the ability to get more fonts from the Microsoft Store. Figure 2-36 shows the Fonts screen.

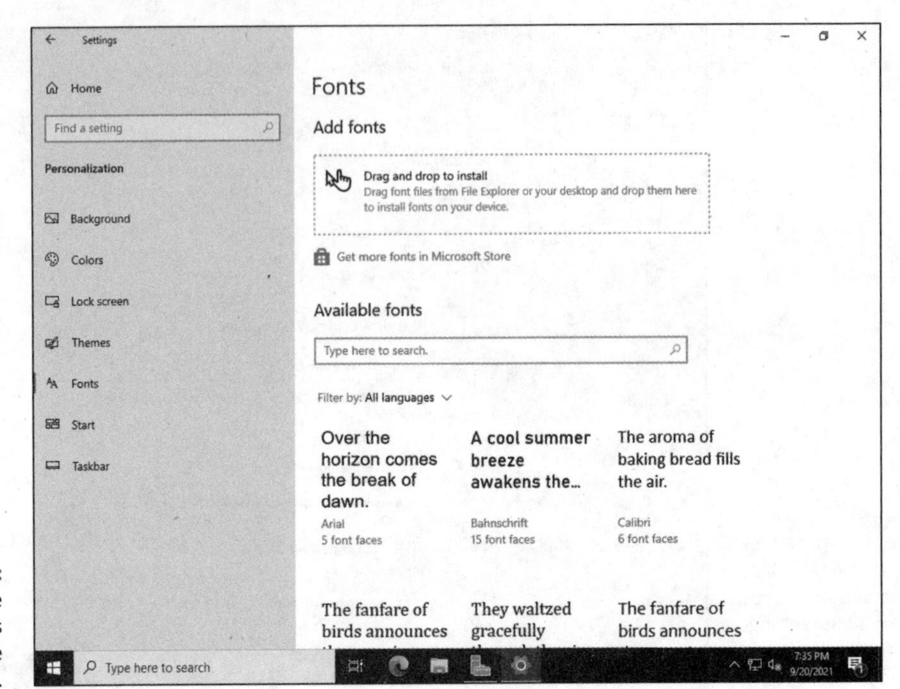

FIGURE 2-36: Samples of the installed fonts available on the server.

» Seeing what's in the Settings menu

Chapter 3
Using the Settings Menu

For many years, the Control Panel was the main management area of Windows Server operating systems. It was where you configured devices, set up time and language settings, changed the appearance of the server, worked with installed programs, and managed Windows Update and setup accessibility options.

Microsoft has made an active effort to replace the aging Control Panel with the newer Settings menu. The Settings menu breaks out settings into categories, similar to what was done in the Control Panel. However, new features for the most part are put into Settings, but not necessarily in the Control Panel. In this chapter, I explain how to access the Settings menu and where to configure items in the Settings menu.

Accessing the Settings Menu

You can access the Settings menu in two ways:

- Click the Start menu and select Settings (the gear icon).
- Right-click the Start menu and then select Settings.

The Settings menu opens and should look similar to Figure 3-1.

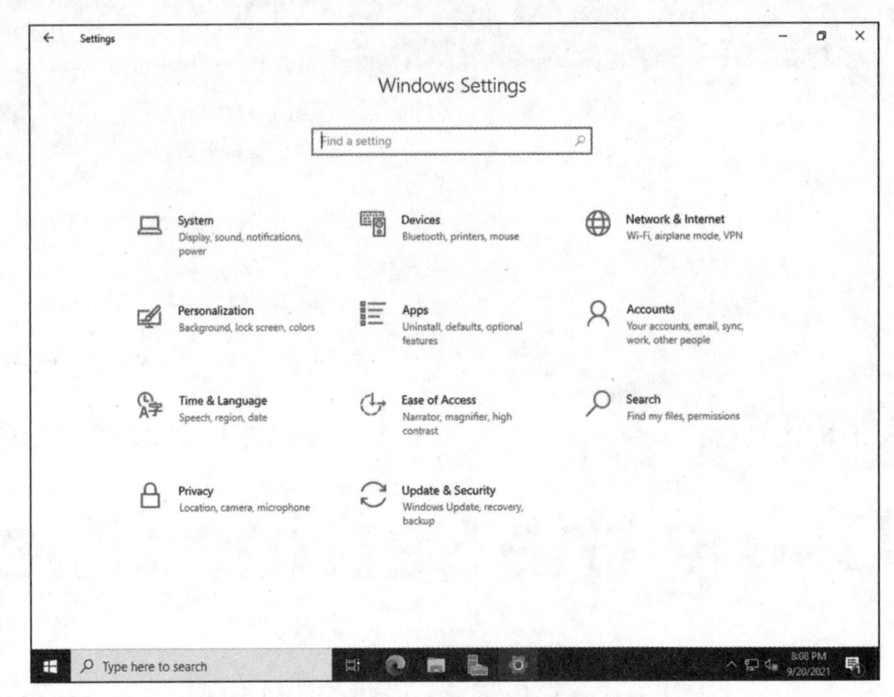

FIGURE 3-1: The Settings menu in all its glory.

Understanding Settings Menu Items

The items in the Settings menu are used to configure functionality on your server. In the following sections, I walk you through the items that are available on a fresh install of Windows Server 2022.

TIP

The Settings menu contains the majority of your administration tools at this point and is where new features are introduced. If you're a new system administrator, I recommend getting familiar with this menu. If you've been around for a while (remember when zip drives were cool?), I recommend transitioning to the Settings menu rather than continuing to use Control Panel at this point.

System

The System menu (see Figure 3-2) shows you many of the things you may want to configure on your system, including the following:

- Changing display settings
- Adjusting your sound settings
- Configuring system notifications

- Adjusting your power and sleep settings
- Accessing disk optimization utilities like disk cleanup
- Handling multiple windows (also known as multitasking)
- Configuring Remote Desktop

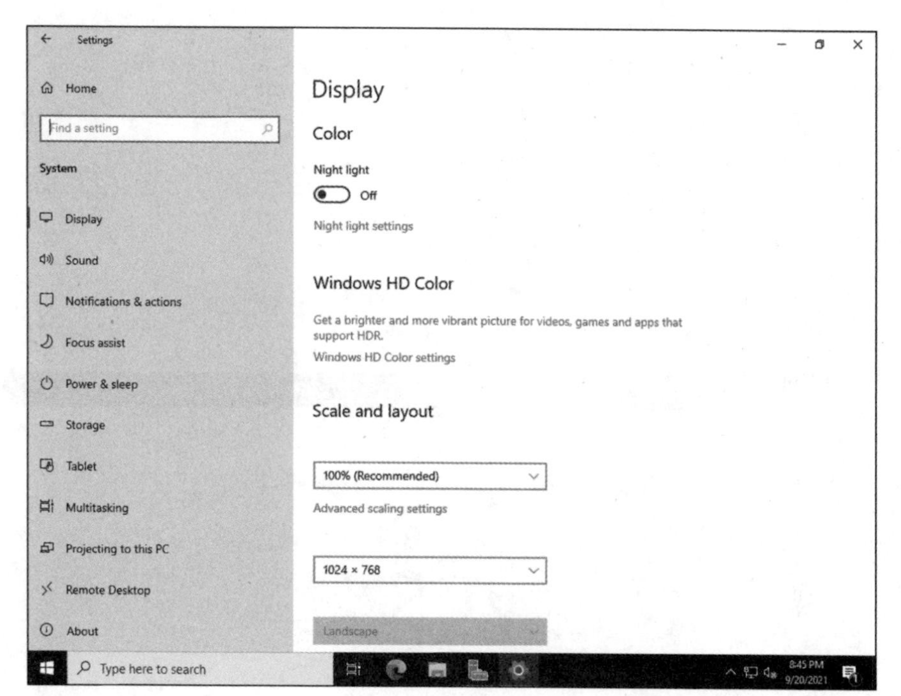

FIGURE 3-2: The System menu contains a lot of configuration choices.

Devices

The Devices menu (see Figure 3-3) is exactly what you would expect. This menu allows you to work with Bluetooth devices, printers, and scanners, as well as keyboard and mouse settings. You can also change autoplay behavior and adjust how the system communicates USB issues.

Network & Internet

The Network & Internet menu (see Figure 3-4) gives you a status on your network connection, as well as the ability to set up or configure ethernet, dial-up, virtual private network (VPN), and proxy connections.

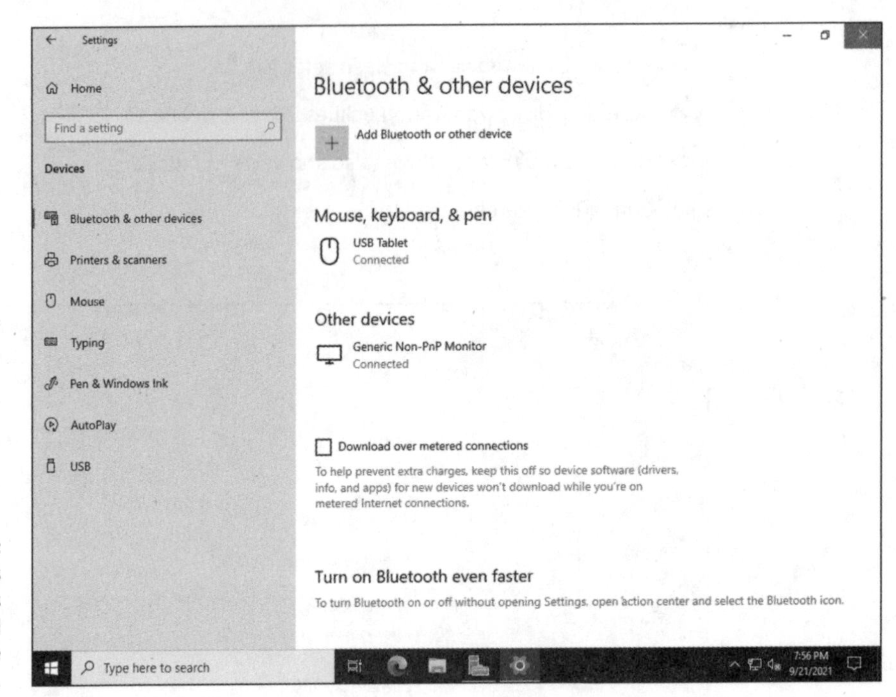

FIGURE 3-3: The Devices menu allows you to install and configure peripherals.

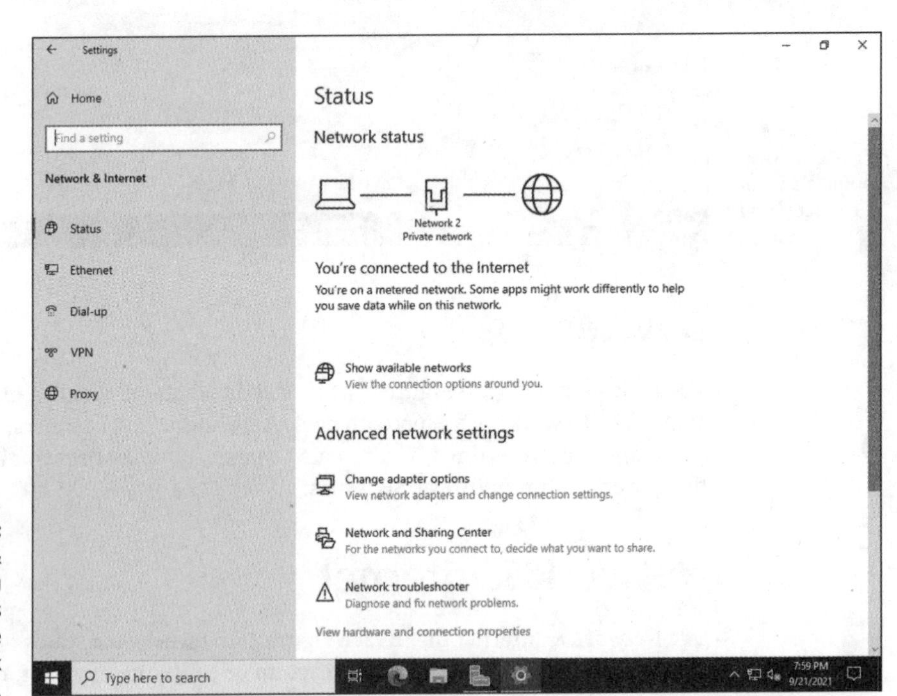

FIGURE 3-4: The Network & Internet menu provides options to configure server network connections.

Personalization

The Personalization menu (see Figure 3-5) allows you to make your server your own. This is where you can adjust background pictures and colors, configure the picture on the lock screen, set themes and manage fonts, as well as customize the Start and Taskbar settings.

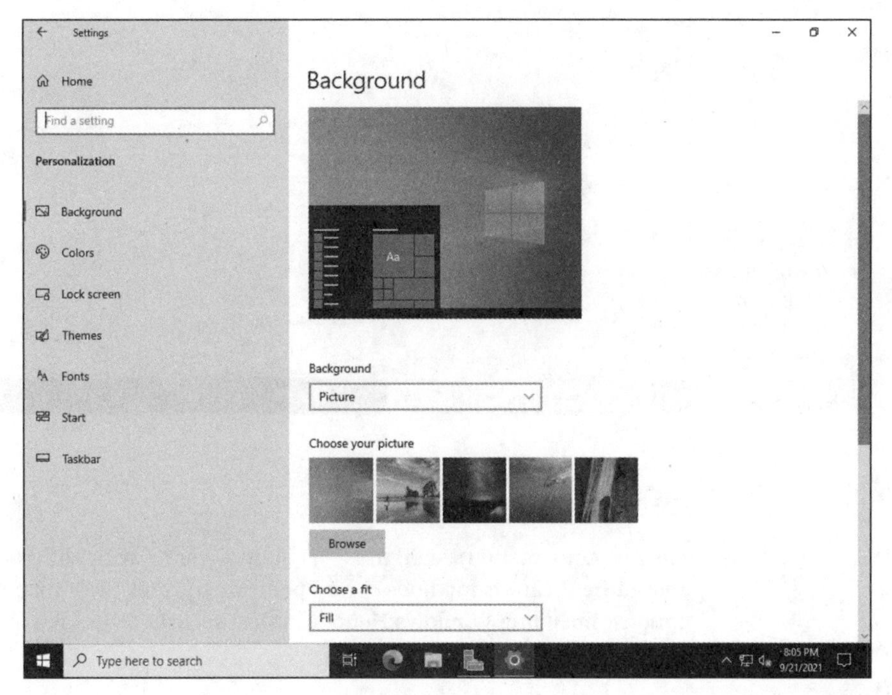

FIGURE 3-5: The Personalization menu gives you many options to make your server your own.

I can hear you saying, "Why would I adjust these things on my server. I don't need backgrounds or custom colors on my server. That's crazy-talk!" I've seen organizations use wallpapers or color schemes to provide a visual indicator that you're on a production (red) system versus a test (green) system. They can be handy when you have multiple Remote Desktop windows open.

Apps

The Apps menu (see Figure 3-6) allows you to specify where you can install programs from, uninstall applications that are no longer needed, and choose which applications are started automatically. Additionally, if you have a website that can open in either an app or a browser, you can adjust this behavior with apps for websites.

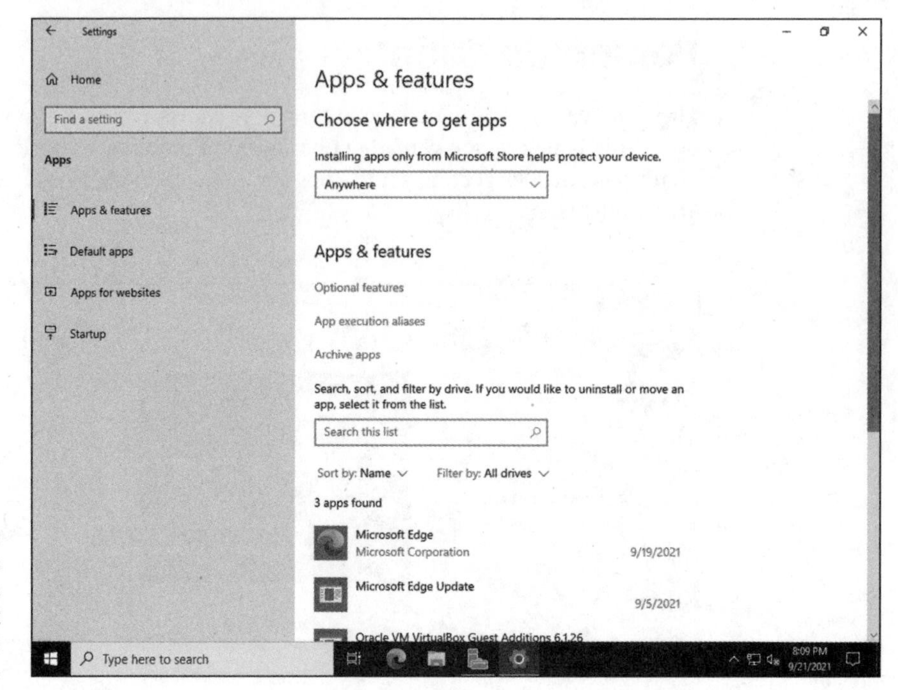

FIGURE 3-6: The Apps menu allows you to work with installed applications.

Accounts

The Accounts menu (see Figure 3-7) allows you to work with local user accounts and authentication methods. The operating system (OS) supports a number of options, including Windows Hello, physical security keys, and, of course, the traditional password scenario.

Time & Language

The Time & Language menu (see Figure 3-8) is where you can adjust the date and time. You can also adjust the region, which changes multiple settings including calendar types and date/time format. You can install new languages and set the default language to use for various services. You can also choose which voice you would like the system to use if you interact with it in that way.

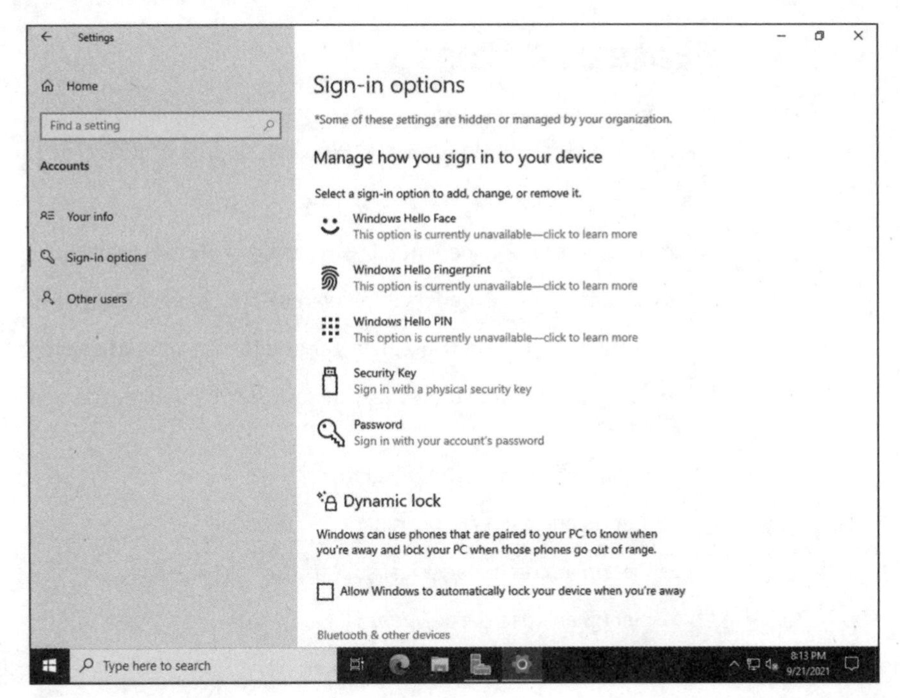

FIGURE 3-7: The Accounts menu allows you to set up various methods of authentication.

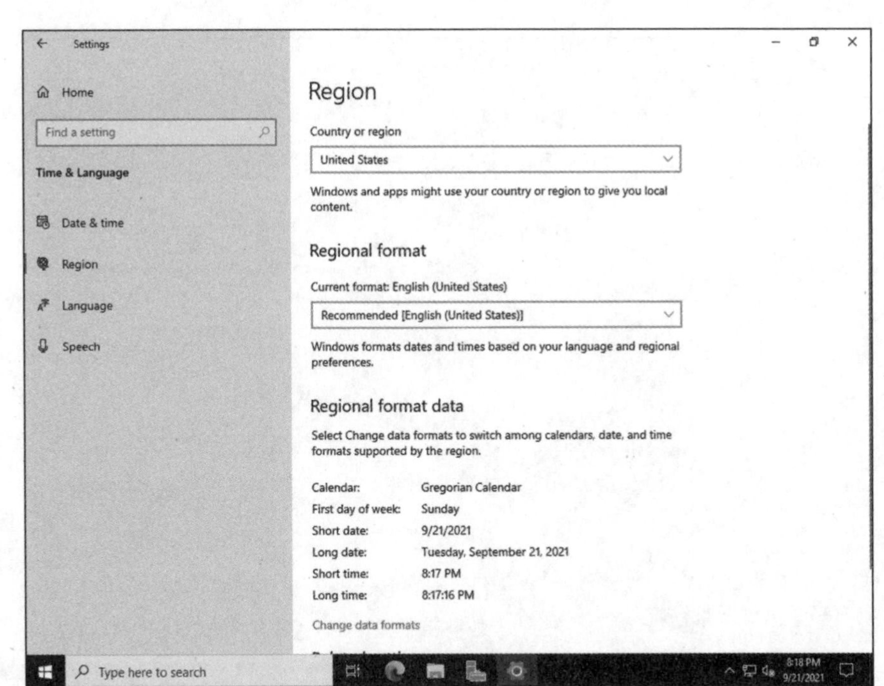

FIGURE 3-8: The Time & Language menu lets you determine how date and time should be formatted.

Ease of Access

The Ease of Access menu (see Figure 3-9) contains all the accessibility tools that are included in the Windows OS. You can do lots of different things here, such as the following:

- » Optimize your experience if you're blind or visually impaired.
- » Optimize your experience if you are deaf or have hearing loss.
- » Configure your mouse with larger or different-colored pointers.
- » Choose different colored text cursors and manipulate the size of the text cursor.
- » Adjust settings on the keyboard.
- » Set up an onscreen magnifier.
- » Set up a screen reader (called Narrator).
- » Set up an onscreen keyboard.
- » Set up high-contrast settings for visibility.

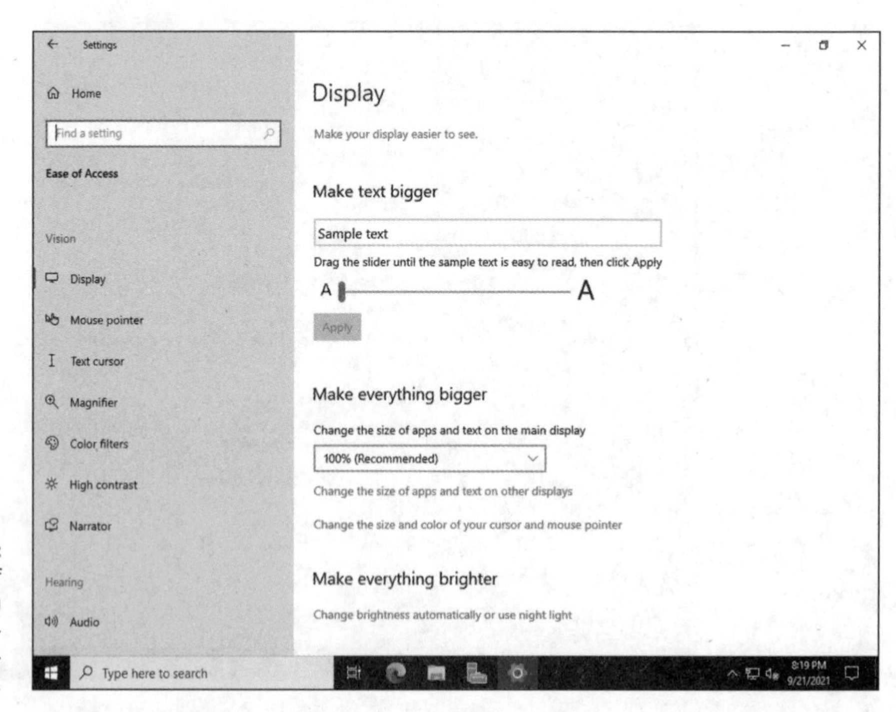

FIGURE 3-9: The Ease of Access menu allows you to configure accessibility options.

Search

The Search menu is relatively small. It allows you to change settings for Safe-Search, which protects you from unwanted images on the Internet, as well as how content is searched on your system and any linked cloud account storage.

Privacy

The Privacy menu (see Figure 3-10) is where you can adjust the permissions granted to both the OS and installed applications. It's pretty straightforward, and there are a ton of settings you can adjust. Be sure to spend some time looking through the submenus.

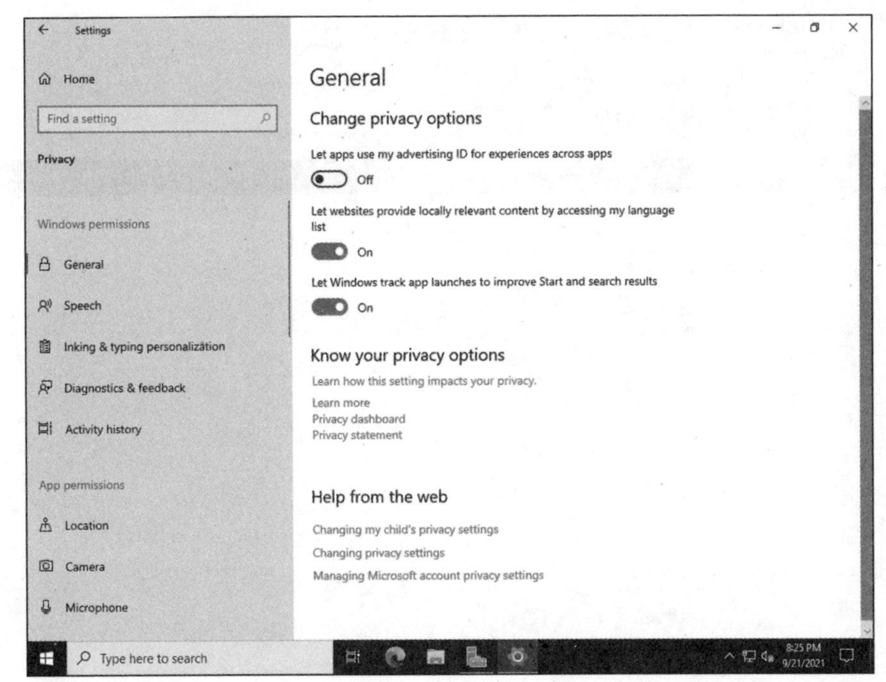

FIGURE 3-10: The Privacy menu allows you to configure tracking and advertisements in both Windows and installed applications.

Update & Security

The Update & Security menu (see Figure 3-11) is where you can configure Windows Updates and manage Windows security settings like antivirus, firewall, app and browser control, and device security. This menu also gives you access to troubleshooting utilities and recovery options. You can activate your copy of Windows and enter into development mode if that's your happy place.

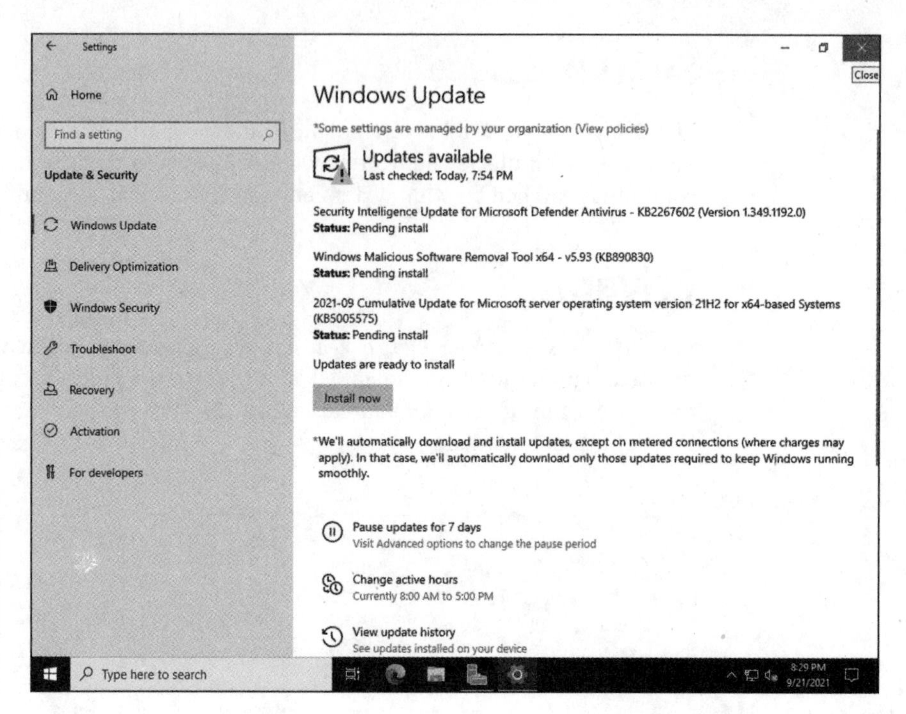

FIGURE 3-11: The Update & Security menu is where you can manage the security-related aspects of your server.

IN THIS CHAPTER

» Defining workgroups

» Weighing the pros and cons of workgroups

» Knowing the difference between centralized and group sharing

» Understanding how to configure a server for a workgroup

» Keeping your workgroup running smoothly

» Examining how Peer Name Resolution Protocol helps workgroup communications

Chapter 4 Working with Workgroups

For some networks, it doesn't make sense to have a domain controller with Active Directory running. In these networks, however, it's still considered desirable to be able to share things from system to system. This is where workgroups come into play.

In this chapter, I explain what workgroups are and how to create them. I also fill you in on the Peer Name Resolution Protocol, what it does, and why it's important for workgroups.

Knowing What a Workgroup Is

A *workgroup* is a peer-to-peer group of computers that share resources but that do not belong to an Active Directory domain. A workgroup may have a central server that it uses to consume various services, or it may simply share data from individual workstations. A workgroup can be as small as two computers, or it can scale up to be larger (though the larger it gets, the more difficult it is to administer). The key point is that there is no Active Directory domain involved.

For a workgroup to work properly, all the systems must share the same workgroup name. By default, in Windows, a non-domain-joined system will belong to a workgroup called WORKGROUP.

The biggest difference to keep in mind between peer-to-peer and client–server relationships like those made in Active Directory is that peer-to-peer are decentralized in every way. You have to explicitly set up access for users and resources on each server that they want to connect to. Systems in a client–server relationship like Active Directory have a centralized database that handles authentication and authorization for access to resources.

TIP

One of the key advantages to workgroups is that they can be very simple to manage as long as they're small. You simply configure a resource for sharing and define who you want to share that resource with. You don't have to worry about complex group policies, because everything is set locally. Workgroups can also be an inexpensive option because you don't need multiple servers to support a workgroup; in fact, you don't technically need servers at all, you could create a workgroup with just client systems. This chapter focuses on workgroups on a Windows Server 2022 system.

WARNING

However, because user accounts are managed on each individual system, security is a concern with workgroups. Passwords may not be changed very often. If they are changed, a user may update his password on a few systems but not on all of them, and then end up out of sync. Plus, some applications may require Active Directory; you won't be able to use these applications if you're using workgroups.

Knowing If a Workgroup Is Right for You

When you're considering creating a workgroup, you need to understand what you can and cannot do with a workgroup. This will determine whether a workgroup is really a good fit for your environment. For instance, if you have a small network with less than 15 systems, a workgroup may be a good route to go. If you have 200 systems, a workgroup probably won't be a good choice because it would result in a very large amount of administrative overhead.

In a workgroup, you can share files, databases, and printers. If this is all you need to do, and you have a small pool of systems (ideally 15 or less), then a workgroup is a good fit. This saves organizations the expense of setting up domain controllers, which will consume another server operating system license, as well as the cost of hiring a system administrator who knows how to take care of Active Directory.

However, you don't have centralized authentication — every user must have an account on the system on which he wants to access resources. You can't manage systems with Group Policy — you have to use local security policies only. Finally, you can't run applications that require Active Directory.

Comparing Centralized and Group Sharing

If you've decided a workgroup is right for you, the next thing you need to decide is whether you want to use a centralized sharing model or a group sharing model.

In the centralized sharing model, all your data is stored on one system —a workstation or a server. All the other members of the workgroup connect to this central location to use the shared resources. The centralized sharing model is the simplest to sustain, because you can set up accounts for the workgroup members on the server and they'll be able to share the resources.

In the group sharing model, all the workstations in the group share different resources. The group sharing model can be far more complex to manage because the users all need to have access to every system on which they need to access a resource.

Configuring a Server for a Workgroup

When configuring a server to be a workgroup server, you need to add groups, users, and whatever resources you want to share. Before you do any of these things, though, you need to change the name of your workgroup.

Changing the name of your workgroup

The Windows default workgroup name is WORKGROUP. You should change this to something that's meaningful to you. Here's how:

1. **In Server Manager, click Local Server.**
2. **Click the WORKGROUP link, below the computer name, shown in Figure 4-1.**

   The System Properties dialog box appears (see Figure 4-2).

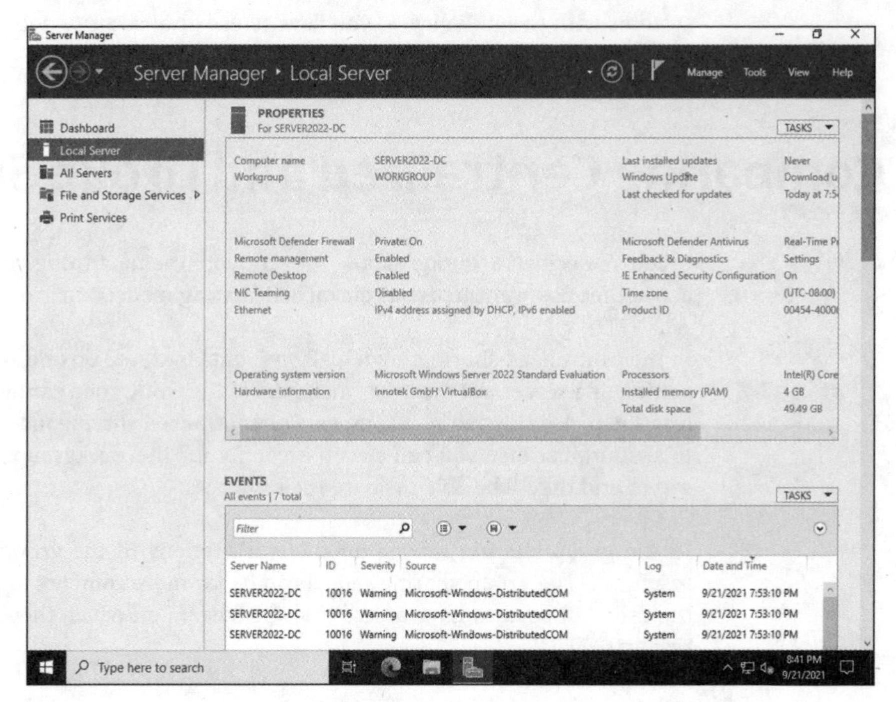

FIGURE 4-1: Configuring the workgroup begins in the Local Server screen in Server Manager.

3. **Click the Change button.**
4. **Change the workgroup name to whatever you want it to be, and click OK.**

   In my example, I changed the name to NOTWORKGROUP (see Figure 4-3).

   You see a message welcoming you to your workgroup.
5. **Click OK.**

   You see a message saying that the server needs to restart.
6. **Click OK.**

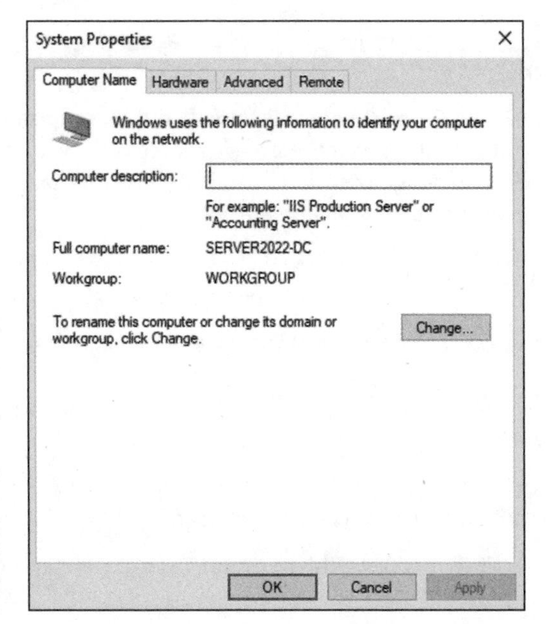

FIGURE 4-2: The System Properties dialog box allows you to set the workgroup name.

7. **In the System Properties dialog box, click Close.**
8. **When prompted to restart the server, click Restart Now.**

Changing the workgroup name is simple, and it follows a security best practice of changing the defaults.

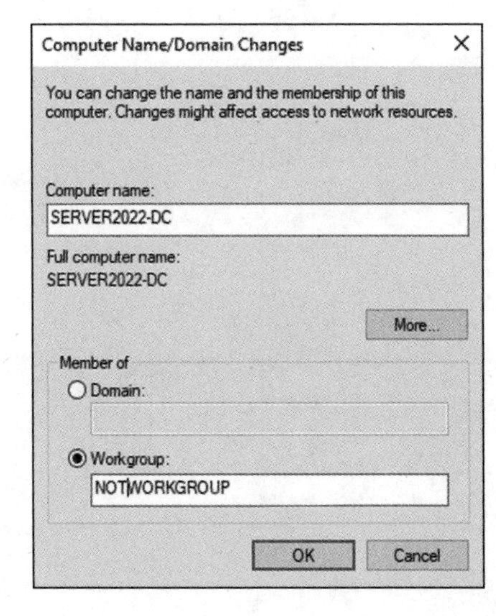

FIGURE 4-3: Setting the workgroup name.

Working with Workgroups

Adding groups

You grant permissions for the shared resources to the group rather than directly to user accounts. This ensures that the users get access to everything they need — and it's far easier to manage than individual user permissions. For this reason, you should create groups before you create users.

Here's how to create a group:

1. **In Server Manager, choose Tools⇨Computer Management.**
2. **Click Local Users and Groups.**
3. **Double-click Groups.**

   Windows Server 2022 includes several groups out of the box. In this example, I'll create a simple group called Workgroup Users.

4. **Click More Actions on the right-hand side of the screen and choose New Group.**

   The New Group dialog box appears (see Figure 4-4).

5. **Enter a name in the Group Name field and click Create.**
6. **Click Close to close the New Group dialog box.**

FIGURE 4-4: Creating a group to allow access to workstation resources.

Creating users and adding users to the group

TIP

In order for a user to access a workgroup, she must have a local account on the system she's trying to connect to. When you create the user account, you can grant permissions directly, but adding the user to a group with the correct permissions is much simpler. Imagine if you have to create five users: You can set permissions for each of them, or simply add them to the group.

Here's how to create a user and add the user to a group:

1. **In Server Manager, choose Tools⇨Computer Management.**
2. **Click Local Users and Groups.**
3. **Double-click Users.**

   Windows Server 2022 includes a few user accounts out of the box. In this example, I'll create a user account called User1.

4. **Click More Actions on the right-hand side of the screen and choose New User.**

   The New User dialog box appears.

5. **Fill in the User Name, Full Name, and Password fields (see Figure 4-5).**

TIP

   In a Production environment, you'll want to leave the User Must Change Password at Next Logon check box selected.

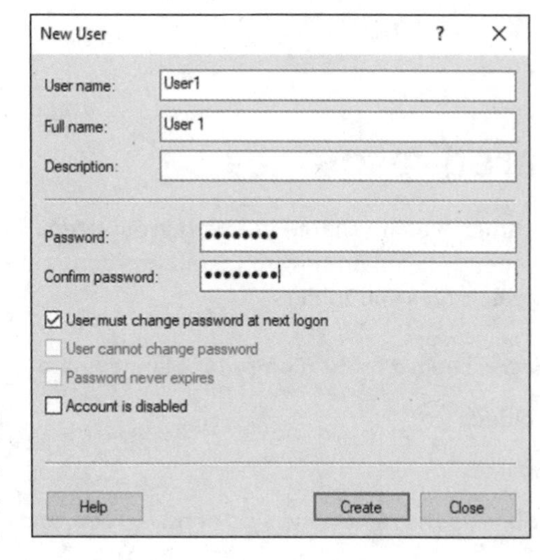

FIGURE 4-5: Creating a user account to allow access to workstation resources.

Working with Workgroups

6. **Click Create.**
7. **Click Close to close the New User dialog box.**
8. **Right-click the newly created user and choose Properties.**

   The *User Name* Properties dialog box appears (where *User Name* is the name of the user whose properties you're looking at).
9. **Click the Member Of tab, and click Add.**

   The Select Groups dialog box appears.
10. **In the Enter the Object Names to Select field, type the name of the group that you created earlier and click Check Names.**

    If the group is found, it will be underlined and will start with the server name, as shown in Figure 4-6.
11. **Click OK to close the Select Groups dialog box.**
12. **Click OK to close the *User Name* Properties dialog box.**

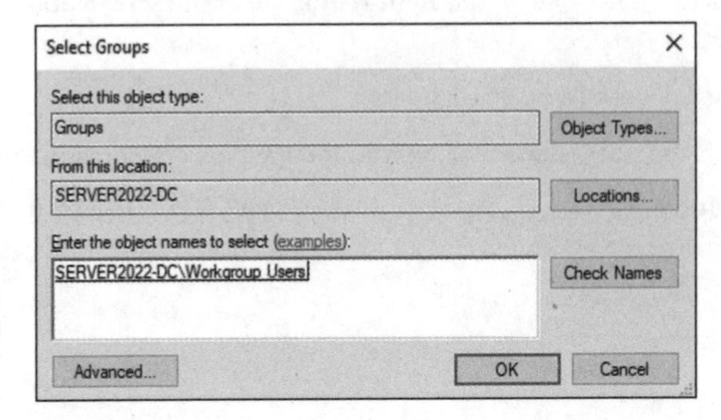

FIGURE 4-6: Selecting the group for your user account.

Adding shared resources

The most common things that are shared in a workgroup scenario are printers and files. In Book 2, Chapter 2, I explain how to install and configure the print server role. Here's how to share files and folders:

1. **In Server Manager, choose Tools⇨Computer Management.**
2. **Click Shared Folders.**
3. **Double-click Shares.**
4. **Click More Actions on the right-hand side of the screen, and choose New Share.**

   The Create a Shared Folder Wizard appears.

5. **On the first screen, called Welcome to the Create a Shared Folder Wizard, click Next.**
6. **On the Folder Path screen, click Browse.**

   The Browse for Folder dialog box appears.
7. **Navigate to the folder that you want to share (see Figure 4-7).**

FIGURE 4-7: Selecting the folder you want to share in the Create a Shared Folder Wizard.

8. **Click OK to close the Browse for Folder dialog box.**
9. **Click Next.**

   The next screen allows you to change the name of the share and set the offline settings. This allows users who are offline to access share files.
10. **After you've selected the options you want, click Next.**

    The next screen allows you to customize permissions for the share.
11. **Select the Customize Permissions radio button, and then click Custom.**

    The Customize Permissions dialog box appears.
12. **Click Add.**
13. **Type the name of the group that you created earlier and click Check Names to ensure that it resolves.**
14. **Click OK.**
15. **In the Permissions for Workstation Users section, set the desired permissions for the share.**

    Typically, you will select the Change and Read check boxes, as shown in Figure 4-8.

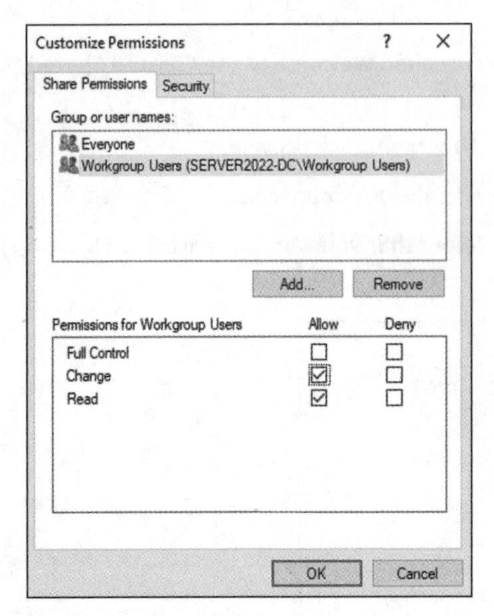

FIGURE 4-8: Customizing the shared folder permissions.

16. **Click OK to close the Customize Permissions dialog box.**
17. **Click Finish.**
18. **On the Sharing Was Successful screen, click Finish again.**

After the share has been added, it will show up in Shares under Shared Folders, as shown in Figure 4-9.

If you ever set up workgroups on an older version of Windows, you know that your next step was to add the group you created to the share permissions on the folder in the file system as well. This was a lot of double work. Lucky for us, Microsoft remedied that. When you share the folder in the Shared Folders area in Computer Management, it sets the share permissions as well. If you want to look at the settings, follow these steps:

1. **Click the folder icon in the taskbar at the bottom of the screen.**

   File Explorer launches.

2. **Click This PC, and then double-click the volume where you create the shared folder.**
3. **Right-click the shared folder, and click Properties.**
4. **Click the Sharing tab, and then click Advanced Sharing, as shown in Figure 4-10.**

   On this screen, you can change the number of users who can connect at the same time, you can change the permissions, and you can change the caching. In this context, *caching* refers to offline access.

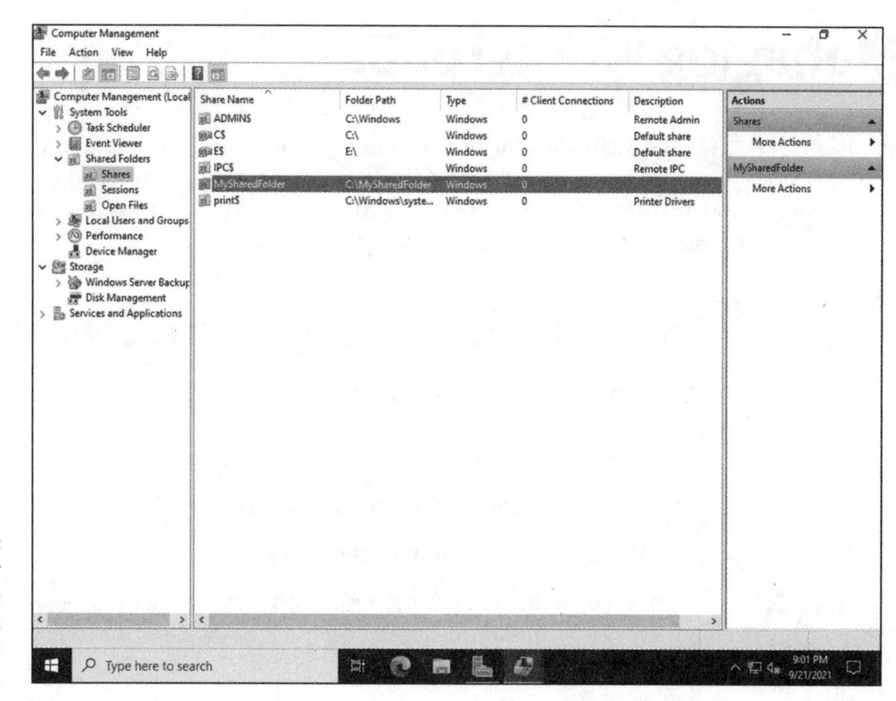

FIGURE 4-9: The new folder share is visible in the Shares section of Shared Folders.

5. **Click OK.**
6. **Click OK again.**
7. **Click Close.**

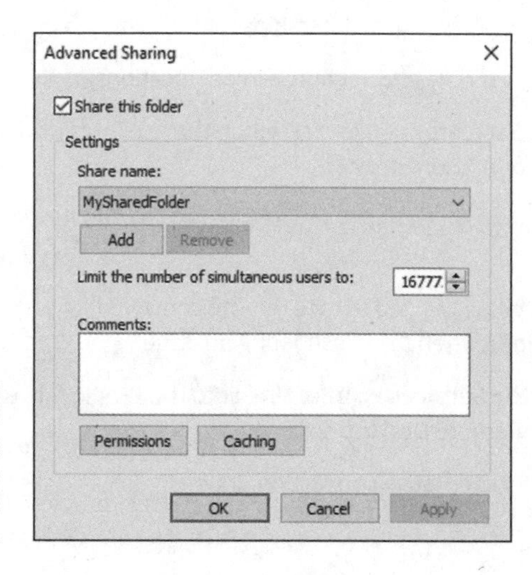

FIGURE 4-10: The Advanced Sharing dialog box allows you to configure the share.

Managing Workgroups

As the administrator of a workgroup, you may need to take care of some ongoing management tasks. Some of the most common are resetting passwords, though you can also change the user's role. In this section, I fill you in on a few different methods to manage user accounts.

The Computer Management console

The Computer Management console is a great place to start. Follow these steps:

1. **From Server Manager, choose Tools⇨Computer Management.**
2. **Double-click Local Users and Groups to expand it.**
3. **Double-click Users to show the user list.**

TIP

If all you want to do is reset the password, right-click the user account and choose Set Password. Click Proceed in the dialog box, and then enter the new password and click OK.

4. **Right-click the account you created earlier, and choose Properties.**

   The Properties dialog box appears. Here's what you can do on each tab of this dialog box:

   - **General:** Set the account to disabled, or require the user to change her password the next time she logs in.
   - **Member Of:** Assign group memberships to the user.
   - **Profile:** Set login scripts and set home folder locations.
   - **Environment:** Set a program to start when the user logs in, and set the desired behavior of client devices.
   - **Sessions:** Set how you want the sessions to be handled for a particular user account.
   - **Remote control:** Allows you to take remote control of a user's session. This is very helpful when troubleshooting an issue.
   - **Remote Desktop Services Profile:** Similar to the Profile tab, except this only applies to Remote Desktop sessions.
   - **Dial-in:** Controls alternative connection options.

The Account window

The Account window allows you to do the same user management functions that the Computer Management console does, just with a nicer looking interface. Here's how to access it:

1. **Click the Start menu and click the Settings icon.**
2. **Click Accounts.**
3. **Click Other Users (shown in Figure 4-11).**
4. **Select the user you created earlier.**

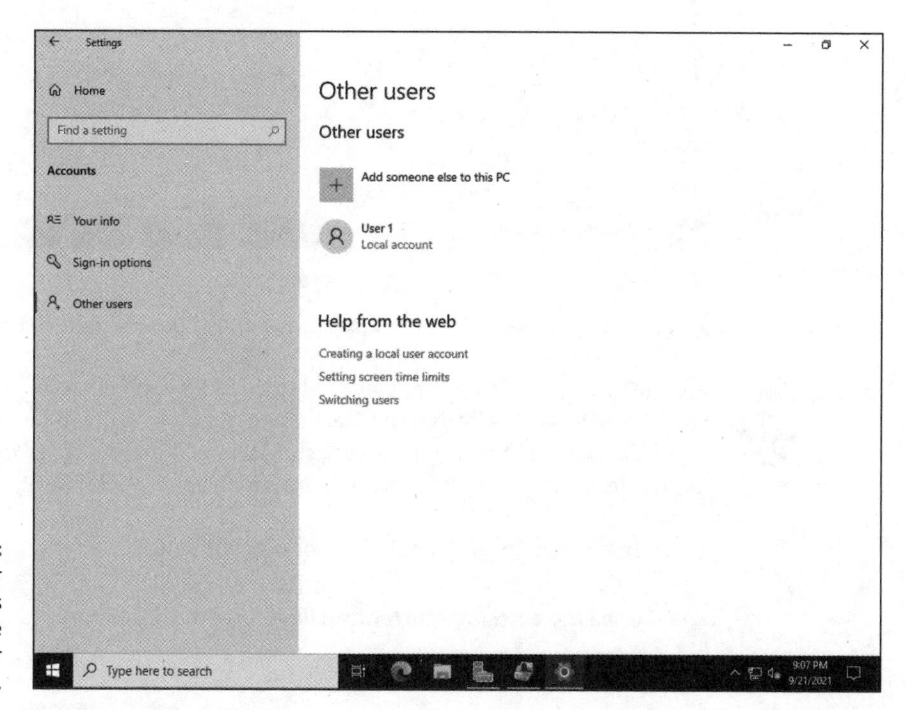

FIGURE 4-11: Clicking Other Users allows you to manage accounts other than your own.

From here, you can change the account type, or delete the account (see Figure 4-12).

PowerShell

You can manage your user accounts with graphical tools, but you can also use PowerShell to do the trick. In fact, there are a few administrative tasks that can only be accomplished by using PowerShell.

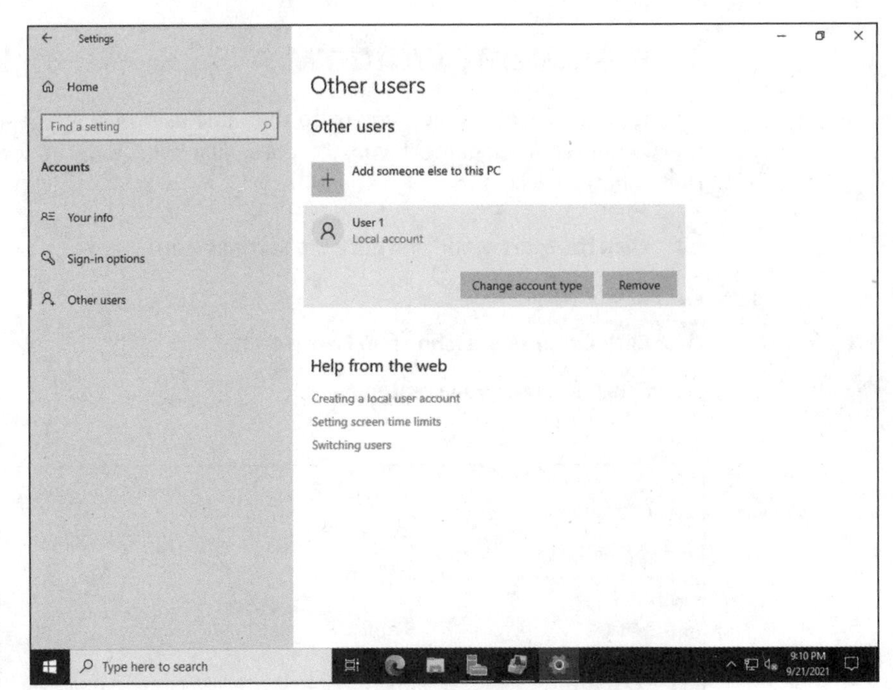

FIGURE 4-12: Managing the account of another user.

To get to PowerShell, click the Start menu and choose Windows PowerShell.

TIP

For the tasks that I'm doing, you need to open PowerShell as an administrator. If you choose PowerShell from the Start menu, that's how it will launch. The majority of the tasks that you'll want to accomplish will require that PowerShell is run as an Administrator. If you want to learn more about PowerShell, see Book 6.

Here are two handy commands for working with local accounts.

- **To see the account's current settings,** type the following:

```
Get-LocalUser -Name "User1"
```

- `Get-LocalUser` queries the system for information on the user account that you specify.
- **To change the user's password,** type the following:

```
$Password = Read-Host -Prompt 'Enter Password'
  -AsSecureString
$UserAccount = Get-LocalUser -Name "User1"
$UserAccount | Set-LocalUser -Password $Password
```

The words with $ in front of them are called *variables.* Think of them as containers for things. For instance, `Read-Host -Prompt 'Enter Password' -AsSecureString` prompts you for a password and then saves it in the `$Password` variable as a secure string. The `Get-LocalUser` cmdlet retrieves the information for User1, and you store User1's information inside a variable named `$UserAccount`. Finally, you take the contents saved in `$UserAccount` and set the password with the secure string you saved earlier into `$Password`.

Figure 4-13 shows an example of the commands having been run.

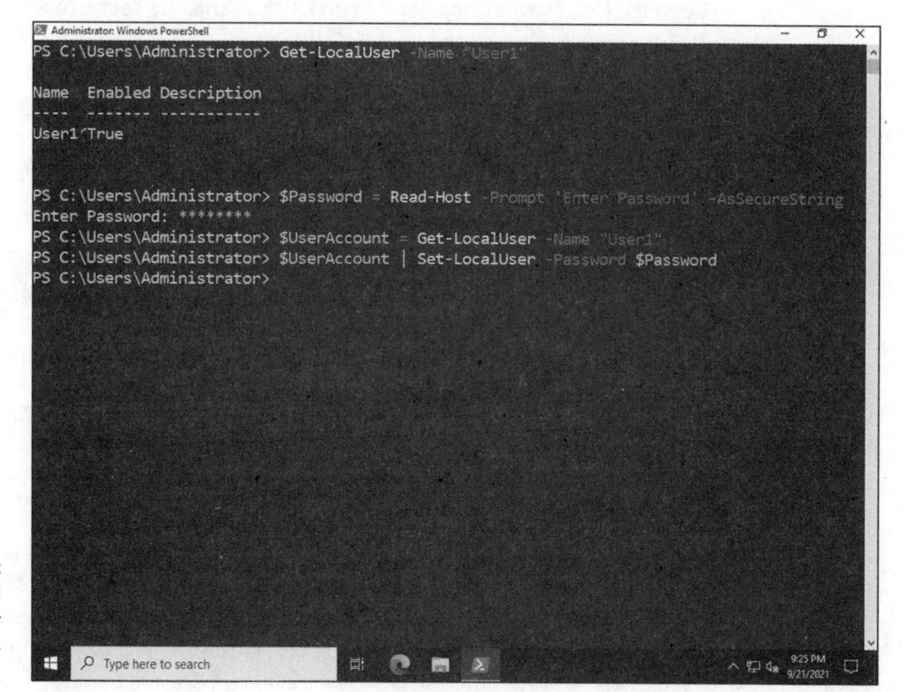

FIGURE 4-13: PowerShell window with user account management code line.

Examining the Peer Name Resolution Protocol

Peer Name Resolution Protocol (PNRP) was first introduced in Windows XP. It allows for name resolution and registration within peer-to-peer networks.

You may have noticed in the file sharing example that the share address started with the server's name. If it weren't for PNRP, you would have to connect with IP addresses because you would have no way to resolve the hostname. I don't know about you, but having to remember a list of IP addresses would be very difficult for me to do, especially if there were multiple systems. You may be thinking, "Why not use Domain Name System (DNS) resolution instead?" There is no harm in using DNS, but you would need a DNS server to support that. PNRP does not require a separate server. You can get name resolution using the infrastructure already in place. Pretty cool, right? You can secure the PNRP traffic with a firewall if you need to provide better security. PNRP uses UDP 3540 to allow for name resolution to occur.

So, how does PNRP work? Let's say that you're on ServerA and you want to talk to ServerB. Here's what happens: ServerA examines its cache to see if it has a PNRP ID for ServerB. There are two possibilities:

- If a match is found, ServerA sends a PNRP request message to ServerB.
- If a match is *not* found, ServerA sends a PNRP request message to whichever system has the closest PNRP ID to ServerB.

The PNRP request is received by whichever node it was sent to. From here, there are three options:

- If the PNRP request was sent to ServerB, and ServerB is up and responding, ServerB will send a reply to the request back to ServerA.
- If the node that received the PNRP request is not ServerB and doesn't know who ServerB is (it doesn't have a PNRP ID for ServerB), the node that received the request will respond to ServerA that it is not able to assist.
- If the node that received the PNRP request is not ServerB, but it either has the PNRP ID for ServerB or knows of a node closer to the PNRP ID of ServerB, it will respond with that information to ServerA. ServerA can then either reach out to ServerB, or reach out to the next closest node and start the search again.

IN THIS CHAPTER

» Understanding Active Directory domains and why you want them

» Preparing to create your domain with Windows Server 2022

» Performing the domain configuration prerequisites

» Configuring your server as a domain controller

Chapter 5

Promoting Your Server to Domain Controller

In the last chapter, I explain how to create and use workgroups. Although workgroups work well for smaller environments with few systems (less than 15 systems), they do not scale well in enterprise environments.

In most enterprise environments, you'll be using a domain. A domain provides one central source of truth for all authentication events. You only need to create a user account once, assign it to the appropriate security groups, and then you're able to assign the security group to whatever resources the user needs to access (assuming, of course, that the resources are on the domain). This simplifies user management significantly and improves your organization's security standing because you're able to enforce settings across the domain with Group Policy (see Book 3, Chapter 2).

Understanding Domains

Before I dive into how to create a domain controller, I think it's important that you understand what an Active Directory domain is, how it's architected, and some of the default groups that are created when you first create an Active Directory domain. I also cover domain controller roles. If you're already familiar with these topics, feel free to skip ahead. If not, read on!

What is a domain?

Let's start with the basic concepts and then build on them. A domain in a Microsoft network is commonly referred to as an Active Directory Domain Services (AD DS) domain. Each AD DS domain has its own database of objects within it. An object could be a user, a group, a computer, or even a printer.

AD domains are organized into a hierarchical structure with the forest being the highest level, followed by domains and organizational units (OUs). Each forest can have multiple domains, and each domain can have multiple OUs.

Forests and domains and OUs, oh my!

I'm throwing around a lot of terminology here, so let me define some things before moving on:

- **Organizational units:** OUs are containers for objects stored in Active Directory. They're typically used to group like objects together — for example, all user accounts for people in the accounting department, or every computer on the second floor of a building. By grouping like items together, it can be simpler to manage these objects because it's likely they'll have similar settings or configurations.
- **Domains:** Think of a domain as a security boundary. A domain contains its own database of the objects it contains. All those objects share that database and any security policies that are assigned within the domain. Additionally, a domain may be set to trust another domain by creating a trust relationship, which may allow users in one domain to access resources in another domain.
- **Domain trees:** A domain tree is simply a collection of domains. For instance, you may have the domain tree namespace set to `sometestorg.com`, and the domains underneath that tree would be something along the lines of `accounting.sometestorg.com` or `hr.sometestorg.com`. The domains share the namespace with the domain tree.

- **Forests:** An Active Directory forest is the top-level object. It stores the entirety of Active Directory including all the trees and domains. The first domain created in a forest is referred to as a *forest root domain.* It will typically share the name of the forest.
- **Schema:** The Active Directory schema is similar to a dictionary in that it contains definitions. It contains the definition of every object class and every attribute that can be used by an Active Directory object.

Depending on the organization where you work, you may have a small instance of Active Directory that has only one forest and one domain, or you may work for a large company with multiple forests and multiple domains.

You may be asking why you would want to have more than one forest. It's more management overhead, right? Yes, it absolutely is. However, if you work in an environment where separating security boundaries is crucial, then multiple forests are your best bet. A common example is separating the corporate network from the production network. The corporate forest is used to manage corporate workstations where email and Internet are accessed. The production forest is used only to manage authentication traffic to servers or network devices. By separating the forests, and by extension the security boundaries of these two networks, you safeguard the production environment from the potentially malware-laced corporate environment.

Understanding privileged domain groups

When you create an Active Directory domain, several administrative groups are created by default. I've known a few administrators who simply put their accounts into the Enterprise Admins group and never looked back. As a system admin, you really want to enforce the concept of least privilege, meaning that people should have the permissions to do what they need to do, but nothing more. The Enterprise Admins group is a very privileged group to be in and shouldn't be the solution to all your permissions woes.

Table 5-1 covers the various groups and what permissions you get when you're placed into them.

Examining Flexible Single Master Operation roles on domain controllers

When Active Directory first came about, there were some definite challenges. Through the years, Microsoft continued to improve on how domain controllers work with Active Directory. It came up with the idea of Flexible Single Master

Operation (FSMO). By using FSMO roles, domain controllers don't fight over who gets to make changes, nor do you have to be as concerned about your main domain controller going down because the roles it was taking care of can be moved to another domain controller.

TABLE 5-1 Default Domain Groups

Group Name	Description
Enterprise Admins	This group is located in the forest root domain. It's a member of the Built-in Administrators in every domain in the forest. Members of the Enterprise Admins group can make changes to the forest, add or remove domains, establish forest trusts, and raise forest functional levels. This group should only be used when needed because it is such a highly privileged group — for example, when you're constructing a forest for the first time, or when you're making forest-wide changes like establishing forest trusts.
Domain Admins	Members of the Domain Admins group are members of the Built-in Administrators group for that domain. They're automatically added to the Local Administrator's group on every system that is joined to their domain.
Administrators	Domain Admins are members of the Administrators group for the domain in which their account resides. Enterprise Admins are members of this group across all domains in their forest. This group is the one that grants them access to the local administrator's group on all domain-joined systems. This group has the ability to manage most of the domain's objects with full permissions, and can take ownership of most every object in the domain.
Schema Admins	Schema Admins is a special group. It exists in the forest root domain similar to the Enterprise Admins group. Schema Admins have permission to manage the schema in Active Directory but nothing else. This makes Schema Admins a less privileged group, but be very careful who you assign to be a Schema Admin. Changes to schema that go poorly can damage Active Directory. This group should only be assigned in the infrequent instance when schema changes need to made.
Additional Administrative Groups	Additional Administrative Groups are created as you install roles and applications. DHCP, DNS, and Exchange Server are just a few example of roles or applications that will create new administrative groups in Active Directory.

Here are the FSMO roles you see in today's domain controllers and what each of them does:

- **Forest wide:**
  - **Schema Master:** Manages the read/write copy of your AD schema.
  - **Domain Naming Master:** Ensures that you don't create domains with the same name as an existing domain.

» **Domain wide:**

- **RID Master:** Every security principal in Active Directory has a security identifier (SID). That SID is made up of the domain SID, which is the same for all systems in the same domain, and the relative ID (RID), which is unique for each security principal in that domain. The RID Master assigns blocks of SIDs to domain controllers so that they can issue them.
- **PDC Emulator:** PDC used to stand for Primary Domain Controller, which was used in environments running Windows NT 4.0 and earlier. The PDC Emulator was introduced to provide support for legacy systems, and it still performs important functions for modern systems. The PDC Emulator is the authoritative domain controller. It handles authentication, password changes, authentication failures and Group Policy Objects and sets the correct time for the entire domain. Next time you're late for a meeting, blame your PDC emulator!
- **Infrastructure Master:** The Infrastructure Master is essentially the translator between domains. It can convert SIDs, globally unique identifiers (GUIDs), and distinguished names (DNs) between domains so that Domain1 understands what Domain2 is requesting. This allows users to access resources on other domains, not just the domain that they exist and authenticate to.

One last important conversation outside of roles is the Global Catalog. A Global Catalog server stores records for all the objects in all the domains in the forest. A domain controller keeps a fully writeable copy of its own Global Catalog, and a read-only copy from the other domains in the same forest. This allows you to search for objects in Active Directory without having to go to each individual domain to do it.

Preparing to Create a Domain

You've decided that you want to create a domain. There are a few tasks that you need to take care of ahead of time.

One of the most important tasks, especially if this is a brand-new server, is to install any available security updates and ensure that you have an up-to-date antivirus software installed. This is going to be your central authentication server after all, so you want to make sure that it's protected.

The task that everybody dreads and sometimes skips is the planning task. I think it gets skipped over because it tends to be less fun than installing the server role and configuring it. Here's the thing, though: If you take the time to plan the installation and configuration ahead of time, the outcome will be much better. You'll be able to ensure that the server will meet your needs and, better yet, you won't have to reinstall it when you find out it doesn't meet your needs. So, sit down for a moment, and map out how you want your Active Directory to look. Will it have multiple domains? What kind of OU structure are you going to use?

Of course, mapping out how you want your AD to look tends to be more for new installations where Active Directory is not currently deployed. But if you're adding a domain controller to an existing Active Directory environment, you should still plan things out. Will this domain controller host any of the FSMO roles? If so, which ones? Is this a newer domain controller being built to replace older domain controllers? If so, what is their current forest functional level and domain functional level?

Functional levels

Functional levels are used to tell domain controllers which features can be enabled at the forest or domain levels. With each new version of Active Directory, new features are added that you may want to take advantage of. If you're creating a new installation of Active Directory, choose the highest functional levels available.

Functional levels can prevent a server from becoming a domain controller if its functional level is too low. Functional levels can be set at the forest level or the domain level. Domains can run with a higher functional level than the forest they're in, but they can't run at a lower functional level than the forest.

When you raise the functional level, you can't add domain controllers that are set to a lower functional level. It's also important to note that the domain functional level can be equal to or higher than the forest functional level, but it can't be below it.

In older server operating systems, raising the functional level was a one-time deal. There was no nice way to back out the change if it caused problems. That is no longer the case. You can revert back to an older version with PowerShell as long as the rollback is supported, and none of the new features has been enabled that require the newer functional level. If you did enable them, you have to disable them before you can roll back.

Rolling back the forest functional level

```
Set-ADForestMode -ForestMode <desired forest level>
```

Forest functional levels include

- Windows2016Forest or 7
- Windows2012R2Forest or 6

Rolling back the domain functional level

```
Set-ADDomainMode -DomainMode <desired domain level>
```

Domain functional levels include

- Windows2016Domain or 7
- Windows2012R2Domain or 6

Forest functional level

In order to raise the forest functional level, you have to be a member of the Enterprise Admin group. You must raise the forest functional level from the domain controller that runs the Schema Master role. You can't do it from the other domain controllers.

To check your current forest functional level, open PowerShell and run the command `Get-ADForest`.

Domain functional level

In order to raise the domain functional level, you have to be a member of the Domain Admin group. You must raise the domain functional level on the domain controller that runs the PDC Emulator role. You can't do it from the other domain controllers.

To check your current domain functional level, open PowerShell and run the command `Get-ADDomain`.

Performing Domain Configuration Prerequisites

You've finished the planning stage. Now comes the fun part! Hopefully, you're installing the Active Directory Domain Services (AD DS) role on a fresh new server. If you are, you can skip ahead to installing Domain Name System (DNS) and Dynamic Host Configuration Protocol (DHCP). If you're installing AD DS on an older server, you need to ensure that there are no unsupported roles or features that need to be uninstalled first.

AD DS requires DNS and DHCP to function properly. If you already have these in your environment, then great — you're all set! If not, you can install them on the same system as AD DS or on separate systems.

Checking for unsupported roles and features

Microsoft doesn't publish a list of roles and features that are incompatible with AD DS. If you're installing AD DS on a system that has other roles and features, there two approaches you can take:

- If the roles and features are no longer in use, uninstall them.
- If the roles and features are in use, back up the system, uninstall the roles and features, install AD DS, and re-install the other roles and features.

REMEMBER

The best and safest practice is to install AD DS on a freshly built server, but if you find that you have no choice (this often happens in smaller companies that can't afford multiple systems), follow the preceding steps.

Installing and configuring Domain Name System

DNS is responsible for resolving easy-to-remember hostnames to not-so-easy-to-remember IP addresses. It's a requirement for Active Directory to function properly. If DNS is not installed on the system on which you'll be installing AD DS, but it is available in your environment, then you're good to go. If DNS does not exist in your environment, you need to install it.

You can install the DNS server role by itself, or you can install it as part of the configuration tasks when you install AD DS. I'm going to show you how to install DNS by itself.

TECHNICAL STUFF

Before you install the DNS role, ensure that your server has a static IP address. If it doesn't, you'll be presented with a warning while installing the role. You'll be allowed to continue with the installation, but your clients could lose connection to the DNS server if the IP address changes.

Installing Domain Name System

Follow these steps below to install DNS:

1. **From Server Manager, click Manage and then click Add Roles and Features.**
2. **On the Before You Begin screen, click Next.**
3. **On the Select Installation Type screen, click Next.**
4. **On the Select Destination Server screen, click Next.**
5. **On the Select Server Roles screen, select DNS Server.**
6. **Click Add Features in the dialog box that pops up.**
7. **Click Next.**
8. **On the Select Features screen, click Next.**
9. **On the DNS Server screen, click Next.**
10. **Click Install.**
11. **Click Close after the installation is complete.**

Configuring Domain Name System

Follow these steps to configure DNS:

1. **From Server Manager, click Tools and then click DNS.**
2. **Click the arrow next to your server's name to expand your options.**

   You see the configuration areas, which should look similar to Figure 5-1.
3. **Right-click Forward Lookup Zones and select New Zone.**
4. **In the Welcome to the New Zone Wizard, click Next.**
5. **On the Zone Type screen, leave Primary Zone selected and click Next.**

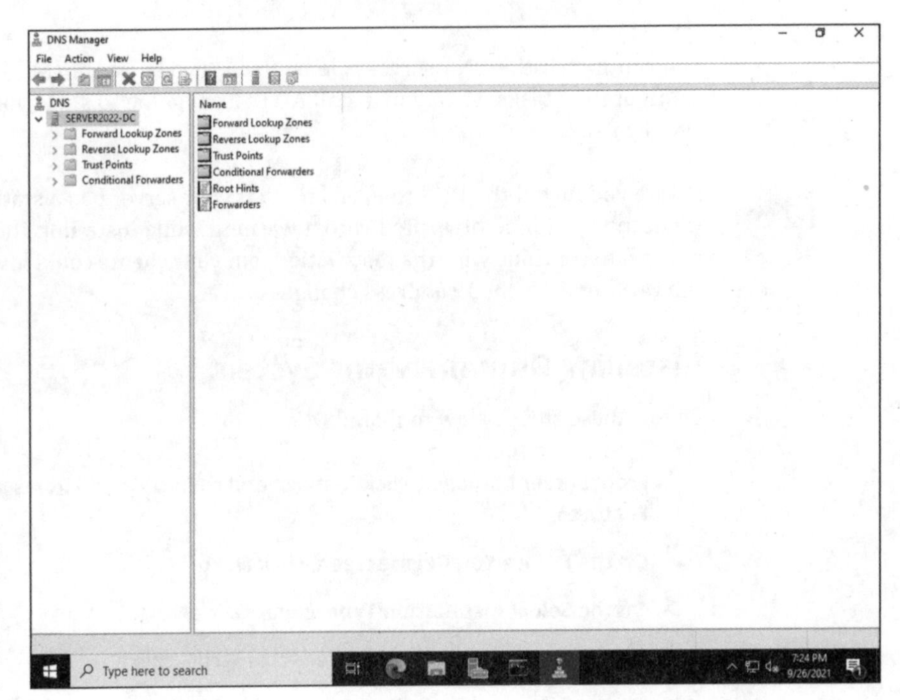

FIGURE 5-1: The DNS Manager screen after installation with no zones configured.

6. **On the Zone Name screen, enter the zone name.**

   This will typically be your domain name. It indicates the namespace in which this DNS server is authoritative. In Figure 5-2, you can see that I've entered sometestorg.com.

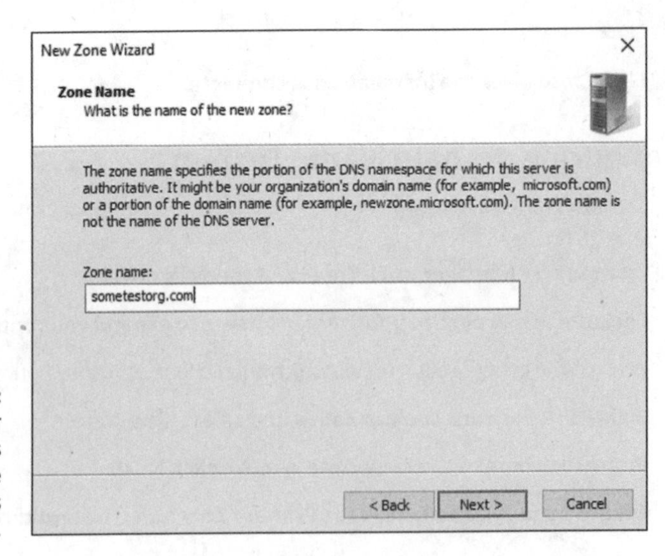

FIGURE 5-2: Naming your zone indicates which zone the DNS server is authoritative for.

7. **Click Next.**
8. **On the Zone File screen, leave the selection on Create a New File with this File Name, and click Next.**
9. **On the Dynamic Update screen, leave the selection on Do Not Allow Dynamic Updates, and click Next.**
10. **In the Completing the New Zone Wizard, click Finish.**
11. **Right-click Reverse Lookup Zones and select New Zone.**
12. **On the Welcome to the New Zone Wizard screen, click Next.**
13. **On the Zone Type screen, leave Primary zone selected and click Next.**
14. **On the Reverse Lookup Zone Name screen, ensure IPv4 Reverse Lookup Zone is selected, and click Next.**
15. **On the next Reverse Lookup Zone Name screen, enter the Network ID.**

    This will be the host bits in the IP address. My DNS server's IP address is 192.168.1.43. Assigning the network ID for this would be 192.168.1, as shown in Figure 5-3.
16. **Click Next.**
17. **On the Zone File screen, leave the selection on Create a New File with this File Name, and click Next.**

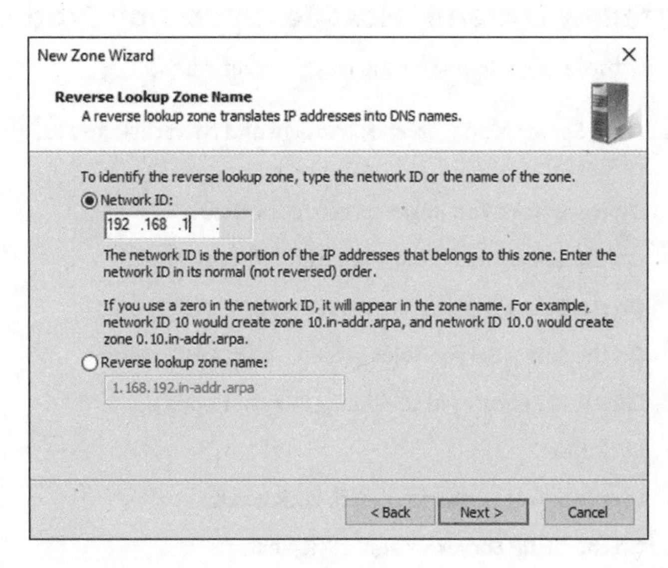

FIGURE 5-3: Assigning the Network ID.

18. On the Dynamic Update screen, leave Do Not Allow Dynamic Updates selected and select Next.
19. In the Completing the New Zone Wizard, click Finish.

That's all there is to it. You now have a functional DNS server. You may be wondering why I told you to leave the dynamic updates off. When Active Directory is installed, you can enable secure dynamic updates. Until then, you have to enable all dynamic updates, which could actually be a very bad idea because any system would be registered in DNS.

Installing and configuring Dynamic Host Configuration Protocol

Installing DHCP is the next step on your journey. DHCP is responsible for assigning available IP addresses to systems on your network. This was a huge improvement over having to manually track IP addresses in a spreadsheet like you used to have to do. The DHCP client is enabled on the majority of systems, workstations, and servers. If a system needs a static IP address, like the infrastructure services I'm discussing right now (AD DS, DNS, DHCP), for example, that IP address should be set as a reservation on the DHCP server so that it doesn't issue an IP address that is already in use.

Installing Dynamic Host Configuration Protocol

Follow these steps to install the DHCP server role:

1. From Server Manager, click Manage and then click Add Roles and Features.
2. On the Before You Begin screen, click Next.
3. On the Select Installation Type screen, click Next.
4. On the Select Destination Server screen, click Next.
5. On the Select Server Roles screen, select DHCP Server.
6. Click Add Features in the dialog box that pops up.
7. Click Next.
8. On the Select Features screen, click Next.
9. On the DHCP Server screen, click Next.
10. Click Install.
11. Click Close after the installation is complete.

Configuring Dynamic Host Configuration Protocol

Follow these steps to configure DHCP:

1. **From Server Manager, click flag, and then click Complete DHCP Configuration.**
2. **In the DHCP Post-Install Configuration Wizard, click Commit.**

   This creates the security groups you need to manage the DHCP server.
3. **Click Close when it's complete.**
4. **From Server Manager, click Tools and then click DHCP.**
5. **Click the arrow next to the server's name, and then the arrow next to IPv4.**

   Your screen should look similar to Figure 5-4.

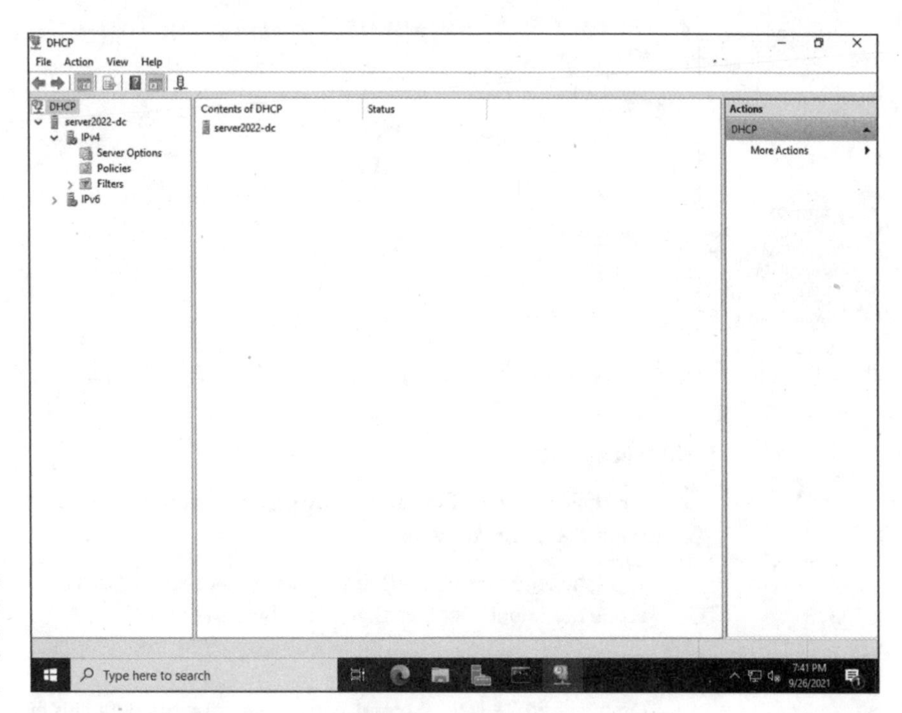

FIGURE 5-4: DHCP Console allows you to configure DHCP services.

6. **Right-click IPv4, and choose New Scope.**

TIP

A DHCP scope defines the IP addresses that are available for lease to the systems that are on the subnet that has been defined. For each subnet on your network, you need a separate DHCP scope defined before you can issue IP addresses for systems that are in that subnet.

7. **On the Welcome to the New Scope Wizard, choose Next.**
8. **On the Scope Name screen, you can leave the description blank and click Next.**

   I'm calling it Demo Scope, just for the sake of this example.

9. **On the IP Address Range screen, enter the pool of addresses you want the DHCP server to manage, as well as your subnet mask (see Figure 5-5).**

   I've created a very small scope for this example. Normally your scope would be much larger, potentially a whole subnet.

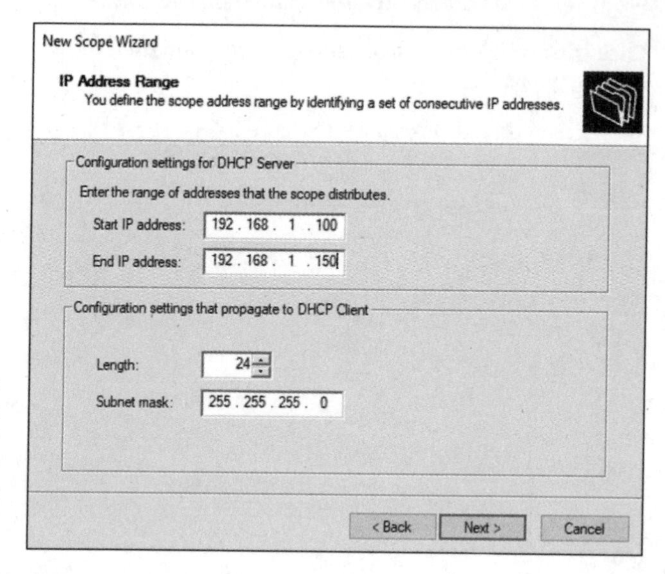

FIGURE 5-5: The IP Address Range screen allows you to specify the address range for the DHCP scope.

10. **Click Next.**
11. **On the Add Exclusions and Delay screen, enter addresses that you don't want the scope to assign.**

    You may enter a router address or other device addresses that exist in this scope but should not be assigned by this server.

12. **Click Next.**

    On the Lease Duration screen, you can see that the default is eight days. In most cases, this will be fine.

13. **Click Next.**
14. **On the Configure DHCP Options screen, leave it set to Yes, I Want to Configure These Options Now and click Next.**
15. **On the Router (Default Gateway) screen, add the IP address for your default gateway and click Next.**
16. **On the Domain Name and DNS Servers screen, enter the domain name you created earlier on your DNS server.**

    You can see my example in Figure 5-6.

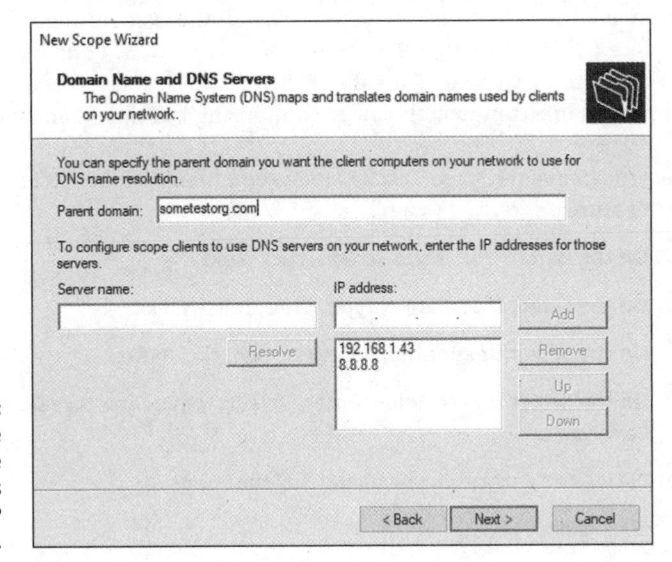

FIGURE 5-6: Setting the domain name and DNS servers in the DHCP Scope Options.

17. **Click Next.**
18. **On the WINS Servers screen, you can optionally enter the name or IP address of a WINS Server on your network, and then click Next.**

    I'm going to leave this blank.

    The last screen asks if you want to activate your scope. The default option is Yes.

19. **Leave it set to Yes, and click Next.**
20. **On the Completing the New Scope Wizard screen, click Finish.**

That's all there is to it.

Configuring the Server as a Domain Controller

To create a Windows domain controller, you need to install and configure Active Directory Domain Services. AD DS installs Active Directory and its database. It creates a central source of authentication and simplifies user management greatly.

Installing Active Directory Domain Services

With most things in Windows, you can install this role through the graphical user interface (GUI) or through PowerShell. In this section, I show you how to install this through the GUI, but I point out to you where you can get the PowerShell to rerun this same command if you're configuring multiple domain controllers.

1. **From Server Manager, click Manage and then click Add Roles and Features.**
2. **On the Before You Begin screen, click Next.**
3. **On the Select Installation Type screen, click Next.**
4. **On the Select Destination Server screen, click Next.**
5. **On the Select Server Roles screen, select Active Directory Domain Services.**
6. **Click Add Features in the dialog box that pops up.**
7. **Click Next.**
8. **On the Select Features screen, click Next.**
9. **On the Active Directory Domain Services screen, click Next.**
10. **Click Install.**

TECHNICAL STUFF

Do you have more domain controllers to build? After you've installed AD DS, the screen has an Export Configuration Settings link at the bottom. Click this link and save the XML file it generates. When you want to use it to create another domain controller, simply use PowerShell to point the installer to the XML file:

```
Install-WindowsFeature -ConfigurationFilePath <path and
  name of file>
```

11. **Click Close after the installation is complete.**

Configuring Active Directory Domain Services

After AD DS is installed, the next natural step is to configure it so that your server can begin its life as a domain controller. Follow these steps:

1. **From Server Manager, click the flag, and then click Promote This Server to a Domain Controller, as shown in Figure 5-7.**

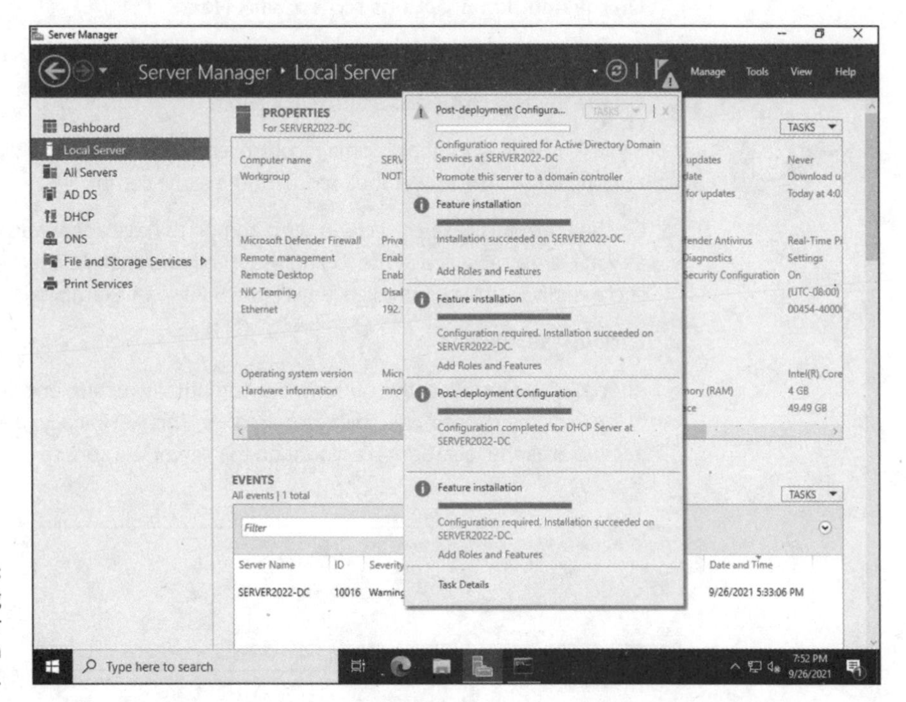

FIGURE 5-7: Promoting the server to a domain controller.

2. **On the Deployment Configuration screen, select Add a New Forest.**

REMEMBER

There are three options when promoting a server to domain controller:

- Add a New Forest is typically used when there is no existing AD infrastructure.
- Add a New Domain to an Existing Forest is used when creating domains other than the forest root domain. As you may imagine, this option is used if there is existing AD infrastructure in place.
- Add a Domain Controller to an Existing Domain does exactly what it says. You aren't creating a forest or a domain; you're just adding a new domain controller.

3. **Enter the root domain name, and click Next.**
4. **On the Domain Controller Options screen, select the functional levels you determined you needed earlier, and then click Next.**

   Leave Domain Name System (DNS) and Global Catalog selected. Enter a restore password and save it somewhere you can get to it; if AD ever has to be restored, you need this password to recover.
5. **On the next DNS Options screen, click Next.**
6. **On the Additional Options screen, click Next.**
7. **On the Paths screen, accept the default locations and click Next.**
8. **On the Review Options screen, click Next.**

TIP

   If you're installing multiple domain controllers, you can click the View script button and see the PowerShell script for the same setup.

   On the Prerequisite Check screen, it's common to have a few warnings. As long as you have a message at the top that says "All prerequisite checks passed successfully," you're good to go. See Figure 5-8 for an example.
9. **Click Install.**

   After AD DS is installed, the server will automatically restart, and you'll be able to log in as the domain administrator account. This will initially use the same password as the local admin account on the server. Be sure to change it.

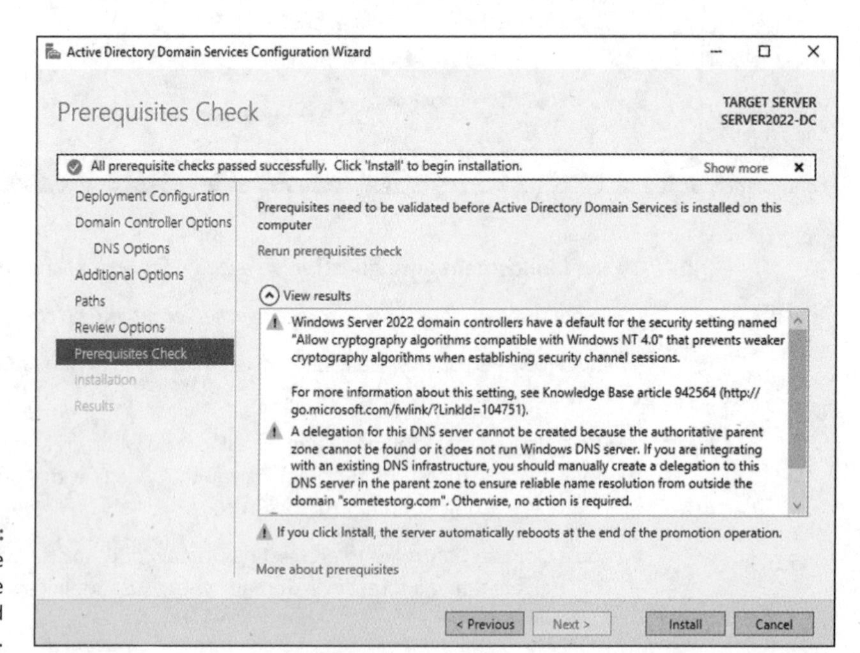

FIGURE 5-8: Checking the prerequisite checks passed before installing.

Converting your DNS Zone to an Active Directory Integrated Zone

When you set up DNS earlier in this chapter, you created the zones but chose manual updates rather than dynamic. Now that the DNS zone is associated with Active Directory, you can go back, integrate the DNS zones with Active Directory, and select secure dynamic updates.

Making your DNS zone and Active Directory integrated zone gives you a multimaster environment where any domain controller that is running DNS can update zones as long as they're an authoritative server for that zone. Zone data is replicated through Active Directory rather than through the traditional zone transfer method. After you've chosen to make your zone and Active Directory integrated zone, you can also enable secure dynamic updates. This allows you to control which systems can update record names, as well as keep unauthorized systems from overwriting names in DNS.

Follow these steps to convert your zone to an Active Directory integrated zone and how to enable secure dynamic updates:

1. **From Server Manager, click Tools and then click DNS.**
2. **Expand the server name, and then expand Forward Lookup Zones.**

   You should see the zone you created earlier.
3. **Right-click the zone and choose Properties.**
4. **On the General tab, under Type: Primary, click the Change button.**
5. **With Primary still selected, check the Store the Zone in Active Directory check box and click OK.**
6. **When you're asked if you want the zone to be Active Directory integrated, click Yes.**
7. **Under Dynamic updates, select Secure Only.**

   See Figure 5-9 for an example of how your screen should look at this point.
8. **Click Apply and then click OK.**
9. **Expand Reverse Lookup Zones.**

   You should see the zone you created earlier.
10. **Right-click the zone and choose Properties.**
11. **On the General tab, under Type: Primary, click the Change button.**

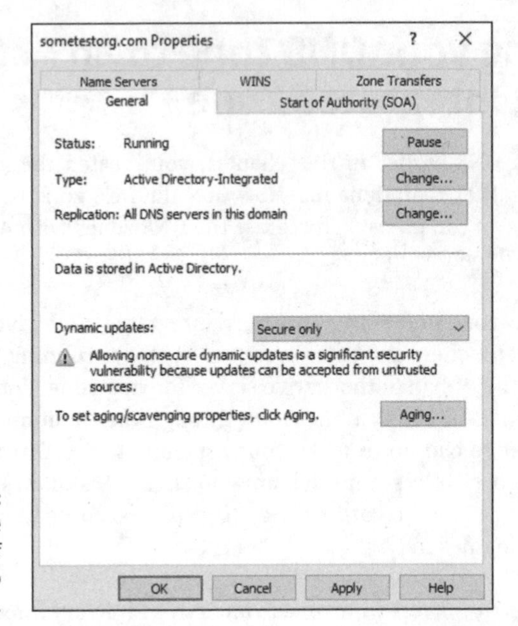

FIGURE 5-9: Setting the properties of your DNS zone to AD integrated.

12. With Primary still selected, check Store the Zone in Active Directory check box and click OK.

13. When you're asked if you want the zone to be Active Directory integrated, click Yes.

14. Under Dynamic updates, select Secure Only.

See Figure 5-9 for an example of how your screen should look at this point. Keep in mind that Figure 5-9 is from the Forward Lookup Zone, however the settings themselves are identical.

15. Click Apply, and then click OK.

16. Under Forward Lookup Zones, create the record for this server.

Future systems should pop up automatically now.

17. Right-click your domain name under Forward Lookup Zones and select New Host (A or AAAA).

Host records are used by the DNS server to map hostnames to IP addresses. A records match hostnames to IPv4 addresses, while AAAA records map hostnames to IPv6 addresses.

18. Under Name, enter the name of your system.

19. Under IP Address, enter the IP address of the system.

20. Check the Create Associated Pointer (PTR) Record check box.

This creates a matching entry in the Reverse Lookup Zone.

21. **Click Add Host.**
22. **Click OK on the confirmation box.**
23. **Click Done in the New Host box.**

Now you're all set. Active Directory is installed, and you're allowing dynamic updates in DNS in a secure fashion. You created a DNS record for your server. You can test at this point to make sure that name resolution is working properly. Here are the steps to test name resolution:

1. **Right-click the Start menu and choose Windows PowerShell.**
2. **Type** nslookup <*servername*> **as shown in Figure 5-10.**

   If DNS is working properly, it should return the IP address of the server name that you put in.

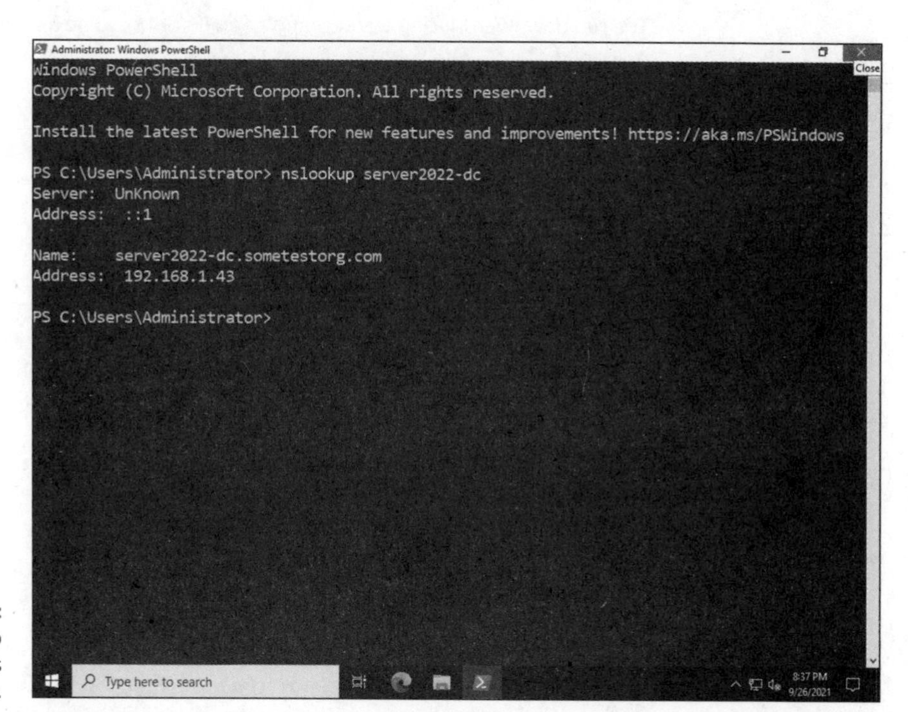

FIGURE 5-10: Using nslookup to verify that DNS is working properly.

As you can see, integrating DNS after the fact can be rather time consuming. You can install it at the same time as Active Directory and save yourself quite a bit of the configuration work. However, now you've seen it done the long way, so the shorter way will be much simpler.

Authorizing your DHCP Server for your Active Directory environment

There's one last thing you need to do before you can use DHCP properly. Earlier, you installed DHCP and then set up a scope. Now that this server is a domain controller, you need to authorize DHCP in Active Directory so it can start handing out IP addresses for domain joined systems. By authorizing your DHCP server, you prevent rogue DHCP servers from appearing on your network. Rogue DHCP servers can cause a denial of service attack. To authorize your DHCP server, follow these steps:

1. **Starting with Server Manager, choose Tools ⇨DHCP.**
2. **Right-click the server's name and select Authorize.**
3. **Click Action and then click Refresh.**

   The red downward-facing arrow that was on IPv4 should be a green check mark now.

Now your DHCP is configured to work with your Active Directory domain. Anytime you stand up new DHCP servers and you want them to work with your Active Directory environment, you need to authorize them.

Configuring the user accounts

Now that Active Directory is installed, you may want to start creating user accounts. The main purpose of Active Directory is for authentication after all! Here's how to create your first user account in Active Directory:

1. **From Server Manager, choose Tools ⇨Active Directory Users and Computers.**
2. **Expand your domain name by clicking the arrow next to it.**
3. **Click Users (this is the Users OU) to select it.**

   You see the built-in accounts and groups.

4. **Right-click Users, and choose New ⇨User, as shown in Figure 5-11.**
5. **Fill in the First Name and Last Name fields.**

   Notice that Full Name fills in automatically with what you typed.

6. **Enter what you want the user's login name to be.**

   In my example, for John Smith, I'm creating a username of jsmith (see Figure 5-12).

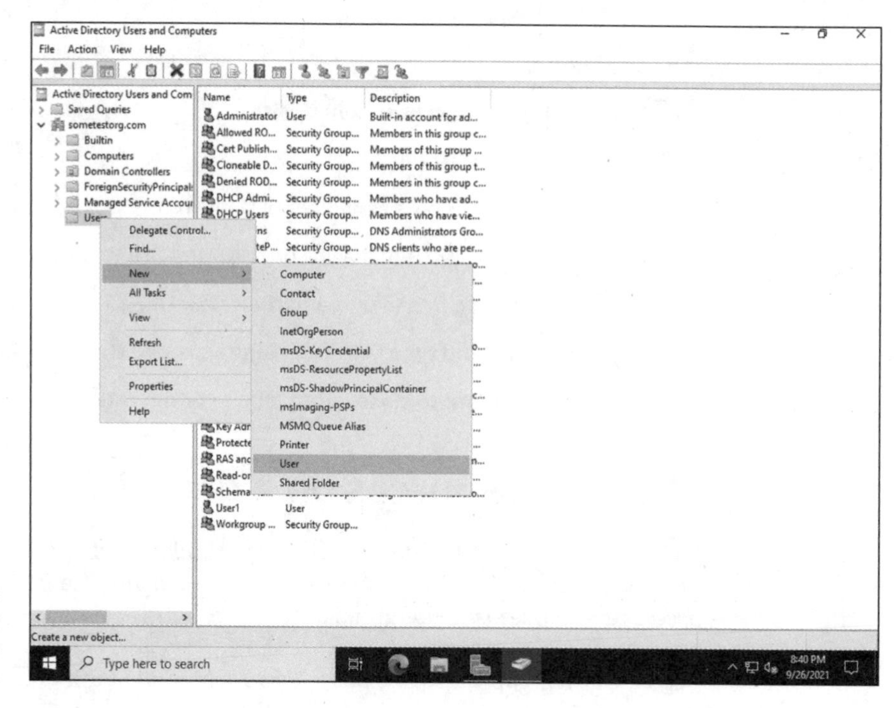

FIGURE 5-11: Creating a new user in Active Directory.

FIGURE 5-12: Creating my first user, John Smith.

7. **Click Next.**
8. **Set the temporary password for the user.**
9. **Type it in again in the Confirm Password field.**
10. **Leave the User Must Change Password at Next Logon box checked and click Next.**

11. **Click Finish.**
12. **Double-click the new user, John Smith.**
13. **Click the Member Of tab.**

    You see that John Smith is a member of Domain Users.

    Let's say that John will be managing our DNS and DHCP servers.

14. **Click Add and type** DnsAdmins **and then click OK.**
15. **Click Add again and type** DHCP Administrators **and then click OK.**
16. **Click OK once more to close out of John's configuration.**

Sharing resources on a domain

Sharing resources on the domain is pretty similar to sharing resources off the domain. When working in an Active Directory domain, the best way to manage access to resources is to use groups.

Let's create a group called File Share Users, add John Smith to it, and then assign it permissions on a folder share. Follow these steps:

1. **Right-click the Users OU, and choose New ⇨ Group.**
2. **Enter a Group Name, and then click OK.**
3. **Double-click the new group (in my case, that's File Share Users) and click the Members tab.**
4. **Click Add.**
5. **Type** John Smith **and then click OK.**
6. **Click OK once more to exit out of the group's configuration screen.**
7. **Click the File Explorer icon in the bottom of the screen (it looks like a folder).**
8. **Click This PC and determine which volume you want to create this folder share on.**
9. **Double-click the volume to open it.**

    In my example, I am just going to use the `C:` volume for the folder.

10. **Right-click the blank space and choose New ⇨ Folder.**
11. **Name the folder Files.**
12. **Right-click the Files folder and choose Give Access To ⇨ Specific People.**

13. **In the box, type the name of the group you created earlier (in my case, File Share Users), and then click Add.**

The security group is added with read permissions.

14. **Click the arrow next to Read, and choose Read/Write, as shown in Figure 5-13.**

15. **Click Share.**

16. **On the Your Folder Is Shared screen, click Done.**

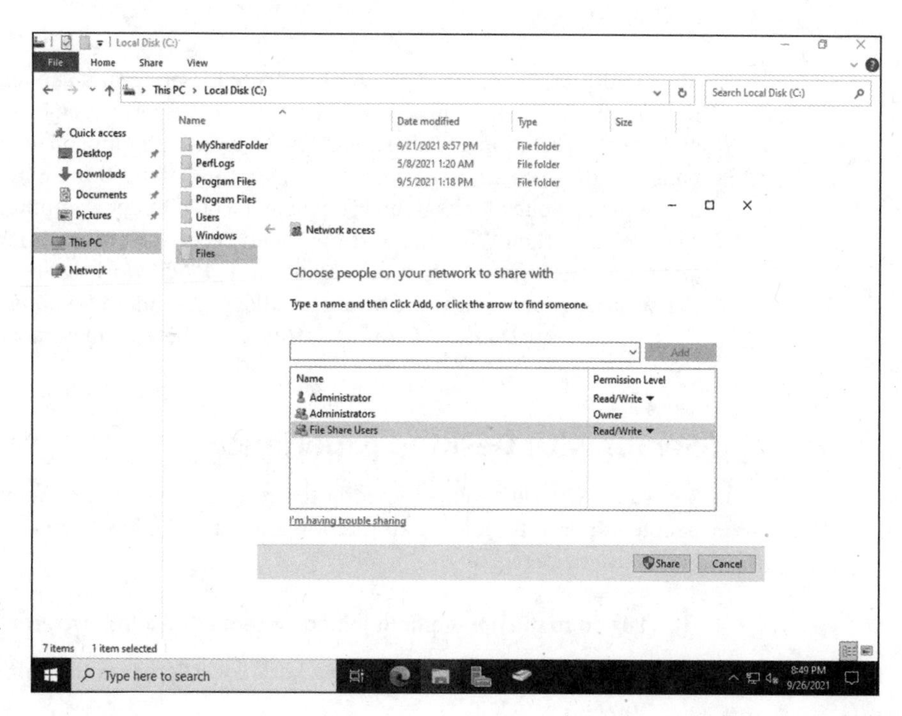

FIGURE 5-13: Sharing a folder to a domain security group.

When a user in the security group wants to access the share that you just created, all he has to do is the following:

1. **Click the Start menu.**

2. **Type** *\\servername\sharename*.

In my case, this is \\server2022-dc\Files.

3. **Press Enter.**

This takes the user to the shared folder.

The great thing about this approach is that new users only need to be added to the security group File Share Users, and they have read/write access to the folder share that you created.

Joining clients to the domain

The last step on this journey is, of course, joining other systems to the domain so that you can then authenticate with domain credentials rather than local credentials. In this section, I cover doing this in both Server with Desktop Experience and Server Core.

By default, any member of the Authenticated Users group can add a workstation to an Active Directory domain. Members in the Authenticated Users group can add up to ten workstations. (This should be changed depending on your organization's policies.) It can be adjusted domain wide in the Default Domain Controllers Policy in the Group Policy Management Console, located under Computer Configuration, then Policies, then Windows Settings, then Security Settings, then Local Policies, and then User Rights Assignment. Double-click Add Workstations to Domain. Add the security group you want to use to allow systems to be added to the domain (I've often seen Domain Administrators used here) and remove Authenticated Users.

Server with Desktop Experience

Server with Desktop Experience is the graphical version of Windows that most people are used to. Joining to the domain with Windows Server is pretty simple. Follow these steps:

1. **Log on to the non-domain-joined system as an administrator.**
2. **When Server Manager launches, click Local Server on the left side of the menu.**
3. **Underneath the server name, click WORKGROUP.**

   The System Properties box, shown in Figure 5-14, appears.
4. **Click the Change button.**
5. **Under Member Of, select the radio button next to Domain, and fill in the name of the domain you want to join.**

   This should look similar to Figure 5-15.
6. **Click OK.**

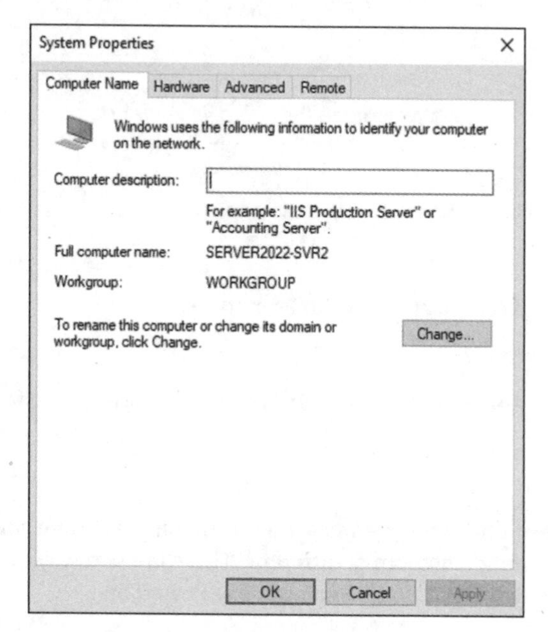

FIGURE 5-14: The System Properties box allows you to change the computer name and the workgroup/domain membership.

FIGURE 5-15: Changing the domain membership of the server.

7. **Enter the username and password of an account that has permissions to join to the domain.**

TECHNICAL STUFF

If you've been following along and you've built your domain from scratch, the domain administrator account will have the same password as the local administrator account did back on the domain controller. If you're joining an established domain, this password may be different.

8. **Click OK.**

   You get the message "Welcome to the <*domain name*> domain.

9. **Click OK.**

   You're told that you need to restart the system.

10. **Click OK.**
11. **Click Close to close out of System Properties.**
12. **Click Restart Now.**

    The system restarts, and when it comes back up it will be joined to the domain.

Server Core

Server Core doesn't have the graphical menus that most admins are used to, so the experience of joining a domain is different. There are two ways to join a domain: You can use the sconfig utility, or you can use PowerShell.

JOINING THE DOMAIN WITH SCONFIG

sconfig is the text-menu-driven utility present in Server Core, which launches by default after you've logged in. Alternatively, from the command line you can access it by typing **sconfig** and pressing Enter. When sconfig starts, you have a screen similar to Figure 5-16.

Let's join the domain using sconfig. For the purposes of instruction, I'll assume you've already typed sconfig and are at the menu. Follow these steps:

1. **Press 1 to select Domain/Workgroup.**
2. **Press D for domain and then press Enter.**
3. **Enter the name of the domain you want to join, and then press Enter.**
4. **Enter a domain admin account.**

   In my case, I type **sometestorg\administrator**.

5. **Enter the password for the account and press Enter.**

   You're asked if you want to change the name of the computer before restart.

6. **Type** no.

   If you need to restart, feel free to select Yes.

   You're prompted to restart the server.

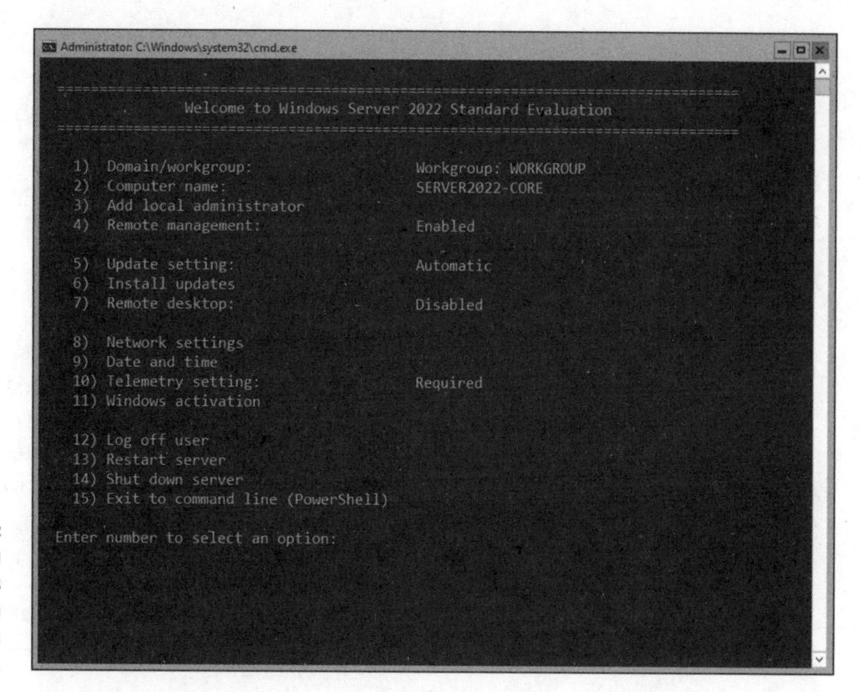

FIGURE 5-16: The sconfig menu in Server Core is the main system configuration area.

7. **Type** yes **to restart.**
8. **When the system comes back up, it will be joined to the domain.**

JOINING THE DOMAIN WITH POWERSHELL

Sconfig is a manual process and, as such, does not work well with server automation efforts. PowerShell can make the domain-joining process much faster than sconfig and can be used as part of a scripted task. Here's the command to join a server to Active Directory with PowerShell:

```
Add-Computer -domainname <domain name> -Credential domain\
   administrator -restart -force
```

See how nice and short and straightforward PowerShell can be? I recommend that you take the time to learn PowerShell. Book 6 is all about PowerShell.

Here are the steps to join Active Directory with the PowerShell command:

1. **Log in to the Server Core system.**
2. **From sconfig, type 15 to exit to** PowerShell.
3. **Using the example command earlier, type the domain name and credentials specific to your test domain, and then press Enter.**

4. **Enter the password, and then click OK.**

   The system will automatically restart. When it comes back up, it's joined to the domain.

Wrapping Things Up

You may be asking yourself at this point, how do all these things tie together? If you worked through this chapter from beginning to end, you installed DNS, DHCP, and Active Directory. You went back and converted the DNS zone to an Active Directory integrated zone and then set up secure dynamic updates. You also went back and authorized your DHCP Server for Active Directory. Then you joined some systems to Active Directory.

If you go back to your DNS, all the systems you joined to the domain are registered. This occurred when you joined them to the domain. You can see my example in Figure 5-17.

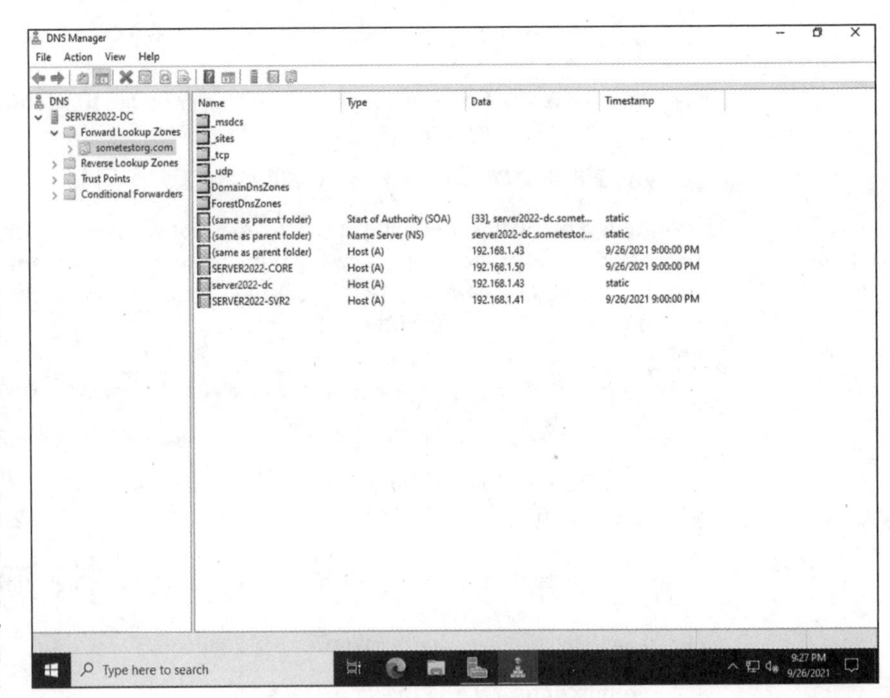

FIGURE 5-17: Active Directory–integrated DNS zones offer improved replication across your domain controllers, and secure dynamic updates ensure that records are added by authenticated users.

It's important that the systems are registered because this means that your users can remember system names instead of IP addresses. With DNS resolving names properly, you can use the name, and the system will know what the IP address needs to be. Pretty cool when you think about it!

Name resolution is great and all, but as you scale, you don't want to keep track of IP addresses manually. Let's check on the DHCP server. Because it was authorized for Active Directory, when you joined your systems to Active Directory, it issued them IP addresses, as shown in Figure 5-18.

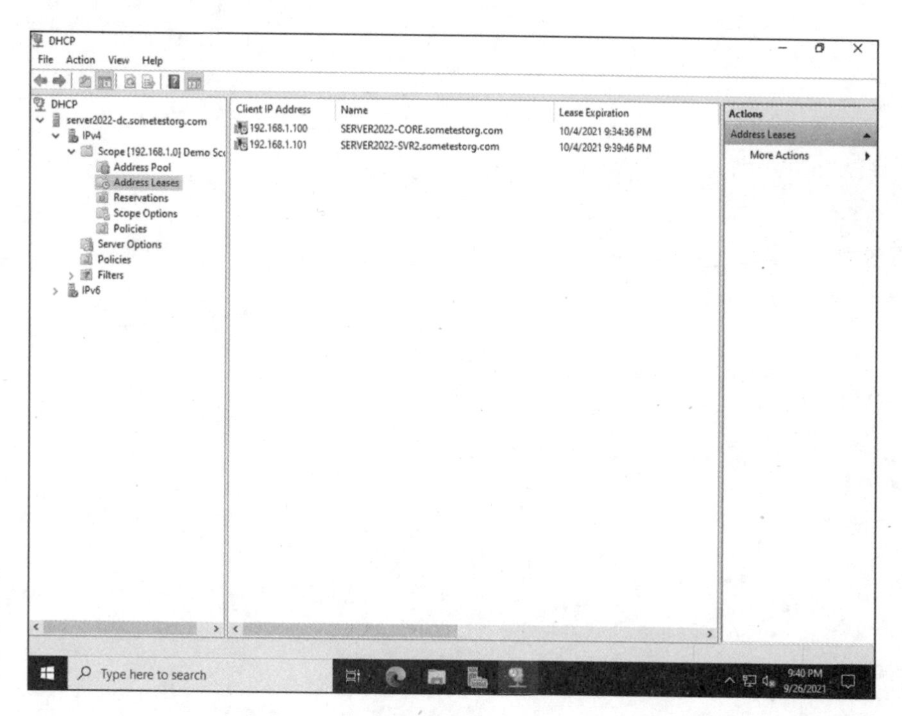

FIGURE 5-18: DHCP address leases for all your domain-joined systems.

You can set reservations to ensure that DHCP does not issue an IP address to another system. You can do this easily by right-clicking a system with a regular lease and choosing Add to Reservation.

Now let's look at Active Directory. Your new domain-joined systems will show up in Active Directory now. Follow these steps:

1. **Starting from Server Manager, choose Tools ⇨ Active Directory Users and Computers.**
2. **Click the Domain Controllers OU.**

You see the system that you set up as a domain controller.

3. **Click the Computers OU.**

   You see the other domain-joined servers.

Now that these systems are in Active Directory, you have a central location to manage these systems. You can view which operating system they're running, and you can delegate permission to them as well.

IN THIS CHAPTER

» Getting IP Address Management up and running

» Setting up IP Address Management the way you want

» Working with IP Address Management

Chapter 6

Managing DNS and DHCP with IP Address Management

When you work for a smaller organization, managing your Domain Name System (DNS) and Dynamic Host Configuration Protocol (DHCP) servers isn't all that bad. You may have one server or just a handful of servers. As your organization grows, however, you may start to feel the pain of managing multiple DNS and DHCP servers. It may get more difficult to keep track of all the zone and scopes.

Microsoft chose to solve for the inevitable sprawl of DNS and DHCP servers with a feature known as IP Address Management (IPAM). The name is super catchy, don't you think? IPAM combines the management of your network services like DNS and DHCP into one application so you can manage both your DNS infrastructure and your DHCP infrastructure all from a central management console.

One of the really great things about IPAM is that it can tell you when a subnet is being very heavily utilized. This can help you keep track of when you may need to add additional subnets so that your users or systems don't run out of usable IP addresses.

In this chapter, I walk you through how to install, configure, and use IPAM.

Installing IP Address Management

Before you begin trying to install IPAM, keep in mind the following requirements:

- IPAM can't be installed on domain controllers.
- IPAM shouldn't be installed on a DHCP or DNS server because it can cause issues with discovery.
- IPAM needs to be installed on a domain-joined system.
- IPAM is Microsoft-centric. You can't manage third-party products like BIND on Linux.

With those simple requirements addressed, you're ready to install IPAM. Follow these steps:

1. **In Server Manager, click Manage and then click Add Roles and Features.**
2. **On the Before you Begin screen, click Next.**
3. **On the Select Installation Type screen, click Next.**
4. **On the Select Destination Server screen, click Next.**
5. **On the Select Server Roles screen, click Next.**
6. **On the Select Features screen, select IP Address Management (IPAM) Server and click Add Features when it pops up.**
7. **Click Next.**
8. **On the Confirm installation selections screen, click Install.**
9. **When installation finishes, click Close.**

Configuring IP Address Management

Installing IPAM is pretty straightforward, you just have to make sure that you don't skip anything. All the tasks that you need to follow to configure IPAM show up as numbered tasks in the Quick Start tile after you've clicked IPAM. Do yourself a favor: For this installation, log in with an account that has domain administration privileges. Then follow these steps:

1. **In Server Manager, click IPAM on the left-hand menu (see Figure 6-1).**

   Notice the tasks in the Quick Start tile. Task 1: Connect to IPAM Server is already complete.

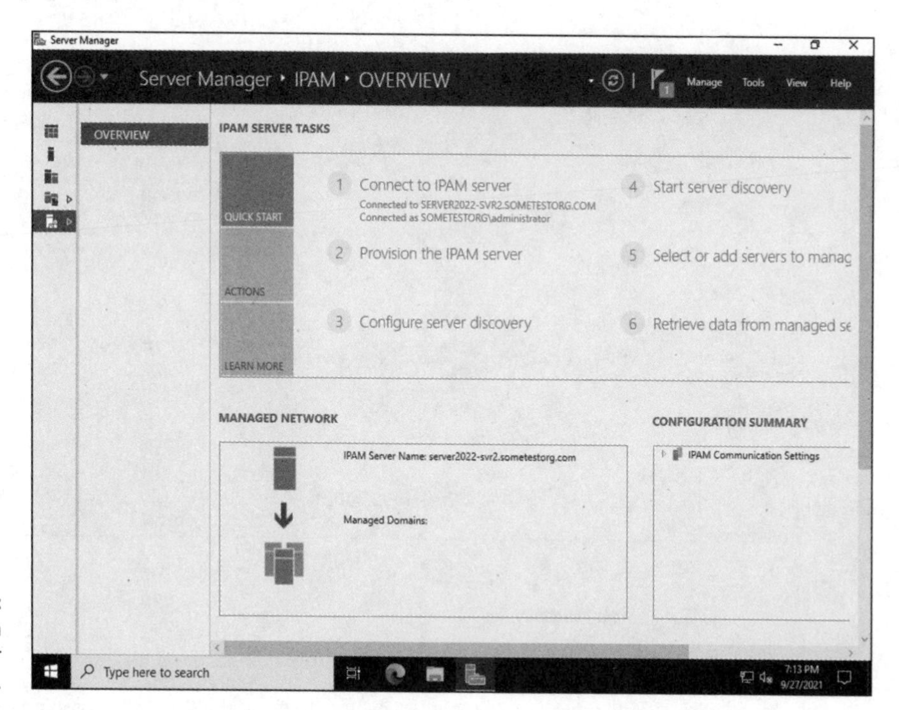

FIGURE 6-1: IPAM shows up in Server Manager after it's installed.

2. **Click Task 2: Provision the IPAM Server.**
3. **On the Before you Begin screen, click Next.**
4. **On the Configure Database screen, accept the default Windows Internal Database, and click Next.**

5. **On the Select Provisioning Method screen, select the Group Policy Based radio button, and enter a Group Policy Object (GPO) name prefix in the GPO Name Prefix field (see Figure 6-2).**

TECHNICAL STUFF

A GPO is a collection of settings that describe how a system will act or how it will look to users. It can be targeted at users and/or computers.

I like the obvious IPAM prefix. It tells me exactly what these GPOs are for.

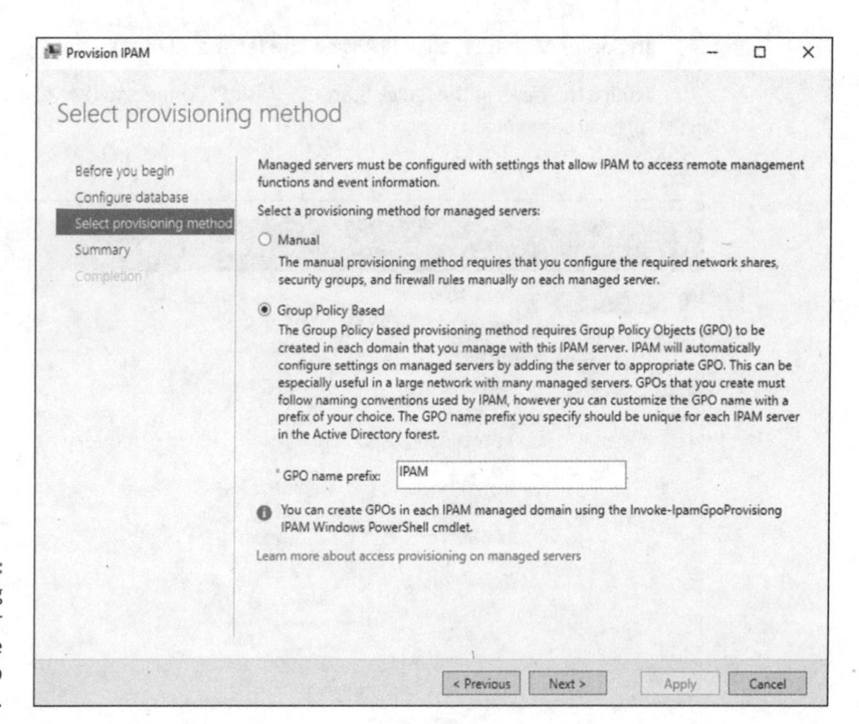

FIGURE 6-2: Provisioning methods for IPAM include manual or Group Policy based.

6. **Click Next.**

7. **On the Summary screen, click Apply.**

   If all goes well, you should be greeted with the IPAM Provisioning Completed Successfully screen.

8. **Click Close.**

   Now you need to push the new group policies out to the domain. You have to do this in every domain that you want IPAM to manage. In this case, I have only one domain.

9. **Right-click the Start menu on the IPAM server and choose Windows PowerShell.**

10. **Run the following command (you must be logged in as a domain admin for this step):**

    ```
    Invoke-IpamGpoProvisioning -Domain <domain name>
    -GpoPrefixName "IPAM" -force
    ```

11. **Confirm that you want to do this three times, once for each policy being created.**

    The policies are: IPAM_DC_NPS, IPAM_DHCP, and IPAM_DNS.

12. **Return to the IPAM Quick Start tile in Server Manager.**

    The next task is Task 3, where you configure server discovery.

13. **Click Configure Server Discovery.**

14. **Click the Get Forests button.**

    A query will be run.

15. **Click OK to close the Server Discovery screen.**

16. **After the job completes, click Configure Server Discovery again.**

    This time, the forest and domain will be filled in.

17. **Next to the domain, click Add.**

18. **Select the server roles you want IPAM to discover.**

    Your screen should look similar to Figure 6-3.

19. **Click OK.**

20. **Click Task 4: Start Server Discovery.**

    This schedules a discovery job. Wait for it to complete. It will let you know when the job finishes.

21. **After the job finishes, click Task 5: Select or Add Servers to Manage and Verify IPAM Access.**

22. **Right-click the server, and choose Edit Server.**

23. **Change the Manageability Status drop-down list from Unspecified to Managed, as shown in Figure 6-4.**

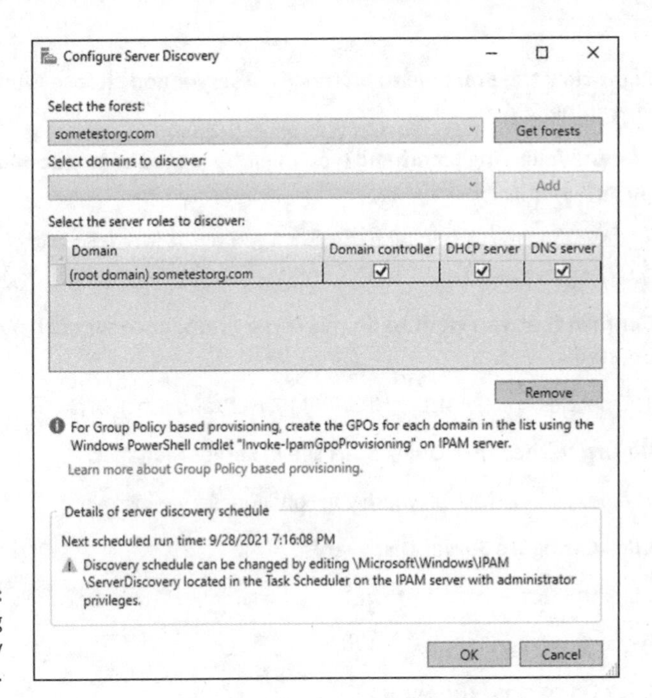

FIGURE 6-3: Configuring server discovery in IPAM.

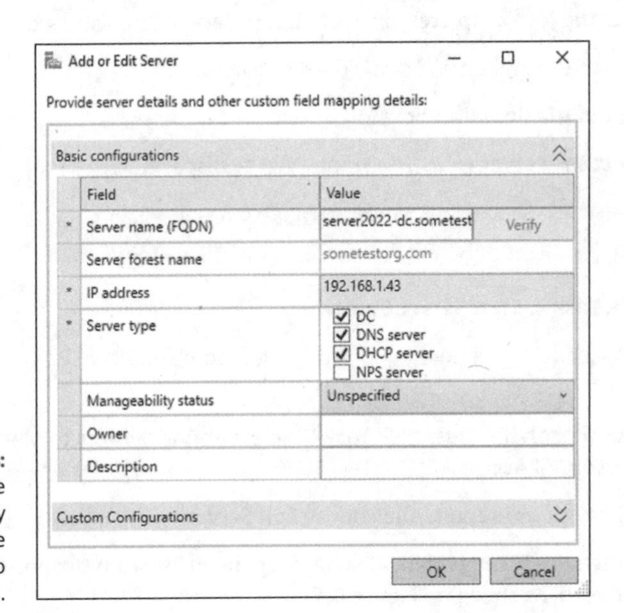

FIGURE 6-4: Setting the manageability status of the server to Managed.

24. **Click OK.**

This adds the managed server to those GPOs you created earlier.

You may see IPAM Access Status blocked at this stage. Typically, this means that the group policies haven't applied on the other system yet (assuming you used Group Policy for your provisioning method). To resolve this, log in to the other system, open a PowerShell window, and type **Invoke-GPUpdate -Force**. Then go back to your IPAM server, right-click the system in question and choose Refresh Server Access Status. If this still doesn't work, you may need to reboot the server that you're wanting to manage.

25. **Right-click the managed server and choose Retrieve All Server Data.**

After the retrieval job is complete, you can start managing the server you added through IPAM. Congratulations! That was the hard part!

Using IP Address Management

You may wonder what you can actually do in IPAM. In the following sections, I walk you through the different areas in IPAM that you can use to configure and manage your DNS and DHCP infrastructure.

Overview

Now that IPAM is configured, go to the Overview section. Remember that Quick Start tile you were using? Instead, click the second orange tile, the one that says Actions. You should see a list of the things you can do similar to Figure 6-5.

Server Inventory

The Server Inventory section should look familiar to you — it may have been the source of much frustration when you were configuring IPAM. The main things that this screen gives you are an overview of the servers you're managing, their IPAM Access Status, and the ability to pull fresh data from the servers outside of the scheduled retrieval task. You can filter by IPv4 and IPv6, and you can also filter by Managed (GPO Provisioned) servers and Unmanaged (Manual) servers. Check out Figure 6-6 for a view of the Server Inventory screen.

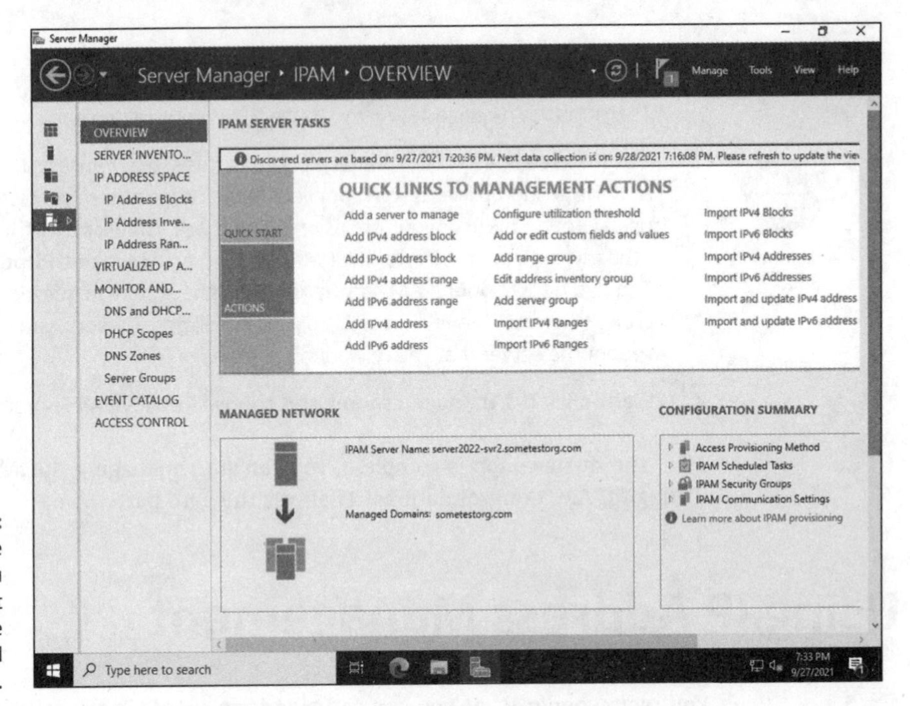

FIGURE 6-5: Some of the actions you can take against servers that are being managed in IPAM.

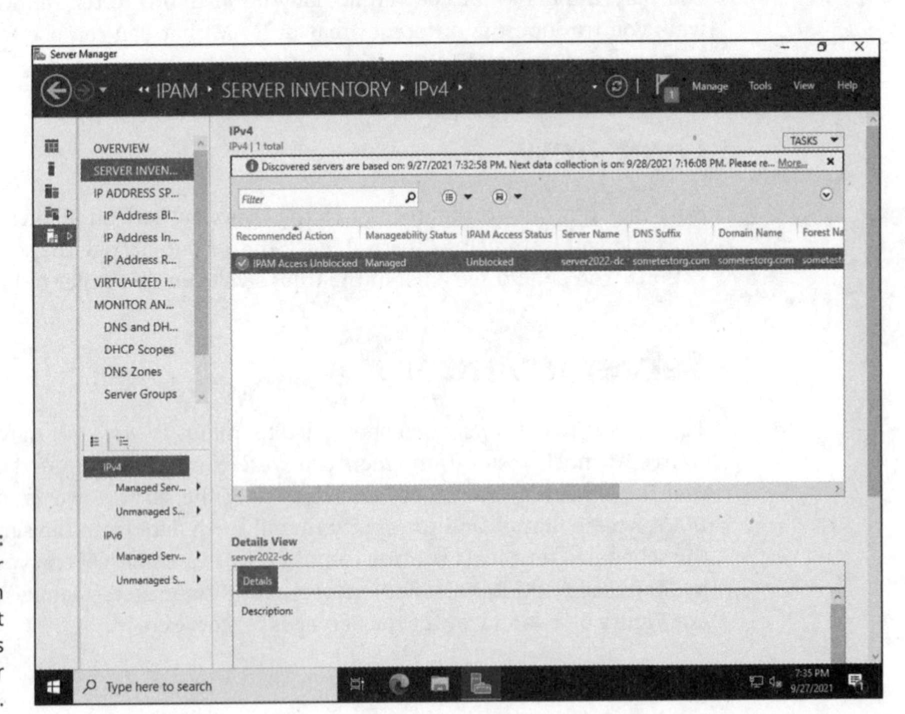

FIGURE 6-6: The Server Inventory screen gives you a list of your servers and what their status is.

IP Address Space

If you click the first selection in IP Address Space, called IP Address Blocks, you see the DHCP scope that was set up earlier. Right now, for Utilization, it says Under, as shown in Figure 6-7. This means that it has plenty of IP addresses available. Before IPAM, you had to track this utilization on a spreadsheet, which could be difficult because the spreadsheet might not be up to date, and the spreadsheet had no way to give you a friendly warning that you were overutilizing your space.

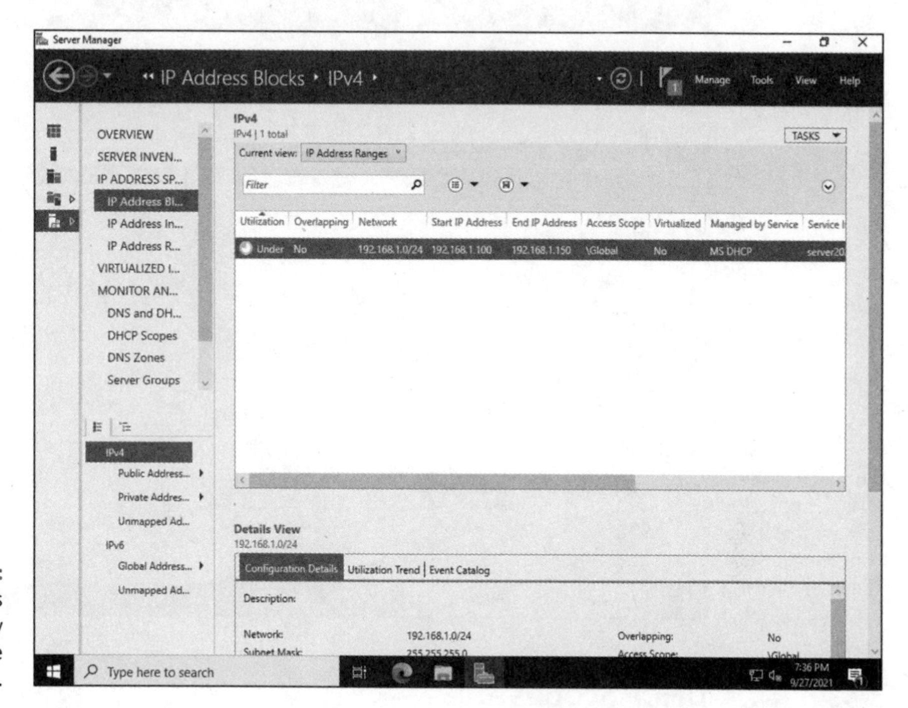

FIGURE 6-7: IP address utilization is easy to see at a glance in IPAM.

The other screens under the IP Address Space category — IP Address Inventory and IP Address Range Groups — give you similar data just with different views. Take the time to click through these and get familiar with what's in each.

Monitor and Manage

Monitor and Manage is where you'll spend a significant amount of your time in IPAM.

DNS and DHCP Servers

When you click DNS and DHCP Servers, you can see the status of the services on every server you're managing through IPAM. If all is well, Server Availability should say Running for both DNS and DHCP. See Figure 6-8 for an example of the status page.

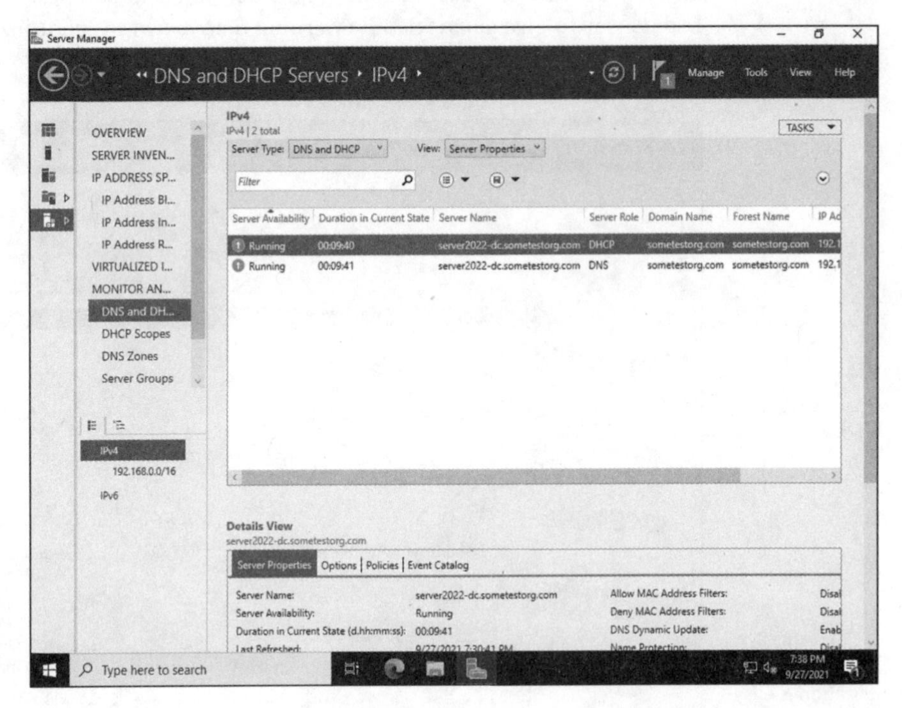

FIGURE 6-8: The status of the DNS and DHCP servers that are being monitored and managed through IPAM.

DHCP Scopes

DHCP Scopes contains all the DHCP scopes that are configured on all the DHCP servers IPAM is aware of. This interface gives you the same utilization metric that you had before in the IP Address Blocks section. It tells you what some of the basic subnet settings are, like the subnet mask and the lease duration. If you right-click over on the existing scope, you see that you're presented with a ton of options to manage the DHCP Scope. See Figure 6-9 for the configuration options available to you.

DNS Zones

In the DNS Zones section, you can see with a quick glance whether the Zone status is good or bad. You can also get some really helpful information at a glance regarding the DNS server that the zone is hosted on. You can select whether you want to look at the forward lookup zones (names to IPs) or if you want to look at the reverse lookup zones (IPs to names). And of course, just like the DHCP Scopes screen, you can right-click the zone and configure it right from IPAM. No need to go to multiple DNS servers anymore! See Figure 6-10 for an example of what the DNS Zones screen looks like.

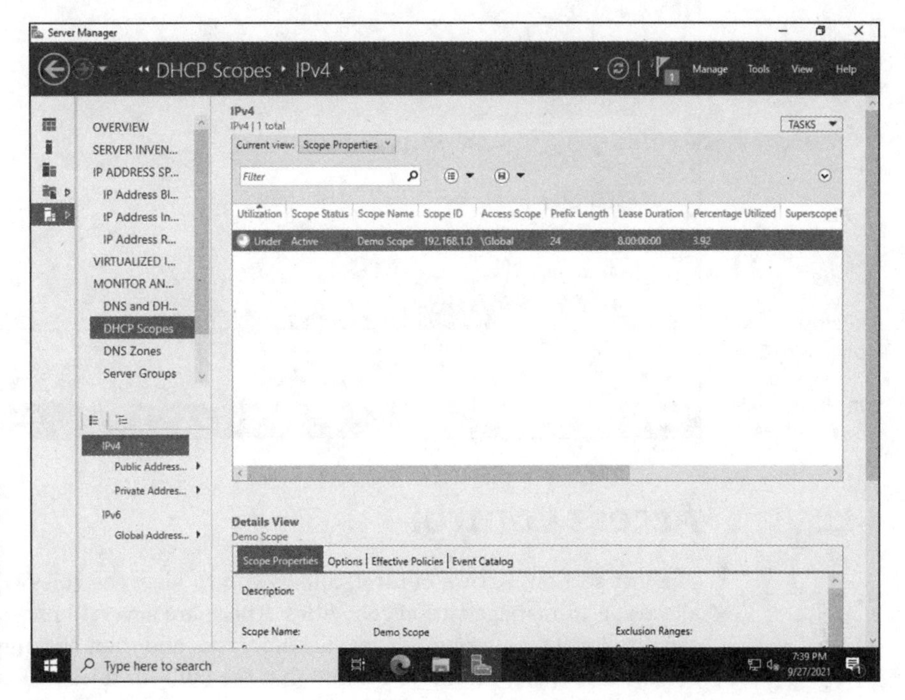

FIGURE 6-9: DHCP Scope configuration can be done across multiple DHCP servers all from IPAM's DHCP Scopes screen.

Server Groups

Server Groups lets you separate out the systems by the type of service running on them. You simply click the Server Type drop-down list and select whichever service you're interested in.

Event Catalog

Event Catalog gathers all the events in Event Viewer that are directly related to IPAM. This can be very helpful if you're troubleshooting why something isn't working properly.

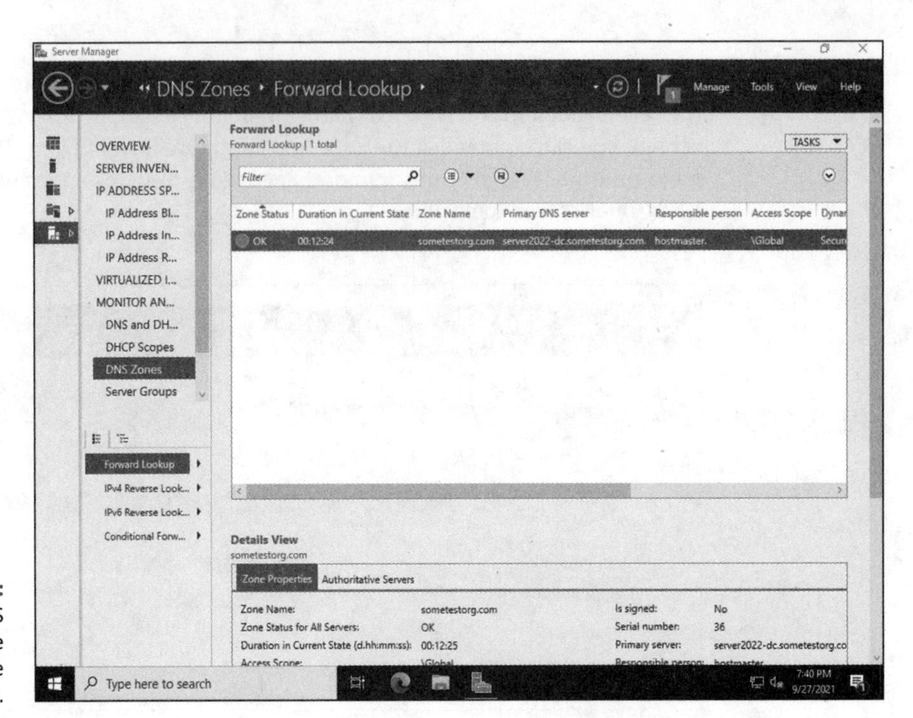

FIGURE 6-10: Configuring DNS zones can be done from inside of IPAM.

Access Control

The last section, Access Control, allows you to view the roles within IPAM that allow you to manage various activities. There are several built-in roles and you can also create your own role by clicking Tasks and then clicking Add User Role. For the most part, I think you'll find that the built-in roles will meet your needs. If you need to create one though, follow these steps:

1. **In the Access Control section, click Tasks and then click Add User Role.**
2. **Name the role and then select the desired permission.**

   In my case, I'm creating a role for a DNS Zone Administrator. See Figure 6-11 for an example of what that looks like.

3. **Click OK.**

   After the role is created, it shows up in the list, and you can tell it isn't a built-in role because it says No under the Built-in Role column. See Figure 6-12 for the final view.

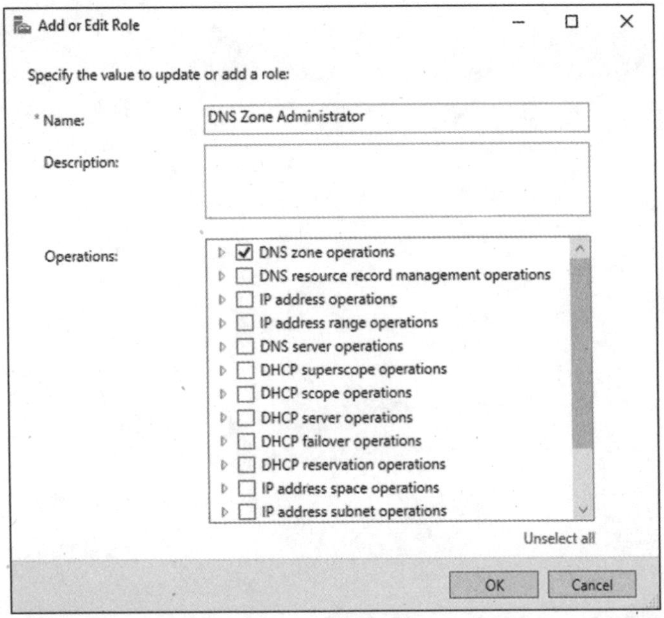

Add or Edit Role

Specify the value to update or add a role:

* Name: DNS Zone Administrator

Description:

Operations:

- [x] DNS zone operations
- [ ] DNS resource record management operations
- [ ] IP address operations
- [ ] IP address range operations
- [ ] DNS server operations
- [ ] DHCP superscope operations
- [ ] DHCP scope operations
- [ ] DHCP server operations
- [ ] DHCP failover operations
- [ ] DHCP reservation operations
- [ ] IP address space operations
- [ ] IP address subnet operations

Unselect all

OK Cancel

FIGURE 6-11: Creating a DNS Zone Administrator is easy given the granular permissions available.

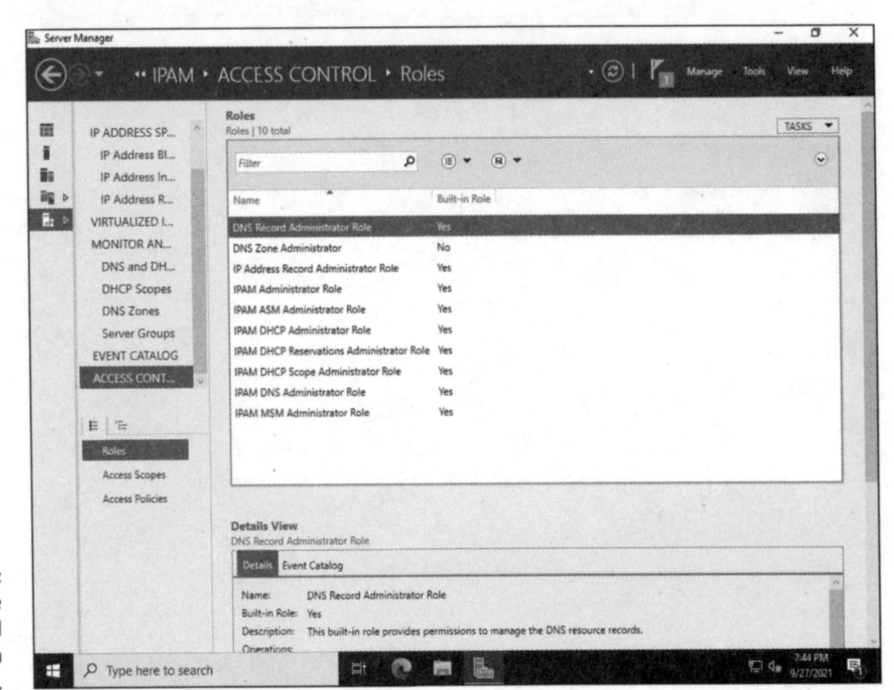

FIGURE 6-12: The new role is created and shows up with the built-in roles.

3 Administering Windows Server 2022

Contents at a Glance

IN THIS CHAPTER

» Accessing the Tools menu in Server Manager

» Working with common administrative tools in Windows Server 2022

» Using Remote Server Administration Tools to manage your servers

Chapter 1
An Overview of the Tools Menu in Server Manager

The majority of the time, when you log in to a server, it's to manage a role, manage a feature, or troubleshoot issues the system may be having. The great thing about the modern server operating systems is that Server Manager launches upon login and is your jumping-off point to all the tools that you'll ever really need to use.

There are some things in the Tools menu that will be there on every server. Disk management utilities, troubleshooting utilities, and other useful system utilities are right at your fingertips. As long as the tool was written by Microsoft, and it has to do with administering the server, including the roles and features you've installed on the server, it will be in the Tools menu.

Accessing the Server Manager Tools Menu

There are a few ways to access the Tools menu. The simplest method is to click Tools from inside Server Manager. You can find this along the top bar of Server Manager. It gives you one-click access to the majority of your administrative tools. See Figure 1-1 for an example of the Tools menu.

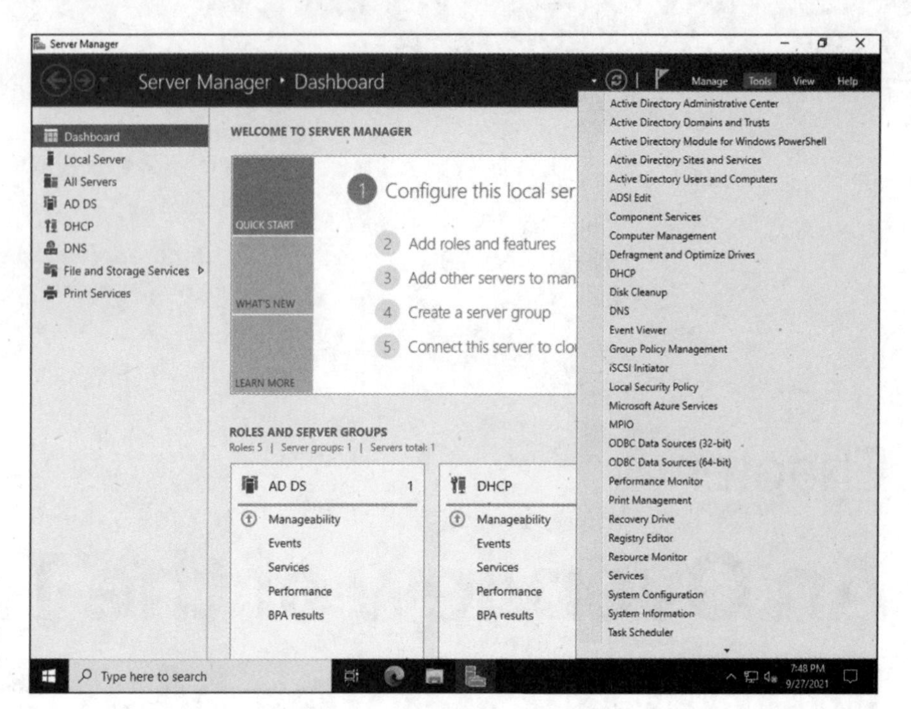

FIGURE 1-1: The Tools menu from within Server Manager.

Notice that there are Active Directory tools in the menu as well. This is due to the fact that we're on a domain controller, which has the administration tools loaded. You can install these tools on your local PC with Remote Server Administration Tools (RSAT), which I cover later in this chapter.

The second method to access the Tools menu is to click the Start menu and scroll down until you reach Windows Administrative Tools. This is the same menu that shows up under Tools in Server Manager. It may be your go-to method of access if you've disabled Server Manager on startup. See Figure 1-2 for the Start menu location of the Tools menu.

TIP

If you've disabled Server Manager on startup, you can click the Start menu and launch it from the Server Manager tile, or simply type **servermanager** into PowerShell, at the Command Prompt, or into the Run box.

The third way to access the Tools menu is the more classic method: Click the Start menu and choose the Control Panel tile. In Control Panel, choose System and Security, and you'll have an option for Administrative Tools. This is the same menu you get with the other two methods. See Figure 1-3 for its location.

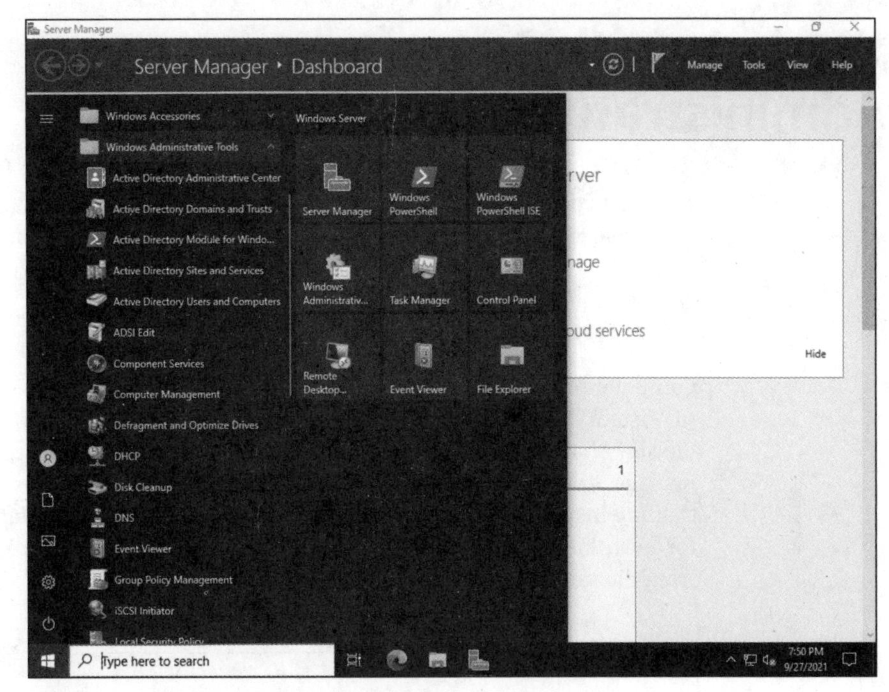

FIGURE 1-2: Windows Administrative Tools in the Start Menu is the same as the Tools menu in Server Manager.

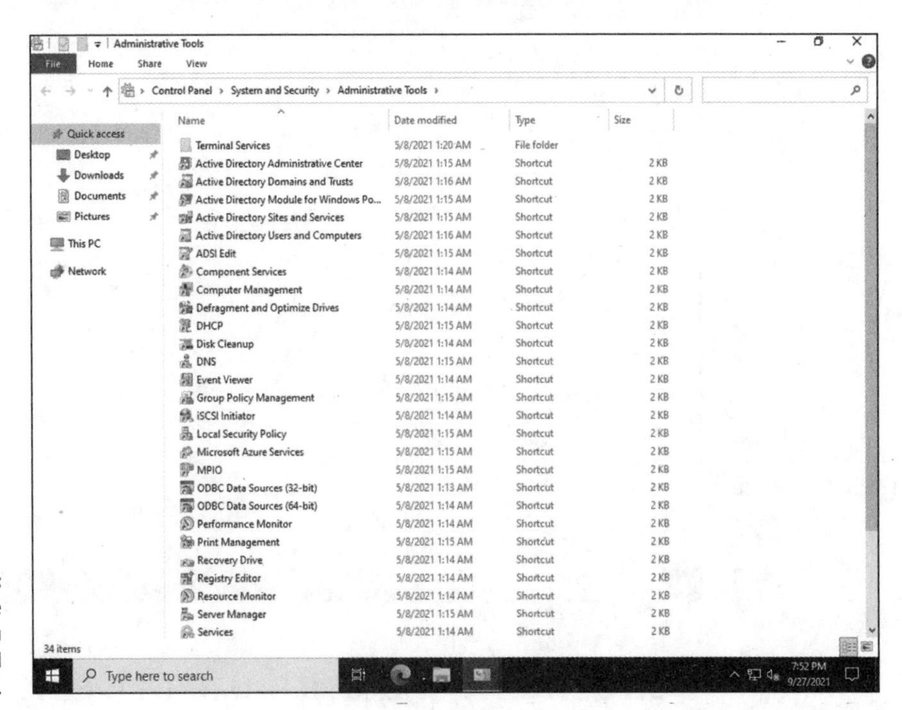

FIGURE 1-3: Accessing the Tools menu through Control Panel.

Working with Common Administrative Tools

Now that you know how to find the Tools menu, I need to fill you in on some of the more common administrative tools that you'll find yourself using.

Computer Management

Computer Management is useful for a number of reasons. It gives you quick access to System Tools, which is a collection of tools such as Task Scheduler, Event Viewer, Shared Folders, Performance, and Device Management. It allows you to manage Storage, and it gives you access to manage the services and applications that are installed on the system. I cover each of these tools in later chapters, but for now, know that Computer Management gives you a way to access them all from the same console, as shown in Figure 1-4.

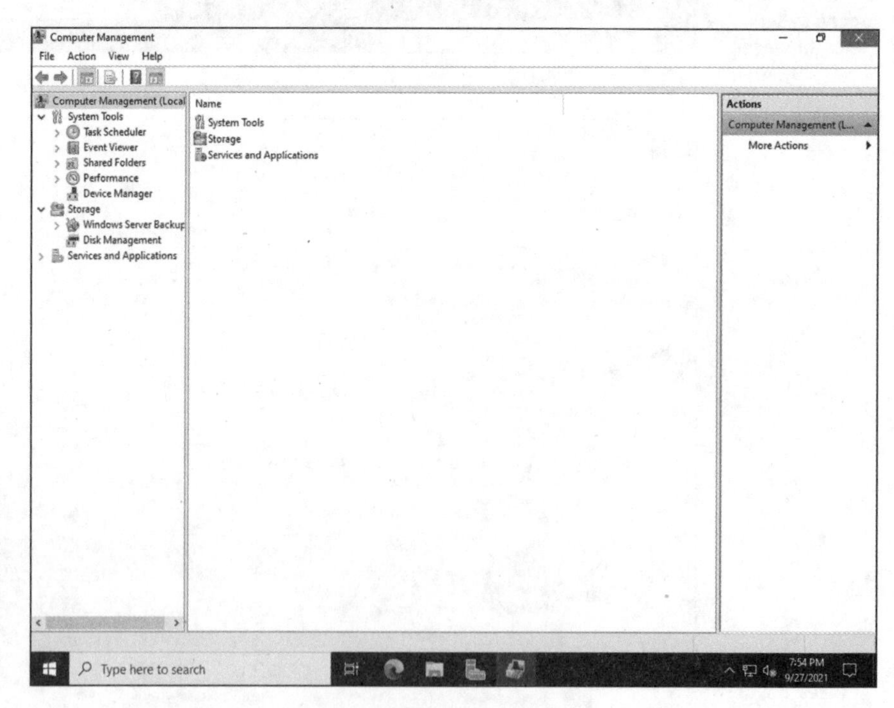

FIGURE 1-4: The Computer Management console gives you a centralized area to use administrative tools.

Defragment and Optimize Drives

When you choose Defragment and Optimize Drives, you're given a menu that will allow you to manually analyze and optimize your drives. Optimization depends on what type of drives you have. A traditional hard disk drive (HDD), for instance, will get defragmented by the optimization process. A solid-state drive (SSD) with TRIM support (which allows the operating system to tell the SSD which blocks are not in use and can be wiped) will be retrimmed. You'll notice that this is scheduled to run automatically once a week by default, as shown in Figure 1-5.

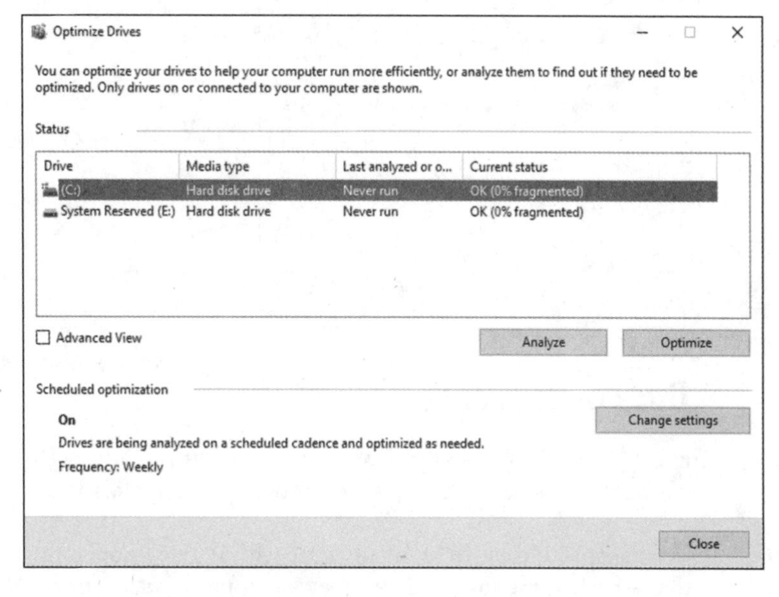

FIGURE 1-5: Defragmenting you hard drive is a scheduled task, but you can choose to run it manually.

This was a pretty good win for administrators, because the process of defragmenting drives used to be a manual one that you had to remember to do. It was typically only done when a system began having performance issues. With defragmentation being done automatically, you don't have to worry about doing it anymore.

Disk Cleanup

The Disk Cleanup tool has been in Windows operating systems for a very long time. When you run it, it asks you which drive you want to run it against, and then it scans for things that you can get rid of safely. This will typically be temporary Internet files, content in the Recycle Bin, and so on. The tool gives you an estimate of how much space it will be able to free up. All you need to do is click OK, and it will remove whatever you have told it to remove. Note the check boxes that are checked in Figure 1-6. These are the items it will remove.

There is a More Options tab you can select that gives you the ability to remove programs you don't use, as well as old copies of System Restore and Shadow Copies if you have them enabled.

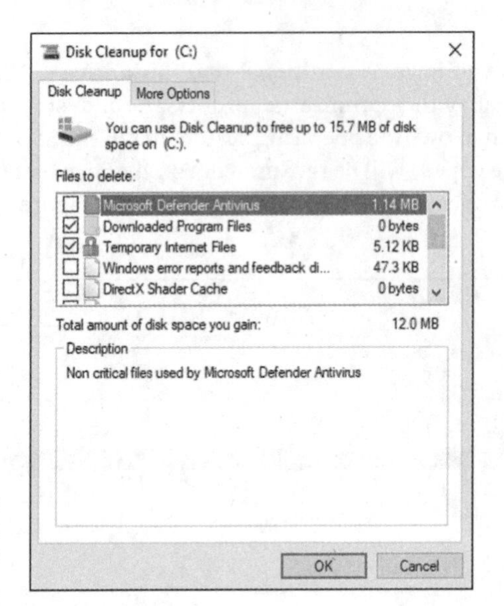

FIGURE 1-6: The Disk Cleanup utility in all its glory.

Event Viewer

Event Viewer is probably one of the most helpful utilities when it comes to troubleshooting issues with hardware, software, and applications.

Most people, if they've spent any time with Windows operating systems, are used to the basic three logs in Event Viewer: Application, Security, and System. There are so many other logs you can leverage, though. Under Custom Views, you can look at logs specific to the roles that you have installed. If you click Applications and Services, then click Microsoft, and then click Windows, you can view logs specific to Windows components. A common scenario, for instance, is when you think the Windows Firewall may be blocking you. The logs for Windows Firewall show up under Applications and Services, then Microsoft, then Windows, and then Windows Firewall with Advanced Security.

If you're looking for something specific, you can choose the log you want to search and then create a filter like the one in Figure 1-7.

I highly recommend digging into Event Viewer and experimenting with filters. The filters will really save time, especially when you're troubleshooting an issue and your boss wants information ASAP!

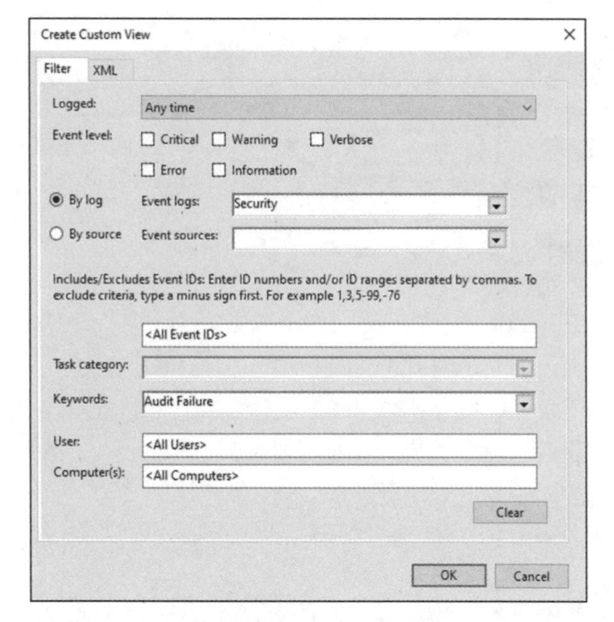

FIGURE 1-7: Using a filter in Event Viewer to find failed logins.

Local Security Policy

Local Security Policy allows you to change many of the settings on your system. If you're in an Enterprise environment, chances are, you won't access this because everything will be set through Group Policy. If you're on a standalone system, or on a computer that is in a workgroup, then Local Security Policy is where you set requirements for security on the system. This may include settings for accounts, auditing requirements, and other security-related settings. You can see the local password policy in Figure 1-8.

Registry Editor

WARNING

The Registry Editor will allow you to make changes to your system's Registry. Use this tool with extreme caution! If you change something that you shouldn't have, you could cause damage to your system.

I cover the Registry in more depth in Book 3, Chapter 3. Be sure to check it out! There, I discuss what each of the hives is for, how to interact with them, and even more important, how to back up the Registry before making changes.

Just about every setting on your system is stored in the Registry somewhere. Figure 1-9 shows you the area of the Registry where system colors are set. These are stored as RGB values. For instance, 255 255 255 is white and 0 0 0 is black.

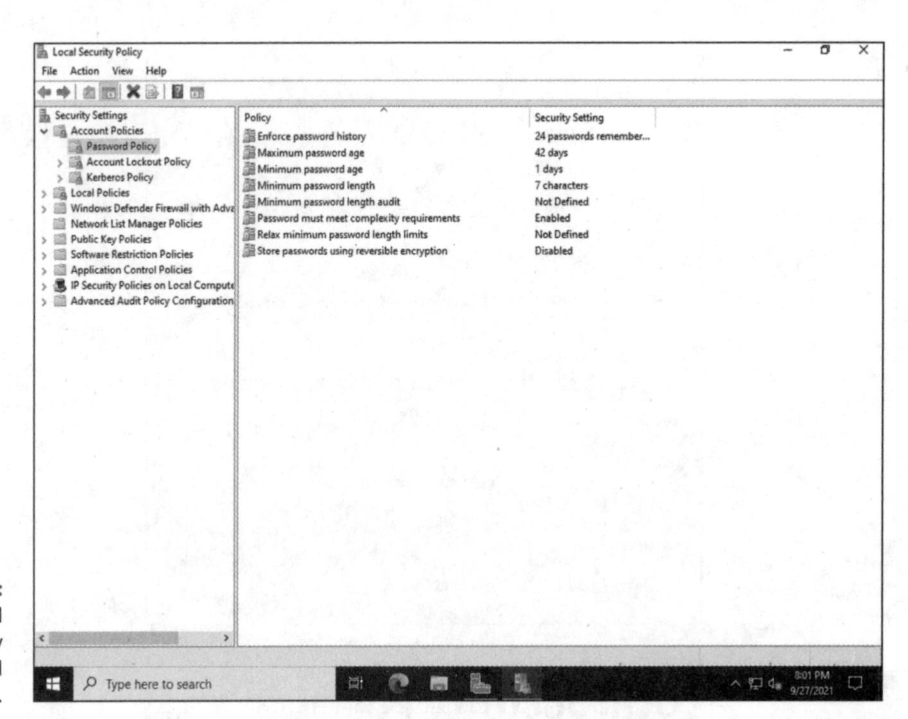

FIGURE 1-8: The local password policy set in Local Security Policy.

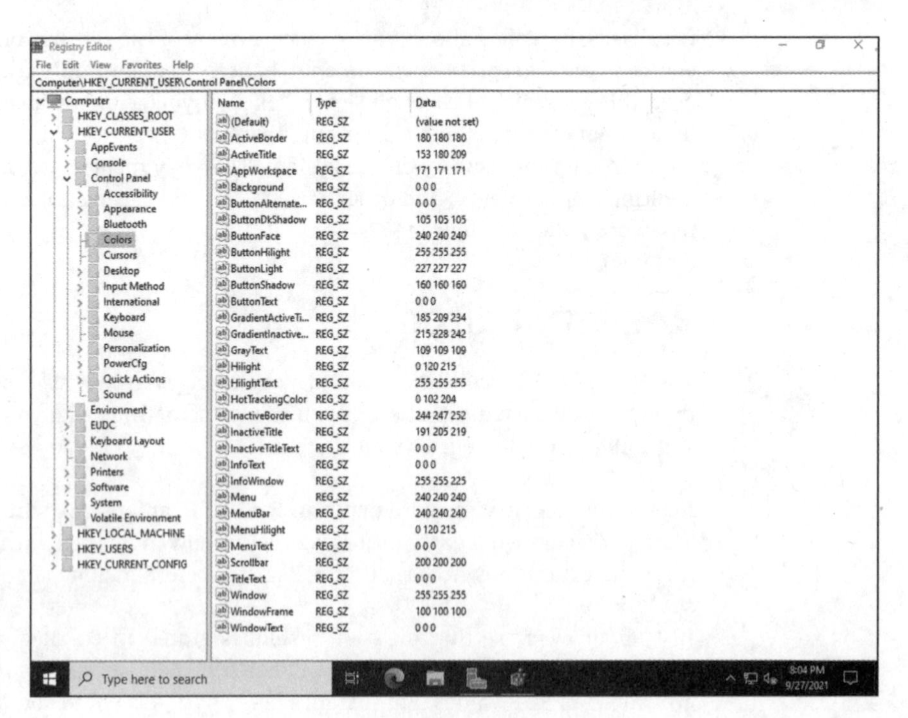

FIGURE 1-9: System colors are stored in the Registry.

Services

Clicking Services brings up the Services management console. Every service on the system shows up in the console. You can see what the status of the service is, what the startup type is set to, and which account each of the services is using to run.

TECHNICAL STUFF

A *service* is an application that runs in the background. You don't normally interact with it directly, although you can stop and start a service and define startup options for it.

If you double-click one of the services, you're presented with several tabs. The General tab allows you to change the startup type. You have your choice of the following:

- **Automatic:** Services whose startup type are automatic will start when the server starts.
- **Automatic (Delayed Start):** If a service has Automatic (Delayed Start) it means that it will start, just not right away. This is typically done for services that have a dependency on another service. It ensures that the dependency is satisfied before the service comes up.
- **Manual:** If a service is set to manual, it means that something has to kick it off. This may be a user action or a call from an application.
- **Disabled:** If a service is disabled, it can't be started.

You can also stop and start the service. On the Log On tab, you can set the account the service will run with. The Recovery tab allows you to specify what action to take if the service runs into an issue. You can set it to restart the service, run a program, or restart the system. The Dependencies tab will tell you if the service is dependent on any other services, and whether any services are dependent on it. You can see the Properties of the Server service in Figure 1-10.

System Configuration

System Configuration allows you to work with boot time services. On the General tab, for instance, you can choose Normal Startup (default), Diagnostic Startup, or Selective Startup. These are helpful for diagnosing issues that happen when the full OS boots. See Figure 1-11 for an example of the General tab in System Configuration.

The Boot tab allows you to specify which drive to boot to, and some boot options like safe boot, No GUI boot, and so on.

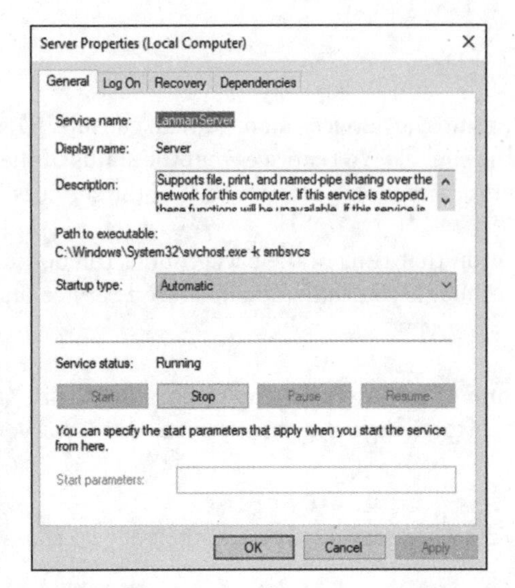

FIGURE 1-10: The Server service Properties page allows you to set startup type and manage the service.

FIGURE 1-11: The General tab in System Configuration lets you set the type of startup you want.

The Services tab allows you to specify which services should be enabled or disabled. You can choose to hide all the Microsoft services so that you can see which third-party services are installed.

The Startup tab still exists, but Startup items are not enabled by default, so the tab is blank unless you enable them.

Last but not least, the Tools tab has many different OS tools. You can select the tool, and it will display the location of the tool. In addition, you can click the Launch button to actually launch the tool.

Task Scheduler

The Task Scheduler allows you to schedule various administrative tasks so that they execute automatically. Many of the Windows components have their own tasks out of the box.

You can create your own tasks to run custom scripts or perform other actions through Task Scheduler as well. You choose what you want the trigger to be. Triggers can be anything from a scheduled time to a logon, an event, and so on. See Figure 1-12 for a view of a custom task in Task Scheduler.

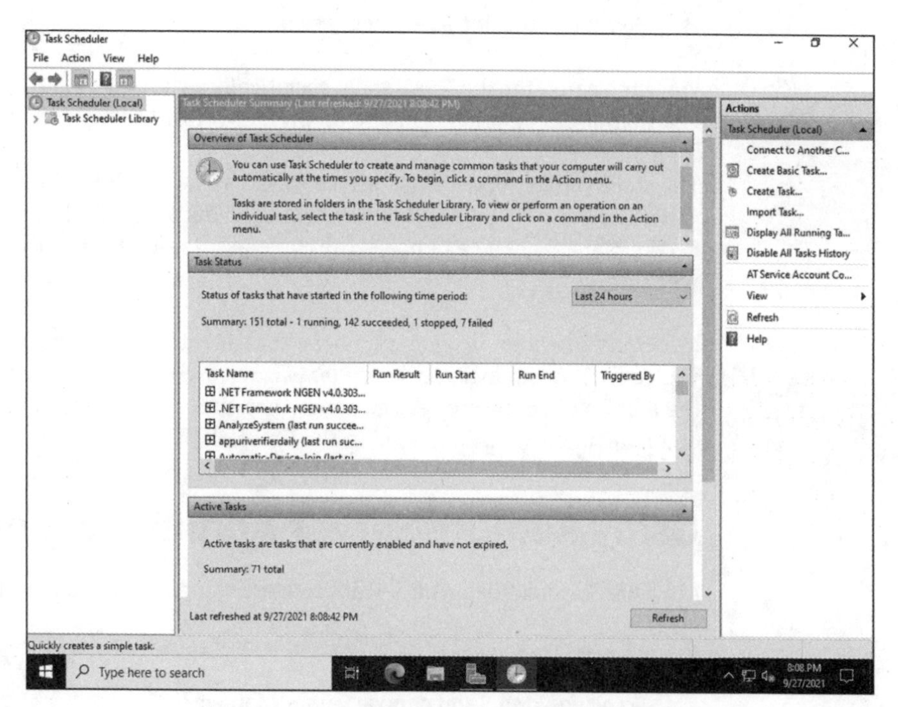

FIGURE 1-12: Creating a custom task in Task Scheduler.

Installing and Using Remote Server Administration Tools

When you install the management features along with roles and features, you are installing components of RSAT. RSAT is often installed on a desktop so that you can manage your servers from your workstation instead of logging on to each individual server. In this section, I walk you through installing RSAT on a Windows 10 system, and show you how to add servers to manage.

Installing Remote Server Administration Tools

As of the 1809 build of Windows 10, RSAT is included as part of the operating system and you only need to enable it. Follow these instructions to install RSAT:

1. **Click Start and then click Settings.**
2. **Select Apps and then select Optional Features.**
3. **Click Add a Feature.**
4. **Select the specific RSAT tools you need.**

You can search for the RSAT tools specifically by typing **RSAT** in the search field and then pressing Enter. They're all prefaced with *RSAT:*.

The 1809 version update was a nice touch. It saves administrators time because you no longer have to go out to download and install the package. Be sure to install Server Manager. The examples I provide assume that you installed this part of RSAT.

Microsoft provides version numbers that identify which build of Windows 10 you're on. For instance, version 1809 was released on November 13, 2018. In 2020, the build numbers were revised somewhat. 20H2 was released in the second half of 2020, and 21H1 was released in the first half of 2021.

Using Remote Server Administration Tools

After RSAT is installed, you'll want to start using it. Let's get started!

1. **Click the Start menu.**
2. **Scroll down to S and choose Server Manager.**
3. **When Server Manager launches, click the Add Other Servers to Manage link on the Quick Start tile.**

   The Active Directory tab allows you to search Active Directory for systems to add, as shown in Figure 1-13.

4. **Select SERVER2022-DC and click the right arrow to select it.**
5. **Click OK.**
6. **After the system is added, the roles and/or features that are installed on that server can now be managed through Server Manager on my Windows 10 client system, as shown in Figure 1-14.**

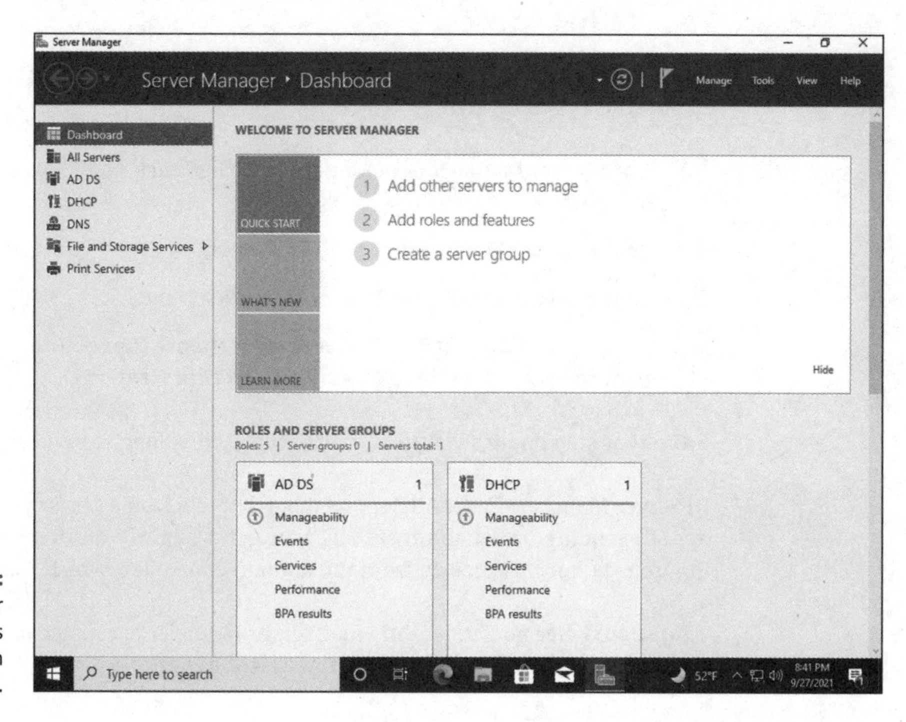

FIGURE 1-14: Server Manager showing the roles from the system I added.

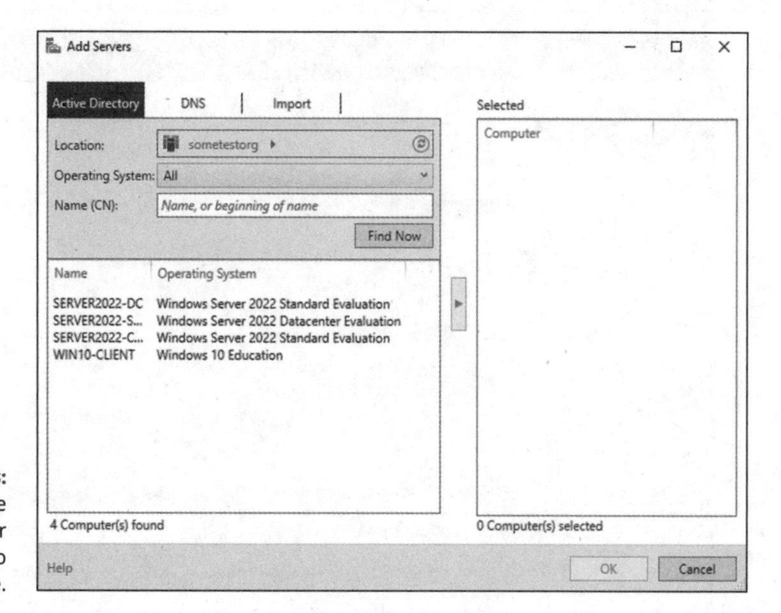

FIGURE 1-13: Searching Active Directory for systems to manage.

You can now add roles and features just as you would if you were on the servers in question. Let's try that.

1. **From Server Manager click Manage and then click Add Roles and Features.**
2. **On the Before You Begin screen, click Next.**
3. **On the Select Installation Type screen, click Next.**
4. **On the Select Destination Server screen, choose the remote server you want to install the role or feature on and click Next.**

From that step on, it's identical to installing while logged into the server.

TIP

In Server Manager on the All Servers tab, right-click on a server you want to manage. Depending on what's installed, you get a slightly different list. The list in Figure 1-15, for instance, is from my domain controller, which I just added.

That should give you an idea of what Server Manager is capable of doing. Install it and make your life as a system administrator easier!

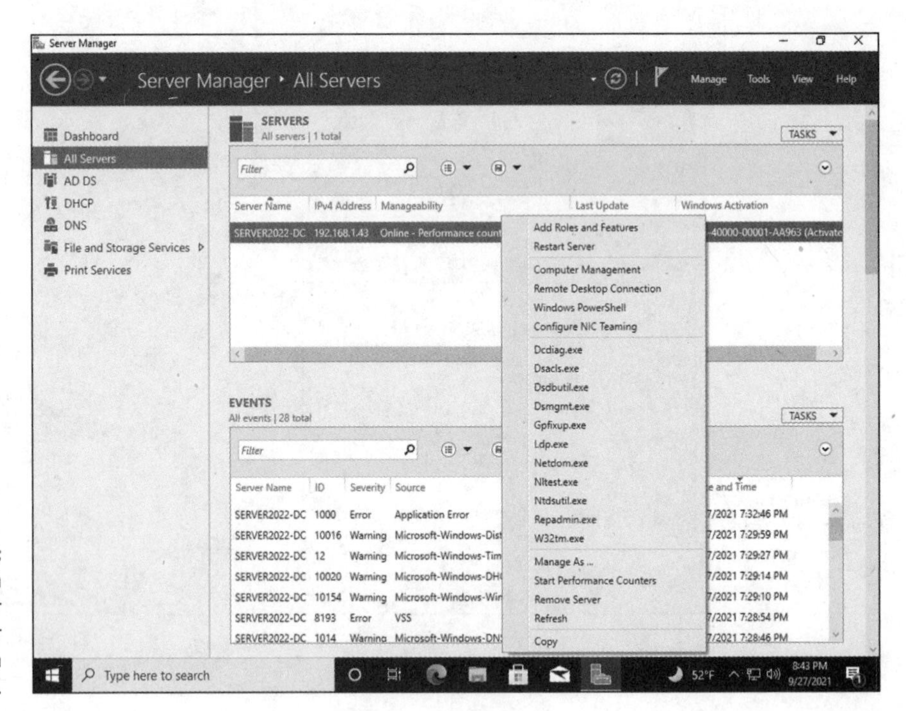

FIGURE 1-15: Managing a remote server with Server Manager on Windows 10.

IN THIS CHAPTER

- » Understanding how Group Policy works in an Active Directory domain
- » Starting the Group Policy Editor to work with Group Policy Objects
- » Performing computer management with Group Policy
- » Achieving user configuration with Group Policy
- » Viewing Resultant Set of Policy to troubleshoot issues with Group Policy

Chapter 2
Setting Group Policy

Making settings changes on one or two systems isn't that bad. But imagine you work in an enterprise with 200 systems. Or 2,000 systems. Suddenly making settings changes on individual machines becomes an impossible task. This is where Group Policy comes into play.

Group Policy allows you create a policy and then target that policy to users or systems within organizational units (OUs), security groups, or even on an individual basis. Group Policy is a very powerful way to enforce your organization's requirements across all Active Directory domain joined systems. Common items that get configured by Group Policy are

- » Password policies
- » User rights assignments (who can do what)
- » User Account Control (UAC) settings
- » BitLocker settings
- » Patch management settings

Group Policy offers hundreds of configuration items that allow you to centrally manage the configuration and security of all your domain-joined systems.

Understanding How Group Policy Works

A Group Policy Object (GPO) is a collection of settings. The GPO can be linked at several different levels, most commonly domain or OU. You can target specific systems or security groups within the GPO as well.

On startup, computer-focused GPOs are applied. These GPOs affect settings on the computer regardless of who logs into the system afterward. User policy objects are applied when a user logs into a system; they are used to make changes that would impact the user such as password policies, lockout policies, and so on. User GPOs follow the user regardless of which system the user logs in to. Remember the following rule: Last applied wins. If you run into an issue with a policy not doing what you're expecting it to do, a later GPO may be overwriting the behavior of a preceding GPO.

By default, Group Policy refreshes every 90 minutes, although there may be a randomized delay of up to 30 minutes. So, if you apply a change, it may take up to 120 minutes for that change to be applied to all systems/users. Some settings do not refresh in this way and require the user to log out and log back in or require a restart of the system. Changes that fall into this category are things like folder redirection, drive mappings, and some file preferences.

Group Policy applies its settings in a set order:

1. Local policies (set by `gpedit.msc`)
2. Site policies
3. Domain policies
4. OU policies

Think of it as a top-down approach to processing: After the local policies have been applied, site GPOs are the most broad, followed by domain GPOs, and then OU GPOs. Most system administrators have experience managing GPOs at the domain and OU level.

A common issue that occurs is when a system administrator makes a change to a domain-level policy, but the change doesn't seem to be applying. The most common culprit is an OU-level policy that is overwriting the setting from the domain policy.

In Group Policy, you configure settings with policies and preferences. Table 2-1 is a comparison of the two to see what the differences are between them.

TABLE 2-1 Group Policy: Policies versus Preferences

Policies	Preferences
When a Group Policy policy is no longer applicable, the setting is removed, and the original value is restored.	When a Group Policy preference is no longer applicable, the setting remains in the registry.
When a Group Policy policy is set to a certain value for an application, the application uses the value set by the policy. If the policy is removed, the application will use the original value.	When a Group Policy preference is set to a certain value for an application, it overwrites the default value for the application. If the preference is removed, the value for the application will remain the same.
When a Group Policy policy is used to enforce settings, users see grayed out options that inform them that the setting is managed by their administrators.	When a Group Policy preference is used to enforce settings, users can change the setting manually. Group Policy will not reapply the configured value after the first time it was applied.

TIP

One last thing before I move on to the next section: Group Policy has been around for a long time, since Windows Server 2000. There is a newer way to configure systems: PowerShell Desired State Configuration (DSC). Group Policy and PowerShell DSC should not be used to make configuration settings on the same systems at the same time. If one setting is different between the two, you'll run into a constant battle between the GP policy applying and overwriting a setting, followed by PowerShell DSC changing the setting back to the previous value. If you're in an enterprise environment that is already using Group Policy, my suggestion is to stick with Group Policy. If you're building your environment from scratch, or if you're in a DevOps-centric environment, PowerShell DSC is the better choice going forward. You can learn more about PowerShell DSC in Book 6, Chapter 5.

Starting the Group Policy Editor

In Server Manager, choose Tools⇨Group Policy Management. Group Policy Management is available on the server itself, assuming you installed the management tools. It is also available through Remote Server Administration Tools (RSAT) so you can manage Group Policy from your Windows 10 desktop.

When you first start the Group Policy Management snap-in, you see an entry for the Active Directory forest you're in. Click the arrow beside that forest to expand it. Click the arrow next to Domains to expand that, and then expand your domain name by clicking the arrow next to it.

With a fresh Active Directory environment, you have a Default Domain Policy. This is linked at the domain level. The Default Domain Policy is typically where you set things that will apply to all users, like passwords and other security settings. You can see in Figure 2-1, that the Default Domain Policy is applied to anyone in the Authenticated Users group, indicated by the presence of this group in the Security Filtering section.

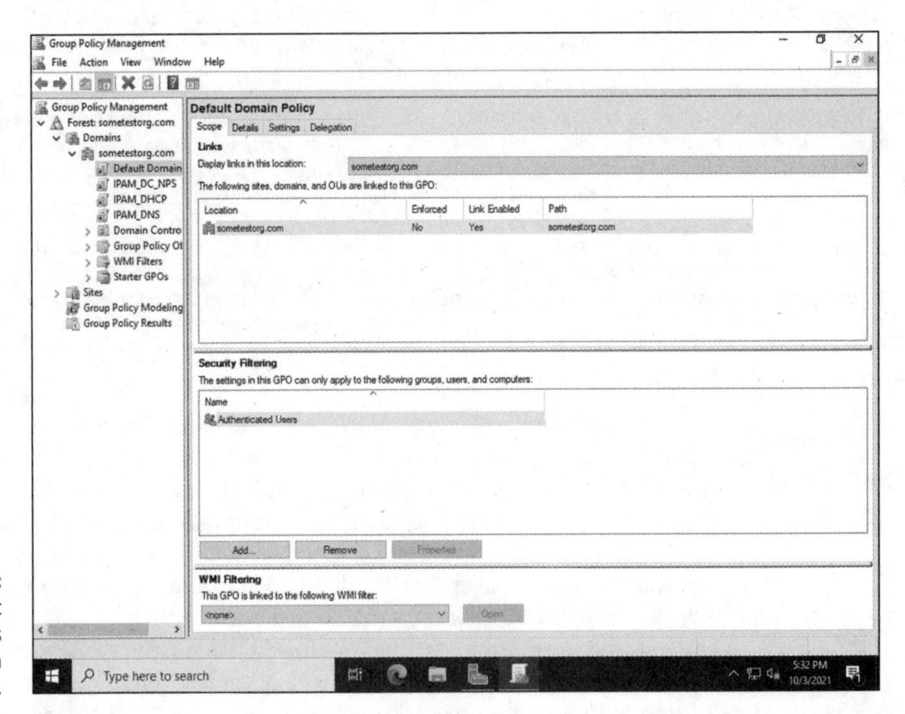

FIGURE 2-1: The Default Domain Policy is set at the domain level.

To open the Group Policy Management Editor, all you need to do is right-click a GPO and choose Edit. The Group Policy Management Editor opens, and you can make your changes from there.

In the following sections, I show you how to make changes to Group Policy. I cover computer management first and then user configuration. I also go over how you can see which policies are getting applied.

Performing Computer Management

Computer management activities can encompass everything from security settings to what applications you want to install on a particular system. You can do quite a few different things with Group Policy. The only real limitation to what you can do with Group Policy is whether there is a setting that will perform the configuration you require.

Follow these steps to create a brand-new GPO:

1. **With Group Policy Management open, and the tree expanded, right-click Group Policy Objects and choose New, as shown in Figure 2-2.**

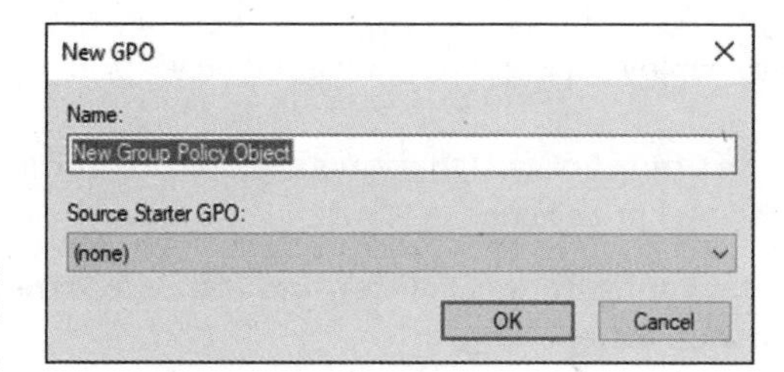

FIGURE 2-2: Creating a new GPO.

2. **Name your GPO.**

   I'll name mine Computer GPO.

3. **Leave Source Starter GPO blank and click OK.**

TIP

   If you want to create GPO templates for common administrative tasks, you can create them in the Starter GPO's folder instead of the Group Policy Objects folder. This allows you to choose them as a source when creating a new GPO and will automatically copy their settings into your new GPO.

4. **With the new GPO selected, click the Add button underneath the Security Filtering section in the main window of the policy.**
5. **Click the Object Types button, select the Computers check box, and click OK.**
6. **Enter the name of the server (or group of servers) you want to target, and click OK.**

   In my case, this will be SERVER2022-DC.

7. **To link the GPO at the domain level, right-click the domain name and choose Link an Existing GPO.**
8. **Select Computer GPO (or whatever you chose to call your new GPO) and click OK.**

That's all there is to getting started with editing Group Policy. Now let's make some actual edits to the Computer GPO that you just created.

Modifying computer software settings

The software settings area in Group Policy allows you to install software on systems to which the GPO is applied. You might use this to deploy software to a particular group of systems, or to all your systems as a whole.

Note: The software you want to deploy needs to be in a shared folder that is accessible to the systems that you're wanting to deploy it to. The systems need to have permission to access the shared folder and its contents. In this case, I've created a shared folder called Software, and that is where I've chosen the package.

Follow these steps to deploy software with Group Policy:

1. **Starting from the Group Policy Management screen, right-click the GPO that you created earlier and choose Edit.**
2. **Expand Computer Configuration, Policies, and Software Settings, and then select Software Installation.**
3. **Right-click the blank space next to Software Installation and choose New⇨Package.**

   The browser box will open looking for Windows installer packages. These typically end in `.msi`.
4. **Select the file and click Open.**
5. **Select Assigned and click OK.**

You can see in Figure 2-3 that I now have my software package assigned to be installed through this GPO, and that the software source is the shared folder I created before.

Modifying computer settings

The Windows Settings area of the Computer Configuration allows you to change certain areas of the system configuration. You can configure Domain Name System (DNS) settings, set Startup and Shutdown scripts, and set up printers and security settings.

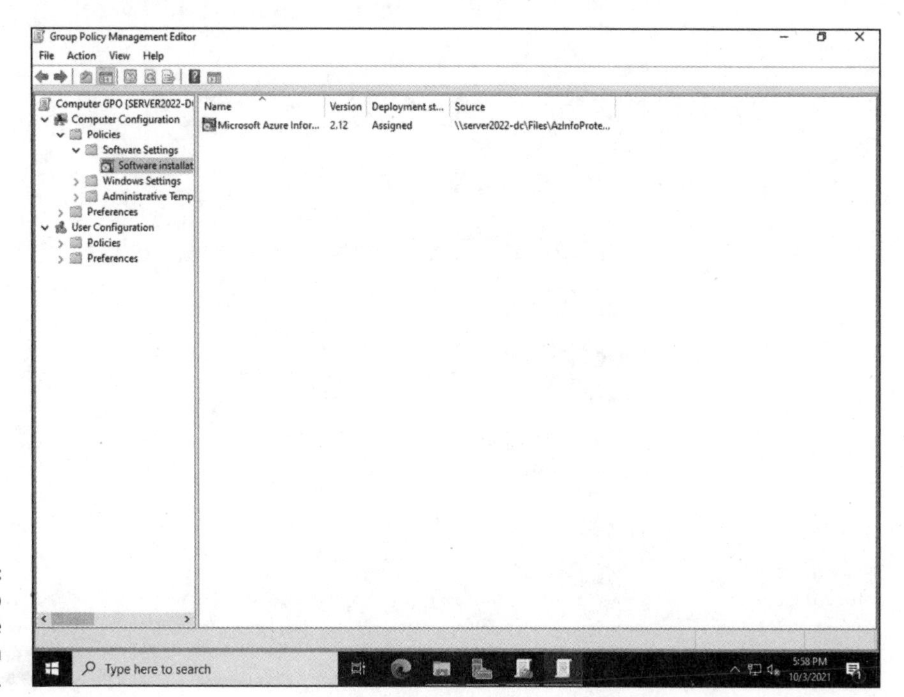

FIGURE 2-3: Setting up software to be deployed via Group Policy.

Here's how to set some of the basic security settings for a computer:

1. **Starting from the Group Policy Management screen, right-click the GPO that you created earlier and choose Edit.**
2. **Expand Computer Configuration, Policies, Windows Settings, Security Settings, and Local Policy, and then select Audit Policy.**
3. **Double-click Audit Account Logon Events.**
4. **Select Define These Policy Settings, choose Success and Failure, and click OK.**
5. **Repeat Steps 3 and 4 for each of the items in Audit Policy.**

When this is completed, your screen should look similar to Figure 2-4.

Take the time to look through the rest of the settings under Local Policies. You can do quite a few things to customize how your users will interact with your systems. For instance, you can block Microsoft accounts so users have to log in with an Active Directory account. You can change the login screen so that it doesn't show who the last logged on user was. There are so many settings available to you that I recommend you look through them and experiment in a home lab or a test lab if you have one.

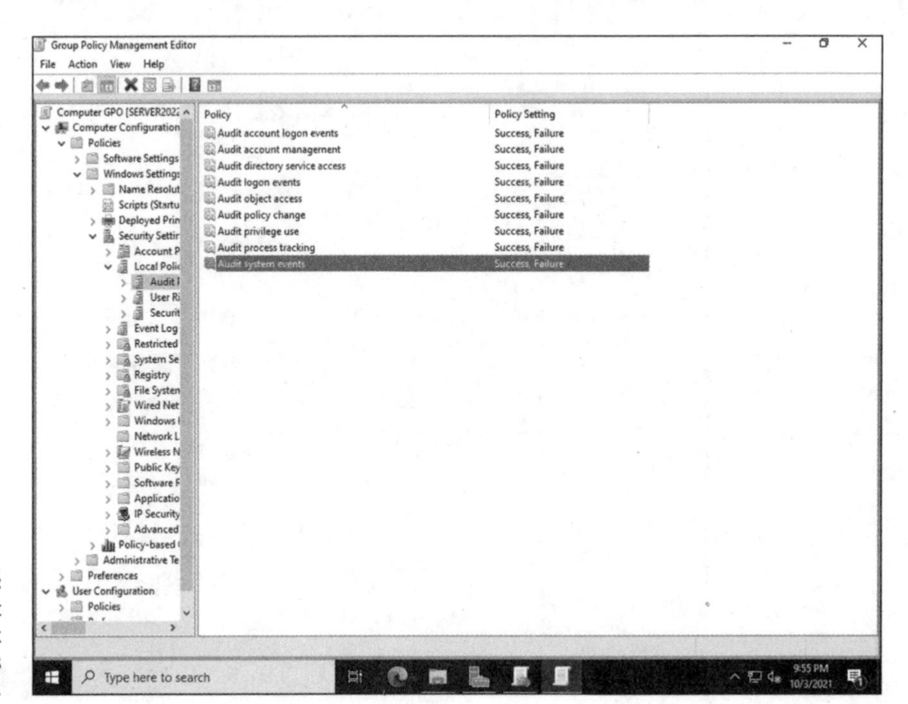

FIGURE 2-4: Setting the audit policy to track both successes and failures.

WARNING

Don't modify Group Policy in a production environment until you've fully tested the change in a test environment. Due to the delay in Group Policy processing, you could break things — badly, and you can't roll back quickly. It could take a few minutes for Group Policy to reapply with the working settings.

Using Administrative Templates

Administrative Templates allow you to further customize how your users will interact with your systems. Here's how to customize the Start menu:

1. **Starting from the Group Policy Management screen, right-click the GPO that you created earlier and choose Edit.**
2. **Expand Computer Configuration, Policies, Administrative Templates, and Start Menu and Taskbar.**
3. **Double-click Do Not Keep History of Recently Opened Documents.**
4. **Choose Enabled and click OK.**

This is just a very small example of what you can do. Look through the rest of the Administrative Template settings that are available to you. You have loads of customization options.

Performing User Configuration

User configuration follows a user rather than a particular computer. User GPOs apply when a user logs in and potentially when a user logs out, depending on what the GPO has been set to do. User GPOs are often used to make account settings or software installations where the user is more important to target than the system. Use them with care, though, because they can increase login times for users.

Modifying user software settings

Similar to the Computer Configuration, software can be installed based on users if configured in User Configuration. Follow these steps:

1. **Starting from the Group Policy Management screen, right-click the GPO that you created earlier and choose Edit.**
2. **Expand User Configuration, Policies, and Software Settings, and then click Software Installation.**
3. **Right-click the blank space next to Software Installation, and choose New⇨Package.**

   The browser box will open looking for Windows installer packages. These typically end in `.msi`.
4. **Select the file and click Open.**
5. **Select Assigned and click OK.**

Your screen should now look similar to Figure 2-5. If it does, you've successfully deployed software to your users. *Note:* You would not typically deploy the same software to computer and user. You would choose one or the other.

Modifying a user's Windows Settings

Similar to Computer Configuration's Windows Settings, the User Configuration's Windows Settings has system-specific settings. The biggest difference is that they apply when the user logs in, rather than when the system starts up. Windows Settings doesn't have anywhere near as many options as the Computer Configuration equivalent does.

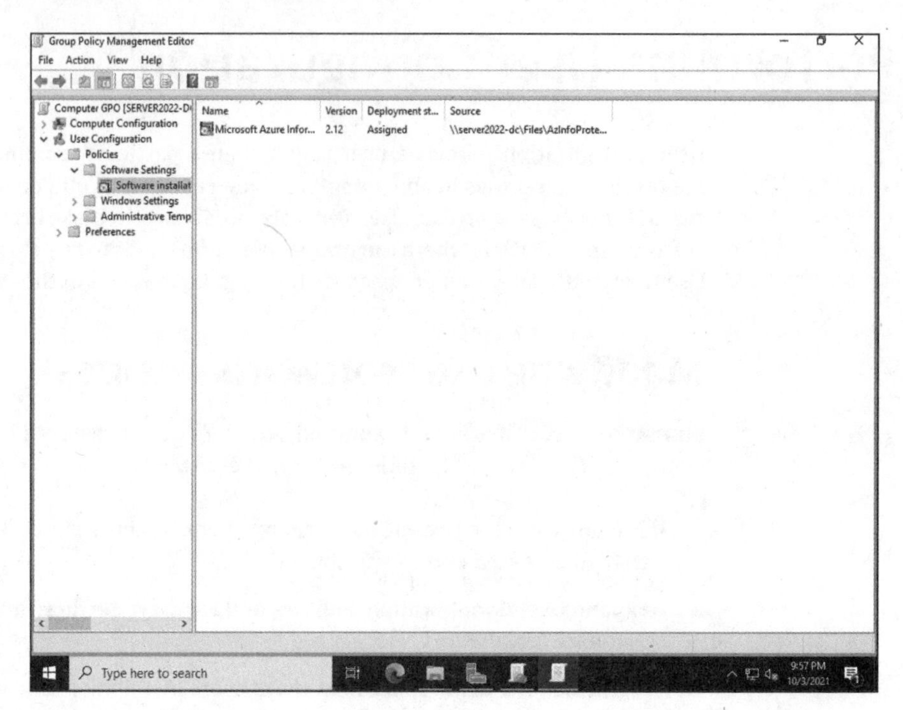

FIGURE 2-5: Software deployed via the User Configuration.

Let's set up folder redirection for the user's Documents folders. This is a common use case because users save things in Documents where they don't get backed up. By turning on redirection, you can save the user's Documents folders on the server where it *does* get backed up. Follow these steps:

1. **Starting from the Group Policy Management screen, right-click the GPO that you created earlier and choose Edit.**
2. **Expand User Configuration, Policies, Windows Settings, and Folder Redirection, and then select Documents.**
3. **Right-click the blank area to the right of Documents, and choose Properties.**

   The Document Properties dialog box appears.
4. **On the Target tab, from the Setting drop-down list, choose Basic.**
5. **Under Target Folder Location, select Create a Folder for Each User under the Root Path.**
6. **Click the Browse button and choose where you want the Documents folders to redirect to.**

   Your screen should look similar to Figure 2-6. Normally the redirection is done to a file server on the network with enough storage to support all the user directories.

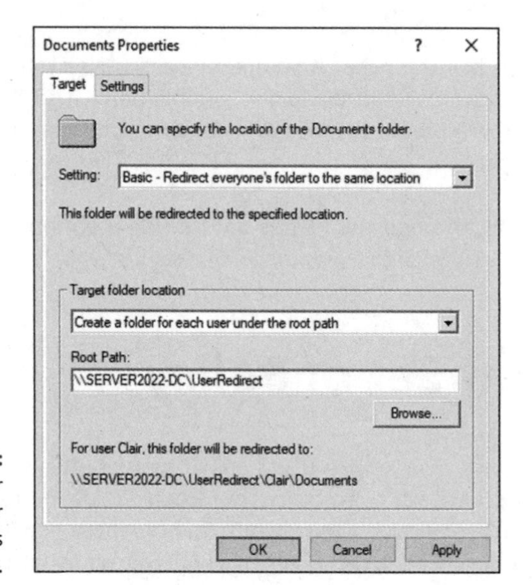

FIGURE 2-6: Setting up folder redirection for the Documents folder.

7. **Click OK.**

   You receive a warning regarding older operating systems not potentially being able to support folder redirection.

8. **Click Yes to continue.**

You've now set up folder redirection for your user's Documents folder. You won't see anything in the Group Policy Management Editor. However, you can verify the redirection is working by having one of the affected users log out and then log back in. She should have her folder created on the server in the share location you specified. Pretty cool!

Using user Administrative Templates

User Administrative Templates are very similar to the Computer versions, except they allow you to customize specific areas in the OS related to users.

1. **Starting from the Group Policy Management screen, right-click the GPO that you created earlier and choose Edit.**
2. **Expand User Configuration, Policies, Administrative Templates, and Desktop, and then choose Desktop.**
3. **Double-click Disable Active Desktop.**
4. **Choose Enable and then click OK.**

If your organization has wallpaper it wants on everyone's systems, you can set that here as well, under Desktop Wallpaper. Simply point it to the shared folder where the wallpaper is stored and — *voilà!* — the next time your user logs in, he'll get the new wallpaper.

I recommend you dig through the menus and see what options are available to configure. You can customize to a great level of detail.

Viewing Resultant Set of Policy

You've configured your GPOs and you've applied them, but something just isn't working right. Your user isn't receiving the Group Policy settings that you're expecting her to, and you aren't sure why. This is where you need to look at Resultant Set of Policy (RSoP). Let's launch RSoP and see what the results look like:

1. **Right-click the Start menu and choose Windows PowerShell.**
2. **Type** rsop.msc **and press Enter.**

   RSoP runs and then displays the results of the settings that are currently applied.
3. **If the software I set up in the GPO hadn't installed, I could click User Configuration, click Software Settings, and select Software Installation.**

   If the software I setup in the GPO is there, then I know the policy applied and the installation must have failed for another reason. If the software I'm applying via policy isn't there, then I know that the policy didn't apply. If this is a user policy, that may mean I need to log out and log back in. If it's a computer policy, a reboot may be needed.

RSoP is a very helpful utility. It shows you only the categories that have settings applied. Keep in mind that User Configuration applies to the currently logged-on user. In Figure 2-7, for example, you can see that I don't have Administrative Templates showing up under my User Configuration even though I set the setting for Active Desktop. This is expected behavior in this case, because I haven't logged out and logged back in since I made that change. If I was troubleshooting an issue for a user and saw this, I would ask the user to log out and then log back in so the new setting could apply.

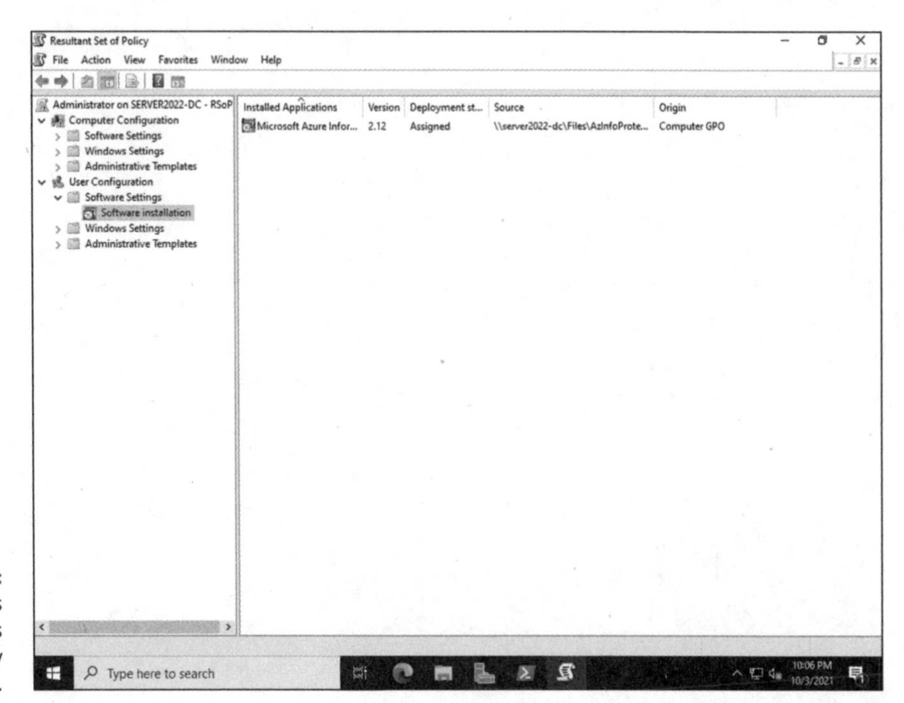

FIGURE 2-7: RSoP shows which GPOs are successfully applied.

Setting Group Policy

IN THIS CHAPTER

» Starting the Registry Editor in Windows Server 2022

» Importing and exporting Registry elements

» Finding the Registry elements you're looking for

» Understanding the different Registry data types you can use

» Understanding the hives and what they do

» Connecting to network registries to configure remote systems

» Setting Registry security to prevent tampering

Chapter 3 Configuring the Registry

There aren't many things that will scare a system administrator, but messing with the Registry is definitely up there. The Registry in a Windows system is like the skeleton of the operating system. Control Panel items are just graphical front ends to the values that exist in the Registry. Every single thing that is installed or has ever been plugged into your system is recorded in your Registry. It's not surprising then, that it can be a very intimidating prospect changing this thing that is so central to the operating system and how it functions. You delete or change the wrong key, and all of a sudden, your system is no longer functional.

In this chapter, I explain what the Registry is, what it's made up of, how to safely configure it, and how to recover it if something goes terribly wrong.

Starting Registry Editor

There are several different ways to start the Registry Editor. If you're in Server Manager, you can simply choose Tools ⇨ Registry Editor. If you aren't in Server Manager, you can right-click the Start menu, choose Windows PowerShell, and then type **regedit.exe**. You need to have administrative privileges to run Registry Editor. Assuming you have the necessary privileges, using either option will open the Registry Editor, shown in Figure 3-1.

When you open the Registry for the first time, all the hives will be collapsed. If you have been in the Registry on the system before, it will typically open to where you were last.

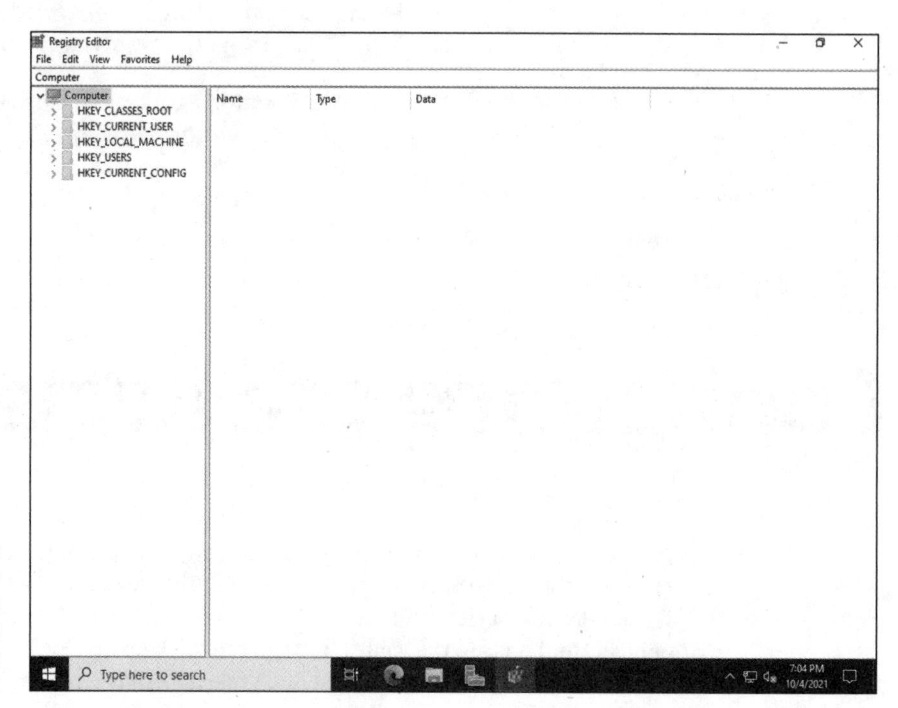

FIGURE 3-1: The Registry Editor allows you to work with the hives and keys in the Registry.

Importing and Exporting Registry Elements

Importing and exporting Registry elements gives you the ability to back up and restore your Registry. The process itself is very simple and really hasn't changed much over the years. This same process will work on older systems, as well as on Windows Server 2022.

Exporting Registry elements

Whenever you're planning on making a change to the Registry, it's always a good idea to get a backup of your settings. This gives you a way to restore your Registry should something go horribly wrong. One of the simplest ways to back up the Registry is to export it to a `.reg` file. The `.reg` is used to identify Registry files.

Here's how to create a backup of the Registry:

1. **From Server Manager, choose Tools ⇨ Registry Editor.**
2. **Select the element that you want to export.**
3. **With the individual element selected, choose File ⇨ Export.**
4. **Select a save location for the `.reg` file.**
5. **Click Save.**

Note that you can export individual keys as well. You don't have to export the whole hive.

Importing Registry elements

Having a backup is great, being able to restore it is even better. To restore your Registry, you need the exported `.reg` file and access to the Registry Editor. There are a couple reasons why you may choose to import a `.reg` file:

- You may have made changes to the Registry that have caused the system to become unstable, so you need to restore from the exported `.reg` file you made before you started your work.
- You may have made customizations to the Registry on one of your servers and you want to more easily apply the same configurations to another server. You can do this quickly by importing the `.reg` file from the configured server into the one that you want to configure.

TIP

If a Registry change has made it to where your system is unable to boot and you can't get to the Registry Editor, your best bet is to restore from your most recent system backup.

Here's how to restore the Registry from your backup file:

1. **From Server Manager, choose Tools ⇨ Registry Editor.**
2. **Choose File ⇨ Import.**

3. **Navigate to the location where your `.reg` backup file is stored.**
4. **Select the `.reg` file and click Open.**

   A progress bar pops up showing you the progress of the restore.
5. **Click OK when the restore is complete.**

Finding Registry Elements

At some point in your career, you'll be charged with making an update to the Registry that will require you to find a specific key within a hive. This can be like finding the proverbial needle in a haystack given the levels of recursion in the hive in addition to the sheer content. Thankfully, the Registry has a search tool that can make finding what you're looking for much easier.

Say, for example, I needed to find the SCHANNEL key because I needed to edit some of the encryption protocols my server supports, but I just can't remember where SCHANNEL is. Here's how the search tool can help me find my key:

1. **From Server Manager, choose Tools ⇨ Registry Editor.**
2. **In Registry Editor, choose Edit ⇨ Find (or just press Ctrl+F).**
3. **Type in your search term in the Find What box, and select what types of elements you want to look for, as shown in Figure 3-2.**

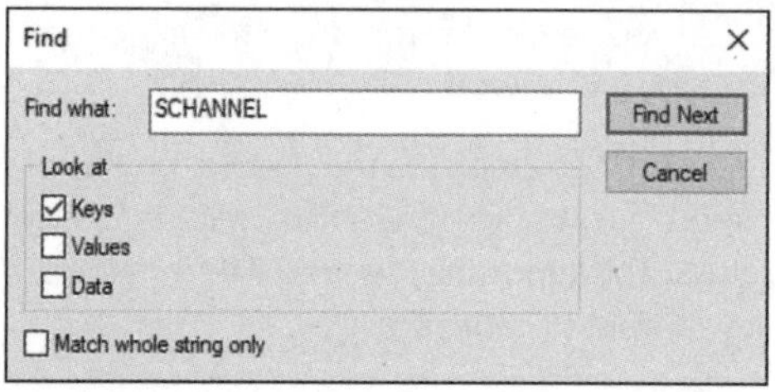

FIGURE 3-2: Searching the Registry for a specific key.

4. **Click Find Next.**
5. **If the first result isn't what you were looking for, choose Edit ⇨ Find Next, or you can simply press F3 until you find the result you want.**

Understanding Registry Data Types

There is quite a bit of new terminology when it comes to the Registry that you may not be used to. *Hives*, for instance, are the highest level in the hierarchy in the Registry and are discussed in the next section.

Within each hive are *keys*, which look like folders in the Registry and are used for organizational purposes. Keys can have multiple levels of keys nested inside of them. Each hive and each key can contain data housed an some kind of data type. It's important that you understand each data type, and what the data type might be used for. A DWORD and a QWORD may sound very similar, for instance, but in general they aren't interchangeable.

Here are the different data types you'll come across in the Registry and what each data type is used for:

- **String Value (REG_SZ):** Contains fixed-length text strings and is the second most common data type used in the Windows Registry.
- **Multi-String Value (REG_MULTI_SZ):** Made up of multiple text strings separated by commas or spaces.
- **Expandable String Value (REG_EXPAND_SZ):** Used to store variables that the operating system and/or applications need to be able to resolve. These may include variables like `\%username%` and are similar to what you would use if you were scripting with environmental variables.
- **Binary Data (REG_BINARY):** Used to store binary data in either decimal or hexadecimal format. This type is most often used in relation to hardware components and information.
- **32-Bit Number (REG_DWORD):** Represents 32-bit numbers. DWORDs are one of the most common data types used in the Registry.
- **64-bit Number (REG_QWORD):** Represents 64-bit numbers.

TIP

A common misconception is that 32-bit programs will write to DWORDs and 64-bit applications will write to QWORDs. That is not the case. If a 64-bit application is writing a value to the Registry that is 32 bits long, it will be stored as a DWORD.

Figure 3-3 gives you an idea of what the data stored in each of these data types might look like. These are just examples, but it should give you a general idea of what to look for.

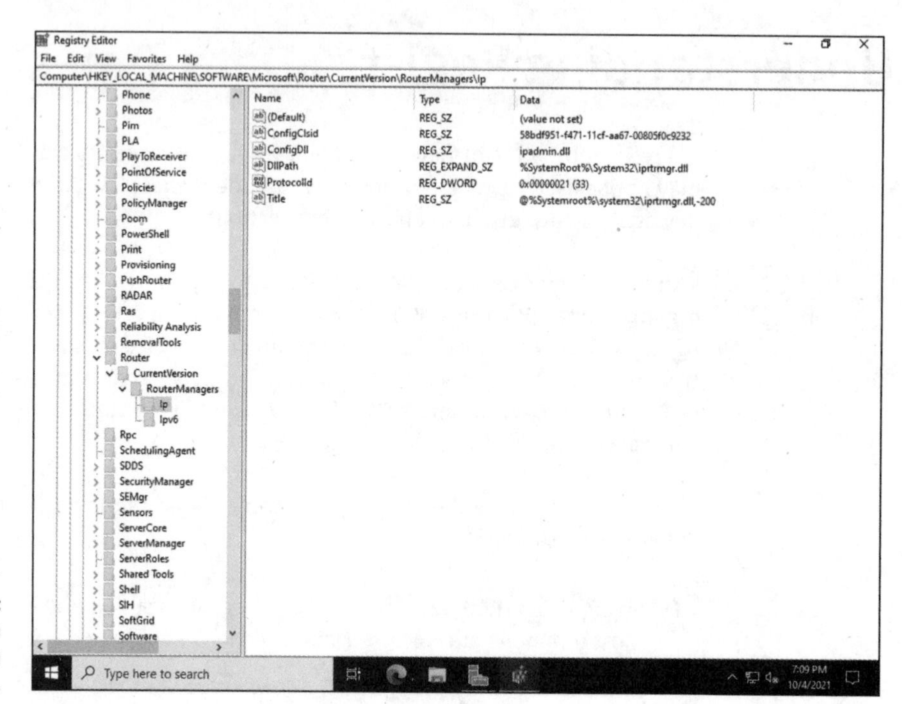

FIGURE 3-3: Data types in the Registry with sample values.

Understanding the Hives

In the context of the Windows Registry, a *hive* is a logical grouping of all the keys, subkeys, and values in the Registry. They're grouped by like settings. For instance, HKEY_USERS has settings that affect all users, while HKEY_CURRENT_USER has settings that affect only the currently logged-in user. These local groupings also contain support files, which I call out in each Registry hive section.

HKEY_CLASSES_ROOT

HKEY_CLASSES_ROOT is commonly abbreviated to HKCR. It's the area in the Registry that keeps track of which file types are associated with which programs, as shown in Figure 3-4. It also holds configuration data for Visual Basic programs and COM objects.

Data in HKEY_CLASSES_ROOT comes from two different sources:

- `HKEY_LOCAL_MACHINE\SOFTWARE\Classes`
- `HKEY_CURRENT_USER\SOFTWARE\Classes`

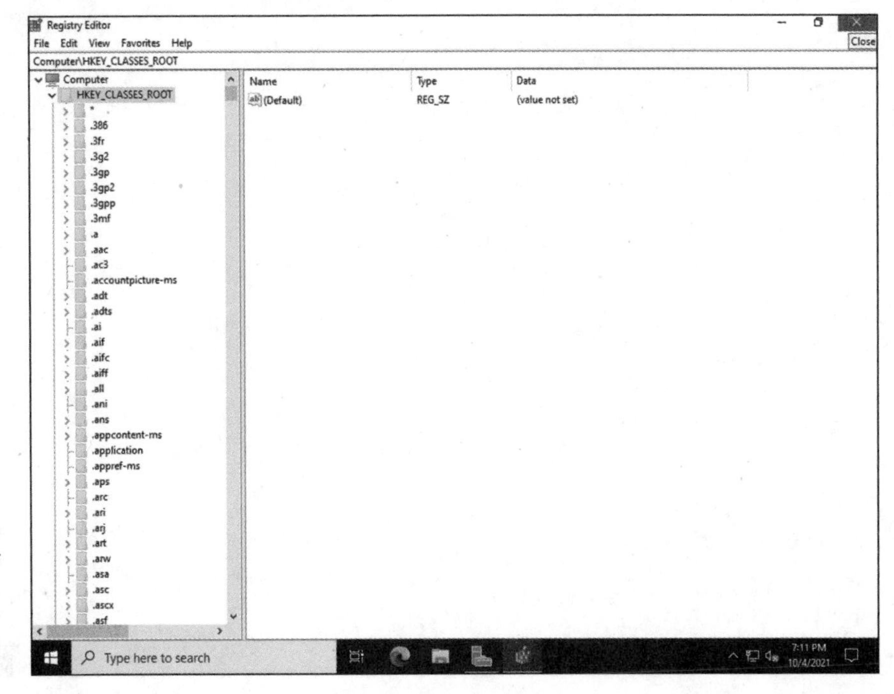

FIGURE 3-4: HKCR keeps tracks of associations between file types and programs.

If there is a subkey created in either location, it also gets created in HKCR automatically.

HKEY_CURRENT_USER

HKEY_CURRENT_USER is often shortened to HKCU. This area of the Registry changes depending on who the actual logged-in user is. Every time a user logs in, this is re-created based on the information stored in the user's `ntuser.dat` file. If the user has never logged in before, then this is created from the default user `ntuser.dat` file. which is stored at `C:\Users\Default\Ntuser.dat`. You can see the HKCU in Figure 3-5.

TIP

If you ever want to look at the default user, you'll need to enable hidden items in your view. Simply navigate to `C:\Users`, click the View tab, and then select the Hidden Items check box. Then you'll see the Default profile.

Note: HKCU does not actually store data. It contains a pointer to the user's actual data, which is housed under HKEY_USERS. Each user is assigned a security identifier (SID), and each user has a key with his SID where his data is stored in HKEY_USERS. If you want to change a user's settings, you can, in fact, change it in HKCU or HKEY_USERS because they're essentially the same space.

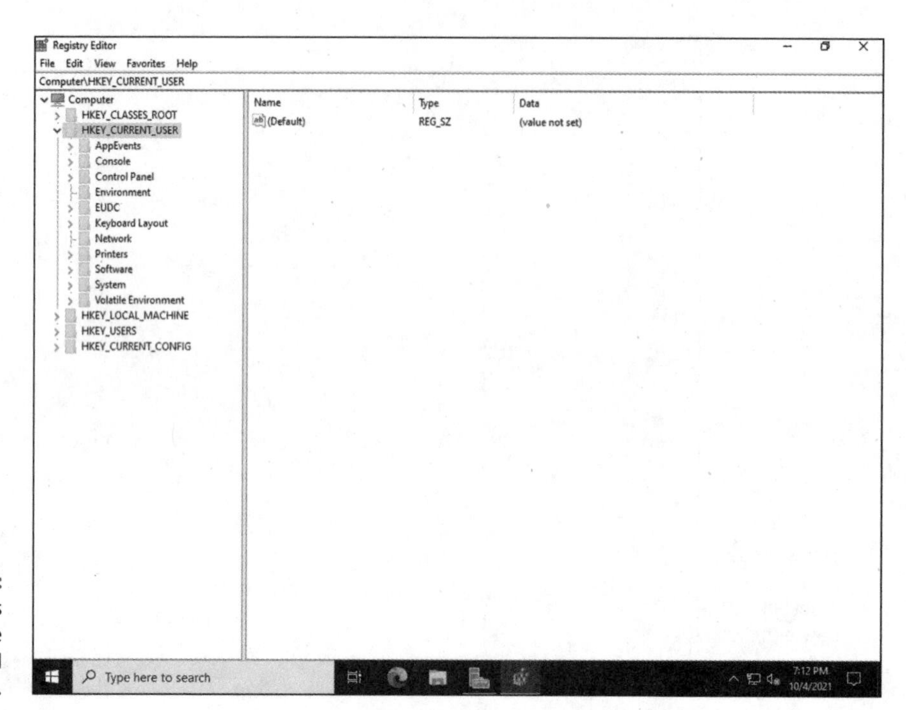

FIGURE 3-5: HKCU contains pointers to the currently logged on user's settings.

HKEY_LOCAL_MACHINE

HKEY_LOCAL_MACHINE is commonly abbreviated to HKLM. It's a treasure trove of information on the system, including information on hardware, operating system, security, drivers, and startup parameters. Here's more information on HKLM:

- **Hardware:** Stores information about hardware that the system has detected.
- **SAM:** SAM stands for Security Accounts Manager. Stores user and group information. Don't ever edit this key directly. If you do, you may block users from being able to log on. If you need to change information in this key, use Active Directory.
- **Security:** Contains security information that is needed by the system and by the network. Don't ever edit this key directly. Stick with Group Policy, Local Security Policy, and Active Directory instead.
- **Software:** The Software subkey stores subkeys for each software program you install, sorted by vendor. It also contains program variables that may be needed to launch or work with the application. These settings are ones that apply to all the users of the system.

- **System:** The System key stores subkeys related to control sets. This includes the current control set and the control sets that have been used at some point in time. If you look at the CurrentControlSet subkey, you can see which control set is actually active at any given point in time. If the entry in the Current value is 1, then ControlSet001 is the active control set.

HKEY_USERS

You'll commonly see HKEY_USERS shortened to just HKU. It contains user-specific settings for each user that has logged on to the system. Each user is represented by his SID, which is unique to each user. In Figure 3-6, you can see a picture of several SIDs. The SID for the default user is just .DEFAULT, while the others have numerical representation.

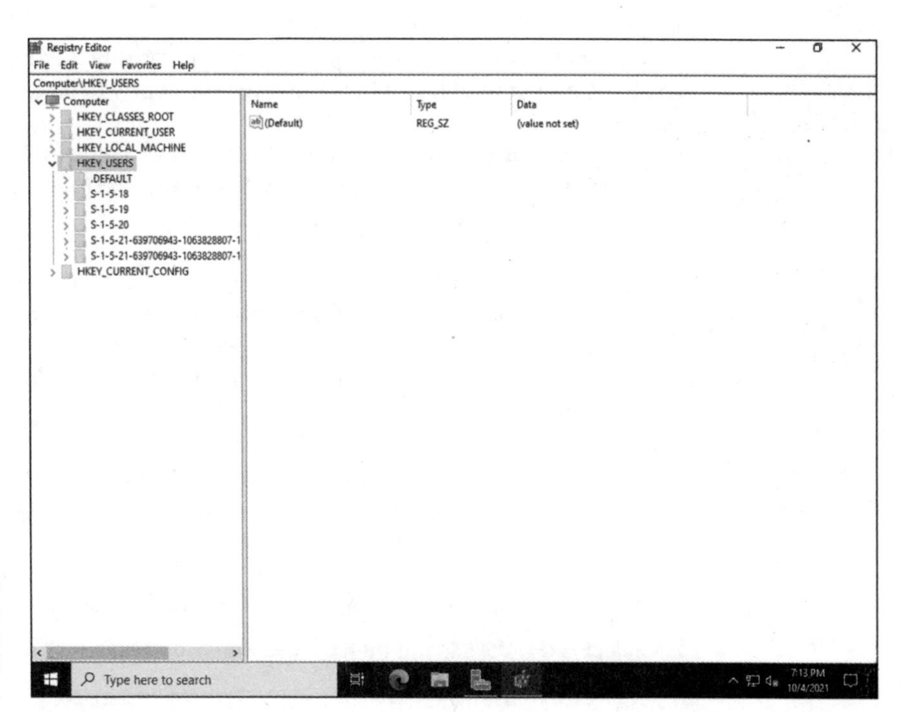

FIGURE 3-6: SIDs for each account on the system are stored in HKU.

The short SIDs you see in the picture are built-in system accounts:

- **S-1-5-18:** Local System
- **S-1-5-19:** NT Authority–Local Service
- **S-1-5-20:** NT Authority–Network Service

Longer SIDs will belong to the accounts of users who have logged into the system. A longer SID with a 500 at the end of it is the administrator account.

HKEY_CURRENT_CONFIG

HKEY_CURRENT_CONFIG is usually written as HKCC. It's similar to HKEY_CURRENT_USER in that it doesn't actually store data; it stores pointers to the data. HKCC has the configuration data pulled from the current hardware profile. The pointer it actually stores points to `HKLM\SYSTEM\CurrentControlSet\Hardware Profiles\Current`.

Loading and Unloading Hives

There may come a time when you want to compare the settings of a user or a machine. The ability to load a hive gives you the ability to do exactly that.

You can only use the Load Hive and Unload Hive commands when you have HKEY_LOCAL_MACHINE or HKEY_USERS selected. With any of the other hives selected, the options will be grayed out.

Let's load a hive and see how this works. To give you some background, I've logged in with the administrator account. I'm going to launch Registry Editor and then compare the settings of this account to the John Smith account I created earlier. If you want to create an account to follow along but you're not sure how to create it, check out Book 3, Chapter 4.

To see the `ntuser.dat` file, you need to have Hidden Items turned on. The simplest way to do that is to open File Explorer, click the View tab, and select the Hidden Items check box.

1. **From Server Manager, choose Tools ⇨ Registry Editor.**
2. **Select HKEY_USERS, and then choose File ⇨ Load Hive.**
3. **Navigate to the `ntuser.dat` file you want to compare to.**

   In my case, the path is `C:\Users\jsmith\ntuser.dat`.
4. **Select `NTUSER.DAT` and click Open.**

   You're prompted to name the key. This key will be created under HKEY_USERS so you can compare the settings you're interested in.
5. **Name it JSMITH, and then click OK.**

You can see in Figure 3-7 that the newly loaded key is present. From here you can expand it and compare its settings to the other user keys.

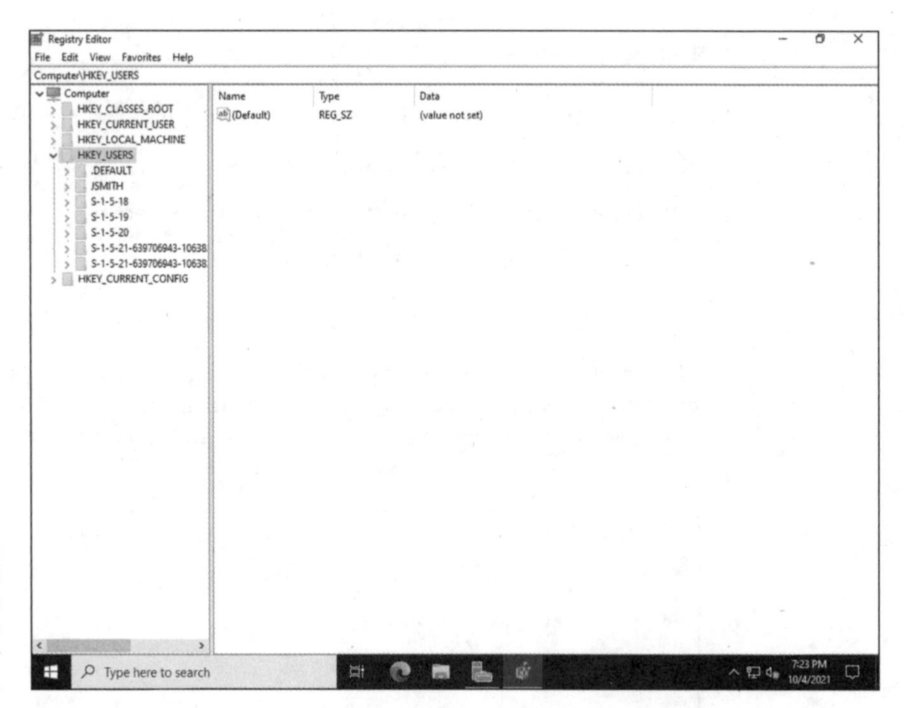

FIGURE 3-7: Loading a hive allows you to compare settings including user settings.

To unload the hive when you're done, follow these steps:

1. **Select the key that you want to remove.**
2. **Choose File ⇨ Unload Hive.**
3. **In the Confirm Unload Hive dialog box, click Yes.**

It's that simple. The key is removed, and you're back to your regular old Registry again.

Connecting to Network Registries

Sometimes you may need to check a setting really quickly or make a small edit to a Registry, but you don't want to log on to that system just to make the change. You can connect to a Remote Registry over the network quite easily. In order for this to work, the remote system has to have remote administration enabled/allowed in the firewall, and the Remote Registry service needs to be running on both the source and destination systems.

Follow these steps:

1. **From Server Manager, choose Tools ⇨ Registry Editor.**
2. **Choose File ⇨ Connect Network Registry.**
3. **Enter the system's name that you want to connect to.**

   In my case, this will be SERVER2022-SVR2.
4. **Click OK.**

The Registry that you wanted to connect to will appear in the list by name. Note that you can only work with HKEY_LOCAL_MACHINE and HKEY_USERS, and those are the only options presented to you. You can see this in Figure 3-8.

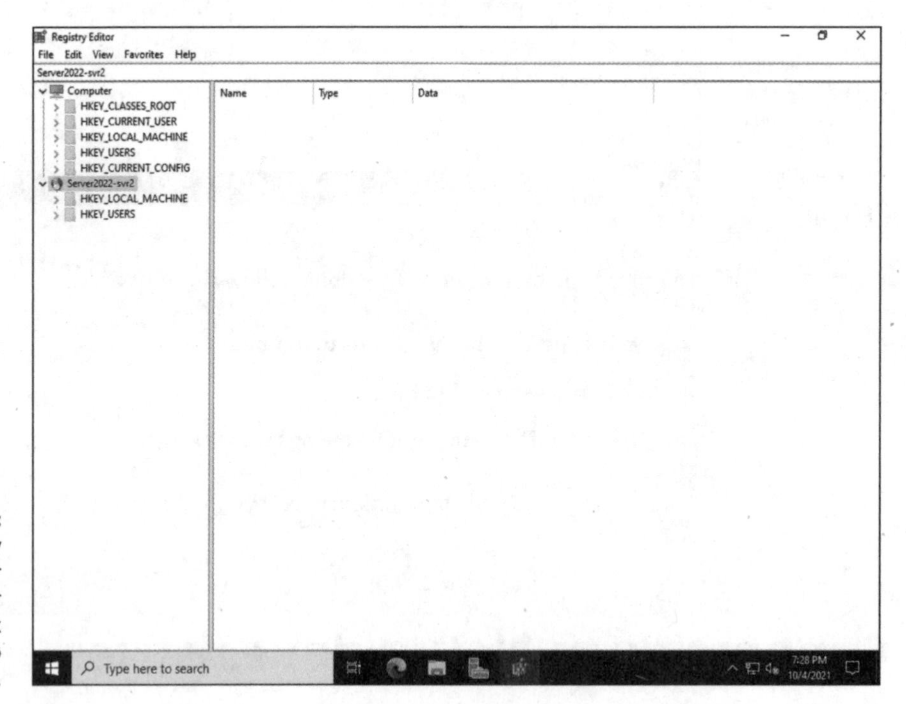

FIGURE 3-8: Remote Registry is useful for comparing or making quick changes to remote systems.

Setting Registry Security

Given that the Windows Registry is at the center of everything that happens in the Windows operating system, it makes sense that you would want to limit access to it and secure it properly.

Setting permissions in the Windows Registry

You can set permissions on the hives and individual keys in the Windows Registry. All you need to do is right-click the hive or key you want to secure and choose Permissions. This brings up a dialog box that most people are familiar with where you can add or remove users and groups and set the desired level of permissions. If you click the Advanced button, you can set much more granular permissions, just as you can in a file system. The Advanced Security Settings dialog box is shown in Figure 3-9.

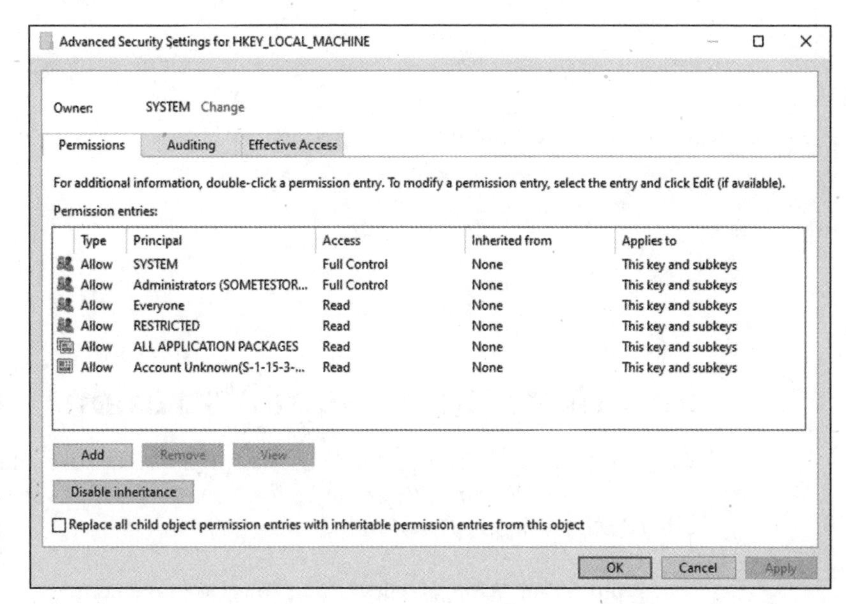

FIGURE 3-9: Setting permissions on the Windows Registry is very similar to setting permissions on file servers.

Disabling Remote Registry access

If you don't need the ability to edit the Registry remotely, the simplest thing to do is disable the Remote Registry service. If this service is not running, then people can't access the Registry remotely; they'll need to log in. Follow these steps to disable Remote Registry access:

1. **From Server Manager, choose Tools ⇨ Services.**
2. **Scroll down to the Remote Registry service and double-click it to open the Properties dialog box.**
3. **Change Startup Type to Disabled, as shown in Figure 3-10.**
4. **Click OK.**

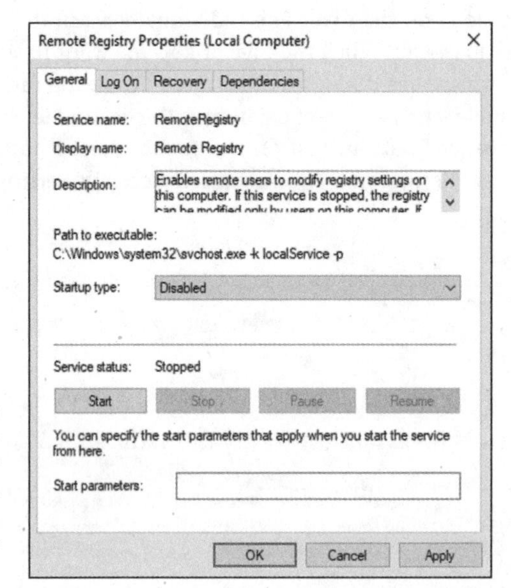

FIGURE 3-10: Disabling the Remote Registry service ensures that nobody can connect remotely to the Registry.

Securing remote administration

You can set which IP ranges or individual IP addresses are allowed to connect remotely to the system with Group Policy. You can push this setting out through Group Policy Objects (GPOs). Follow these steps:

1. **From Server Manager, click Group Policy Management.**
2. **Expand Group Policy Objects.**
3. **Select the GPO you want to edit, right-click it, and choose Edit.**
4. **Navigate through Computer Configuration, Policies, Administrative Templates, Network, Network Connections, Windows Defender Firewall, and Domain Profile.**

5. **Double-click Windows Defender Firewall: Allow Inbound Remote Administration Exception.**

6. **Choose Enabled and then add the address or subnet you want to allow remote connections from into the text box underneath Allow Unsolicited Incoming Messages from These IP Addresses (see Figure 3-11).**

   You can put an asterisk (*) there, but this allows connections from anywhere and is not a good practice.

7. **Click OK.**

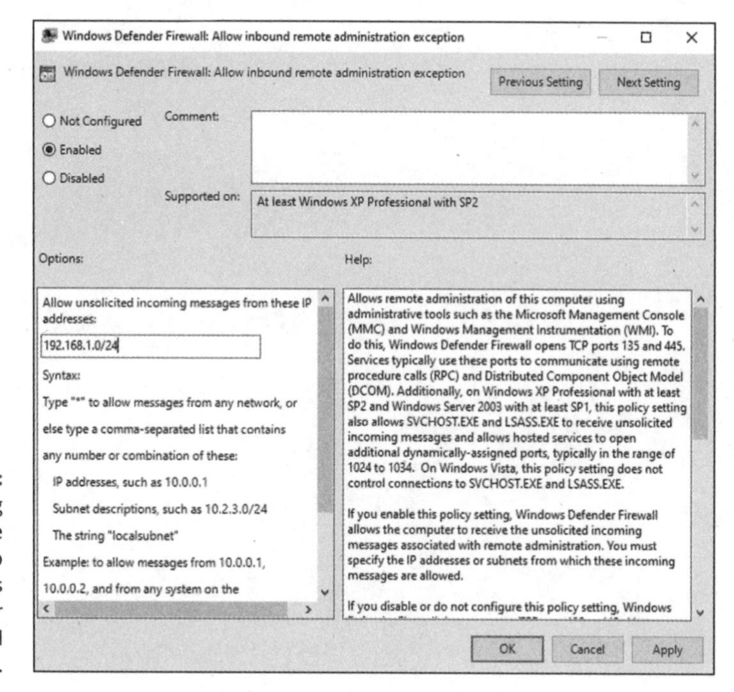

FIGURE 3-11: Restricting remote administration to specific subnets is a great layer of additional security.

» Configuring Active Directory objects in Windows Server 2022

Chapter 4
Working with Active Directory

Most system administrators have worked with or will at some point in their careers work with Active Directory. Most become familiar with Active Directory Users and Computers, because they use it in their daily work. But that's such a small part of what Active Directory contains.

In this chapter, I fill you in on what Active Directory is, what it does, and what the different components in Active Directory are.

Active Directory 101

Active Directory is a directory service. It's used to store user accounts, computer accounts, group policies, and all kinds of records.

An Active Directory forest is the top level in the hierarchy. Each forest serves as a security boundary and is created by default with a root domain. Other domains can be created within the forest, though they must share the same namespace as the forest (this namespace matches the underlying DNS zones). For instance, if your forest's namespace is `sometestorg.com`, the domains in that forest can be things like `hr.sometestorg.com` or `accounting.sometestorg.com`.

Domains are the next tier down from forests in the Active Directory hierarchy. Active Directory domains are a collection of objects that share the same database in Active Directory. Those objects are things like users, groups, workstations, and servers.

Within domains, organizational units (OUs) are used to organize systems or users that have something in common. This helps if you want to apply a Group Policy only to accounting users, because you can put them in their own OU and then link the Group Policy Object to that OU. Common criteria for organizing systems and users are things like their geographic location, their building, or their department.

Users in Active Directory are assigned to security groups to gain access to resources. The security groups are given permissions to the resources, and when the user is added to the security group, they're able to access any of the resources granted to her by the security group. In Active Directory, security groups come in three types, referred to as scopes:

- **Universal:** A universal security group is able to grant permissions for any domain in the same forest, or for other forests where there is a trust in place (trusts are discussed later in this chapter).
- **Global:** A global security group is able to grant permissions to resources for any domain in the forest, or for domains and forests when there is a trust in place.
- **Domain local:** A domain local security group is able to grant permissions for resources as long as they're in the same domain.

Configuring Objects in Active Directory

Active Directory has a few different components. Each component serves a different purpose, but all the components support Active Directory in some way. I know plenty of system administrators who have only ever worked in Active Directory Users and Computers. They've never had a reason or the inclination to explore the other areas. But that's not going to be you! In this section, you explore the different components of Active Directory and what each of them controls.

Using Active Directory Domains and Trusts

Small organizations often may have only one domain, but larger organizations will end up with multiple domains. To simplify administration and the user experience, you can set up trusts between domains so that an authenticated user in one domain can access resources in another domain without having to authenticate with a separate set of credentials. There are a few terms you should know before we discuss the type of trusts:

- **Transitive trust:** A transitive trust can take advantage of trust relationships formed by other domains. For instance, say Domain 1 trusts Domain 2, and Domain 2 trusts Domain 3; if the trust is set as a transitive trust, Domain 1 will also trust Domain 3.
- **Nontransitive trust:** A nontransitive trust means that trust relationships made with other domains do not automatically apply to all other domains. For example, Domain 1 trusts Domain 2, and Domain 2 trusts Domain 3; however, Domain 1 does not trust Domain 3 because it does not have a direct trust relationship.
- **One-way trust:** A one-way trust establishes trust in one direction only. Set this way, Domain 1 trusts Domain 2, but Domain 2 does not trust Domain 1.
- **Two-way trust:** A two-way trust is a bidirectional trust relationship. If Domain 1 trusts Domain 2, then Domain 2 also trusts Domain 1.

Identifying types of trusts

You'll work with four types of trusts in Windows Server 2022. Each has its pros and cons and different use cases:

- **Shortcut trust:** Shortcut trusts are used on Windows Server domains that reside in the same forest, where there is a need to optimize the authentication process. This may happen when a user on Domain A frequently needs to authenticate to Domain B. They're transitive and they can be created as one-way or two-way trusts.
- **Realm trust:** A realm trust allows you to create a trust between a Windows Server domain and a non-Windows (think Linux, Unix, or MacOS Server) Kerberos realm. Realm trusts have a lot of flexibility; they can be transitive or nontransitive and created as one-way or two-way trusts.
- **External trust:** External trusts connect a Windows Server domain in one forest to another Windows Server domain in a different forest. External trusts are nontransitive and can be established as one-way or two-way trusts.

» **Forest trust:** Forest trusts create a trust relationship between two Windows Server forests. They're transitive and can be established as one-way or two-way trusts.

Creating a domain trust

Creating a domain trust is pretty simple. Let's look at a use case and build it out.

Let's say you have your domain in its own forest, and you want to connect to another Windows domain in another forest. You know that you need to set up an external trust, given the types of trusts you can create. Here are the steps to set that up:

1. **From Server Manager, choose Tools⇨Active Directory Domains and Trusts.**
2. **Right-click your domain name and select Properties.**
3. **Click the Trusts tab, and then click the New Trust button.**
4. **On the Welcome to the New Trust Wizard screen, click Next.**
5. **On the Trust Name screen, type the name of the domain you want to establish the trust relationship with.**

   In my case, it's `someotherorg.com`, see Figure 4-1.
6. **Click Next.**

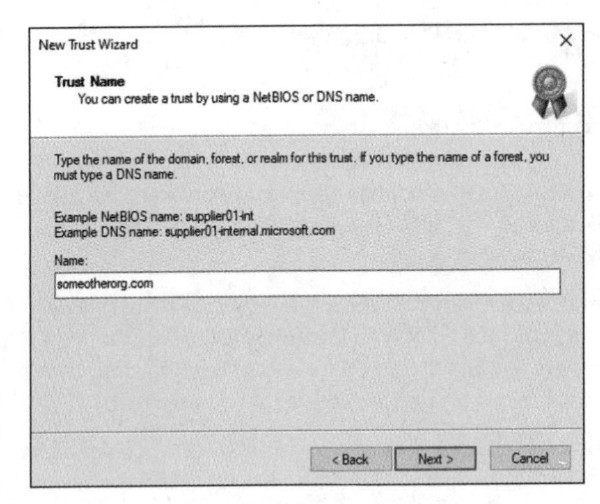

FIGURE 4-1: Setting the trust name for the new trust.

7. On the Trust Type screen, assuming that you're creating a trust between the root domains of each forest, select External Trust (as shown in Figure 4-2), and click Next.

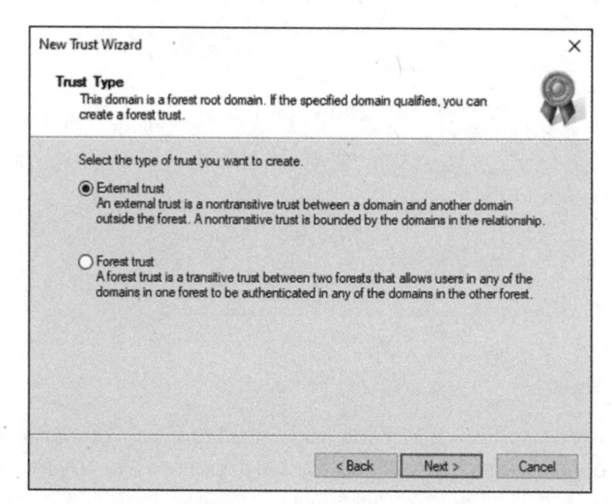

FIGURE 4-2: Setting the type of trust.

8. On the Direction of Trust screen, you can set whether you want the trust to be two-way or one-way; in this case, select Two-Way (shown in Figure 4-3), and click Next.

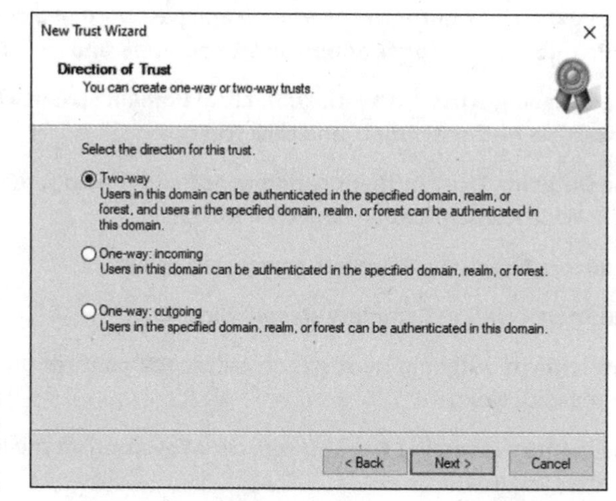

FIGURE 4-3: Setting the direction of the trust.

9. On the Sides of Trust screen, you need to set whether you're setting the trust on your domain or on both domains; for this example, select Both This Domain and the Specified Domain (as shown in Figure 4-4).

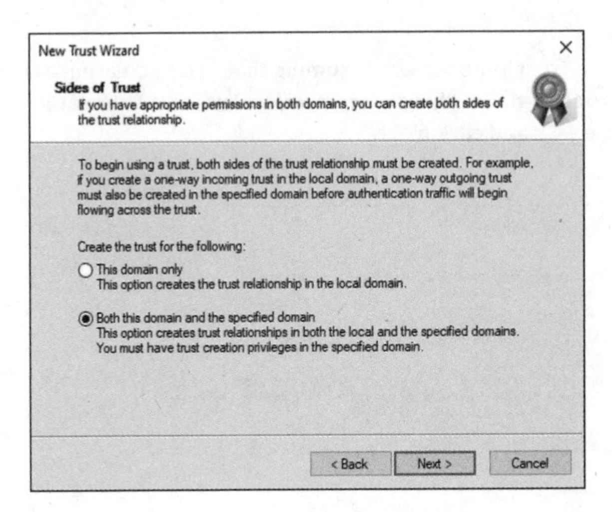

FIGURE 4-4: Setting where you want to set the trust at.

10. Click Next.

TECHNICAL STUFF

You may be wondering why you would need to specify your local domain or both domains after you said you wanted to create a two-way trust. If you have the credentials for both domains (via universal or domain local security groups), then you can choose to create the trust in both. Say you recently acquired a business partner but you don't know the credentials for a domain admin or enterprise admin account. You'd have to choose This Domain Only, and then one of their admins would need to repeat the process on their side to establish a two-way trust.

11. On the next screen, enter the username and password of a domain administrator or enterprise administrator account, and click Next.

12. On the Outgoing Trust Authentication-Local Domain screen, choose Domain-Wide Authentication, and click Next.

13. On the Outgoing Trust Authentication-Specified Domain screen, choose Domain-Wide Authentication, and click Next.

14. On the Trust Selections Complete screen, click Next.

15. On the Trust Creation Complete screen, click Next.

16. On the Confirm Outgoing Trust screen, select Yes, confirm the outgoing trust, and click Next.

17. On the Confirm Incoming Trust screen, click Yes, confirm the incoming trust, and click Next.

If all goes well, you see the Completing the New Trust Wizard screen, shown in Figure 4-5.

18. Click Finish.

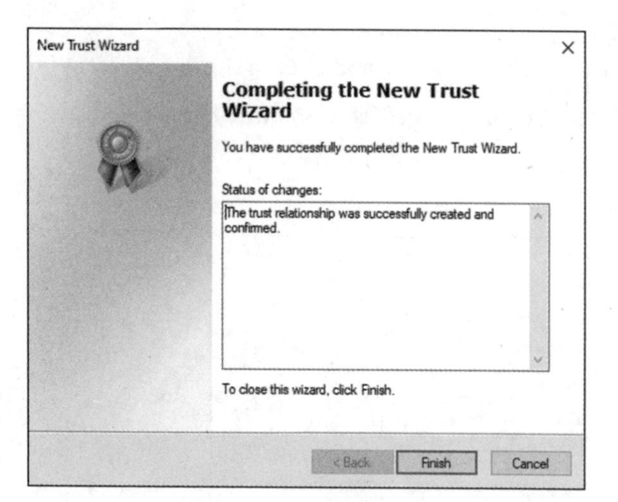

FIGURE 4-5: The successful completion of the New Trust Wizard.

Using Active Directory Sites and Services

Active Directory Sites and Services is very useful when you need to identify separate sites inside your Active Directory domain. This is typically done when a domain is used across multiple physical geographically separate locations.

Understanding what an Active Directory site is

A site in Active Directory is a representation of a physical site. Each site gets associated with an Active Directory domain. Although domains can contain many sites, a site can't be joined to more than one domain.

Sites are useful with geographically distant locations as they can tell clients that are connecting which domain controller is the closest. This minimizes the amount of traffic that needs to flow over the network and can also improve performance since Active Directory replication is able to occur with the closest domain controller.

Creating a site

Now that you know what a site is, you probably want to create your own. Here's how to create a site in Active Directory Sites and Services:

1. **With Server Manager open, choose Tools⇨Active Directory Sites and Services.**
2. **Right-click Sites and choose New Site.**

3. **Fill in the name of the site and select a site link object.**

   In this case, choose the default, as shown in Figure 4-6.

4. **Click OK.**

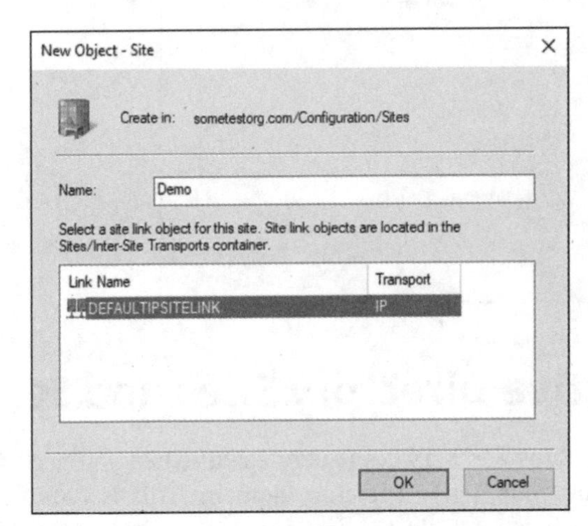

FIGURE 4-6: Creating a site in Active Directory.

Of course, a site is not useful without a subnet assigned to it, so let's do that next:

1. **Right-click Subnets and choose New Subnet.**
2. **Enter the prefix for the site.**

   For my example, I've entered 192.168.178.0/24.

3. **Select the site you just created (in my case, Demo).**

   Your screen should look similar to Figure 4-7 at this point.

4. **Click OK.**

That's all there is to it. You've created your first site in Active Directory!

Using Active Directory Users and Computers

Active Directory Users and Computers is the component of Active Directory that most people are familiar with. With Active Directory Users and Computers, you can manage the users, groups, and other objects that reside inside the Active Directory database. User accounts created in Active Directory Users and Computers are referred to as domain users. By having a domain account, users can log in to multiple resources with the same account. This is in stark contrast to local accounts, where the user would need a separate account on every system and/or application. In this section, I show you a couple of the most common use cases.

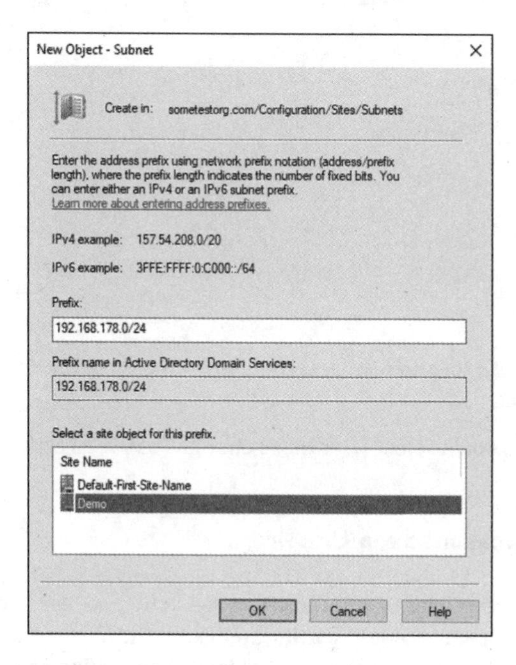

FIGURE 4-7: Configuring a subnet for the new Active Directory site.

Creating users

Creating users is arguably the most common task a system administrator will do in relation to Active Directory. Thankfully, Microsoft has made the process pretty painless:

1. **From Server Manager, choose Tools⇨Active Directory Users and Computers.**
2. **Right-click the OU or container that you want to create the user account in.**

   For this example, I use the default Users container.
3. **Select New and then User.**
4. **Fill in the First Name and Last Name (Full Name will automatically populate with what you use), and select a logon name (in my case, ksmith, as shown in Figure 4-8).**
5. **Click Next.**

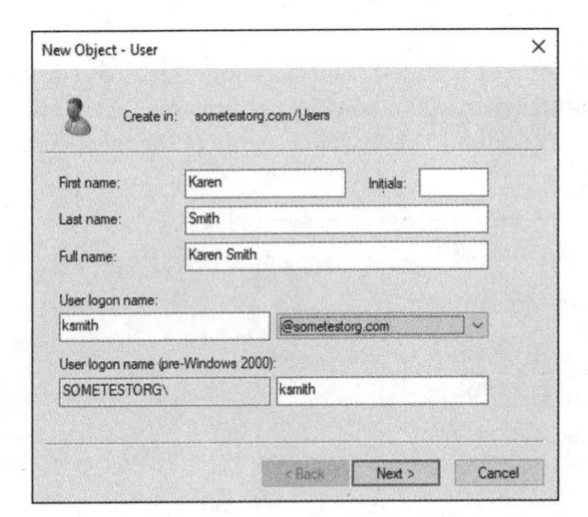

FIGURE 4-8: Creating a new user from within Active Directory Users & Groups.

6. **Type a password for the user and then type it again in the Confirm Password field.**
7. **In most cases, you'll select User Must Change Password at Next Logon.**
8. **Click Next.**
9. **On the confirmation screen, click Finish.**

Now if you look in the Users container, you see your new user account. You can reset the password on the account by right-clicking the user's name and choosing Reset Password. You can disable the account by right-clicking the user's name and choosing Disable Account. Finally, you can configure the more advanced options for the user account by right-clicking the user's name and choosing Properties. This gives you a multitude of selections, like setting organizational information, setting group memberships, and other pertinent information (see Figure 4-9).

Creating groups

Having a domain user account is great because it gives your users the ability to log in to multiple systems or applications with the same account. Giving each user account direct access to resources quickly becomes a nightmare, however, because it's very difficult to keep up. When a user leaves, for example, you have to remove them from every location they had access to. If you use a domain group instead, you can still ensure they get the access they need, but when they leave the organization, you can simply remove them from the group and disable their account.

1. **From Server Manager, choose Tools⇨Active Directory Users and Computers.**

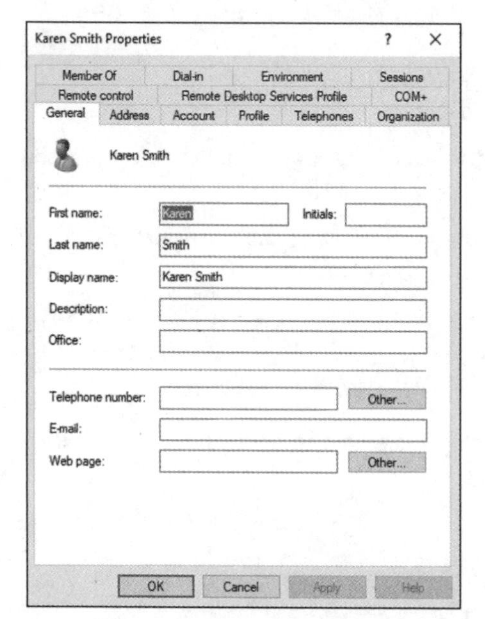

FIGURE 4-9: The Properties screen for the user account gives you more options to configure.

2. **Right-click the OU or container that you want to create the user account in.**

   For this example, I'll use the default Users container.

3. **Select New and then Group.**
4. **Fill in the Group Name field.**

   I simply entered New Group.

5. **Choose the Group Scope.**

   Your options are as follows:

   - **Domain local:** Groups that are only visible within their own domain.
   - **Global:** Visible forest-wide but can only contain accounts and groups from the same domain.
   - **Universal:** Visible forest-wide and can contain accounts and groups from across the forest.

   For this example, I chose Domain Local.

6. **Choose the Group Type.**

   Your options are as follows:

   - **Security:** Used to manage user or computer access to resources on the network.
   - **Distribution:** Used for email lists; can't be used to assign permissions to objects.

   In this case, I chose Security, shown in Figure 4-10.

7. **Click OK.**

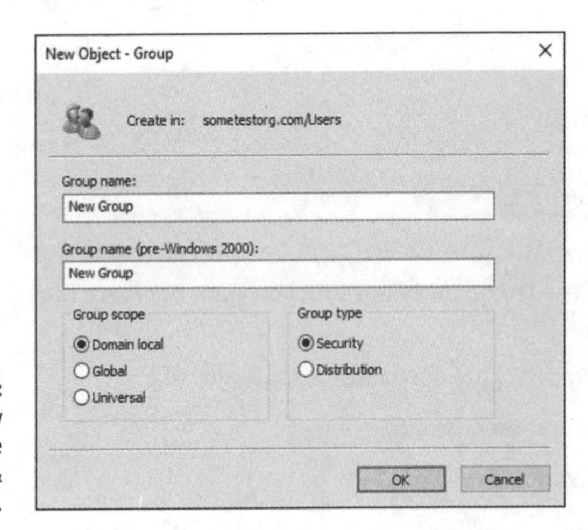

FIGURE 4-10: Creating a new group in Active Directory Users & Groups.

Groups don't really do much without users, so let's assign the new user account you created to the new group that you just created:

1. **Double-click your new group.**
2. **Click the Members tab.**
3. **Click Add, and type the name of the user and click OK.**

   In my case, this is Karen Smith.

4. **Click OK to close the dialog box.**

Using Active Directory Administrative Center

Active Directory Administrative Center is your central management point for Active Directory configuration tasks. You can run queries and build users and groups; you can also change functional levels and other settings. In the following sections, I walk you through a few of these settings.

Enabling Active Directory Recycle Bin

Before the Active Directory Recycle Bin was introduced, there was no nice way to restore a user or a group after it was deleted. You could do a restore from a backup of Active Directory (no small feat), or you could re-create the object you deleted. Restoring Active Directory entailed stopping AD DS, restoring from a system state backup that was taken before the object was deleted, marking the objects as authoritative with the `ntdsutil.exe` utility, and then restarting AD DS.

TECHNICAL STUFF

ntdsutil.exe is a command line tool that allows you to work with Active Directory at a database level.

Microsoft realized that this was a pain point for Active Directory administrators everywhere, so it introduced the Active Directory Recycle Bin. But it isn't enabled by default, so you must plan ahead and enable it. After it's enabled, you can restore deleted items to Active Directory with no downtime. Deleted items aren't kept forever; by default, they're retained for 180 days. A system administrator can adjust this value if desired.

To enable Active Directory Recycle Bin, follow these steps:

1. **From Server Manager, choose Tools⇨Active Directory Administrative Center.**
2. **Select your domain from the left menu.**
3. **Click Enable Recycle Bin from the menu on the right, as shown in Figure 4-11.**

   You see a dialog box warning you that Recycle Bin can't be disabled after it's enabled.
4. **Click OK.**

   You see a dialog box telling you to refresh Active Directory Administrative Center.
5. **Click OK.**
6. **Click the circle with the two arrows inside of it at the top of the screen to refresh.**

   Enable Recycle Bin is now grayed out.

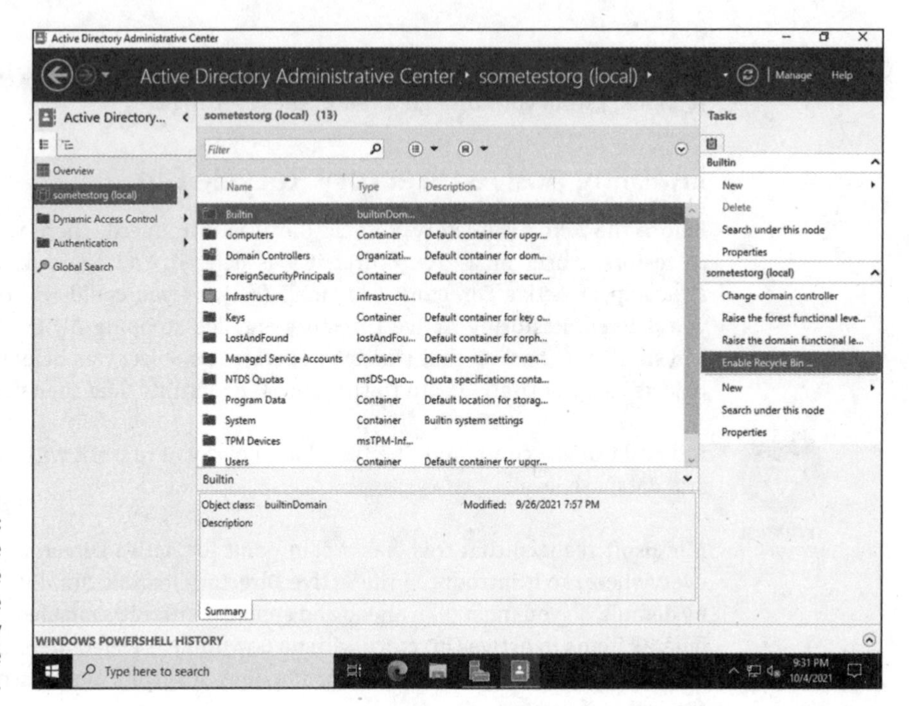

FIGURE 4-11: Enable the Active Directory Recycle Bin through the Active Directory Administrative Center.

Managing users and groups

Managing users and groups within Active Directory Administrative Center is not difficult. Most of the options become available when you select your domain name after you've opened Active Directory Administrative Center. For instance, if you right-click the Users container, you have the option to select New, and then User or Group. If you select User, you find that the user creation screen has a lot more options to work with. Figure 4-12 shows an example of user creation in Active Directory Administrative Center.

Being able to customize all these settings when creating a user is really nice. In the past, when using Active Directory Administrative Center, for example, you had to create the user. When the user was created, you could go into the User Properties screen and make the changes to the organizational tabs. Active Directory Administrative Center's user creation utility lets you make all these settings all at once.

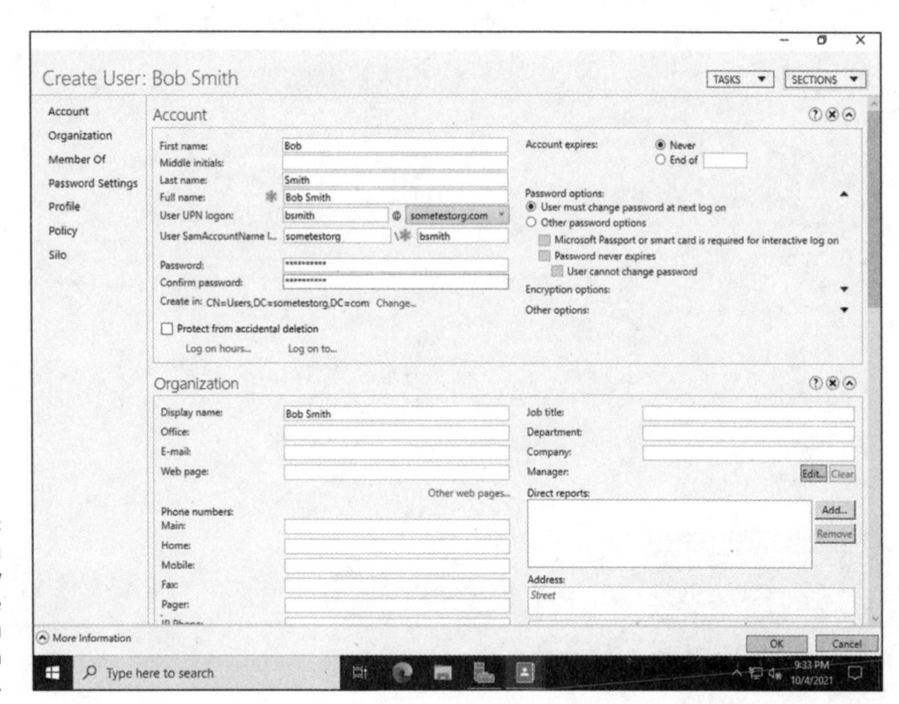

FIGURE 4-12: Creating a user in Active Directory Administrative Center gives you more options in the beginning.

IN THIS CHAPTER

» Activating Windows Server 2022

» Adding and removing standard applications on the server

» Measuring reliability and performance on Windows Server

» Protecting system data from loss or corruption

» Working with Remote Server Administration Tools for server administration tasks

» Managing servers with Admin Center

Chapter 5 Performing Standard Maintenance

This chapter covers the everyday tasks that a system administrator would be expected to know how to do. From simple tasks like activating the license on a new server to working with the brand-new Windows Admin Center, there's a little something in this chapter for everyone.

Activating Windows

When you first install a new operating system on a server, one of the first things you do is activate it. If you're lucky enough to be on a domain with a Key Management Server (KMS), the activation of the server is automatically done for you. You don't need to find the product key and type it into the server. The KMS manages the product keys (it uses Multiple Activation Keys [MAKs]) and activations for you. If you aren't lucky enough to have a KMS, then you'll need to manually activate the server. Here's how to activate a server via the graphical user interface (GUI) and via the command line.

Through the graphical user interface

Activating Windows Server 2022 from the GUI is very simple. It starts in my favorite place: Server Manager! Here's how to activate Windows Server 2022 through the Server Manager console:

1. **Starting in Server Manager, click Local Server in the left-hand side of the menu.**
2. **To start the activation process, click the Not Activated hyperlink next to Product ID.**

   A dialog box will launch automatically asking for the product key.
3. **Enter your product key and click Next.**

   You're prompted to activate Windows.
4. **Click Activate.**

   You get a confirmation that Windows has been activated.
5. **Click Close.**

   You're left on the Activation screen shown in Figure 5-1, where you see that your version of Windows is now activated.

Through the command line

You can activate your copy of Windows Server via the command line as well. You may want to do this because you prefer to be a Command Prompt ninja, or maybe because you have Server Core installed and don't have a choice.

Either way, first, you have to install the key. You do this with the Windows Server License Manager script, slmgr.vbs.

To use slmgr.vbs to install the key, you use the command with the parameter `-ipk`. Just replace *<productkey>* with your 25-character license key, including the dashes:

```
slmgr.vbs -ipk <productkey>
```

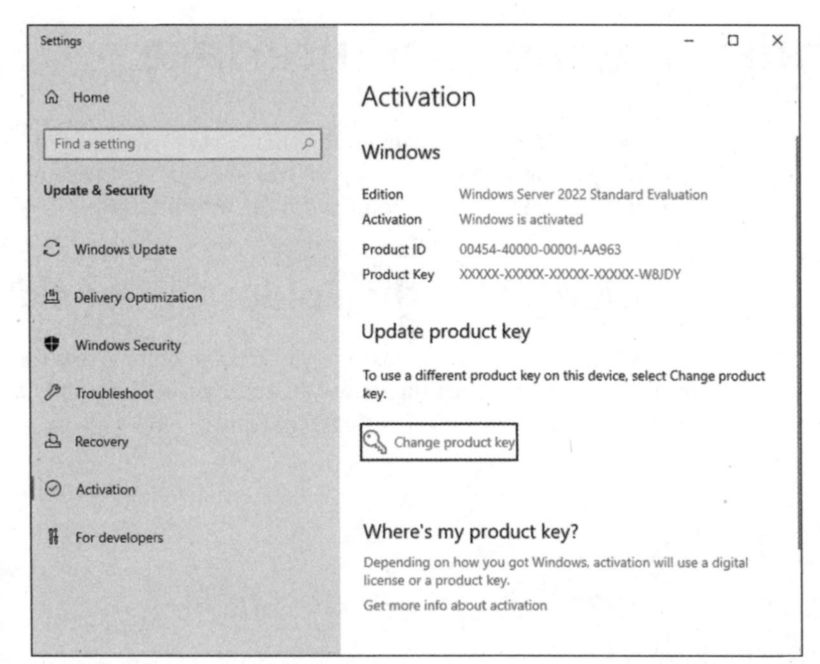

FIGURE 5-1: The activation window showing that Windows Server 2022 is activated.

You get a dialog box that tells you the product key installed successfully. Click OK.

After the license key is installed, you use the same script with the `-ato` parameter to do an online activation of your copy of Windows. You do that with the following command:

```
slmgr.vbs -ato
```

If the activation was successful, you get a dialog box that says the product was activated successfully, as shown in Figure 5-2.

Windows Script Host

Activating Windows(R), ServerStandardEval edition (c1a197b6-ba5e-4394-b9bf-b659a6c1b873) ...
Product activated successfully.

OK

FIGURE 5-2: Using slmgr.vbs to activate Windows Server.

Configuring the User Interface

The user interface can be something that is very particular to a person. You may like things to be set a certain way. In this section, I cover some of the configuration options you have for working with the user interface.

Working with the Folder Options dialog box

The Folder Options dialog box lets you change quite a few things. To access it, open File Explorer, click the View tab, and then click Options and select Change Folder and Search Options (as shown in Figure 5-3).

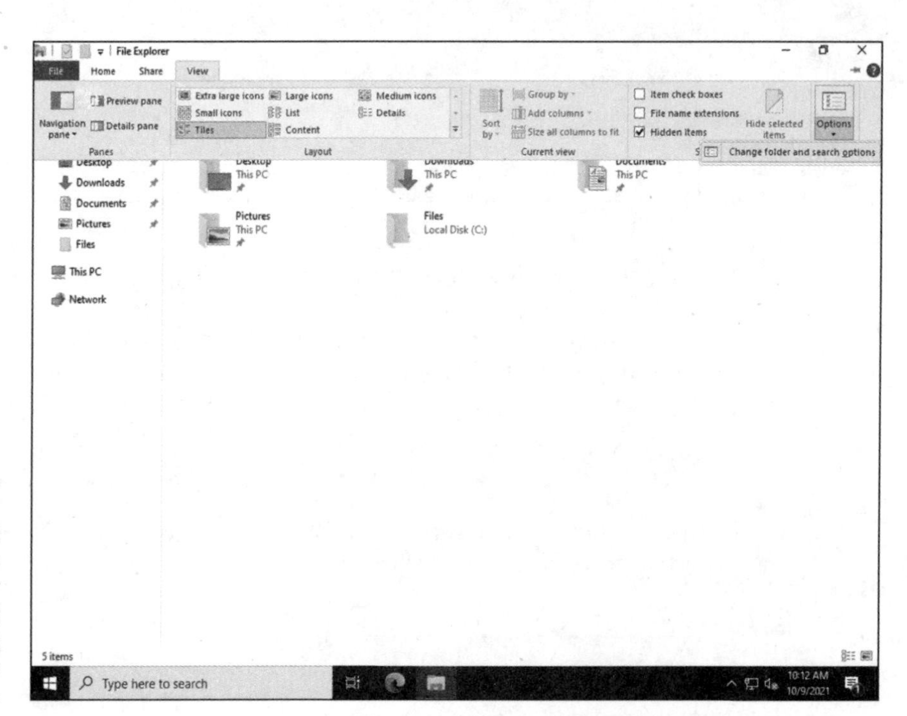

FIGURE 5-3: You can access the Folder Options dialog box through the View tab in File Explorer.

The Folder Options dialog box has three tabs: General, View, and Search.

The General tab

The General tab (shown in Figure 5-4) lets you adjust the general behavior of folders. You can set File Explorer to open by default to Quick Access (default) or to This PC. You have the choice when you open folders to make them open in the same window (default) or a new window. Tired of double-clicking to open a folder? Change it to a single-click. You can even set Quick Access to display frequently used folders (default), frequently used files, or both.

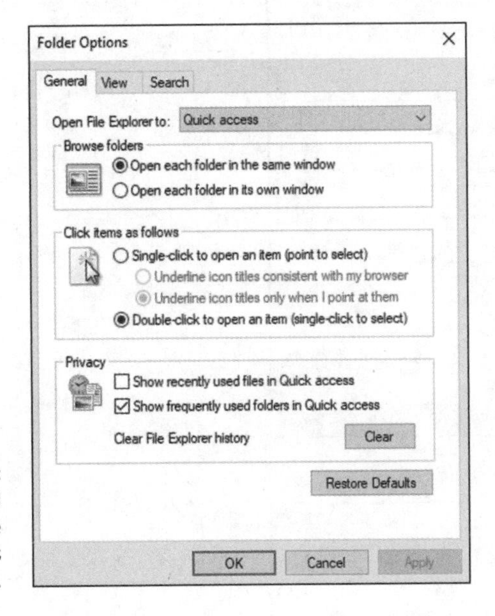

FIGURE 5-4: The General tab of the Folder Options dialog box.

The View tab

On the View tab (shown in Figure 5-5), you can make changes to how folders, and the items in folders, display. One of the most common changes I make on the View tab is to select Show Hidden Files, Folders, and Drives and Hide Extensions for Known File Types. With these two options set that way, I can view hidden folders like the Default user, and hidden files like `ntuser.dat`. And by not hiding extensions, it makes it easier to differentiate between application and config files when they have the same name.

Notice in the background of Figure 5-5 that you can see the file extensions and that the hidden file, which was previously not visible (in Figure 5-4), is now visible with a transparent icon.

The Search tab

The Search tab (shown in Figure 5-6) controls the behavior of the search function when searching through folders and files on the operating system. By default, the only choice selected here is Include System Directories in the When Searching Non-indexed Locations section. Other options include searching compressed files, file names, and contents. And you can tell Windows Search not to use the index during its search. Use that last option with caution — the process could take a while to complete depending on how much data it needs to go through.

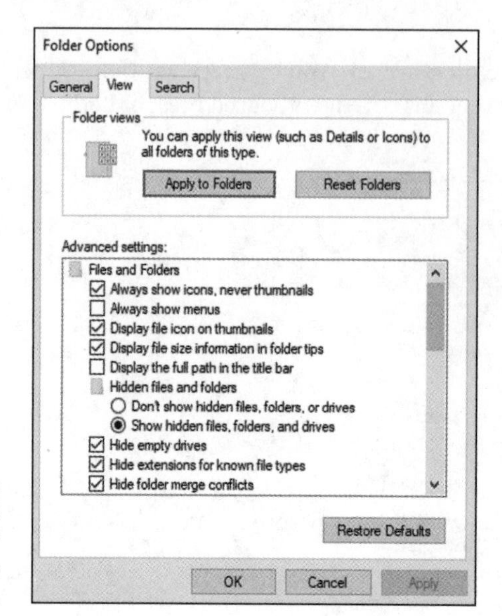

FIGURE 5-5: The View tab of the Folder Options dialog box.

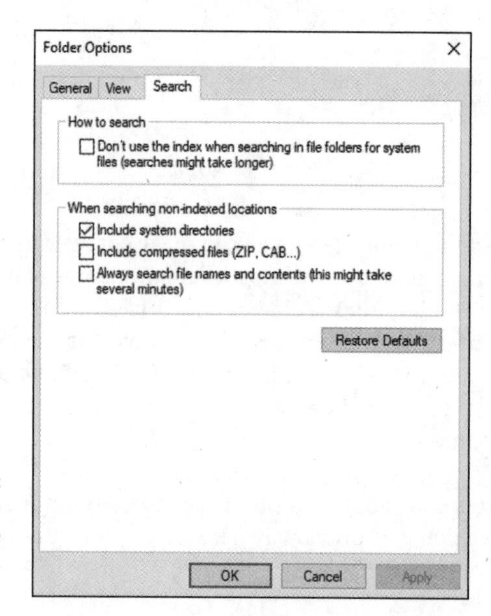

FIGURE 5-6: The Search tab of the Folder Options dialog box.

Setting your Internet Options

You can set your Internet Options in two ways: through the Control Panel, or you through the Network & Internet screen.

I'll start with the Control Panel method. Follow these steps:

1. **Click Start, click Windows System, and then click Control Panel.**
2. **Click Network & Internet, and then click Internet Options.**

   The Internet Properties dialog box, shown in Figure 5-7, appears.

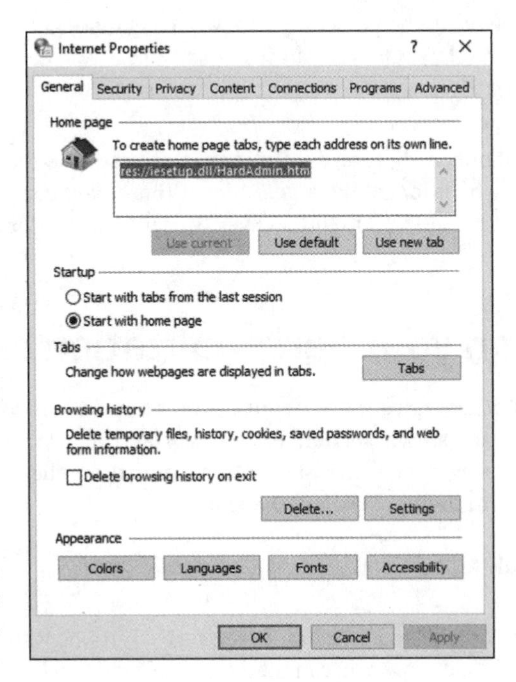

FIGURE 5-7: The Internet Properties dialog box.

There are seven tabs on the Internet Properties dialog box:

- **General:** The General tab allows you to set your own home pages, specific the startup behavior for Internet Explorer, define how you want web pages to be displayed in tabs, delete your browsing history, and tweak items related to appearance.
- **Security:** The Security tab allows you to set the security level for individual zones — the Internet, the Local Intranet, Trusted Sites, and Restricted Sites.
- **Privacy:** Allows you to specify if websites are allowed to track your location, how you want the pop-up blocker to function, and whether you want InPrivate mode to disable toolbars and extensions.
- **Content:** The Content tab focuses on various content-related features. You can work with certificates, control how AutoComplete functions, and look at the settings for any RSS-related feeds.

- **Connections:** Allows you to set up an Internet connection and specify a proxy within the local area network (LAN) settings area.
- **Programs:** The Programs tab lets you set default behaviors for clicking links and editing HTML files. It also lets you manage any add-ons you have installed and set programs that you want to use for things like email. You can set the file types that you want Internet Explorer to open as well.
- **Advanced:** The Advanced tab gives you a way to set more granular settings within Internet Explorer (IE). You can customize how IE will react to scripts, change which cryptographic settings you want it to use, and more.

The newer method of setting Internet Options is accessed via the Start menu, and the Settings icon, which looks like a gear. From there, you can select Network & Internet from the new Settings menu. I cover this area in greater detail in Book 4, Chapter 1.

Focusing on your Personalization settings

Personalization settings have gotten a lot of attention in Windows Server 2022. In fact, they have their own section in the new Settings menu. To access the Personalization settings, click the Start Menu, and then click the gear icon to access Settings. From there, click Personalization.

There are several different locations you can customize:

- **Background:** This section allows you to set the background with a picture, a solid color, or a slide show. You can set the background to fill, fit, stretch, tile, center, or span.
- **Colors:** In this section, you choose what your accent colors will be for things like your Start menu. You can choose specific colors, or you can select Automatically Pick an Accent Color from My Background, and new colors will be set anytime the background color changes. You can set where the accent color shows up, and you can select a light or dark app mode.
- **Lock Screen:** Here, you can set a custom image on your lock screen. You can choose from stock photos or choose your own.
- **Themes:** Much like in Windows 10, you can set themes in Windows Server 2022. These themes affect the background, colors, sounds, and what your cursor looks like.

- **Fonts:** The Fonts section allows you to search, view, and install new fonts. If you like a particular font, you can click it to see the font faces that are available, as shown in Figure 5-8 with the Courier New font.
- **Start:** The Start section allows you to customize what appears on your Start menu. You can choose to show more tiles or apps, or even let the Start screen take over the entire screen (as in Windows 8).
- **Taskbar:** The Taskbar section allows you to change the behavior of the taskbar. You can lock it or hide it, and you can choose to replace Windows PowerShell with the Command Prompt when you right-click the Start menu.

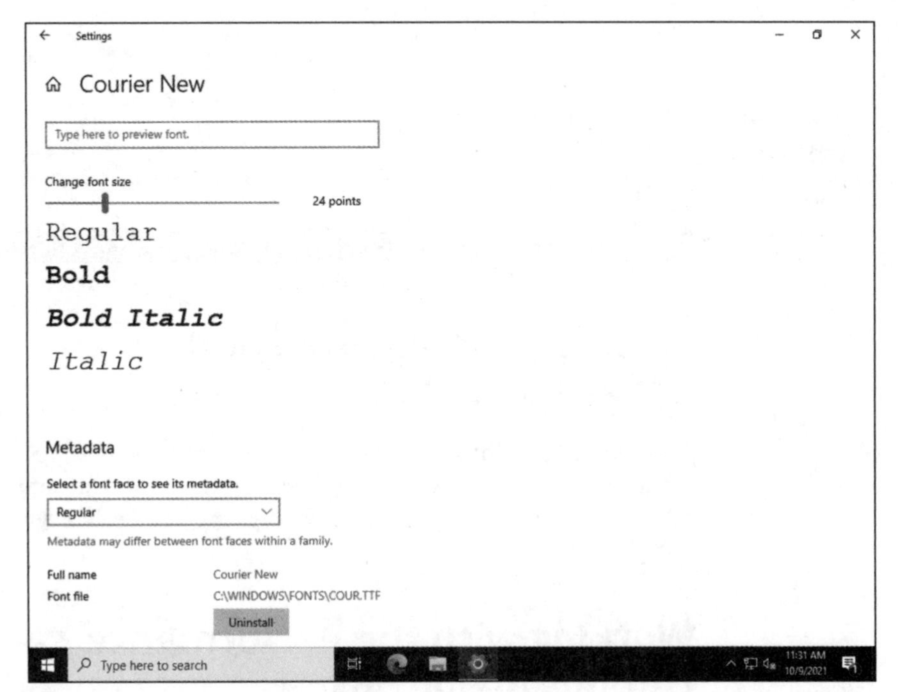

FIGURE 5-8: Viewing the available font faces for the Courier New font.

Reporting problems

The troubleshooting utility is still located in the Control Panel. To access the Control Panel, click Start, click Windows System, and then click Control Panel. From there, click System and Security, and then click Review Your Computer's Status and Resolve Issues (in the Security and Maintenance section). Potential issues are broken out into Security and Maintenance issues, as shown in Figure 5-9.

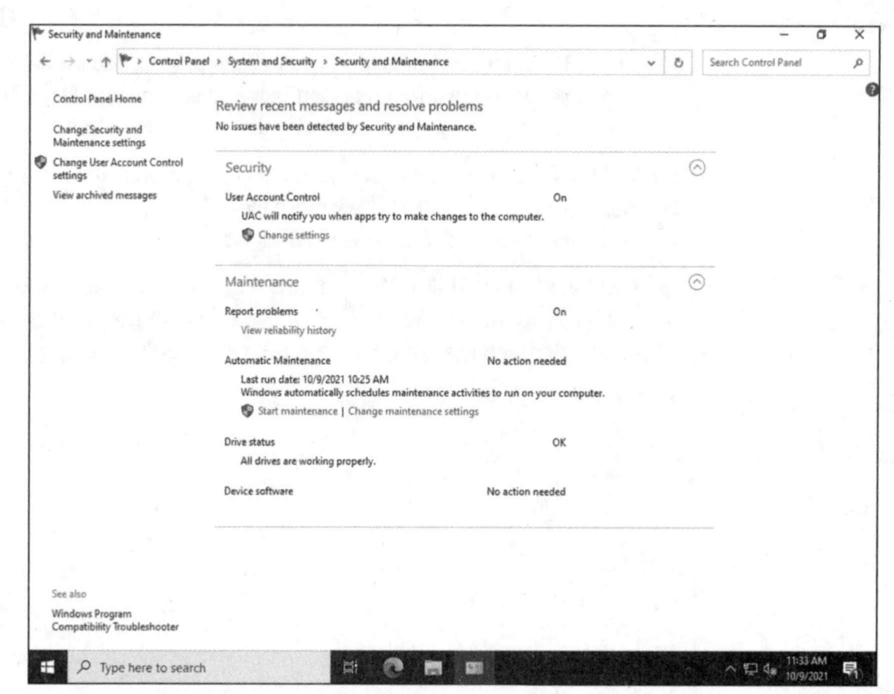

FIGURE 5-9: The troubleshooting utility can notify you of issues and possible resolutions.

Setting your Regional and Language Options

The Regional and Language Options are set in the new Settings menu under Time & Language. You can reach this by clicking the Start menu and then clicking the gear icon to access Settings. This allows you to set the date and time, the region, the language, and speech settings.

Working with the Performance Options dialog box

The Performance Options dialog box is pretty straightforward.

There are two ways to get there:

- Click Start, then click Windows System, and then click Control Panel. From there, choose the System and Security section, and then click System. Click Advanced System Settings. On the Advanced tab, click the Settings button under Performance.

» Click the Start menu, and then click the gear icon to access the Settings menu. Click System, click About, click System Info, and click Advanced System Settings. On the Advanced tab, click the Settings button under Performance.

Either way, you get a screen similar to Figure 5-10.

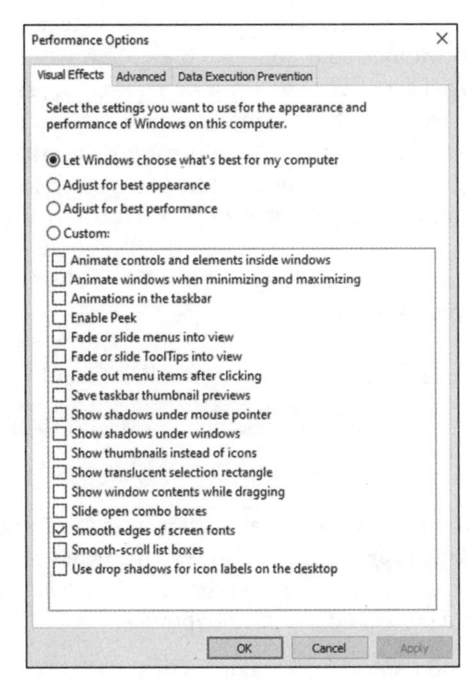

FIGURE 5-10: The Performance Options dialog box.

There are three tabs in the Performance Options dialog box:

» **Visual Effects:** Allows you to adjust the amount of visual effects that Windows uses. On a server, this should normally be set for performance rather than flashy screen transitions.

» **Advanced:** Allows you to set whether performance should target better performance for programs or background services, and allows you to adjust the amount of virtual memory being used.

» **Data Execution Prevention:** Data Execution Prevention (DEP) is a great feature for protecting your system against malicious code, but it can cause performance issues with some applications. Typically, your vendor will let you know if DEP needs to be disabled, and you can whitelist that program.

Understanding How User Access Control Affects Maintenance Tasks

User Account Control (UAC) was designed to protect your system against unauthorized changes. It requires certain types of changes to get administrator approval before they can execute. This ensures that malware and other rogue applications can't run without your consent. It can also be incredibly annoying when you're simply trying to get your work done. It prompts you for approval anytime a program wants to make a change on your system. If you're logged in as a standard user, you must provide an administrator's username and password or PIN. If you're logged in as an administrator, you must select yes or no. Most maintenance tasks require some level of privilege elevation, so you'll most likely run into this at some point. You can tell if you'll get prompted by looking at the application you want to run. If the program's icon has a shield icon in its lower-right corner, it will prompt for elevation.

There are four levels of settings within UAC, ranked from least secure to most secure:

- **Never Notify:** This is essentially turning UAC off. Don't do this.
- **Notify Me Only When Apps Try to Make Changes to My Computer (Do Not Dim My Desktop):** This notifies you when programs are trying to make changes to your system. You can continue to do things in the background without acknowledging the UAC prompt.
- **Notify Me Only When Apps Try to Make Changes to My Computer (Default):** This will notify you when programs are trying to make changes to your system. You must answer yes or no (assuming you're logged in as admin) before you'll be allowed to continue. If you aren't logged in as admin, you need to use an administrator account and password at the prompt.
- **Always Notify:** This is the most secure setting, but it's also the most annoying because it will prompt you any time an application or user wants to make a change that needs administrator permissions.

If you want to make changes to UAC on an individual system, you can go through the Control Panel to System and Security, Security and Maintenance, and then Change User Account Control Settings, shown in Figure 5-11. If this is being controlled via Group Policy, the options in the image may be grayed out.

For more information on UAC, see Book 5, Chapter 3.

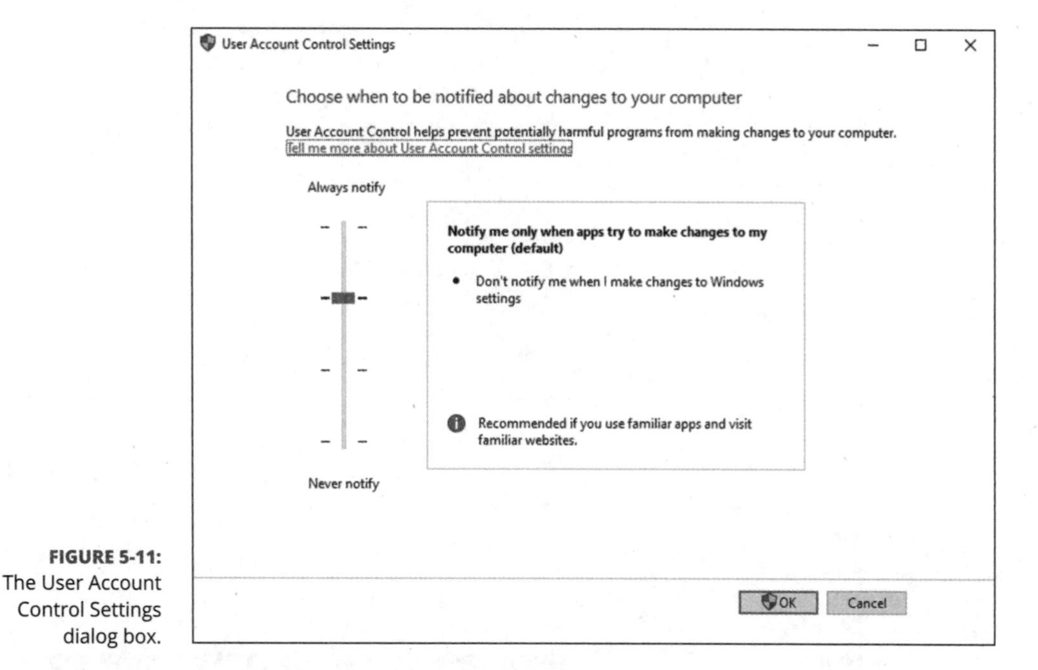
User Account Control Settings
Choose when to be notified about changes to your computer
User Account Control helps prevent potentially harmful programs from making changes to your computer.
Tell me more about User Account Control settings
Always notify
Notify me only when apps try to make changes to my computer (default)
Don't notify me when I make changes to Windows settings
Recommended if you use familiar apps and visit familiar websites.
Never notify
OK
Cancel

FIGURE 5-11: The User Account Control Settings dialog box.

Adding and Removing Standard Applications

Windows Server 2022 much like Windows 10 can take advantage of the Windows App Store for installing applications. You can also install them in the traditional way (by double-clicking the installer package that you downloaded or got from disc or flash media).

Removing apps can be done from one central place. Click Start, and then click the gear icon to access Settings. From there, click Apps. To remove an application, all you need to do is select it, and then click the Uninstall button, as shown in Figure 5-12.

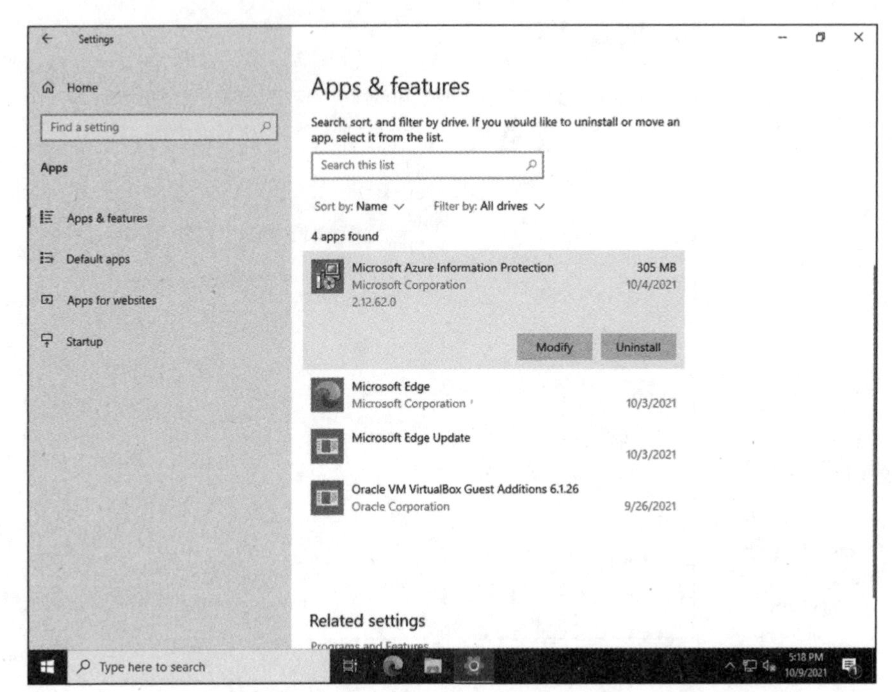

FIGURE 5-12: Uninstalling a program in the Apps & Features section of the Settings menu.

Measuring Reliability and Performance

As a System Administrator, you'll have many days that start with complaints of "The server is sooooo slow." If this happens, always blame the networking team. Just kidding.

Windows Server 2022 like many of its predecessors has several utilities that allow you to measure how well your system is performing and find potential bottlenecks or resource constraints. The most common tools are Performance Monitor, Resource Monitor, and Task Manager.

Performance Monitor

Performance Monitor allows you to gather information in real time and collect that information into a log for later review. It looks at what impact running applications on your system have to your system's overall performance. You can access Performance Monitor from Server Manager by clicking Tools and then clicking Performance Monitor.

Performance Monitor is the perfect tool to track basic performance metrics across your central processing unit (CPU) and random access memory (RAM), although it contains quite a few more counters than just CPU and RAM. You can monitor performance for most of the hardware, software, and services running on your system.

In Figure 5-13, I've started Performance Monitor with the default % Processor Time and have added Interrupts/sec.

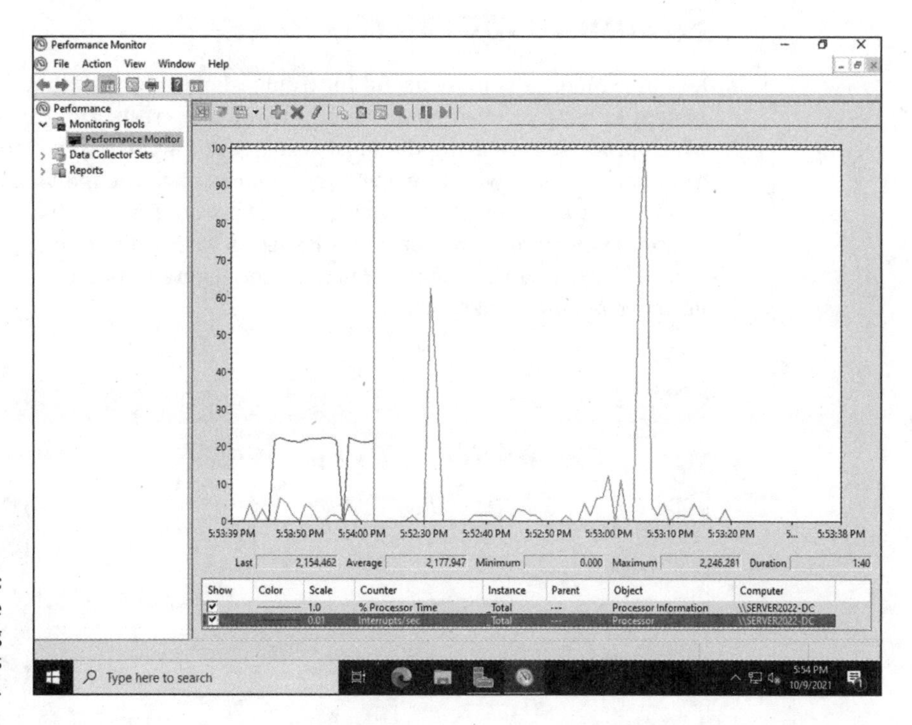

FIGURE 5-13: Performance Monitor running on Windows Server 2022.

One of the best things about Performance Monitor is that you can schedule collection. Say, for instance, that there is always a slowdown on your system between 10 p.m. and midnight. You can schedule performance monitor to run and analyze the data to determine where the issue lies and then see the results when you come in the next morning.

One of the downsides of Performance Monitor is how granular the counters are. It can be difficult to locate the counter that will give you the information that you need, so you may need to play with it a bit to find the right counter.

To add a counter to Performance Monitor, follow these steps:

1. **Open Performance Monitor and click Performance Monitor.**
2. **Click the plus (+) sign in the menu bar.**

3. **On the Add Counters screen, expand Processor and select Interrupts/Sec.**
4. **With Interrupts/Sec highlighted, click the Add button and then click OK.**
5. **Double-click your new counter, select a color different enough from the default counter's color, and then click OK.**

Resource Monitor

Resource Monitor is more useful for troubleshooting issues that are causing your system to hang or even potentially causing it to crash. You can see at a glance which processes are using the most resources or are using the largest amount of disk, network bandwidth, or memory. You can click the individual tabs for CPU, Memory, Disk, or Network to get more details on each. I actually like using this utility to troubleshoot performance issues as well. I find that it presents performance issues in a very helpful manner. See Figure 5-14 for an example of what Resource Monitor looks like.

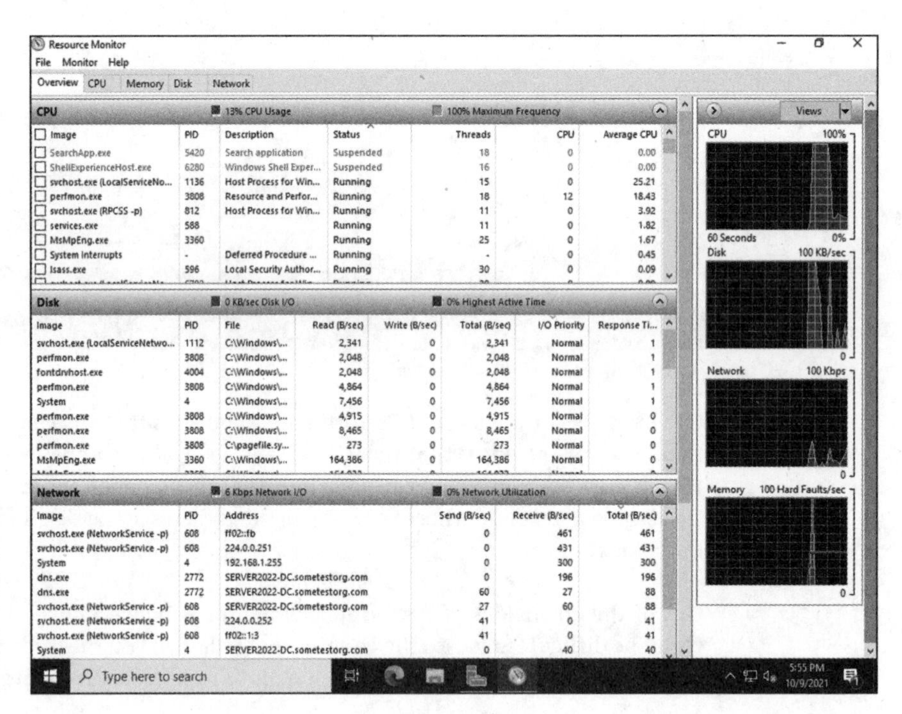

FIGURE 5-14: Troubleshooting resource contention is simple with Resource Monitor.

Task Manager

Task Manager has been around in the Windows operating system for a very long time. It provides a quick glance at CPU, network, and memory utilization, and it allows you to stop running processes or applications ungracefully if they aren't responding.

There are two ways to access Task Manager:

- Right-click the taskbar and choose Task Manager.
- Press Ctrl+Alt+Del and choose Task Manager from that screen.

After you open Task Manager, you see a screen similar to Figure 5-15.

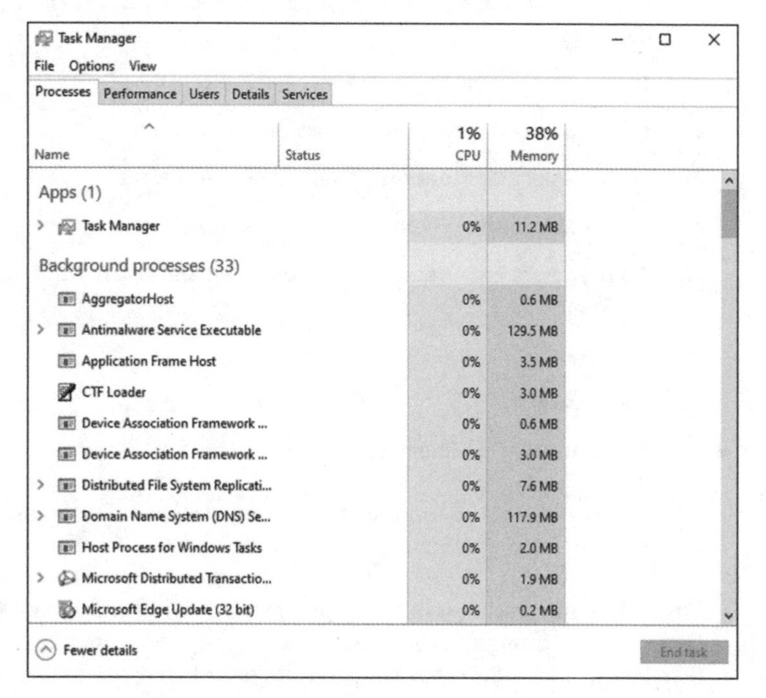

FIGURE 5-15: Task Manager in Windows Server 2022.

Most System Administrators come into Task Manager with one thing in mind. They have an application or a process that isn't responding and they need to terminate it. Here's how to do that in Task Manager:

1. **With Task Manager open, select the unresponsive application.**
2. **Click End Task.**

   The application or process will be stopped.

WARNING

Terminating programs through Task Manager is an ungraceful shutdown. It may lead to data loss and should be used as a last resort.

Protecting the Data on Your Server

One of the many jobs a System Administrator is tasked with is protecting a company's critical assets. One of the most critical assets is almost always going to be data. Windows Server 2022 provides an optional utility that can be used to back up and restore data, known as Windows Server Backup.

Before you can use Windows Server Backup, you have to install the feature. Follow these steps:

1. **From Server Manager, click Manage and then click Add Roles and Features.**
2. **On the Before You Begin screen, click Next.**
3. **On the Select Installation Type screen, click Next.**
4. **On the Select Destination Server screen, click Next.**
5. **On the Select Server Roles screen, click Next.**
6. **On the Select Features screen, scroll down and select Windows Server Backup.**
7. **Click Next.**
8. **Click Install.**
9. **When installation finishes, click Close.**

Windows Server Backup (once installed) can be accessed through Server Manager. Simply choose Tools⇨Windows Server Backup.

TECHNICAL STUFF

There is an interesting issue with the install of Windows Server Backup that has been around through several revisions of the server operating system: You've installed Windows Server Backup, but it doesn't show up in the Tools menu. The fix: Go back to Add Roles and Features, and on the Select features screen, expand Remote Server Administration Tools and choose Network Load Balancing. I have no idea why installing the Remote Server Administration Tools for Network Load Balancing makes this show up. I experimented with several features and Network Load Balancing is the only one that resolved this issue.

System Backup

After you've installed Windows Server Backup, you can launch the console from Server Manager. Choose Tools⇨Windows Server Backup. You're be presented with a screen similar to Figure 5-16.

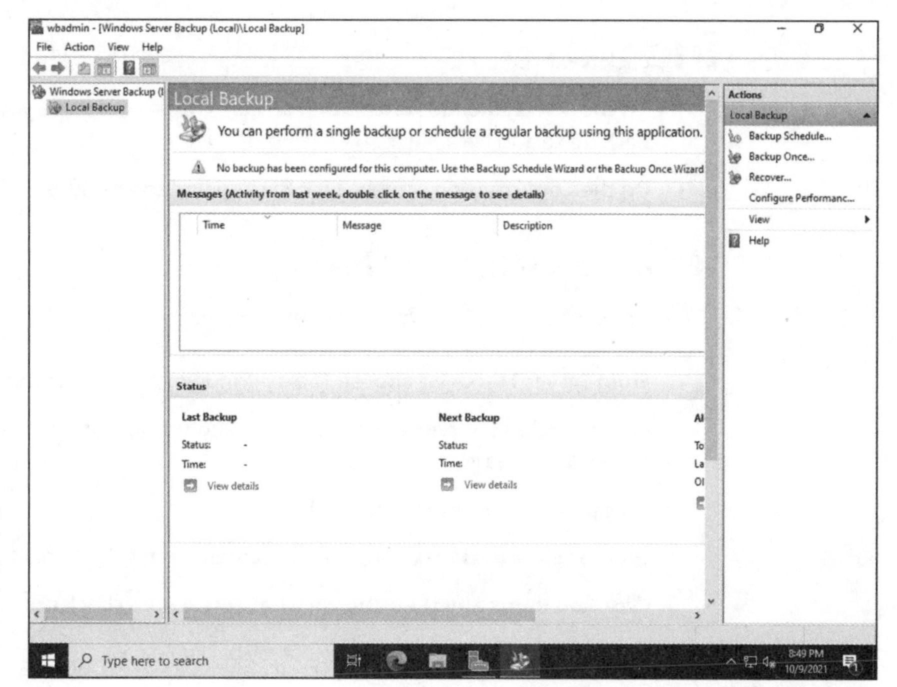

FIGURE 5-16: Windows Server Backup allows you to create and run backups, once or on a schedule.

Creating a one-time backup

There may be times when you want a recent backup because you're about to do some maintenance or upgrade some software. In these instances, the scheduled backup may not be recent enough to suit your purposes. If this happens, a one-time backup is the best solution. Follow these steps:

1. **From Server Manager, choose Tools⇨Windows Server Backup.**
2. **In the Windows Server Backup console, choose Backup Once on the right side of the screen.**
3. **On the Backup Options screen, select Different Options and click Next.**
4. **On the Select Backup Configuration screen, select Custom and click Next.**
5. **Click Add Items and then choose what you want to back up.**

   I'm going to choose everything except my backup drive.
6. **Click OK and then click Next.**
7. **On the Specify Destination Type screen, you can select any of these — just don't save the backup to the same drive you're backing up.**

   I'm going to choose my local F: drive, so I'll choose Local Drives.

8. **Click Next.**
9. **On the Select Backup Destination screen, choose the drive you want to back up to and click Next.**
10. **On the Confirmation screen, click Backup and the backup will start.**

Creating a scheduled backup

To create a scheduled backup, follow these steps:

1. **From Server Manager, choose Tools➪Windows Server Backup.**
2. **In the Windows Server Backup console, choose Backup Schedule on the right side of the screen.**
3. **On the Getting Started screen, click Next.**
4. **On the Select Backup Configuration screen, select Custom and click Next.**
5. **Click Add Items and then choose what you want to back up.**

   I'm going to choose everything except my backup drive.
6. **Click OK and then click Next.**
7. **On the Specify Backup Time screen, select how often you want the backups to run.**

   I'm going to select Once a Day and 9 PM.
8. **Click Next.**
9. **On the Specify Destination Type screen, you can select any of these — just don't save the backup to the same drive you're backing up.**

   I'll select Backup to a Hard Disk That Is Dedicated for Backups.
10. **Click Next.**
11. **On the Select Destination Disk screen, select the disk that you want to back up to.**

    If it isn't visible, click Show All Available Disks and select it from there.
12. **After you select your backup disk, click Next.**

    You're presented with a warning that the disk will be reformatted when you finish the wizard.
13. **Click Yes.**
14. **On the Confirmation screen, click Finish.**

System Restore

Restoring items is nice and simple through Windows Server Backup. Here's how to recover a file that was deleted:

1. **From Server Manager, choose Tools⇨Windows Server Backup.**
2. **On the right-hand menu, click Recover.**
3. **On the Getting Started screen, select where the backup is stored.**

   In my case, I'll choose This Server.
4. **Click Next.**
5. **Choose the available backup that you need.**

   This will be done with a calendar that has available dates, and a drop-down box with times.
6. **Click Next.**
7. **Select what you want to recover.**

   In my case, I deleted a file and a folder, so I'm going to choose Files and Folders.
8. **Click Next.**
9. **Browse the tree until you find what you're looking for.**

   In my case, that looks like Figure 5-17.
10. **Click Next.**
11. **On the Specify Recovery Options screen, select the recovery destination and how the restore should react if it finds items with the same name in the destination.**

    In my case, I'll choose Original Location and Overwrite Existing Versions.
12. **Click Next.**
13. **On the Confirmation screen, click Recover.**

    You can watch the progress of your recovery.
14. **When the recovery is complete, click Close.**

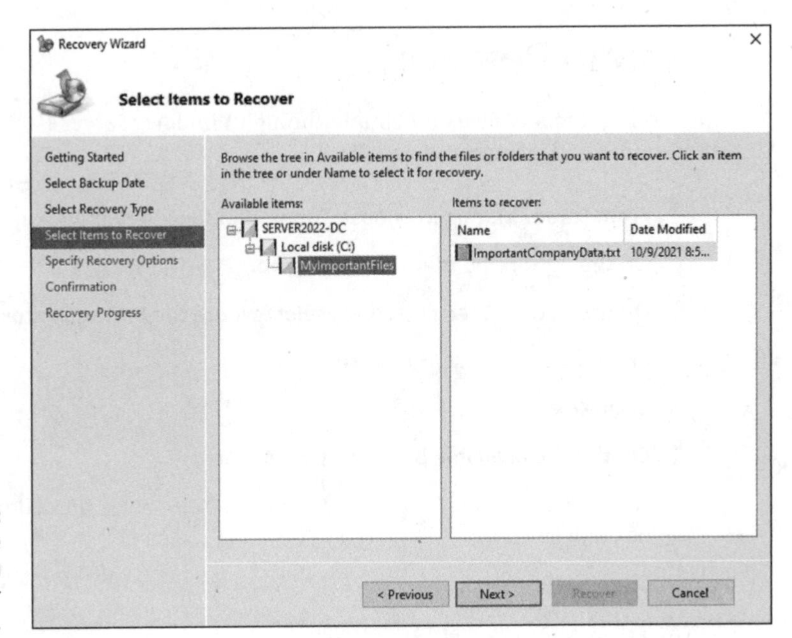

FIGURE 5-17: Restoring some files through Windows Server Backup.

Performing Disk Management Tasks

Managing server storage is an important task. So much so that larger organizations will typically have storage administrators in addition to system administrators. As a system administrator, you should have a good working grasp of how to manage and maintain the storage on your servers.

Managing storage

Much of the storage management tasks in Windows Server 2022 can be performed from the File and Storage Services area in Server Manager. From here, you can manage the volumes, disks, and storage pools associated with your server. If you would like to learn more about configuring and managing storage, check out Book 2, Chapter 2.

Managing disks

Managing the disks on your system is done through the File and Storage Services area in Server Manager. After you click File and Storage Services, just click Disks and you can manage the physical and logical disks, as well as the related volumes. By right-clicking the drive, you can create a new volume, bring an offline disk online, or take an online disk offline, or you can reset the disk. If you right-click the volume, you have a lot of options similar to the old Disk Management utility (which you could still use if you wanted to). You can see these options in Figure 5-18.

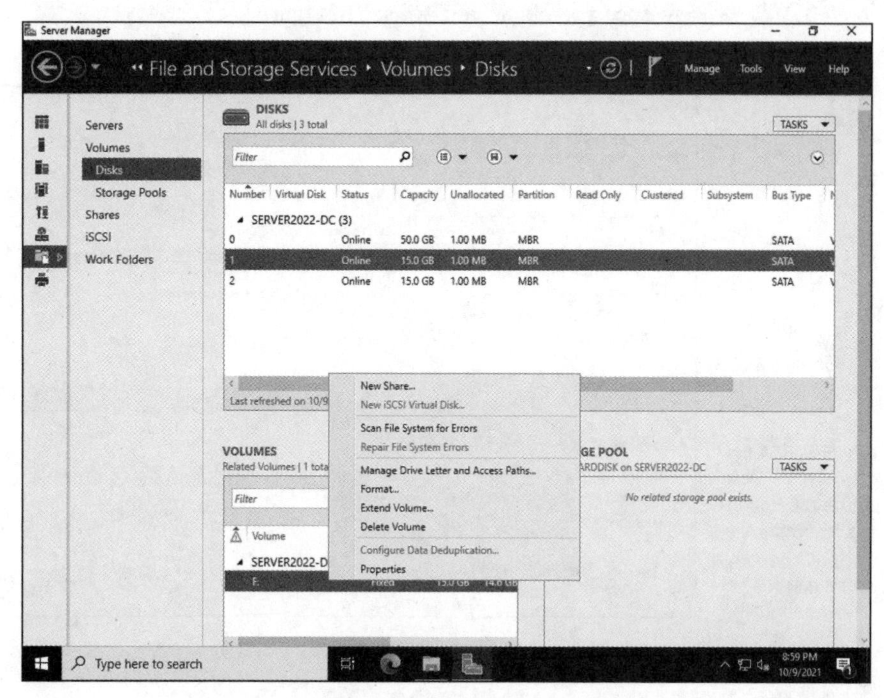

FIGURE 5-18: Managing a disk volume in File and Storage Services.

Defragmenting drives

As a disk is used and data is created and deleted, data on the disk becomes fragmented. This can cause performance issues on hard disk drives (HDDs). You'll need to defragment your disk to resolve this issue. By defragmenting your HDD, you're taking all the scattered data and combining it so that it occupies a single contiguous space. Solid-state drives (SSDs) can benefit from the optimization as well if they support TRIM, which allows the operating system to tell the SSD drive when blocks of data are no longer needed. To defragment or optimize your drive, follow these steps:

1. **From Server Manager, choose Tools➪Defragment and Optimize Drives.**
2. **Select the drive that is fragmented and click Analyze.**

   The analysis will return with recommendations on whether you need to optimize it.

3. **If needed, select Optimize and the drive will be defragmented.**

Disk defragmentation used to be something that a System Administrator had to do manually, but in recent server operating systems, it's scheduled automatically to run once a week, as you can see in Figure 5-19.

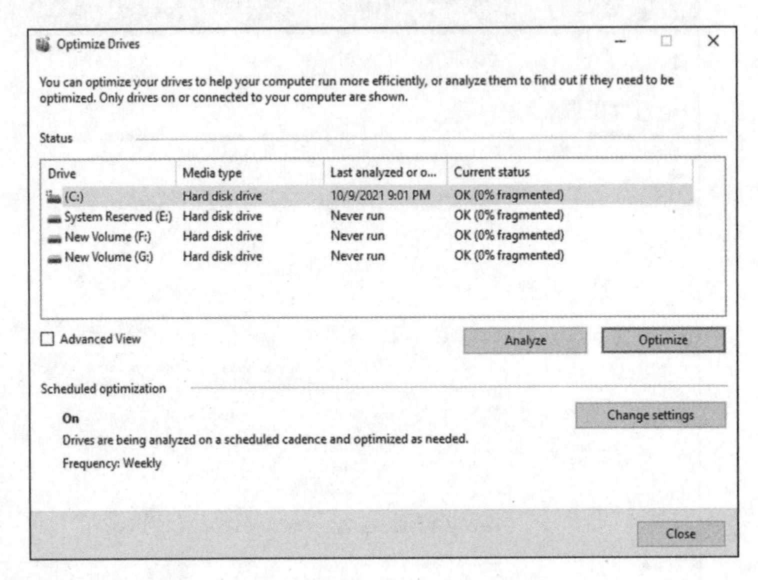

FIGURE 5-19: Disk defragmentation runs automatically once a week, but it can still be run manually.

Automating Diagnostic Tasks with Task Scheduler

For the System Administrator, some routine tasks need to get done, but they're either inconvenient to manually perform or monotonous. For those kinds of tasks, Task Scheduler is a great way to automate maintenance tasks so the you no longer have to remember to log on to a system to do them.

Getting to Task Scheduler is simple. Server Manager launches when you log on to the system. From Server Manager, choose Tools⇨Task Scheduler. The screen that launches looks similar to Figure 5-20.

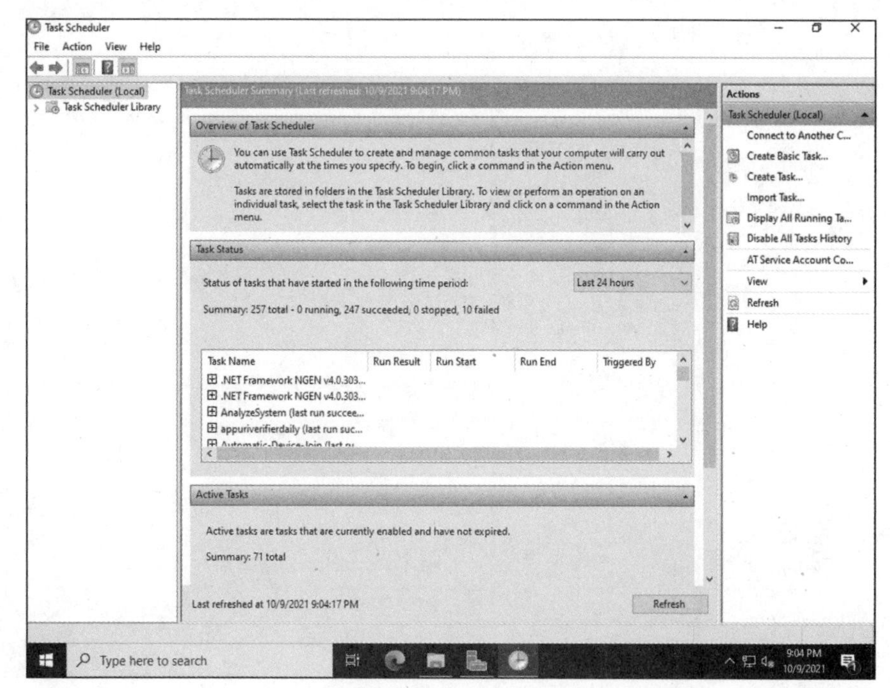

FIGURE 5-20: Task Scheduler is the central area for managing automated/scheduled tasks.

Discovering task status

If you want to know what the status of a task is, you can check the Status column after selecting Task Scheduler Library or any of the folders contained within the Task Scheduler Library. Here are some of the statuses that you may encounter:

- **Ready:** The scheduled task is ready for its next scheduled execution.
- **Running:** The scheduled task is currently running.
- **Disabled:** The scheduled task will not run.

Using preconfigured tasks

Windows Server 2022 comes with quite a few predefined tasks that you can choose from. For instance, Disk Cleanup runs automatically when the system starts to run low on free space. You can see in Figure 5-21 that the Status is Ready, and it gives you the last run time as well.

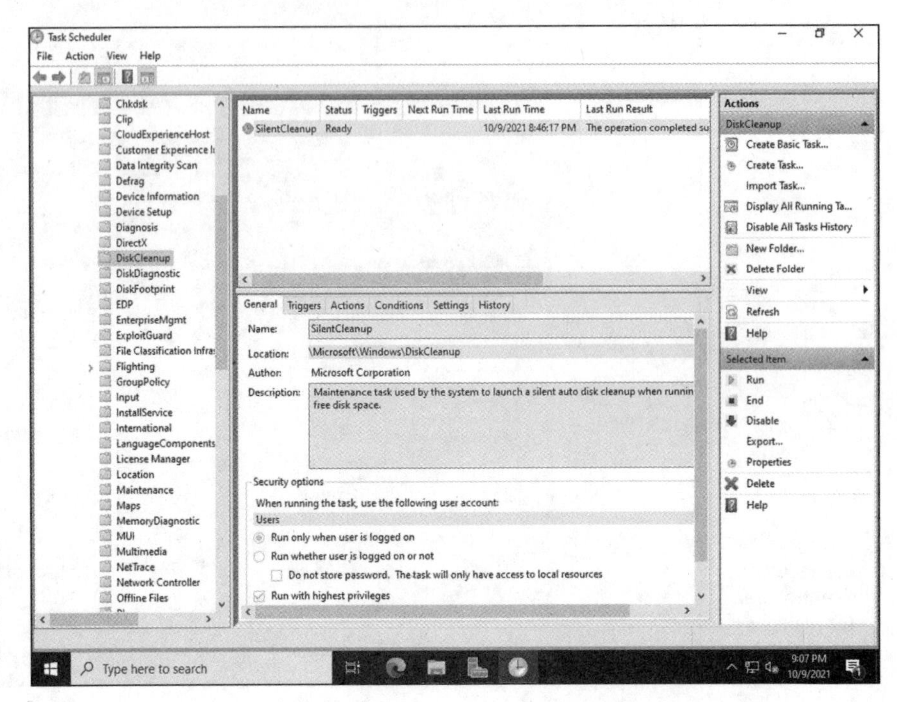

FIGURE 5-21: Configuring one of the Windows Server preconfigured tasks.

You can change the schedule that the task runs on or even delete the task if you want. There are several tabs that you should understand:

- **General:** The General tab contains information about the task, including who authored it and what account it will run under.
- **Triggers:** This tab is used to automate the task. Here, you can create a schedule if you want the task to run on a certain day at a certain time, or if you want the task to kick off when other events occur like logon, startup, or when a certain event occurs.
- **Actions:** The Actions tabs contains what you want the task to actually do. Typically this will be starting a program of some kind, though you can still use the deprecated options of sending an email or displaying a message.
- **Conditions:** The Conditions tab can place conditions on when the task will be allowed to run. For instance, you may require that the task only be able to run after the system has been idle for a predetermined amount of time.
- **Settings:** The Settings tab allows you to tweak settings related to how the task can run and what it should do if it runs into a failure.
- **History:** The History tab tells you if the task has been running successfully. If it runs into an error, it can give you valuable information on what caused the error.

To edit a preconfigured task, simply double-click the task and make your changes, and then click OK. For example, say I want to schedule the SilentCleanup task to run every evening. I would follow these steps:

1. **From Server Manager, choose Tools⇨Task Scheduler.**
2. **Expand the Task Scheduler Library.**
3. **Go to the DiskCleanup folder and double-click SilentCleanup.**
4. **Click the Triggers tab and then click New.**
5. **Under Begin the Task, select On a Schedule.**
6. **Change the settings area to Daily and select the time you want it to run at.**
7. **Under Advanced settings, select Stop Task if It Runs Longer Than and select a reasonable time frame.**
8. **Click OK to save the trigger.**

Now the task will run every day at 7:00 p.m. in my case. You can see what this looks like in Figure 5-22.

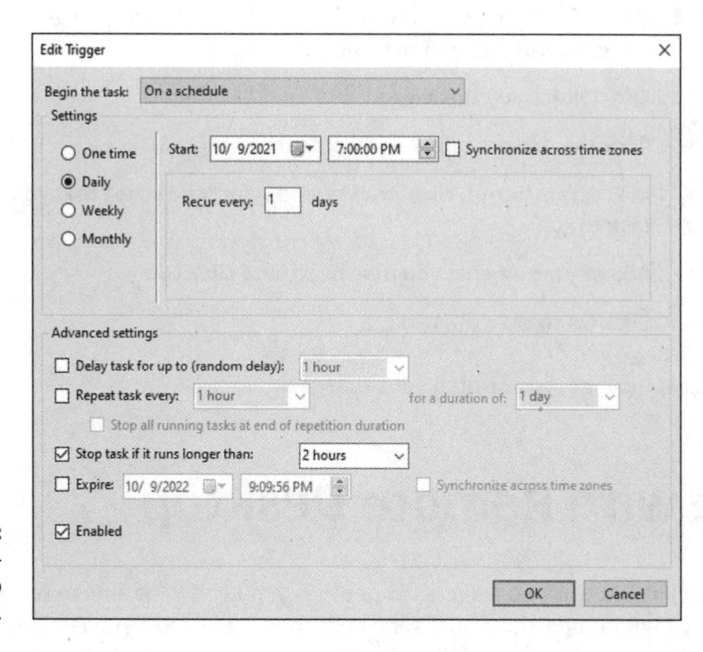

FIGURE 5-22: Changing a preconfigured task to run every day.

Creating your own tasks

Maybe the preconfigured tasks aren't what you need to do the job. Maybe you have a custom script that you need to run on a schedule. This is where creating your own tasks can be very helpful.

To create your own task, select the Task Scheduler Library. You have the choice of creating a basic task or simply creating a task. Here are the steps to create a task:

1. **From Server Manager, choose Tools⇨Task Scheduler.**
2. **Click Task Scheduler Library.**
3. **Select Create Task from the right-hand menu.**
4. **On the General tab, give your task a name, select which user account you want it to run under, and select which OS you want to configure it for.**
5. **Click the Triggers tab and then click New.**
6. **Under Begin the Task, select On a Schedule.**
7. **Change the settings area to Daily and select the time you want it to run at.**
8. **Under Advanced settings, select Stop Task if It Runs Longer Than and select a reasonable time frame.**
9. **Click OK to save the trigger.**
10. **Click the Actions tab and select Start a Program.**
11. **In Program/Script, click Browse and select the script that you want the task to run.**
12. **Add any arguments you may need, and click OK.**
13. **Click OK to save the task.**

There you go, you created your first task!

Working with Remote Desktop

When you work on servers, sometimes you need to be able to see what's going on by logging into the actual server. In these cases, Remote Desktop is the tool you want. Remote Desktop allows you to connect remotely to the console of the Windows Server system. There are several configuration options you can make that impact how the Remote Desktop session works.

On the server on which you want to use Remote Desktop, all you need to do is click the Start menu, click Windows Accessories, and choose Remote Desktop Connection. Enter the name of the system that you want to connect to and click Connect, shown in Figure 5-23.

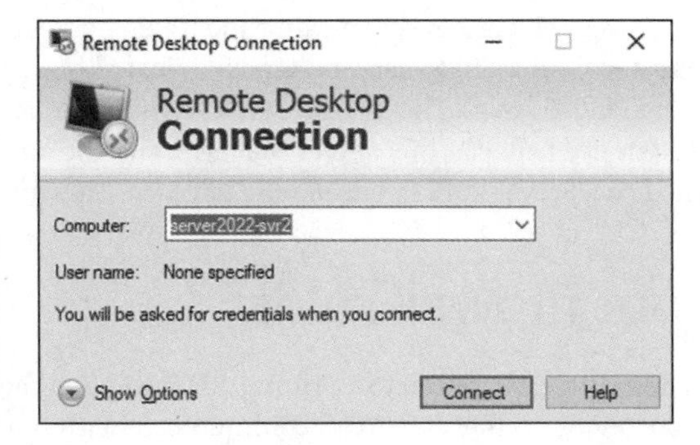

FIGURE 5-23: Connecting to another system via a Remote Desktop Connection.

You'll be prompted for credentials to log on to the other server. Enter a username and password and click OK.

If this is a system that you connect to often, you can save the connection settings and even tell it which username you want to use. With the Remote Desktop connection window open, click the Options arrow. Fill in the username, and then choose Save As. This will create an RDP file that you can double-click to automatically launch a future Remote Desktop session with your saved settings. Here are some of the other settings you can create, which can also be saved into this RDP file:

- **Setting the display configuration:** With Remote Desktop Connection open, and the Options bubble expanded, click the Display tab. From this tab, you can set the size of the remote window, you can enable support for multiple monitors, and you can choose how many colors you want. By default, it's set to 32 bit, which is the highest quality setting.
- **Accessing local resources:** With Remote Desktop Connection open, and the Options bubble expanded, click the Local Resources tab. On this tab, you can configure audio, how the keyboard will react with the remote session, and which local devices you want to pass through the remote connection like printers, drives, and the clipboard.
- **Optimizing performance:** With Remote Desktop Connection open, and the Options bubble expanded, click the Experience tab. By default, this is set to detect the connection quality automatically. You can set it to whatever type of connection you're using, including modem, low-speed broadband, satellite, high-speed broadband, WAN, or LAN. You can also choose whether you want the connection to reconnect if it's dropped.

Working with Remote Server Administration Tools

Remote Server Administration Tools (RSAT) allows you to manage all the various roles and features on your server. The great thing about RSAT is that you can install it on your desktop system and manage servers remotely, just as you would if you were connected to the individual server. The interface is the familiar Server Manager interface. The main difference between this Server Manager and the one on the server is that instead of the Local Server option, you simply have the All Servers option.

Figuring out firewall rules

To use RSAT, the remote servers need to be running WinRM and the Windows Firewall must allow port 5985 so that RSAT can communicate with the remote server.

Connecting to the server

To manage a server with RSAT, you need to connect to it first. Follow these steps to add a server to Server Manager on a Windows 10 PC:

1. **Click Start, open Server Manager, and click All Servers.**
2. **Right-click All Servers and choose Add Servers.**
3. **Search for the name of the system.**
4. **Select the server(s) that you want to add and click the arrow to move the selected server(s) to the Selected box.**
5. **Click OK.**
6. **Right-click the newly added server and choose Start Performance Counters.**

After you follow these steps, your server will show Online under manageability (shown in Figure 5-24), and you can now manage your server remotely through the RSAT tools.

Managing your servers

To manage servers that have been added to the RSAT version of Server Manager, simply right-click the server and choose what you want to do. There are a ton of management options available to you from the Server Manager interface. You can add roles and features, restart the server, or run AD tools if the remote system is a domain controller. You can even run a remote PowerShell window. See the available options in Figure 5-25.

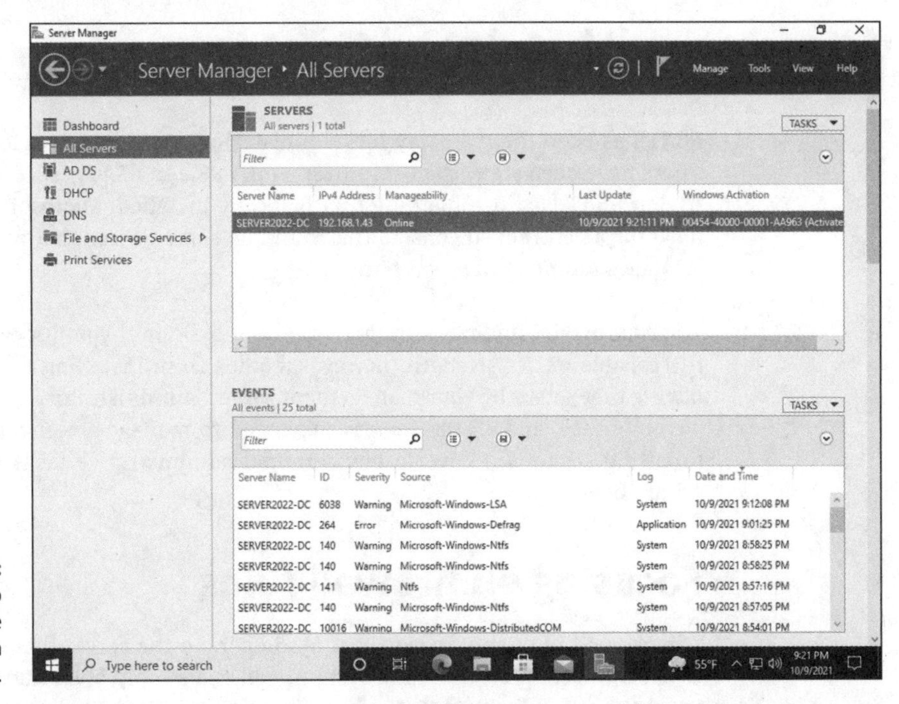

FIGURE 5-24: Using RSAT to manage remote servers from a Windows 10 client.

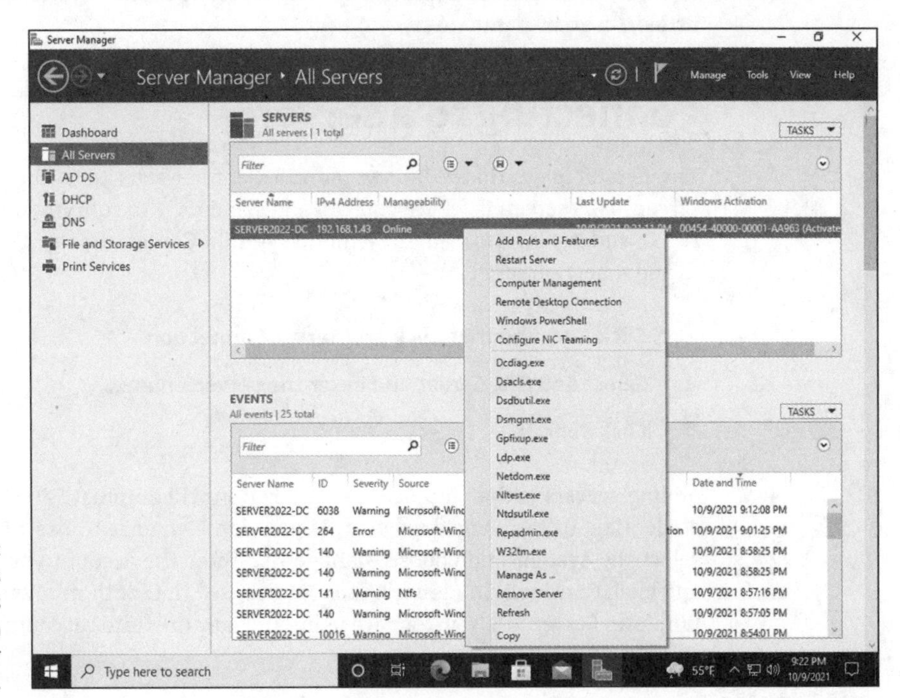

FIGURE 5-25: There are lots of options to manage a remote server using RSAT.

Working with Admin Center

RSAT has been the mainstay for administering remote servers for quite a while. There have been significant improvements made to RSAT, especially in Windows 10, but RSAT is still something that is locally installed. Microsoft recognized the need for a centralized console that could be served through a web browser, and Windows Admin Center was born.

You can install Windows Admin Center on a desktop if you just want to see what it's capable of. Its strength, however, comes from installing it on a server and having that server be your management server. Administrators can connect to the web interface, and all the servers they need to manage are already there. In the following sections, I cover a few common administrative tasks using Windows Admin Center.

Focusing on firewall rules

Windows Admin Center needs port 6516 open on the server that it's running on. When connecting on the same system, you can simply open Microsoft Edge and type **https://localhost:6516**. If accessing it from a remote system, you need to type **https://<server name>:6516**.

Connecting to a server

The default page that Windows Admin Center opens to is the All Connections screen. This screen is where you can add systems. I'm running this on a Windows 10 PC, and my PC shows up in Windows Admin Center already. Follow these steps to add a new server:

1. **Click Add, and then click Add Server Connection.**
2. **Select Add One Server and enter the server's name.**
3. **Click Submit.**

Adding servers is that simple. The connection will automatically be set to use the credentials of the logged-in user. If you don't want it to use those credentials, select the system and choose Manage As. Enter the account you do want to use and click Continue. In Figure 5-26, you can see that both my Windows 10 PC and Windows Server 2022 system are using the administrator account.

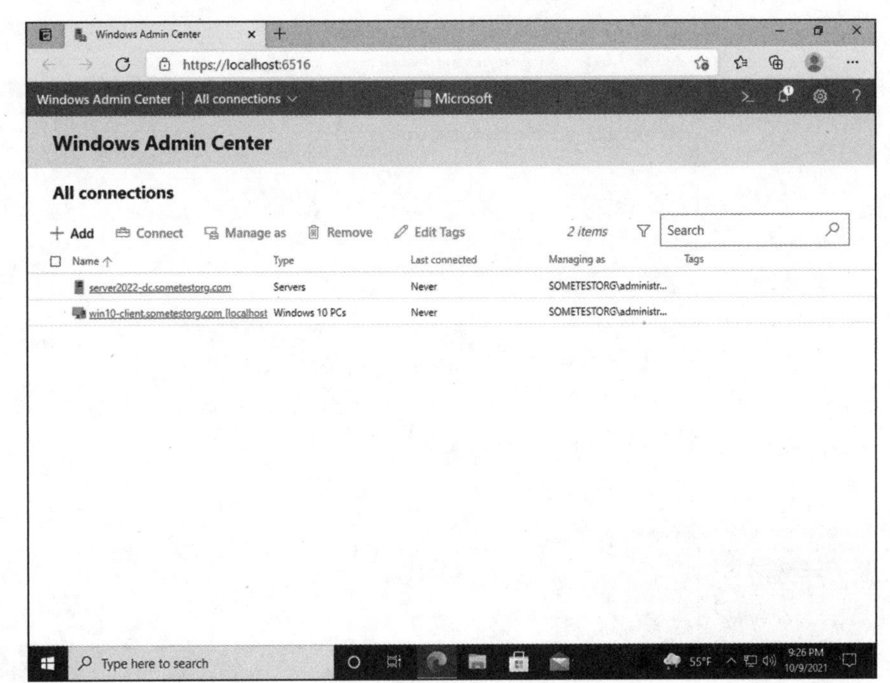

FIGURE 5-26: Connecting to a server with Windows Admin Center.

Using Windows Admin Center to manage your servers

From Windows Admin Center, click the name of the system that you want to manage. This changes your view to the Server Manager view, shown in Figure 5-27.

From here, you can manage the system, check its resource utilization, and more. You can do everything you could do in the traditional Server Manager, with several new additions. The newest version of Windows Admin Center includes lots of new Azure functionality, including the ability to manage Azure Kubernetes clusters, Azure Backups, Azure Security Center, and more.

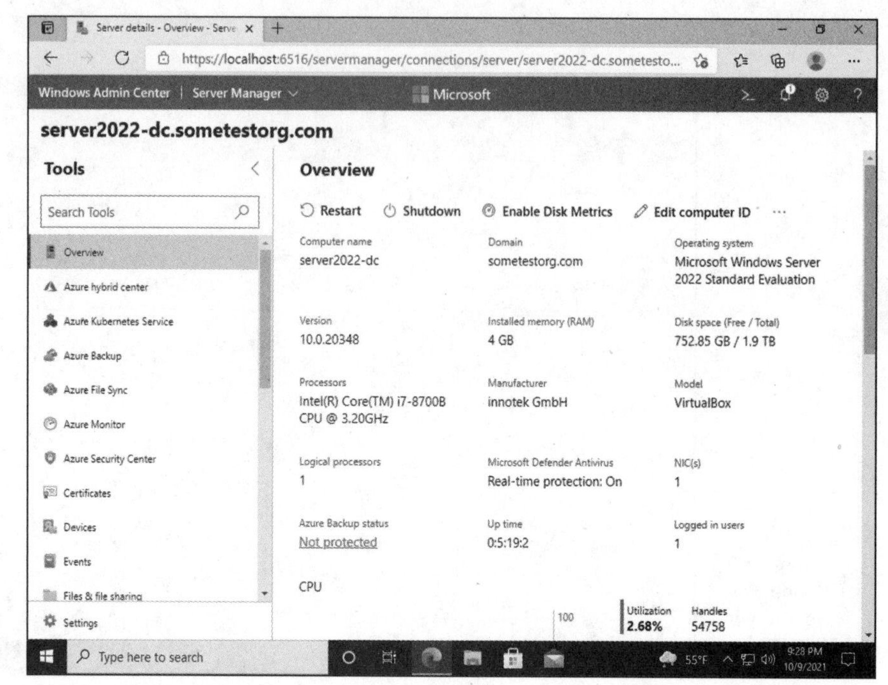

FIGURE 5-27: The Server Manager view in Windows Admin Center.

Creating a Windows Recovery Drive

A Windows Recovery drive can be very helpful if you ever have to do a bare metal restore. This can recover the operating system, custom settings, and configurations, as well as drivers. Follow these steps to create a recovery drive:

1. **Insert a blank USB flash drive into the system.**
2. **Click the Start Menu, scroll down to Windows System, and choose Control Panel.**
3. **Change View by Category to View by Large Icons.**
4. **Select Recovery, and then click Create a Recovery Drive.**
5. **On the Create a Recovery Drive screen, click Next.**

   The system takes a moment, but it should find the USB flash drive.

6. **When the system finds the driver, click Next.**

   On the final Create the Recovery Drive screen, you're warned that everything on the destination drive will be deleted.

7. **Click Create.**

IN THIS CHAPTER

» Configuring the command line for use

» Setting up user and session environmental variables

» Getting additional help at the command line

» Understanding command line symbols and what they do

Chapter 6

Working at the Command Line

Just about every system administrator has worked with the Command Prompt at some point in their career. The Command Prompt is a launching point for a number of diagnostic utilities and a great resource for gathering information, in addition to being a utility to automate repetitive tasks.

This chapter discusses working with the command line — from the basics on how to use the command line to how to customize it to your liking.

Opening an Administrative Command Prompt

The simplest way to get to the Command Prompt in Windows Server 2022 is to click the Start menu, scroll down to Windows System, and click Command Prompt. This will run the Command Prompt in an unprivileged state.

TIP

For some commands to work, you must run the Command Prompt as an administrator. To do this, right-click Command Prompt in the Start menu, click More, and click on Run as Administrator, as shown in Figure 6-1. When the Command Prompt opens, the top bar will read "Administrator: Command Prompt." This is a quick visual way to verify that you're running as administrator.

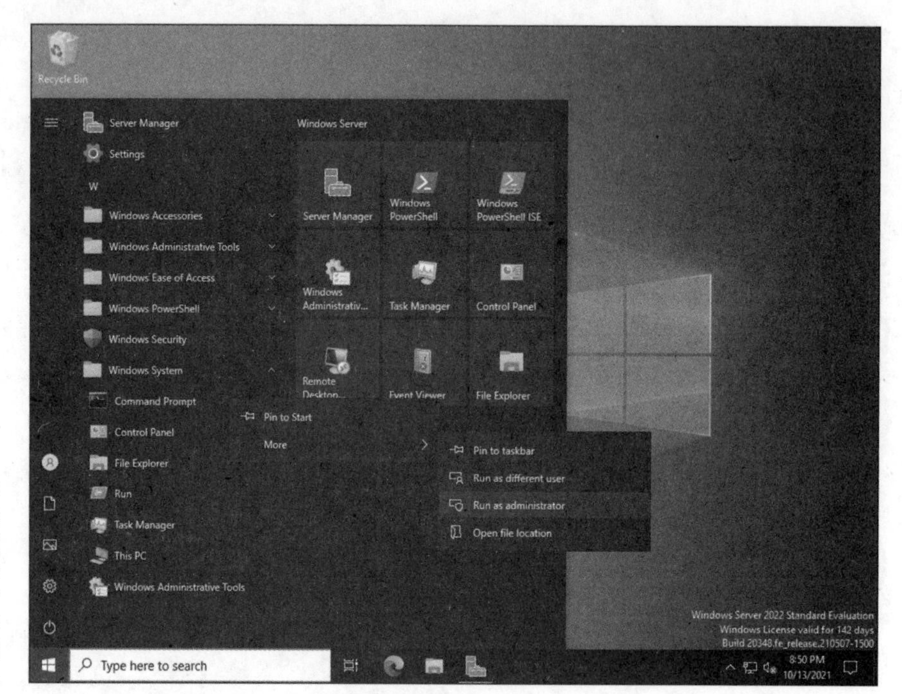

FIGURE 6-1: Running the Command Prompt as administrator allows you to run more privileged commands.

Configuring the Command Line

You can customize the Command Prompt quite a bit. If you right-click the menu bar and choose Properties, you can set the customizations that you want. These customizations will last for as long as you have the session open. If you want the settings to be saved, right-click the menu bar and choose Defaults. The Properties and Defaults menus are nearly identical.

In this section, I give you a look at the different customizations you can make.

Customizing how you interact with the Command Prompt

The Options tab (shown in Figure 6-2) is where you can set things like the size of the cursor, how many commands to keep saved in the buffer, editing options, and text selection. Refer to this screenshot as I walk you through each section of the Option tab.

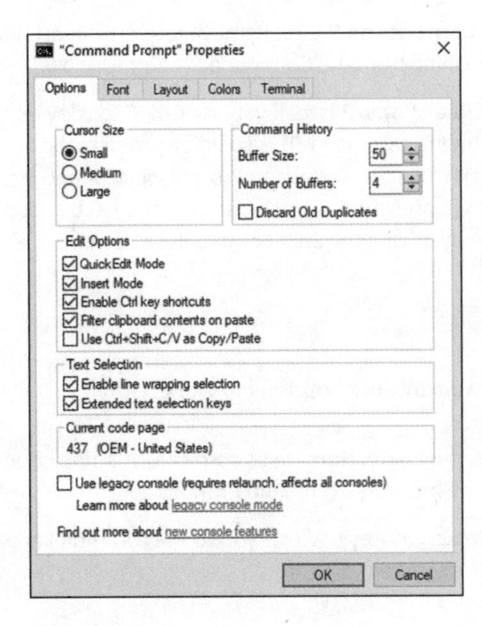

FIGURE 6-2: The Options tab allows you to customize how you interact with the Command Prompt.

Cursor Size

Changing the cursor size makes the cursor wider and easier to spot. This setting can be very helpful for someone who is visually impaired.

Command History

The command history allows you to simply press the Up key to back through old commands, which can save you from having to retype commands if you're doing something repetitive. The default buffer size is 50, but you can increase or decrease that as you like.

Edit Options

In this section, you a few choices in controlling how you can edit things within the Command Prompt:

- **Quick Edit Mode:** Allows the use of the mouse to copy and paste text into and from the Command Prompt window.
- **Insert Mode:** Allows you to type wherever the cursor is at. If Insert Mode gets disabled for some reason, you'll overwrite existing text starting at the location of the cursor.
- **Enable Ctrl Key Shortcuts:** Lets you use Ctrl key shortcuts like Ctrl+C to copy or Ctrl+V to paste.
- **Filter Clipboard Contents on Paste:** Remove tabs and smart quotes from pasted material coming from the Command Prompt window.
- **Use Ctrl+Shift+C/V as Copy/Paste:** If you use Ctrl+C to copy text, you can't use it to stop a running command. This could be problematic because your only other option is close the Command Prompt window and start again. If you check this box, it preserves the original use of Ctrl+C but also allows you to copy and paste by adding the Shift key into the command.

Text Selection

In this section, you two additional options to work with:

- **Enable Line Wrapping Selection:** Helps correct formatting issues when copying and pasting from the Command Prompt.
- **Extended Text Selection Keys:** Allows you to use common keyboard shortcuts inside the Command Prompt.

Current Code Page

Current Code Page is not an adjustable field. It's simply passing on information regarding which character code you're using. I'm in the United States, so I'm assigned 437 (which is a holdover from the old IBM PC days) and the additional identifier of United States.

Use Legacy Console

Legacy console removes a lot of the newer features that were added to the Command Prompt. If you select the Use Legacy Console check box, some of the customization options that I've mentioned will no longer show up on the Options tab for you. You may want to do this for compatibility issues, or for someone who is used to the way that the Command Prompt used to work and finds it disruptive to work with it with the newer options.

Changing the font

The Font tab (shown in Figure 6-3) is pretty straightforward. It allows you to select the size of the font and which font you want to use. It even gives you a preview of what your selection will look like in the Window Preview. In Figure 6-3, for instance, I've chosen a font size of 24 in the Consolas font. You can see in the bottom box the preview of what my font will look like if I keep this setting.

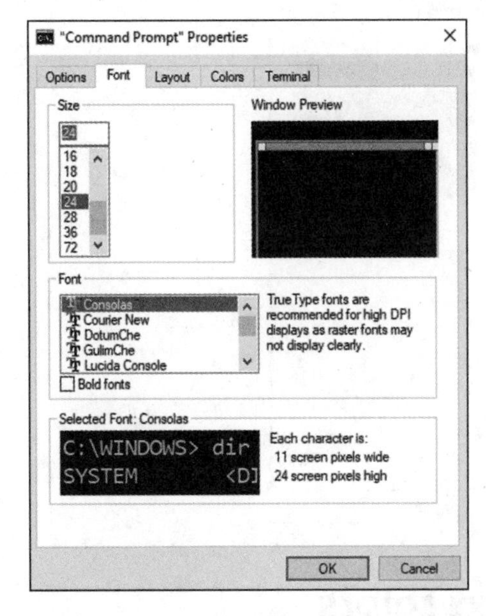

FIGURE 6-3: The Font tab allows you to change the font and how it displays in the Command Prompt.

Choosing your window layout

On the Layout tab (shown in Figure 6-4), there are three configurable options:

- **Screen Buffer Size:** The number of characters you can see on a single line is controlled by the Width adjustment. The Height adjustment determines how many lines will be stored in memory. Select the Wrap Text Output on Resize if you would like for the text to adjust whenever you resize the Command Prompt window. This is useful if the output of a command is difficult to read because the window size was small. You can adjust the Command Prompt window and the text will resize automatically, instead of your having to resize the window and then rerun the command.
- **Window Size:** The Width and Height adjustments change the size of the actual Command Prompt window.

» **Window Position:** If you uncheck the Let System Position Window check box, you can adjust how far away you want the Command Prompt to open from the top and left edges of the screen. If you don't select the check box, you can't change the Left and Top settings.

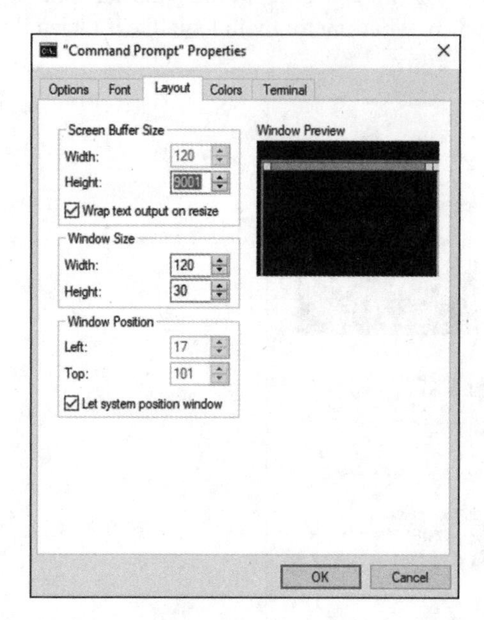

FIGURE 6-4: The Layout tab configures what the window looks like.

Defining text colors

The Colors tab (shown in Figure 6-5) lets you set the background and text colors that are used in the Command Prompt. You can adjust the text and background colors for the Command Prompt screen, as well as for popup boxes. By default, the Command Prompt has a black background and off-white text. You can change that to whatever you like.

You can also adjust the opacity of the Command Prompt using the Opacity slider. The slider is normally set to 100%. However, if you adjust the slider downward, you can see what's behind the Command Prompt.

Making the Command Prompt your own

A new tab introduced is the Terminal tab which allows for further customization of the Command Prompt. This is shown in Figure 6-6 with four brand new options.

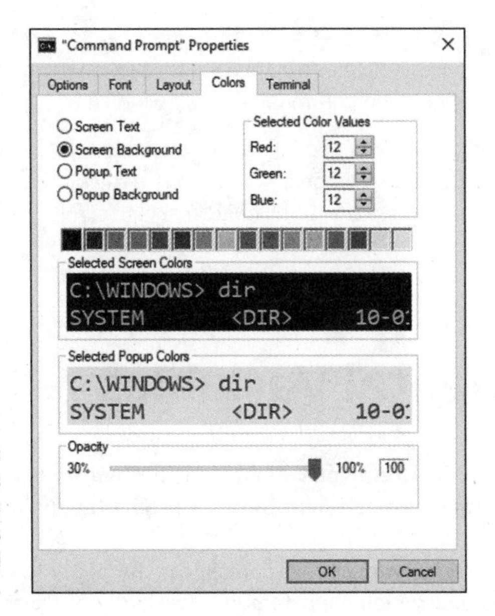

FIGURE 6-5: The Colors tab lets you customize colors in the Command Prompt.

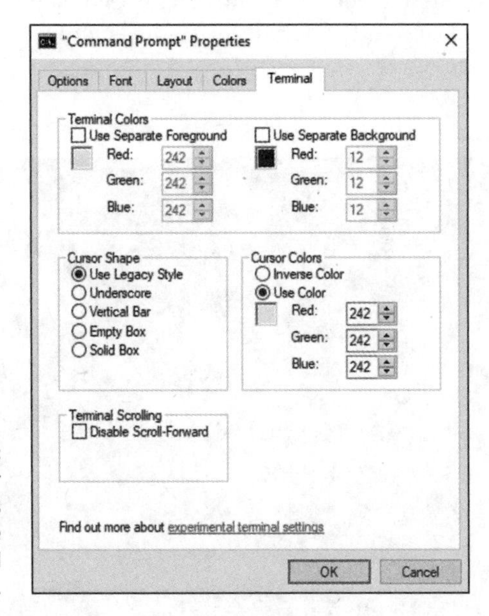

FIGURE 6-6: Further customizations that are available in the Command Prompt.

- **Terminal Colors:** You can make changes in the Terminal Colors section to both the background and the foreground of the Command Prompt.
- **Cursor Shape:** In the Cursor Shape section, you can adjust the cursor to appear as the legacy cursor, an underscore, a vertical bar, an empty box, or a solid box.

» **Cursor Colors:** As you might imagine, this section allows you to adjust the color of the cursor within the Command Prompt window.

» **Terminal Scrolling:** By default Scroll-Forward is enabled. This allows you to scroll through the command history that is stored in the buffer. If you check this and disable Scroll-Forward you will only be able to scroll to the last line output.

Setting Environmental Variables

There are two types of environment variables. User environment variables apply to an individual user, and system environment variables apply to all users.

There are quite a few environmental variables, far too many to cover in a single book. If you want to know which environmental variables are available to you and what their current settings are, you can do this on the Command Prompt window. Just type **Set | More** and you get the output for all the environmental variables on the system, as shown in Figure 6-7.

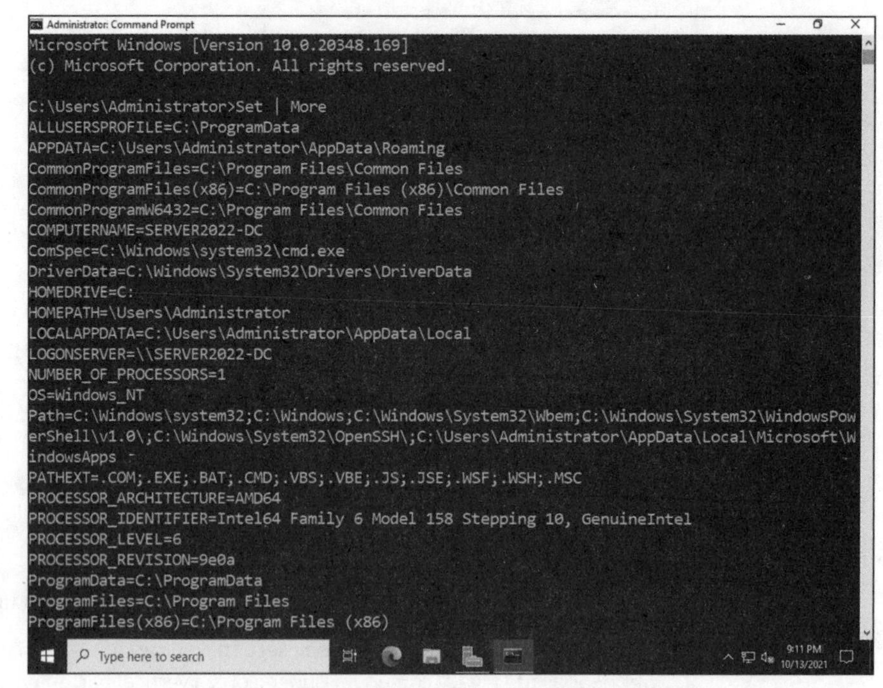

FIGURE 6-7: Displaying environmental variables and their current settings is possible with the Set command.

One of the most common environmental variables to edit is the PATH variable. You can edit the PATH variable when you want to include a directory in your path so that you can simply run programs in that directory without having to actually navigate to that directory. The syntax is simple. For instance, to add a folder named Tools to my path, I type **SET PATH=%PATH%;C:\Tools**. This appends my Tools folder to the existing path variables, as you can see in Figure 6-8. I can verify that my new entry is in my path by typing **echo %PATH%**.

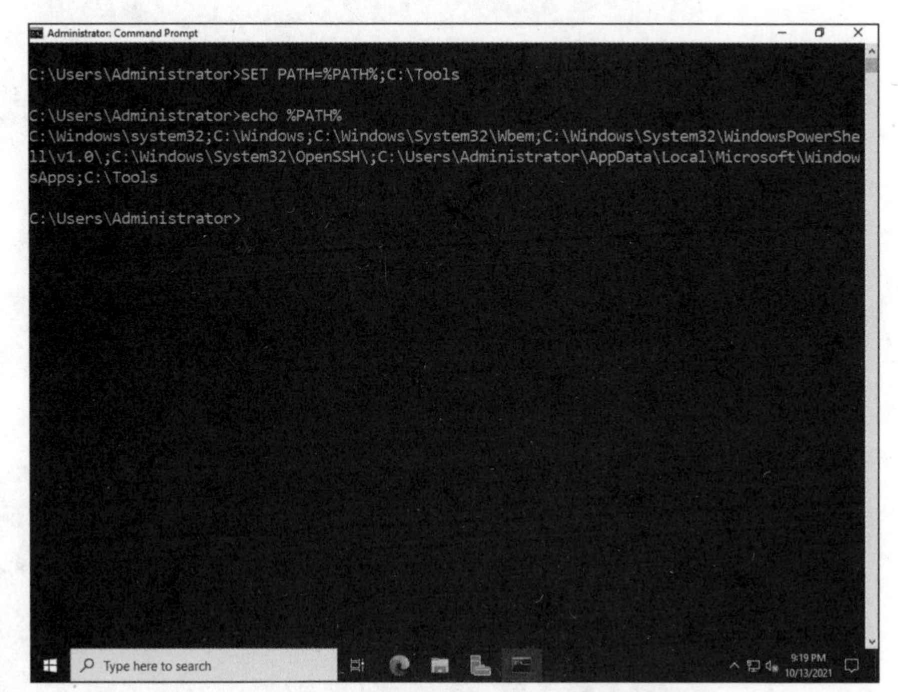

FIGURE 6-8: The view after appending a folder to my PATH environment variable.

Getting Help at the Command Line

Getting help at the command line is pretty easy. To get help with a specific command, you can type the command followed by **/?**. For example, in Figure 6-9, you can see that I wanted to get help with the nslookup utility, so I typed **nslookup /?** and I was presented with the options for that command.

If you want general help at the Command Prompt, such as guidance as far as what you can do, simply type **help** and press Enter. You get a list of all the commands you can run at that point in time, shown in Figure 6-10.

FIGURE 6-9: Using the Command Prompt help to get more information on a command.

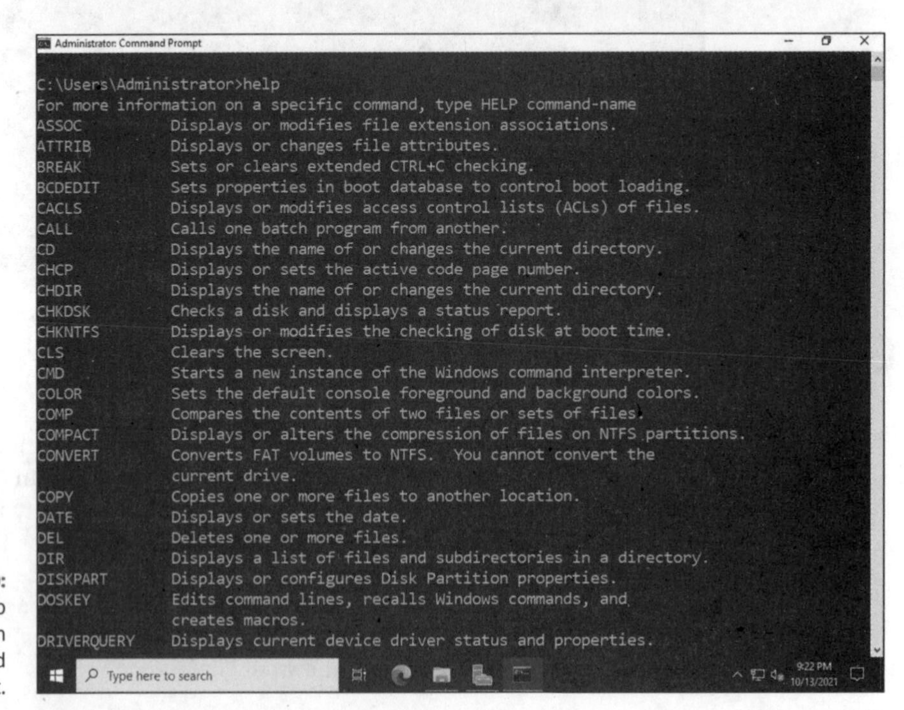

FIGURE 6-10: Using the help command in the Command Prompt.

Understanding Command Line Symbols

Symbols can extend the usefulness of the Command Prompt. They can allow you to send output to a file or combine commands. Table 6-1 provides a list of symbols that you can use on the Command Prompt window.

TABLE 6-1 **Command Line Symbols**

Symbol	Example	Description
>	*command* > file.txt	Writes the output of the command to the filename you specify. If the file does not exist, it will be created. If the file does exist, the file will be overwritten.
<	*command* < file.txt	Runs a command and inserts the contents of the file after the command.
>>	*command* >>file.txt	Similar to the single > symbol except this command appends if the file exists, rather than overwrites the file.
|	*command* A | *command* B	Sends the output of command A to the input of command B.
&	*command* A & *command* B	Runs command A and then command B.
&&	*command* A && *command* B	Executes command B only if command A finishes successfully.
||	*command* A || *command* B	Executes command B only if command A does *not* finish successfully.
@	@echo off	Typing the at symbol (@) will suppress whatever comes after it. It's not truly a command; it's actually an optional flag that can be used to suppress whatever comes after it. This can prevent a line of code from showing up in your server's logs. You commonly see it used with @echo off to turn off the output of commands that are running in a script, or on the console.

» Configuring and customizing your PowerShell session

» Setting environmental variables with PowerShell

» Obtaining assistance from within the PowerShell window

» Understanding PowerShell punctuation and how to use it

Chapter 7

Working with PowerShell

The Command Prompt has been a staple for many years, but Microsoft has been making a big push toward PowerShell. And it's not hard to see why. PowerShell can run the same utilities and things that can be run in the Command Prompt, but it can also run much more than that. By importing modules, you can expand the things that PowerShell can do.

PowerShell is a very flexible option for system administrators. The console can run the legacy commands that were available in the Command Prompt, as well as the newer PowerShell commands and scripts. PowerShell improves on the ability to support automation across platforms, including on-premise datacenters, Azure, Amazon Web Services (AWS), and with PowerShell Core, even Linux and macOS!

As you work in PowerShell, you'll discover how easy it is to type in longer commands as PowerShell uses tab complete. This allows you to type the first few letters of a cmdlet and then press the Tab key. If a cmdlet matches what you've typed so far, it will be displayed. If it isn't the right cmdlet, you can continue to press the Tab key until the correct cmdlet is displayed. This makes administering from the PowerShell window very efficient.

One of the things that I love most about PowerShell over the Command Prompt is the common language that is used with PowerShell cmdlets. PowerShell cmdlets utilize a verb-noun format. When you're using a PowerShell cmdlet, there is a well-documented set of "verbs" that you can use. The most common PowerShell verbs that you see are `Get`, `Set`, `New`, and `Invoke`. Many more verbs are available — you can see them on the Microsoft site along with examples of when they would be used; go to `https://docs.microsoft.com/en-us/powershell/scripting/developer/cmdlet/approved-verbs-for-windows-powershell-commands?view=powershell-7.1`.

Nouns in the context of PowerShell cmdlets are what you want to take action against. Consider the cmdlet `Get-Date`. `Get` is the verb; you're telling PowerShell you want to query for some information. `Date` is the noun; you're asking PowerShell to retrieve the date.

This chapter serves as a brief introduction to PowerShell. If you want to learn more about PowerShell, check out Book 6.

Opening an Administrative PowerShell Window

Microsoft has been making a pretty major push to get more system administrators to embrace PowerShell, given its flexibility and utility. In Windows Server 2022, when you right-click the Start menu, you no longer see the Command Prompt by default; instead, you see Windows PowerShell, as shown in Figure 7-1.

You have two options when in this view:

- You can choose Windows PowerShell, which opens a non-elevated PowerShell window. This window will allow you to perform PowerShell tasks that don't require administrative privileges.
- You can choose Windows PowerShell (Admin), which opens an elevated PowerShell window. Much of the work you do as a system administrator will require administrative access, so you'll want to choose Windows PowerShell (Admin).

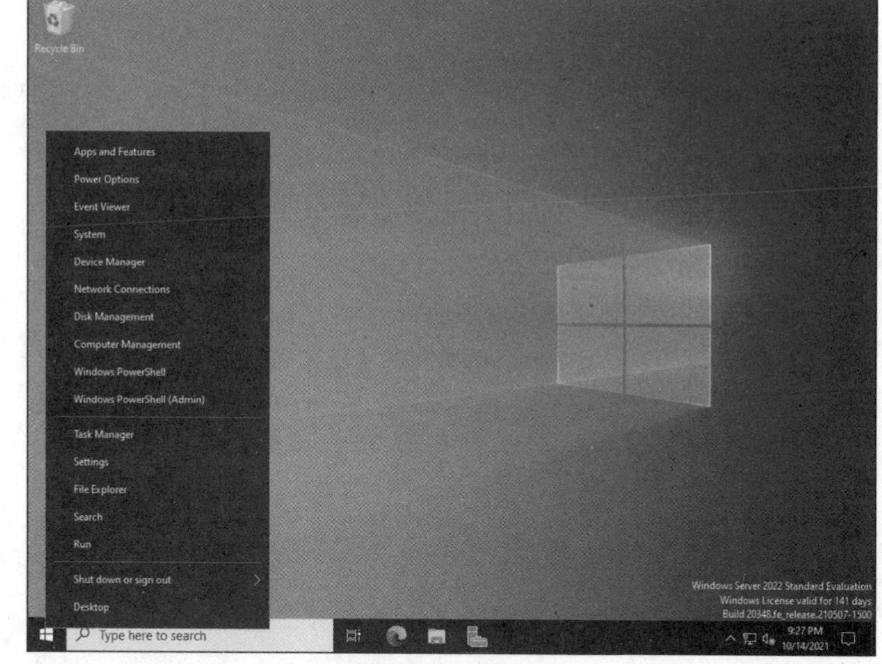

FIGURE 7-1: Windows PowerShell now resides by default in the menu that you access by right-clicking the Start menu.

Configuring PowerShell

You can configure the PowerShell window in the same way that you can configure the Command Prompt window. You can customize the Window one time or set defaults so that settings will load every time. The one limitation with making the settings through the Defaults or Properties selection in the menu is that the colors that PowerShell uses for commands and other things are not affected by the Properties settings. If you want to affect the color of the commands and other components in PowerShell, you need to use a profile script. If this is what you want to do, see "Using a Profile Script" later in this chapter.

To configure the PowerShell window, launch Windows PowerShell, right-click the Windows PowerShell title bar, and then choose Properties. The Windows PowerShell Properties dialog box appears. In the following sections, I walk you through this dialog box tab by tab.

Options

The Options tab (shown in Figure 7-2) is where you can set things like the size of the cursor, how many commands you want to be able to recall, editing options, and text selection. The following sections walk you through each of the sections of the Options tab.

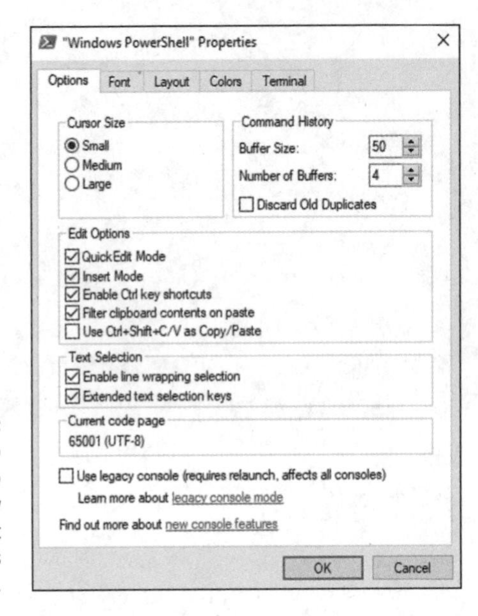

FIGURE 7-2: The Options tab allows you to customize how you interact with Windows PowerShell.

Cursor Size

Changing the cursor size makes the cursor wider and easier to spot. This setting can be very helpful for someone who is visually impaired. I personally prefer this setting to be on Large for both the Command Prompt and Windows PowerShell.

Command History

The Command History settings allows you to press the up arrow key to go back through previous commands. This can save you from having to retype commands if you're doing something repetitive. The default Buffer Size is 50, but you can increase or decrease that as you like. The Number of Buffers setting is used to specify how many processes are allowed to have their own individual buffer. The Discard Old Duplicates check box is optional; I usually don't check it out of personal preference. It removes duplicate commands, which can make it simpler to find an old command because you won't have to go through a lot of repeated commands.

Edit Options

The Edit Options section gives you some options to control how you can edit things within the PowerShell window:

- **Quick Edit Mode:** Checking this box enables you to use the mouse to copy and paste text into and from the Windows PowerShell window.
- **Insert Mode:** Checking this box enables you to type wherever the cursor is. If this option is disabled, you'll overwrite existing text depending on where your cursor is located.
- **Enable Ctrl Key Shortcuts:** Checking this box enables you to use Ctrl key shortcuts like Ctrl+C to copy or Ctrl+V to paste.
- **Filter Clipboard Contents on Paste:** Checking this box removes formatting from pasted material coming from the Windows PowerShell window.
- **Use Ctrl+Shift+C/V as Copy/Paste:** If you're using Ctrl+C to copy text, you can't use that shortcut to stop a running command. This could be a problem if you have a hung command, because you'll have to close the Windows PowerShell window. If you enable this option, it will allow the normal usage of Ctrl+C (to stop a running command) but will still allow you to copy and paste by adding the Shift key into the shortcut.

Text Selection

The Text Selection section gives you two additional options to work with:

- **Enable Line Wrapping Selection:** Checking this box can correct formatting issues when copying and pasting from Windows PowerShell.
- **Extended Text Selection Keys:** Checking this box allows the use of common keyboard shortcuts inside the PowerShell window.

Current Code Page

Current Code Page is not an adjustable field. It's letting you know which character code you're using. In my PowerShell window, you can see that I'm using a UTF-8 character set.

Use Legacy Console

Checking the last option, Use Legacy Console, removes a lot of the newer features that were added to PowerShell. If you enable it, some of the customization options that I've discussed will disappear from the Options tab for you. I do not recommend checking this box in a Windows PowerShell window.

Font

The Font tab (shown in Figure 7-3) is a simple tab with just a few settings. It allows you to select the font size you want to use and which font to use. The Window Preview section gives you a preview of what your selection will look like. In Figure 7-3, for example, I've chosen a Size of 24 and a Font of Consolas. You can see in the bottom box (Selected Font: Consolas) the preview of what my choices will look like if I choose to keep my changes.

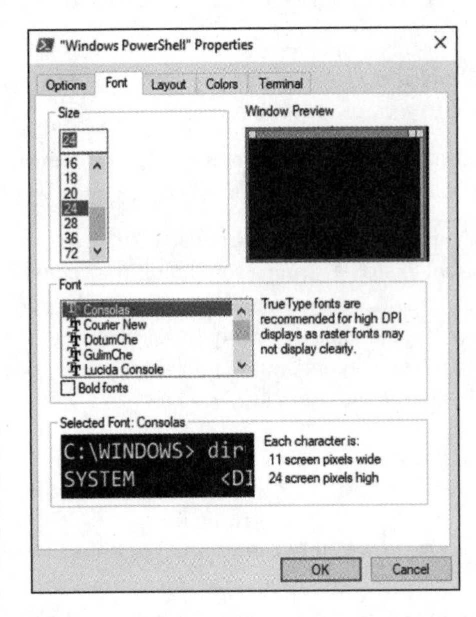

FIGURE 7-3: The Font tab allows you to change the font and how it displays in PowerShell.

Layout

The Layout tab (shown in Figure 7-4) has three configurable options:

- **Screen Buffer Size:** The Width adjustment controls the number of characters that can fit on the screen. The Height adjustment determines how many lines will be stored in memory. Check the Wrap Text Output on Resize box to allow the text on the screen to automatically adjust itself when you resize the PowerShell window.
- **Window Size:** The Width and Height adjustments in this section change the actual size of the PowerShell window.

» **Window Position:** If you uncheck the Let System Position Window check box, you can adjust how far away you want the PowerShell window to be from the top and left of the screen. (If you leave the check box selected, the Left and Top options are grayed out.)

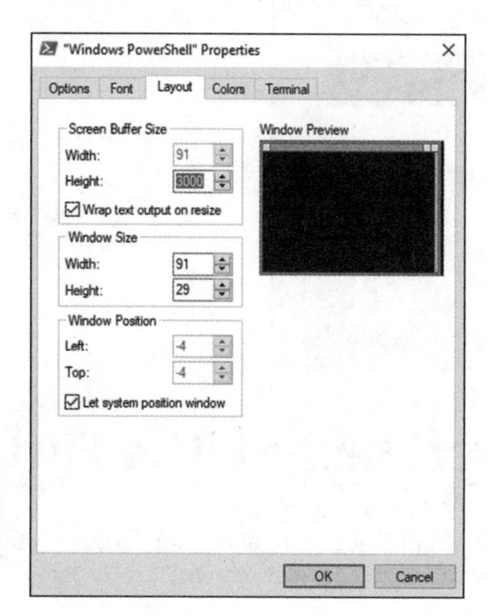

FIGURE 7-4: The Layout tab configures what the PowerShell window will look like.

Colors

The Colors tab (shown in Figure 7-5) lets you set the background and text colors that are used in the Windows PowerShell window. You can adjust the background and text colors for the PowerShell window, as well as for any popup boxes that appear. Select the radio button of the option you want to change (for example, Screen Background), and then choose the color you want either using the Red, Green, and Blue drop-down lists or by clicking one of the colored boxes below.

By default, Windows PowerShell has a blue background and off-white text. You can change that to whatever you like. In all of my screenshots in this chapter, I'm using a black background with white text, mainly because it prints better.

You can adjust the opacity of the PowerShell window as well, using the Opacity slider. The slider is normally on 100%, which makes it solid so you can't see through it. However, if you slide the slider to the left, you can see what's behind the PowerShell window. This can be fun, but personally, I find it distracting; I recommend leaving the Opacity slider on 100%.

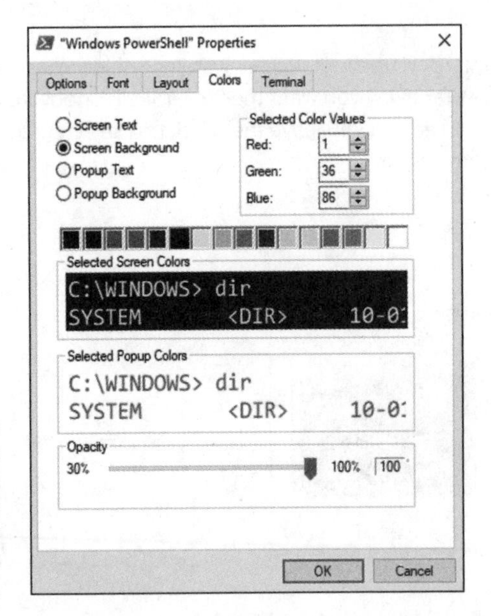

FIGURE 7-5: The Colors tab allows you to customize screen colors in Windows PowerShell.

Customizing PowerShell a Little Further

The Terminal tab (shown in Figure 7-6) allows for further customization of the PowerShell window with four brand-new options.

- **Terminal Colors:** You can make changes in the Terminal Colors section to both the background and the foreground of the PowerShell window.
- **Cursor Shape:** In the Cursor Shape section, you can adjust the cursor to appear as the legacy cursor, an underscore, a vertical bar, an empty box or a solid box.
- **Cursor Colors:** This section allows you to adjust the color of the cursor within the PowerShell window.
- **Terminal Scrolling:** By default, Scroll-Forward is enabled. This allows you to scroll through the command history that's stored in the buffer. If you check this box and disable Scroll-Forward, you'll only be able to scroll to the last line output.

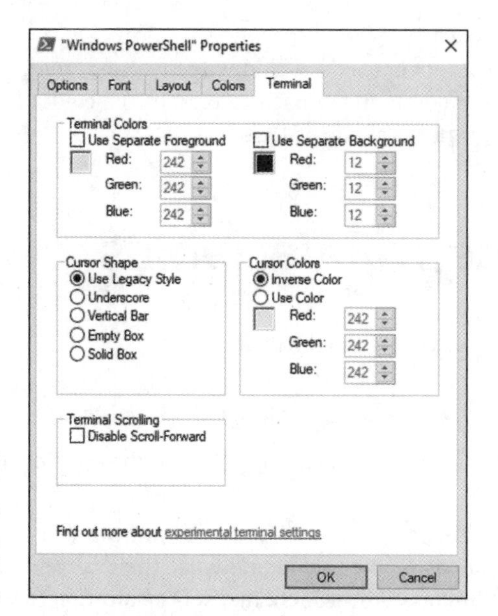

FIGURE 7-6: Further customizations that are available in the Command Prompt.

Using a Profile Script

Before you create your script, you need to create a WindowsPowerShell folder in your Documents folder. The profile script should be named `profile.ps1` and should be placed inside the WindowsPowerShell folder.

To set the colors for the various components that appear on the screen like commands, variables, strings, and so on, you need to build out a script that sets the desired color for each. Say I wanted all the components to just be white on my black background (which works out great for print). My script would look like this:

```
$colors = @{}
$colors['String'] = [System.ConsoleColor]::White
$colors['Variable'] = [System.ConsoleColor]::White
$colors['Comment'] = [System.ConsoleColor]::White
$colors['None'] = [System.ConsoleColor]::White
$colors['Command'] = [System.ConsoleColor]::White
$colors['Parameter'] = [System.ConsoleColor]::White
$colors['Type'] = [System.ConsoleColor]::White
$colors['Number'] = [System.ConsoleColor]::White
$colors['Operator'] = [System.ConsoleColor]::White
$colors['Member'] = [System.ConsoleColor]::White
Set-PSReadLineOption -Colors $colors
```

You can see each component is set individually, and then at the end the Set-PSReadLineOption command is used to read in the colors from the variable you created at the beginning called $colors.

Setting Environmental Variables

There are two types of environment variables:

- **User:** User environment variables apply to individual users.
- **System:** System environment variables apply to all the users on a system.

There are quite a few environmental variables. If you want to know which environmental variables are available to you and what their current settings are, you can check this from Windows PowerShell. Just set the location to the environmental variables (it's treated like a drive) by typing the following:

```
Get-ChildItem Env:
```

You receive the output for all the environmental variables on the system, shown in Figure 7-7.

One of the most common environmental variables to edit is the PATH variable. This is done when you want to include a directory in your path so that you can simply run programs in that directory without having to actually be in that directory. The syntax is simple. For instance, to add a folder named Tools to my path, I would type the following:

```
[Environment]::SetEnvironmentVariable("DEMO","C:\Tools","User")
```

This would append my Tools folder located in my C: drive to the existing path variables, as you can see in Figure 7-8. I can verify my new entry is in my path by typing the following:

```
Get-ChildItem Env:
```

TIP

You'll need to close PowerShell and reopen it after adding an environment variable to see it with the Get-ChildItem Env: command. If you've just added one and you don't see it, make sure you didn't forget this step.

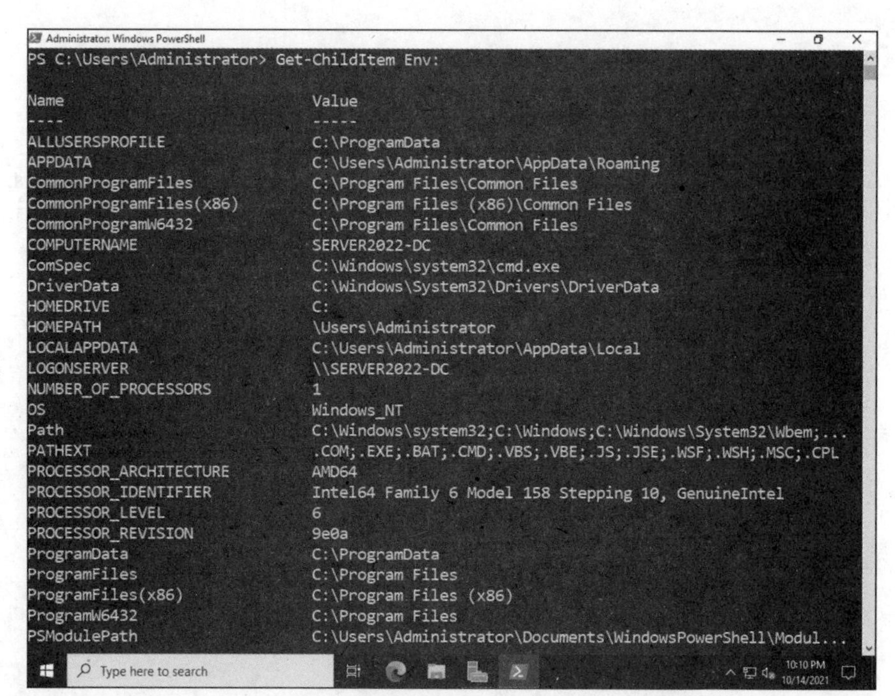

FIGURE 7-7: Displaying environmental variables and their current settings is easy with the `Get-ChildItem` command.

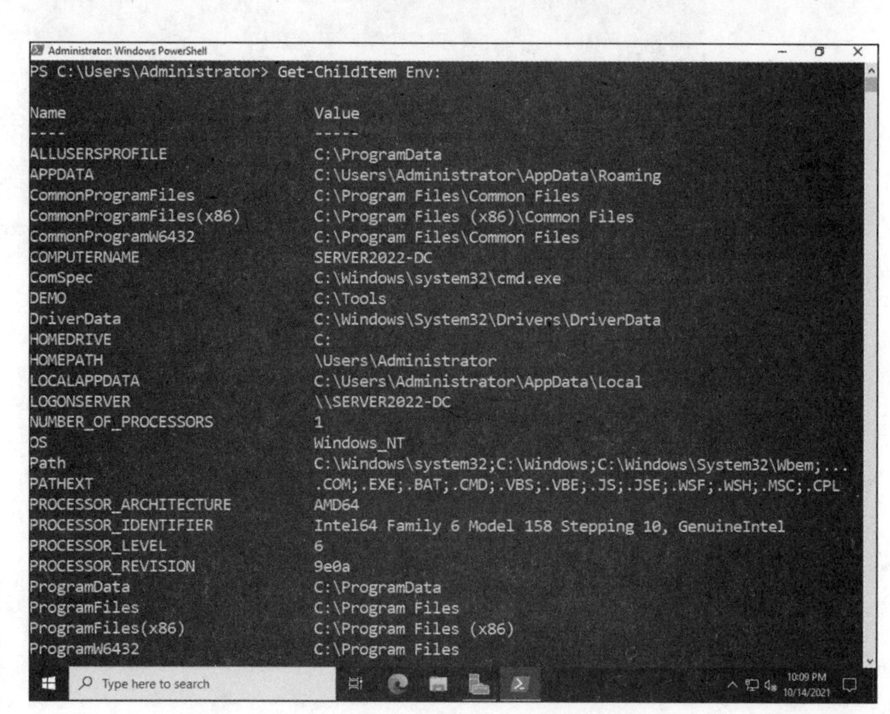

FIGURE 7-8: The view after creating a new environment variable.

Getting Help in PowerShell

You may need help with the syntax of a particular cmdlet. For instance, the `Get-Command` cmdlet has syntax that you need to follow to get specific and relevant information. To get help on the `Get-Command` cmdlet, simply type **Get-Help Get-Command**, as shown in Figure 7-9.

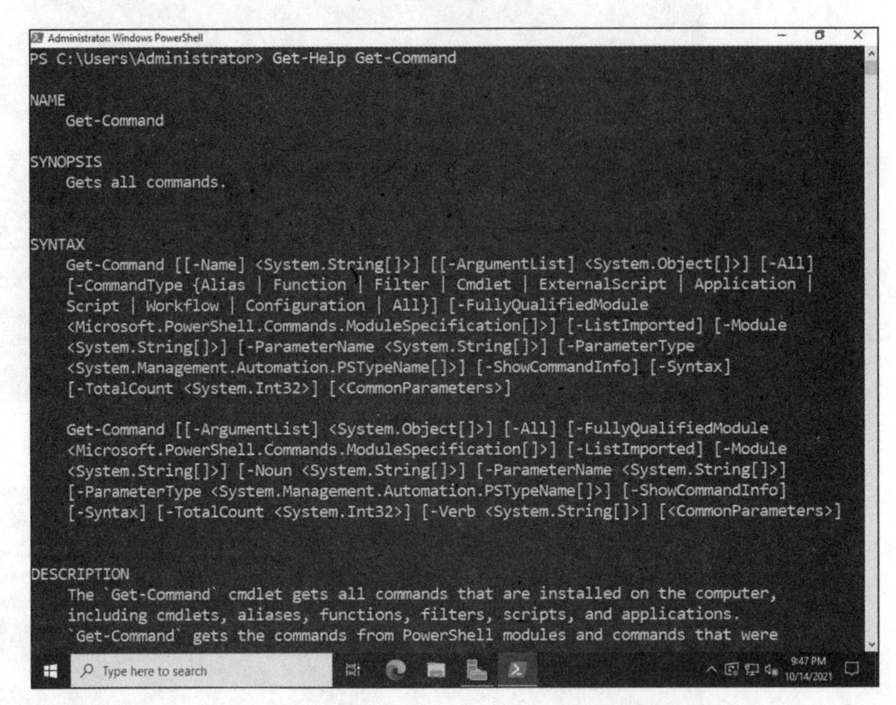

FIGURE 7-9: Using the PowerShell Help to get more information on the `Get-Command` cmdlet.

If you want general help from within Windows PowerShell, such as guidance as far as the syntax of the Help command, simply type **help** and press Enter. You get the Help page for Windows PowerShell that explains what it is and gives some examples, as shown in Figure 7-10.

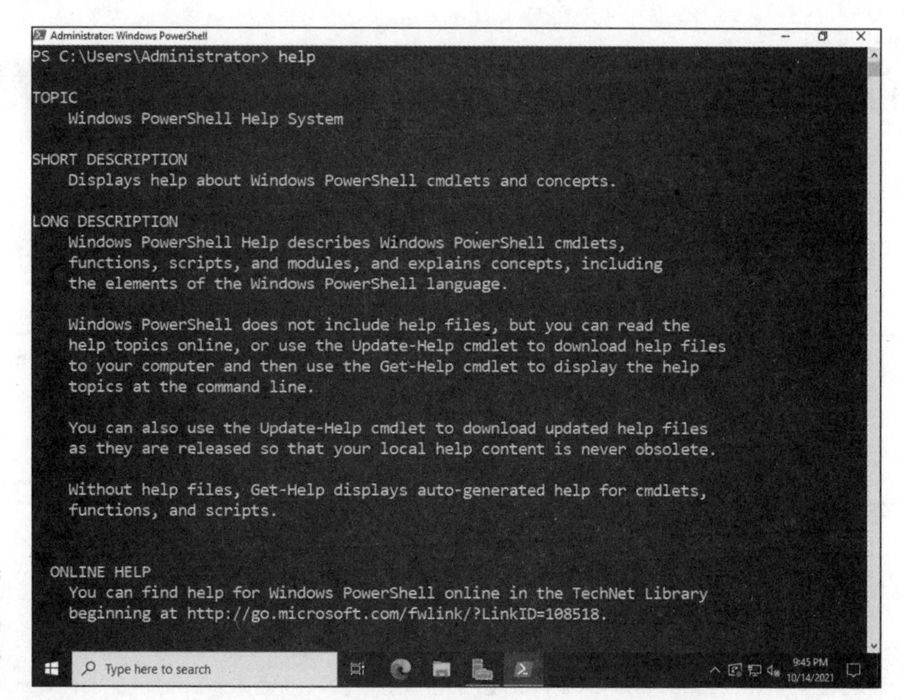

FIGURE 7-10: Using the `help` command in Windows PowerShell.

Understanding PowerShell Punctuation

Where the Command Prompt was all about symbols, Windows PowerShell is all about punctuation. Table 7-1 lists punctuation types that you can use in the Windows PowerShell window.

TABLE 7-1 Windows PowerShell Punctuation

Punctuation	Example	Description
#	`#This is a comment`	Identifies comments. Comments are used in PowerShell scripts to document what sections of code are meant to do, and other information. All text entered after the # on a line is considered a comment.
$	`$myvariable`	Declares variables.
=	`$myvariable=1`	Assigns a value to a variable.
|	`Get-ChildItem | Get-Member`	Takes the output of the first command and passes it into the input of the second command.
"	`"My value is $var"` (The result might be: My value is 5.) This assumes that the variable `$var` was previously set to 5.	Encapsulates text; variables show the appropriate value.

(continued)

(continued)

Punctuation	Example	Description
'	`'My value is $var'` (The result would be: My value is `$var`.) Even if the variable `$var` is set to 5, `$var` will be printed as `$var`.	Encapsulates text; treats text literally so variables are treated as text.
( )	sometext.ToLower() 100/(5+4)*6	Provides arguments for cmdlets and groups items like numbers.
[ ]	$fruit=[apple,orange] -like [sometext]	Typically used for arrays and like comparisons.
{ }	Invoke-Command -ScriptBlock {*cmdlets*}	Used to enclose blocks of code.

PowerShell ISE

The PowerShell Integrated Scripting Environment (ISE) has been an excellent resource for those learning PowerShell or for those wanting to troubleshoot PowerShell scripts. It has two panes for working with PowerShell scripts. The top pane is the text editor, which does color coding and performs command completion suggestions for you with a feature known as IntelliSense.

The second pane is a traditional PowerShell window. You can type commands into it directly, though typically I use it to see what the output of an individual line of code was when I'm troubleshooting why a script is not running properly.

The right side of the screen gives you a Command Explorer of sorts. It lists all the commands you can use and will help you complete the command.

PowerShell ISE is supported through Windows PowerShell 5.1. Starting with PowerShell 6, it's no longer supported and Microsoft's recommendation is to transition to using Visual Studio Code if you want a graphical tool for working with PowerShell scripting. The majority of the PowerShell examples in this book have been done in Visual Studio Code for exactly this reason, though it should be noted that Windows Server 2022 at the time of this writing does ship with PowerShell 5.1.

4
Configuring Networking in Windows Server 2022

Contents at a Glance

IN THIS CHAPTER

» Examining the Network and Sharing Center

» Controlling your network settings

» Configuring TCP/IP on Windows Server 2022

» Understanding DHCP and why you need it

» Understanding DNS and how it makes your life easier

Chapter 1

Overview of Windows Server 2022 Networking

A server can't do its fundamental job without a solid and reliable network to support it. The network you'll support will most likely be an Ethernet network. Ethernet networks use an unshielded twisted pair (UTP) cable. There are different categories of Ethernet cable, but the most common ones are Category 5e (Cat5e) and Category 6 (Cat6). As you go up in category, you gain speed and, in some cases, distance.

If your organization occupies a single building or a small office space, you'll probably be supporting a local area network (LAN), and your organization will most likely own all the network components. If your organization is larger and more geographically dispersed, you may be supporting a wide area network (WAN); in this case, your organization will own some of the network equipment, but an Internet service provider (ISP) will likely own some of the copper or fiber that your traffic is crossing.

In this chapter, I explain how to network in Windows Server 2022. I introduce you to the Network and Sharing Center, and fill you in on how to configure TCP/IP, DNS, and DHCP (all of which I explain in greater detail in the pages that follow).

Getting Acquainted with the Network and Sharing Center

In Windows Server 2022, the Network and Sharing Center gives you a central location to start from for all your networking needs. This utility has been in previous versions of Windows, so if you're a long-time system administrator, this interface will be comfortable for you.

To access the Network and Sharing Center, right-click the Start menu and choose Network Connections. On the Status page, scroll down to Network and Sharing Center.

On the Network and Sharing Center screen (shown in Figure 1-1), you get access to a few of the useful utilities all in one spot.

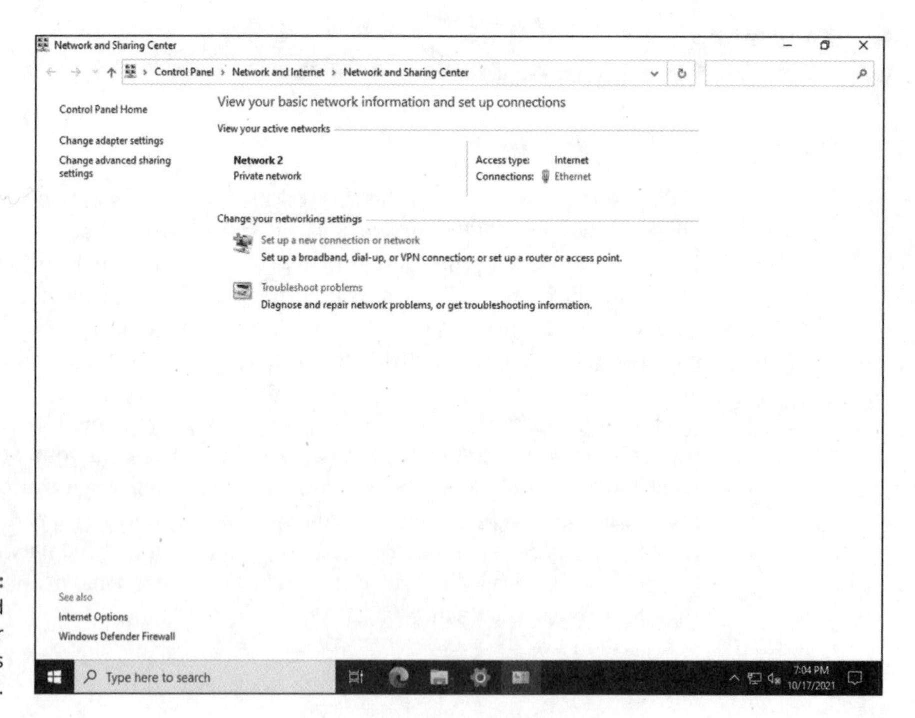

FIGURE 1-1: The Network and Sharing Center in Windows Server 2022.

In the "View Your Active Networks" section, you can see at a glance if your connection is enabled and whether you have Internet connectivity. If you don't have Internet connectivity, the Access Type will say, "No Internet."

On the left side of the screen, you see the Change Adapter Settings link. When you click that link, you're presented with a list of all the network adapters present on your system. Right-click the adapter and choose Properties (see Figure 1-2).

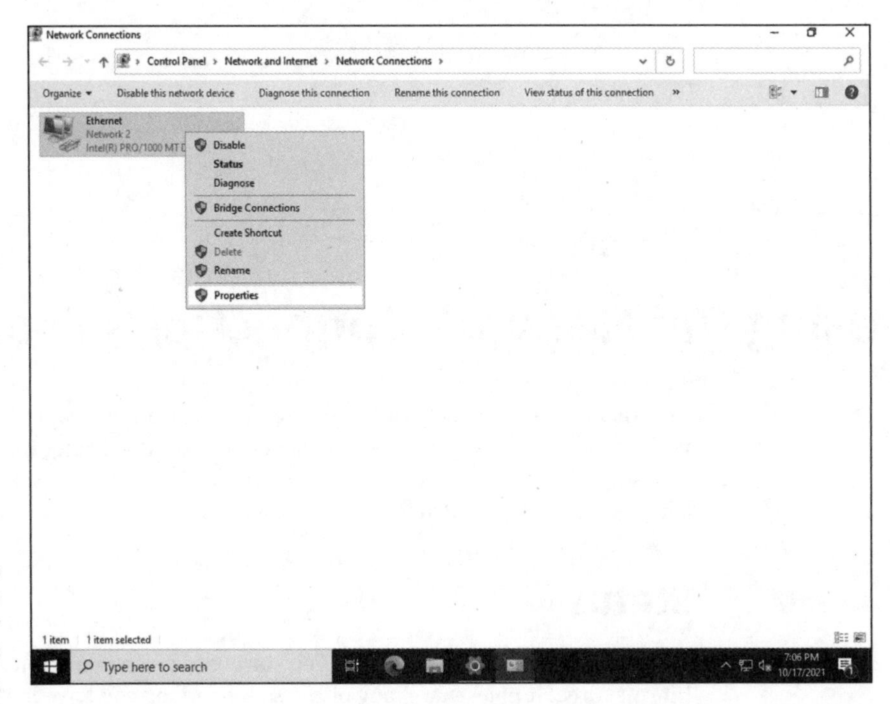

FIGURE 1-2: Getting to the network adapter properties screen.

The Properties screen is where the majority of the TCP/IP configuration takes place. For more on this subject, turn to the "Configuring TCP/IP" section, later in this chapter.

The final section of the Network and Sharing Center that I want to draw your attention to is the Troubleshoot Problems utility. When you click the Troubleshoot Problems link (refer to Figure 1-1), you're taken to the Troubleshoot area. From here, you can click Internet Connections (see Figure 1-3) and get a wizard-based utility that can help you identify and resolve issues.

If the system doesn't think there is a problem, you may need to click Additional Troubleshooters to get the option you need. From there, you can select Internet Connections or choose from a plethora of other troubleshooting options.

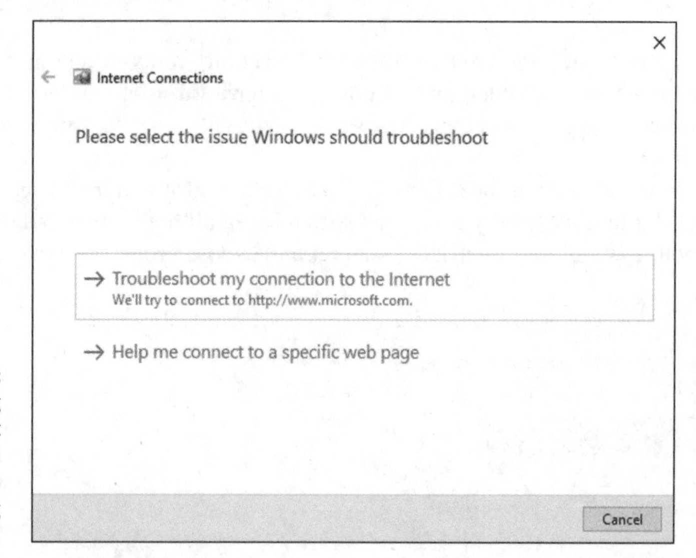

FIGURE 1-3: Troubleshooting your Internet connection with the built-in troubleshooting utility.

Using the Network Connections Tools

In Windows Server 2022, there is a section that allows you to control all your network settings. You can access these tools by right-clicking the Start menu and choosing Network Connections. Alternatively, you can click Start, click Settings (the gear icon), and then click Network & Internet.

Status

The Status page is the default page you start with when you get into the Network & Internet area. It gives you many of the same options you have in the Network and Sharing Center, as well as the status of your network connection (see Figure 1-4).

TECHNICAL STUFF

At the bottom of the page, you can see the "Change Adapter Options" link and the "Network Troubleshooter" link, just as you did in the Network and Sharing Center. My guess is that this screen will eventually replace the Network and Sharing Center because the toolset on this screen is identical.

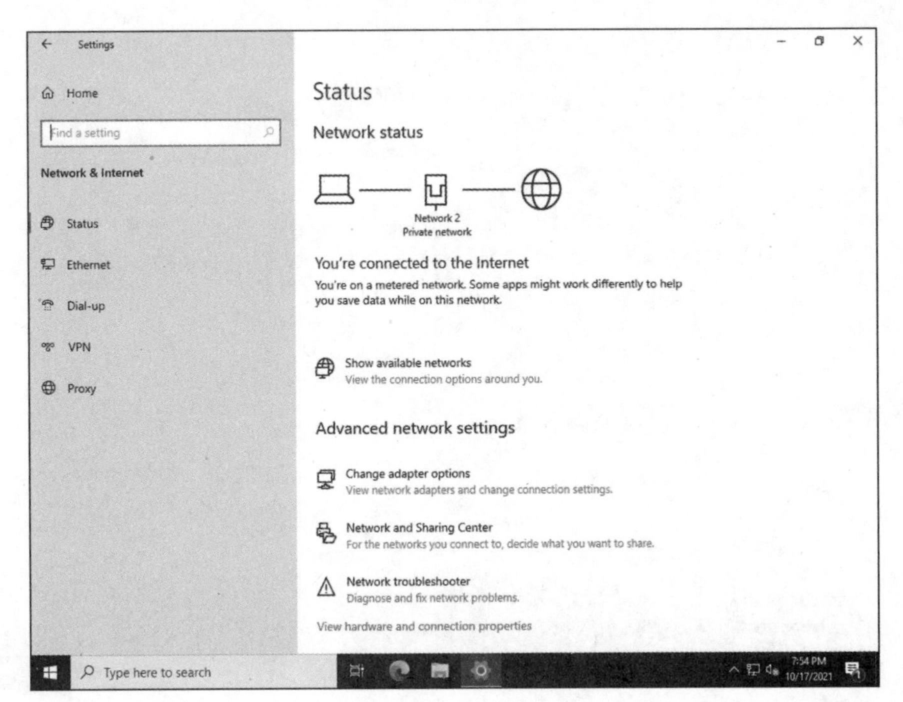

FIGURE 1-4: The Status screen in the Network & Internet section of Settings.

Ethernet

When you click the Ethernet link on the left-hand menu of the Network & Internet area, you're presented with options specific to the Ethernet connection (see Figure 1-5):

- **Change Adapter Options:** If you click this link, you get a list of all the adapters installed on the system. You can choose which adapters you want to work with from there.
- **Change Advanced Sharing Options:** If you click this link, you can change the network discovery and file and printer sharing settings for your network profiles.
- **Network and Sharing Center:** Clicking this link opens the good old-fashioned Network and Sharing Center.
- **Windows Firewall:** Clicking this link opens the newer Firewall & Network Protection screen, shown in Figure 1-6. From here, you can allow specific applications through the firewall, enable or disable the firewall for the different profiles, and tweak the notifications the firewall will make. You can also enter the Advanced Settings area, which lets you specify more granular rules by Internet Protocol (IP), port number, and so on.

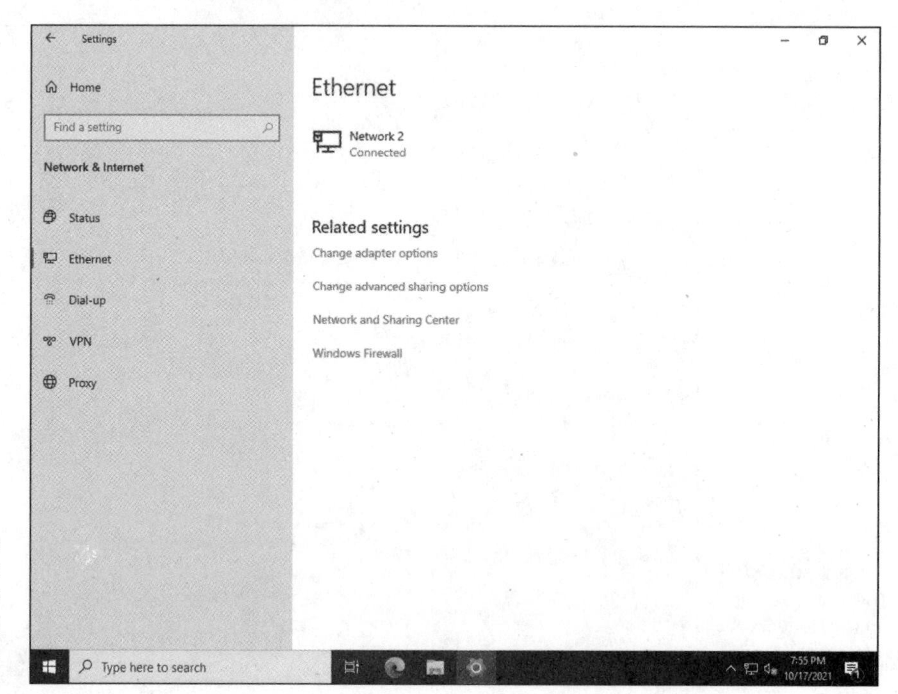

FIGURE 1-5: The Ethernet screen in the Network & Internet section of Settings.

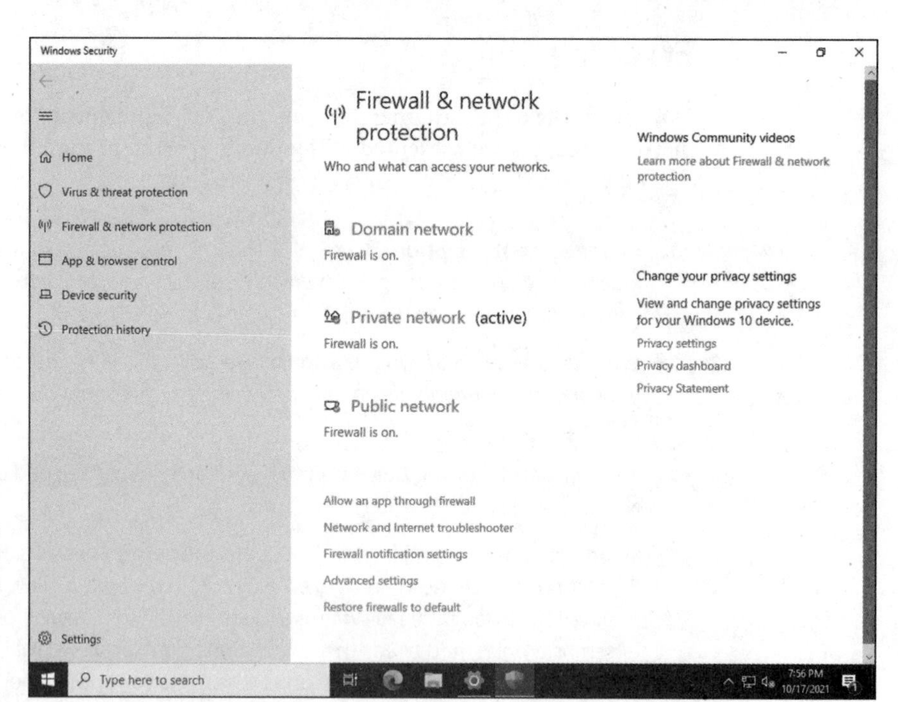

FIGURE 1-6: The Firewall & Network Protection screen in the Network & Internet section of Settings.

Dial-up

When you click the Dial-up link on the left-hand menu of the Network & Internet area, you can create a new connection if you have a modem attached to your system. A wizard guides you to set the number you need to dial out to for service.

VPN

When you click the VPN link on the left-hand menu of the Network & Internet area, you can create a virtual private network (VPN) connection (see Figure 1-7). By default, the only VPN provider available is built into Windows. You need to name the connection and then tell it the address of the VPN server that you're connecting to.

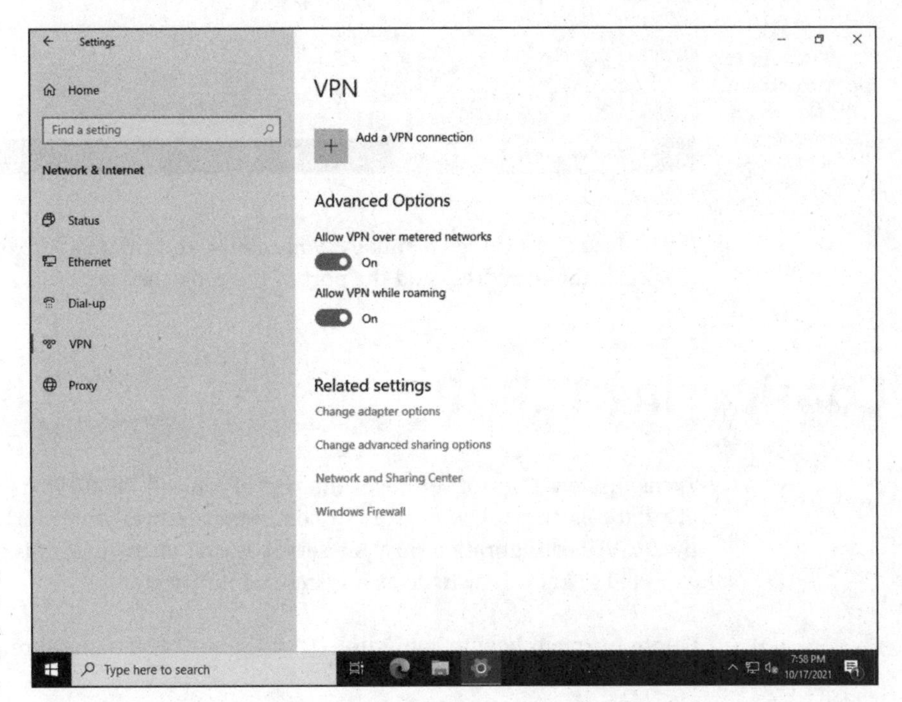

FIGURE 1-7: The VPN screen in the Network & Internet section of Settings.

Proxy

When you click the Proxy link on the left-hand menu of the Network & Internet area, you can set up the proxy settings — which you'll need to do if your organization uses a proxy server (see Figure 1-8). If you're using an automatic configuration script, you can turn on the Automatically Detect Settings switch and the Use Setup Script switch.

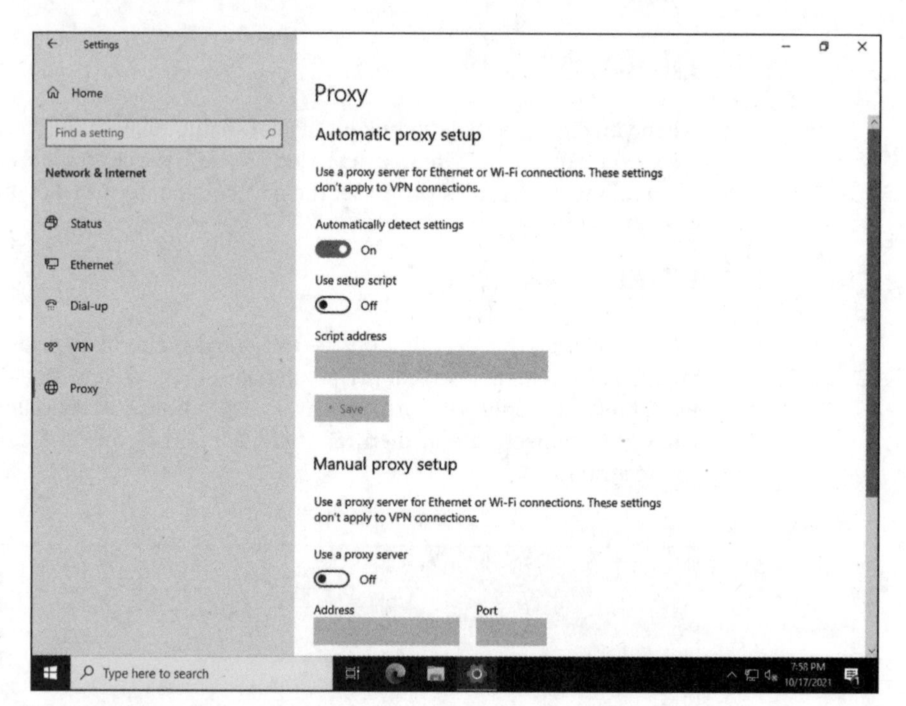

FIGURE 1-8: The Proxy screen in the Network & Internet section of Settings.

If you need to set things manually, you can use the "Manual Proxy Setup" section to specify the IP address and the port of the proxy server.

Configuring TCP/IP

Transmission Control Protocol/Internet Protocol (TCP/IP) is a whole suite of protocols that allow devices to communicate over a network. Working with the TCP/IP configuration on your server is one of the basic tasks that you'll be expected to know how to do as a system administrator.

Before I explain how to configure TCP/IP, I need you make sure you understand a few terms:

- **IP address:** An IP address is a number that uniquely identifies a system on a network. There are two versions of IP addresses:
  - **IPv4:** A 32-bit address that identifies a system on an IPv4 network (for example, 192.168.10.10)
  - **IPv6:** A 128-bit address that identifies a system on an IPv6 network (for example, FE80:0000:0000:0000:0202:B2EF:FC4B:5749)

» **Domain Name System (DNS):** Translates hostnames to IP addresses using forward lookup zones, and IP addresses to hostnames with reverse lookup zones.

» **Windows Internet Name Service (WINS):** Responsible for converting NetBIOS names to IP addresses. WINS was primarily used in older versions of Windows (Windows 2000, Windows XP, and Windows Server 2003).

Now that you know the key terms, let's assign an IP address to a system. Follow these steps to walk through the settings you can change:

1. **Right-click the Start menu, and choose Network Connections.**
2. **Click Change Adapter options.**
3. **Right-click one of the adapters and choose Properties.**

   The Properties dialog box for the adapter appears.

4. **Select Internet Protocol Version 4 (TCP/IPv4) and click the Properties button (as shown in Figure 1-9).**

   The Internet Protocol Version 4 (TCP/IPv4) Properties dialog box appears. By default, this is set to obtain an Internet Protocol (IP) and Domain Name Server (DNS) address automatically. You can see in Figure 1-10 that the server has static addresses set. This is because it's a DNS server and domain controller.

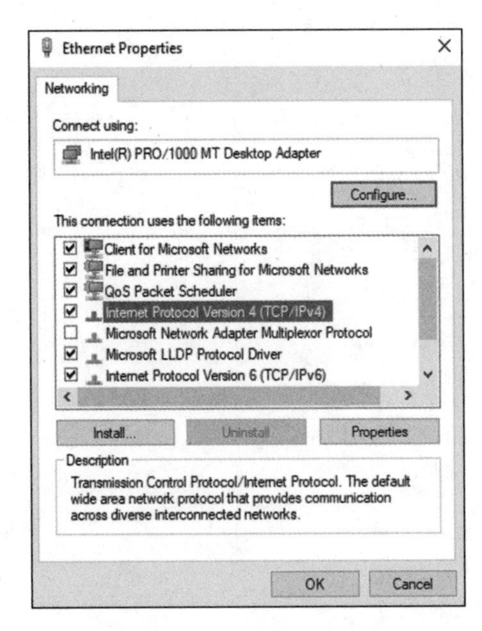

FIGURE 1-9: Selecting which network protocol you want to work with in the adapter's Properties dialog box.

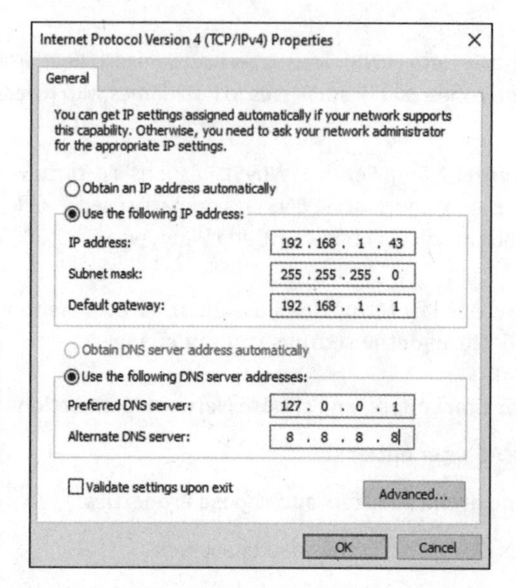

FIGURE 1-10: Statically set IP and DNS server settings are common on servers serving critical infrastructure services.

You can configure additional, more advanced settings as well. In the Internet Protocol Version 4 (TCP/IPv4) Properties dialog box (refer to Figure 1-10), click the Advanced button. The Advanced TCP/IP Settings dialog box has three tabs that allow you to make more advanced configurations in relation to your IP addressing, DNS servers, and WINS servers, shown in Figure 1-11.

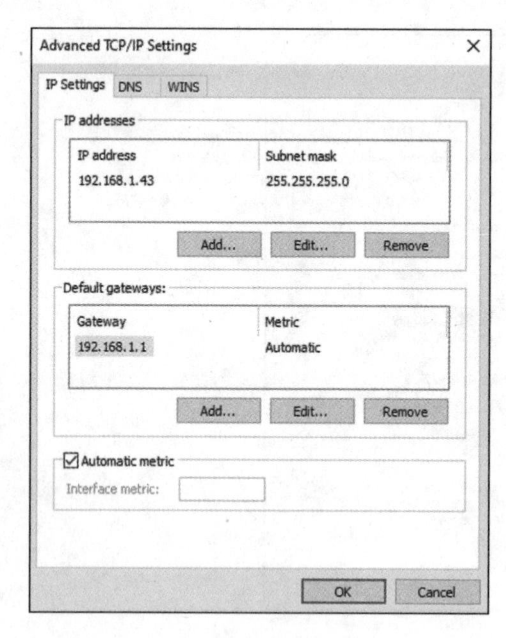

FIGURE 1-11: The Advanced TCP/IP Settings dialog box lets you perform more advanced configuration tasks.

The IP Settings tab can be used to add, edit, or remove IP addresses or gateways. The DNS tab allows you to add, edit, or remove DNS servers. (The basic configuration screen allows two servers to be configured, but from this tab you can configure additional servers if you need to.) If you're using WINS, the WINS tab allows you to specify WINS servers that are available in the network.

Understanding DHCP

Dynamic Host Configuration Protocol (DHCP) makes your life easier by automatically assigning an IP address to a system. It manages addresses that are in use and ensures that duplicate IP addresses are never issued. By default, an address is leased for eight days, at which point the lease can be renewed or the IP address can be reassigned to another system.

DHCP shifts the burden of assigning and tracking IP addresses from the system administrator and a manual spreadsheet to a more automated process. Especially for large organizations, DHCP ensures that the available IP address space is utilized more efficiently. If a device with a lease is removed from the network or is offline, the lease is removed when it expires, and the IP address is made available to another device.

Automating the provisioning of IP addresses is desirable, but there will be instances when you need to set a static IP address that won't change. Systems that host major infrastructure services like DNS, DHCP, and Active Directory should have static IP addresses. You can still manage the static IP addresses in DHCP by setting a reservation so that the IP address is accounted for but DHCP will not re-issue it.

DHCP can provide configuration other than simply issuing IP addresses. By using DHCP options, you can set things like the default gateway, the name servers for a network, imaging servers that are available, and more.

So, how does DHCP work? One of the easiest ways to remember is with the *DORA* acronym; *DORA* stands for Discover, Offer, Request, and Acknowledge:

1. **Discover.**

   A DHCP client requests an IP address by sending a DHCPDiscover message out to its local subnet as a broadcast.

2. **Offer.**

   The DHCP server makes an offer to the client using a DHCPOffer message, which contains the IP address and configuration information, including the lease time.

3. **Request.**

   The DHCP client broadcasts a DHCPRequest to indicate that it has accepted what was sent.

4. **Acknowledge.**

   As a last step, the DHCP server broadcasts a DHCPAck message, which lets the client know that the lease has been finalized.

DHCP uses ports UDP/67 and UDP/68. UDP/67 is used as the destination port on the DHCP server, and UDP/68 is used by the DHCP client.

Defining DNS

Domain Name System (DNS) is the service that is used to map human-friendly names like `www.dummies.com` to an IP address, which is how a computer addresses locations.

You need to understand a few terms to understand how DNS breaks down addresses:

- **Top-level domain:** The top-level domain is used to indicate the country of origin or the type of organization. For example, a commercial organization might use `.com`, or a website in Brazil might use `.br`. Common top-level domains include
  - `.com` (commercial)
  - `.edu` (educational institutions)
  - `.org` (usually used by not-for-profit organizations)
  - `.net` (an alternative to `.com`)
  - `.gov` (government sites)
  - `.mil` (military sites)
  - Country codes like `.us`, `.br`, `.tk`, `.cn`, and so on
- **Second-level domain:** A second-level domain is registered to either an individual or an organization. For instance, `dummies.com` is a second-level domain.
- **Subdomain:** Subdomains are additional names that an organization chooses to register. An example of a subdomain would be the `www` in `www.dummies.com`.

Zones in DNS are used to separate administrative boundaries within a common DNS namespace (like sometestorg.com). Multiple subdomains can exist within the same zone, and multiple zones can exist on the same DNS server. For example, I may have three subdomains: hr.sometestorg.com, sales.sometestorg.com and legal.sometestorg.com. My HR and Sales subdomains are managed by the same group of people, so they are in the same zone. My Legal domain is managed by a different group of people, so it is in its own zone. Each zone stores its information in a DNS zone file. The DNS zone file contains all of the records that exist in the zone.

DNS records are stored in DNS zones. There are several types of DNS records you may find in a zone, and the record type defines what kind of record you're using. Subdomains, for example, are typically defined by an A record. Table 1-1 lists common DNS record types.

TABLE 1-1 DNS Record Types

Record Type	Description
SOA	Start of Authority defines the primary DNS server name, refresh intervals, and time-to-live settings.
A or AAAA	A records are host records for IPv4 addresses; AAAA records are host records for IPv6 addresses. This record provides a mapping of a hostname to an IP address.
PTR	Maps an IP address to a hostname and is used for reverse DNS lookups.
NS	Defines name servers for the DNS zone.
MX	Defines the mail exchange server's DNS record.
CNAME	CNAMEs are used to create an alias record. For example, you might have `server1.example.com`, but you want people to use the name `www.myawesomesite.com`. You can accomplish that by creating a CNAME with the desired URL and point the CNAME to the A record for `server1.example.com`.

Now that you know all this, you may be asking, "But how does DNS work?" The simple answer: DNS queries. When a DNS client needs to resolve a record, it sends a DNS query to a local DNS server. If that server knows the address, it responds with the IP address. If it doesn't know the address, it can query another server.

Let's use the following example: You want to reach `www.dummies.com` because you're no dummy. You type the address into your browser and behind the scenes, this is what happens:

1. **The DNS client queries the local DNS server if it knows who `www.dummies.com` is.**
2. **The local server does not know, so it sends a query to the root server to get an authoritative DNS server for `.com`.**

   It receives a referral for the `.com` DNS servers.
3. **The local server queries the `.com` servers for `www.dummies.com`.**
4. **The `.com` server may not know the address, but it provides the address in a referral for the DNS server for `dummies.com`.**
5. **From there, the local server can query the `dummies.com` DNS server for `www.dummies.com`, and will receive a valid IP address back as an answer to the original query.**

This all happens within the span of the few seconds that it takes for you to reach the website. The answer is cached by the DNS client for however long the time to live (TTL) is set, so future requests can be answered by the cache on the local system instead of having to go through the queries again.

DNS uses port 53 to communicate. Regular DNS queries are made over UDP/53; however, larger queries like IPv6 and DNSSEC queries need TCP/53. TCP/53 is also used for zone transfers, which is why it was historically blocked by organizations at the firewall. However, if your organization plans on using IPv6 or DNSSEC, you need to allow it.

Creating a DNS zone

At some point in your career, it's highly likely that you'll be asked to create a zone. When I was explaining zones earlier, I had mentioned the use case of having one zone with two subdomains: a Sales subdomain and an HR subdomain. In that example, I had mentioned needing a new zone for the Legal subdomain. Here are the steps involved in setting up the new zone:

1. **From Server Manager, choose Tools⇨DNS.**
2. **Right-click Forward Lookup Zones and click on New Zone.**
3. **In the Welcome to the New Zone Wizard, click Next.**
4. **On the Zone Type screen, select the radio button next to Primary Zone, and click Next.**

On the Zone Type screen, you're prompted to check the Store the Zone in Active Directory check box. You must be on a writeable domain controller to select this option. I suggest selecting this to get the fault tolerance of Active Directory integrated zones.

5. **On the Active Directory Zone Replication Scope screen, select the radio button next to To All DNS Servers Running on Domain Controllers in This Domain: <*domain_name*>, shown in Figure 1-12.**
6. **Click Next.**
7. **On the Zone Name screen, enter the name of the zone that you want to create and click Next.**

   In this example, I'll enter legal.sometestorg.com.

   On the Dynamic Update screen, you have three options. If you opted to store the DNS zone in Active Directory all three options will be available. If you chose not to store the DNS zone in Active Directory, you won't be able to select Allow Only Secure Dynamic Updates.

8. **Select Allow Only Secure Dynamic Updates and click Next.**
9. **On the Completing the New Zone Wizard screen, click Finish.**

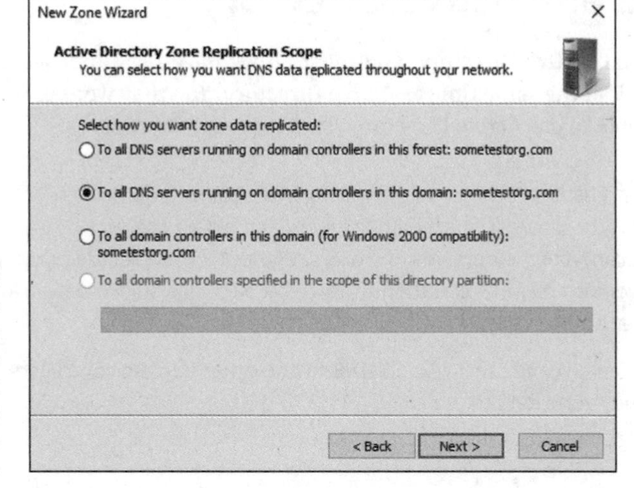

FIGURE 1-12: Selecting the replication scope for the new zone is important; in this case, the scope is at the domain level.

If you've followed along with these steps, your screen should look similar to Figure 1-13. You can see the new zone for the subdomain of legal.sometestorg.com.

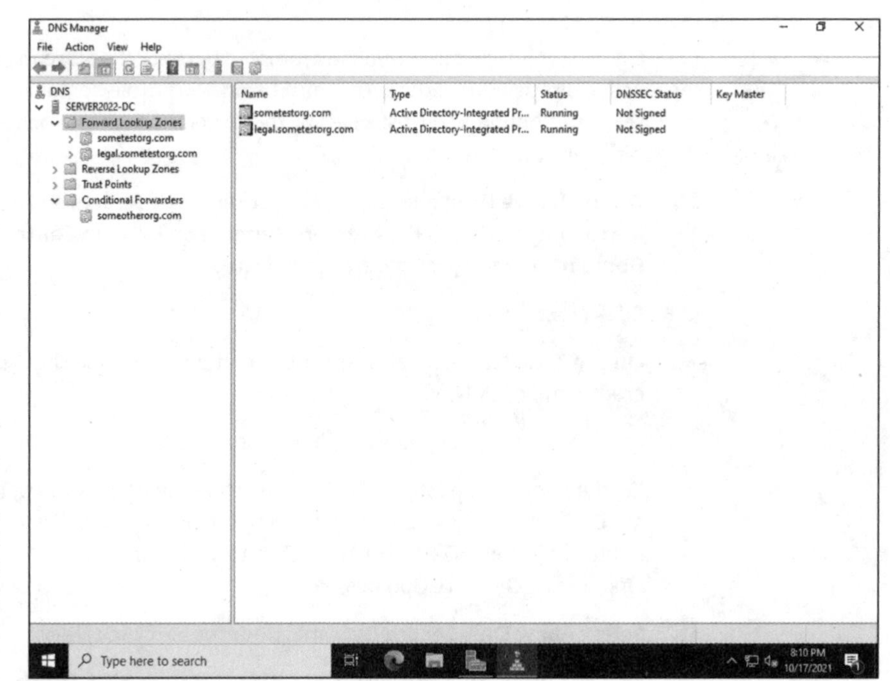

FIGURE 1-13: The legal.sometestorg.com zone provides a separate administrative boundary from the sometestorg.com zone.

DNS and Active Directory

To install Active Directory, you must either have DNS installed ahead of time or install it at the same time as Active Directory. You may wonder why you need DNS to be able to use Active Directory. There a couple of reasons:

- DNS assists systems on the network in locating a domain controller by using locator records, which help the workstations and servers by providing the location of the domain controllers. Without locator records, your systems wouldn't be able to authenticate, because they wouldn't have a location to send authentication traffic.
- Domain controllers rely on DNS to find other domain controllers to replicate their zone data to.

Making DNS fault tolerant

When you create a DNS server, you specify whether it will be the primary DNS server or the secondary DNS server for a zone. There can only be one primary DNS server for a zone, but you can have multiple secondary DNS servers on a zone. The primary DNS server for the zone serves queries that come in and, more important, accepts changes and additions to zone records. Secondary DNS servers can service queries but can't accept additions or changes to the zone records. They contain a read-only copy of the zone that is copied from the primary DNS server. If the primary DNS server for the zone goes down, you can promote the secondary DNS server for the zone to the primary. This is not an automatic process — as the server administrator, you must initiate it.

If you're using Active Directory Integrated DNS zones, you're automatically fault tolerant. DNS zones that are Active Directory integrated store their records in Active Directory. The DNS servers that are considered authoritative for those zones exist in a multi-master configuration. If one of the DNS servers was to go down, the other DNS servers would continue to service queries without any issues. Just remember that to be fault tolerant, you need at least two DNS servers that are authoritative for the Active Directory integrated zone.

IN THIS CHAPTER

» Viewing network properties on Windows Server 2022

» Connecting to other networks and using virtual private networks

» Managing and configuring network connections in Windows Server

Chapter 2
Performing Basic Network Tasks

System administrators are expected to know the basic ins and outs of how to configure networking on a Windows server. You may need to change the IP address on a server for instance. Certain servers should have their addresses set statically. These are generally going to be critical infrastructure systems like Active Directory Domain Services (AD DS), Domain Name System (DNS), and Dynamic Host Configuration Protocol (DHCP) servers.

In this chapter, I cover how to do the basic configuration on a network interface card and delve into some other neat things that you can do with networking in Windows Server 2022.

Viewing Network Properties

Looking at the properties of your network adapter gives you a quick and simple way to see how your system is configured to communicate on your network. To look at your network properties in Windows Server 2022, follow these steps:

1. **Click the Start Menu, and then click the gear icon to open the Settings menu.**
2. **Click Network & Internet.**
3. **Choose Ethernet from the left-hand menu.**
4. **Click Change Adapter Options.**

When you select the network adapter, you have several options that appear in a bar across the top of the screen, as shown in Figure 2-1:

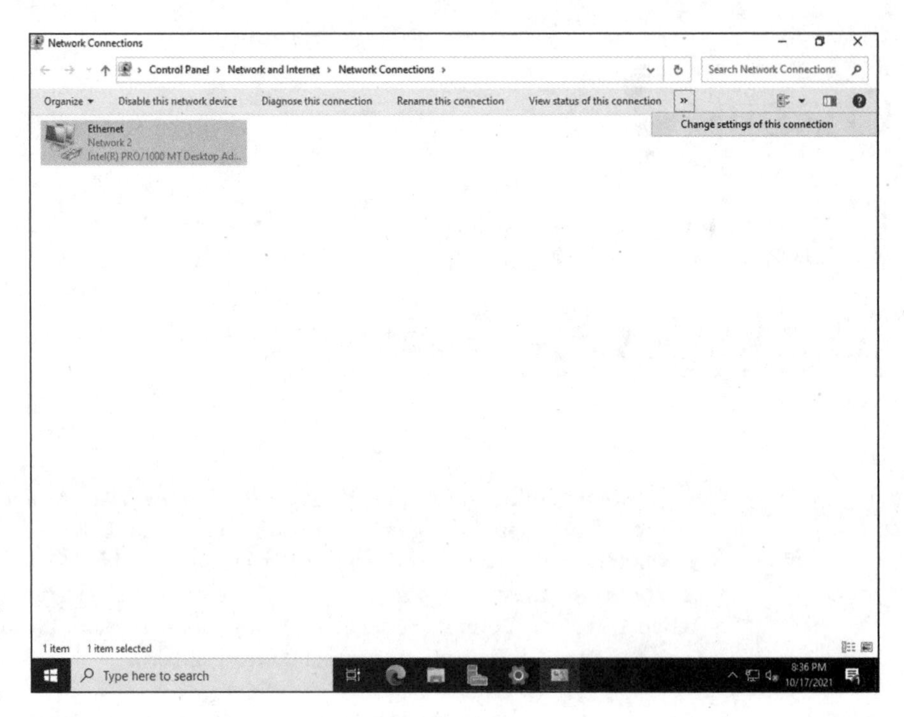

FIGURE 2-1: Working with the properties of a network adapter.

- **Disable This Network Device:** Disables the selected network adapter.
- **Diagnose This Connection:** Used to troubleshoot connection issues related to network connectivity.
- **Rename This Connection:** Allows you to rename the connection. This is helpful if there are multiple network adapters and you need to keep track of what each one is doing.
- **View Status of This Connection:** Shows you the status of the network connection.
- **Change Settings of This Connection:** Bring up the Properties dialog box for the network adapter that was selected.

There are also options if you right-click the network adapter. The majority of the options match what appeared on the top bar when you selected the network adapter, with the exception of Bridge Connections, which allows you to bridge two network adapters so that the operating system sees them as a single network adapter.

Assuming that you want to configure the network adapter, you can either right-click on the adapter and choose Properties or select Change Settings of This Connection from the top bar when the network adapter is selected. After you've opened up the Properties dialog box for the network adapter, you should see something that looks like Figure 2-2.

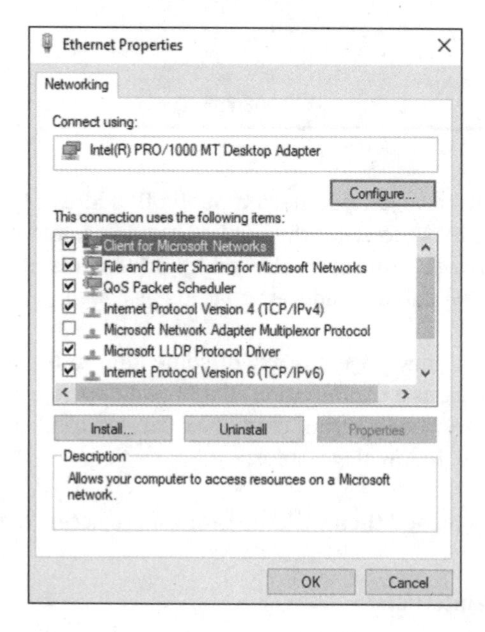

FIGURE 2-2: The Properties dialog box for the network adapter allows you to change the network adapter's configuration.

Some of the items in the network adapter Properties dialog box can only be uninstalled, while others can be configured via the Properties button. Plus, you can install new network features by clicking the Install button.

Connecting to Another Network

Connecting to your local network is a great first step, but in most cases you really want to be able to connect to another network. The best and most common example of this is connecting to the Internet. The Internet is an entirely different network from the one that your computer is on, and you need to be set up properly to access it.

Connecting to the Internet

In some networks (like a home network, for example), you connect your computer to a router or a switch attached to a cable modem, and setup is automatic. You're able to access the Internet within minutes. If you were to check your IP address, you would have one of the non-routable internal IP addresses, from one of the ranges shown in Table 2-1.

TABLE 2-1 **IPv4 Private Address Ranges**

Subnet	Range
10.0.0.0/8	10.0.0.0–10.255.255.255
172.16.0.0/12	172.16.0.0–172.31.255.255
192.168.0.0/16	192.168.0.0–192.168.255.255

To get to the Internet however, you need a public IP address. The cable modem you lease from your ISP is receiving the public IP address, and it's usually doing the "translation" between your internal IP address (which is not routable on the Internet) to the routable public IP address that it's assigned.

In an organization, you may have a proxy of some kind in between you and the Internet. Proxies can act as a combination of a firewall and a web filter; they can protect your system from dangerous traffic, as well as block known malicious sites. To set up a proxy, follow these steps:

1. **Click the Start menu and then click the gear icon to access the Settings menu.**
2. **Click Network & Internet.**
3. **Click Proxy.**

   In an organization, if you're using a proxy, you'll most likely have a setup script.
4. **Click the Use Setup Script toggle, and enter the Script address, including the name of the** `.pac` **file.**
5. **Click Save.**

After you hit save, your server's settings should look similar to Figure 2-3. The location of the proxy script will be different, of course.

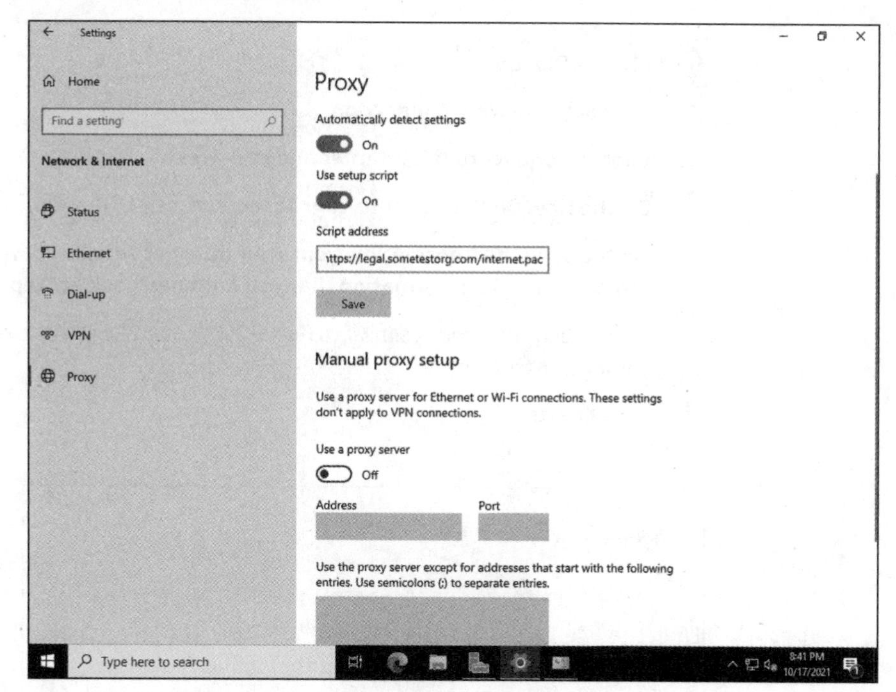

FIGURE 2-3: Setting up a proxy script for Internet access on a Windows server.

Setting up a dial-up connection

Okay, I know you probably giggled when you saw this header. Who uses dial-up anymore, right? Believe it or not, dial-up is still around, though it's certainly becoming less common. Why might you use dial-up? Well, there are a few reasons why it might be a good solution depending on your use case:

- Maybe you have a traditional phone line but DSL is not offered in your area.
- Maybe you need a reliable, consistent connection, and you're really just checking emails or doing simple Google searches.
- Maybe you need Internet access, but you need it to be as inexpensive as possible (dial-up is approximately $20 a month).

Whatever the reason, if you need to use a dial-up connection, here's how to set it up:

1. **Sign up with a dial-up service and get the service number from them.**

   This is the number you will call to connect.

2. **Click the Start menu, and click the gear to access the Settings menu.**
3. **Click Network & Internet.**

4. **Choose Dial-up.**
5. **Click Set Up a New Connection.**
6. **Choose Connect to the Internet and click Next.**
7. **On the How Do You Want to Connect screen, select Dial-up.**
8. **On the Type the Information from Your Internet Service Provider (ISP) screen, enter the information that you obtained back in Step 1.**

   Your screen should look similar to Figure 2-4, though for obvious reasons the entries won't match.

9. **Click Create.**

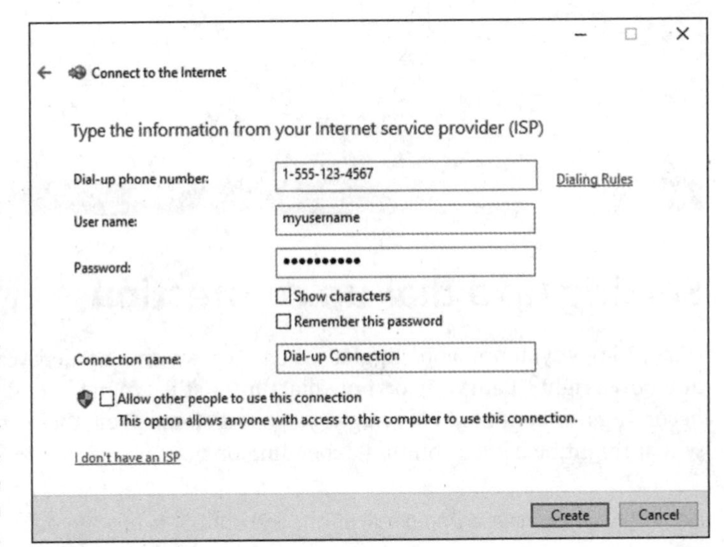

FIGURE 2-4: Configuring a dial-up connection in Windows Server 2022 is simple if you have the information from your ISP.

Connecting to a virtual private network

You can use a virtual private network (VPN) to gain remote access to a network. The great thing about VPNs is that they allow you to work as if you were actually on your work network.

If your workplace uses a VPN to connect, you can use the built-in Windows VPN client. Here's how to configure that:

1. **Click the Start menu and then click the gear icon to access the Settings menu.**
2. **Click Network & Internet.**

3. **Select VPN.**
4. **Click Add a VPN Connection.**
5. **For VPN Provider, select Windows (Built-in).**
6. **Enter a connection name, and the address of the VPN server that you want to connect to.**

   Your screen should look similar to Figure 2-5, though your fields will have different data in them.

7. **Enter your username and password, and click Save.**

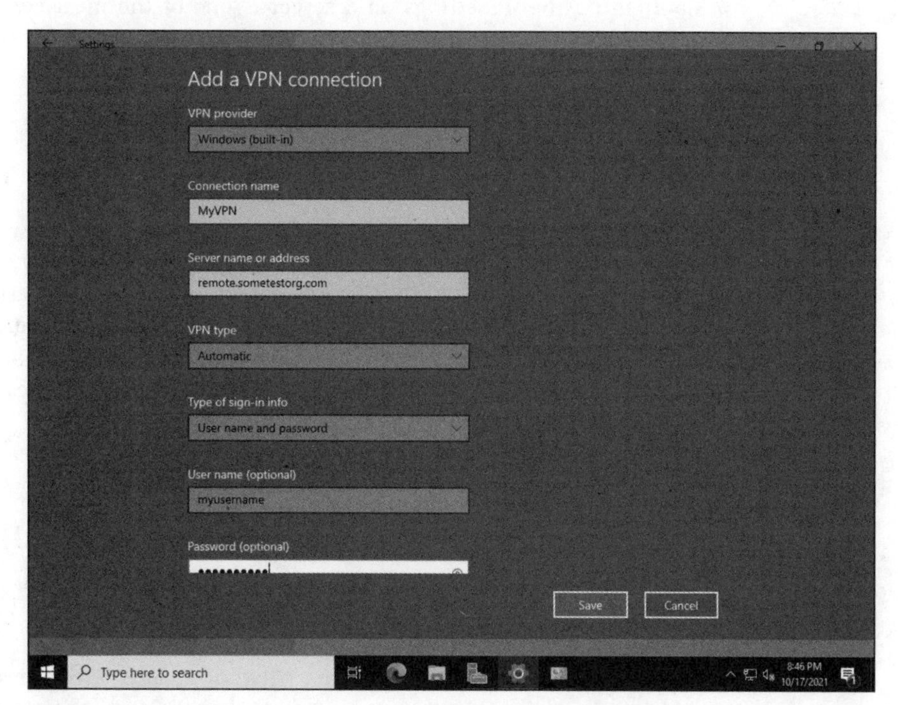

FIGURE 2-5: Setting up a VPN for remote access with the built-in Windows VPN client.

Managing Network Connections

Changing the IP address is pretty standard activity, but some of the other options in the network adapter's Properties dialog box aren't so straight forward. In this section, I walk you through each of these options, because these are the ones that you'll most likely work with.

Understanding the Client for Microsoft Networks feature

Client for Microsoft Networks has no configuration that you can do on it. You can uninstall it, but that's it. The Client for Microsoft Networks feature is required to allow a client to remotely access files, printers, and other shared resources on a Windows server. It's installed by default, and it should not be uninstalled.

Configuring the Internet Protocol

Configuring the IP settings on a server is one of the most common tasks that system administrators will do on a server. Before we move on to configuring, let's look at a little terminology first.

The *IP address* is the address given to a system. It's how other systems on the network will address your system.

The *subnet mask* identifies which part of the address is a network address and which part of the address is a host address. For instance, 172.22.0.0/16 has a subnet mask of 255.255.0.0. The /16 is referred to as Classless Inter-Domain Routing (CIDR) notation and tells me how many bits the network portion of the address takes up. Other systems on this same network will all have IP addresses that start with 172.22.x.x, and the x refers to the host part of the address that will differ from system to system.

The *default gateway* is the IP address used by a system to reach systems in other networks.

Let's get a little background on IPv4 and IPv6 and then look at how you can configure each.

To start with, you need to open the network adapter Properties dialog box for whichever network adapter you want to work with. Here are the steps to get to the Properties dialog box:

1. **Click the Start menu, and then click on the gear icon to open the Settings menu.**
2. **Click Network & Internet.**
3. **Choose Ethernet from the left-hand menu.**
4. **Click Change Adapter Options.**
5. **Right-click the desired network adapter and choose Properties.**

IP Version 4

IP Version 4 (IPv4) addresses are 32-bit addresses. Each number represents 8 bits in binary — for instance 255 is represented as 1111 1111.

To configure a static IPv4 address, select Internet Protocol Version 4 (TCP/IPv4) and click Properties. By default, this is set to Obtain an IP Address Automatically and Obtain DNS Server Address Automatically. You can select Use the Following IP Address and fill in the IP address, the subnet mask, and the default gateway in the top half of the dialog box. Then select Use the Following DNS Server Addresses and fill in the preferred DNS servers at the bottom, and click OK. See Figure 2-6 for an example.

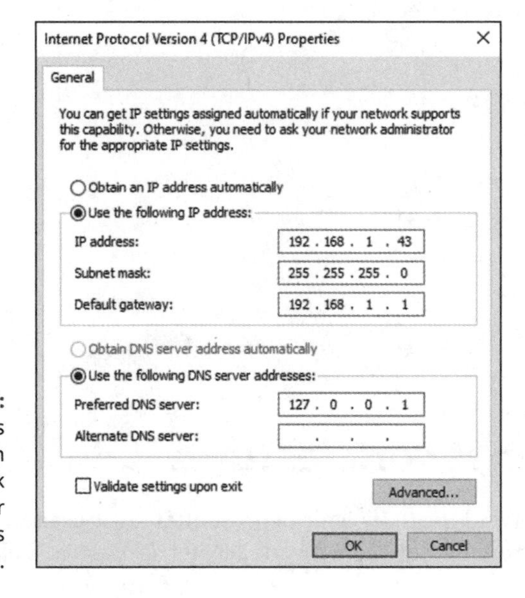

FIGURE 2-6: IPv4 properties are set through the network adapter Properties dialog box.

IP Version 6

IPv4 addresses have been in use for quite some time and it was recognized that eventually there would be no more public IPv4 addresses available. The last available public IPv4 address available from the American Registry for Internet Numbers (ARIN) was issued in September 2015. IP Version 6 (IPv6) was created to address the issue of running out of IPv4 addresses. Instead of a short 32-bit address, IPv6 addresses have a much longer 128-bit address. The address comprises 8 groups of 16 bits, separated by colons.

To configure a static IPv6 address, select Internet Protocol Version 6 (TCP/IPv6) and click Properties. By default, this is set to Obtain an IP Address Automatically and Obtain DNS Server Address Automatically. You can select Use the Following IPv6 Address and fill in the IP address, the subnet mask, and the default gateway in the top half of the dialog box. Then select Use the Following DNS Server Addresses and fill in the preferred DNS servers in the bottom of the dialog box, and click OK. See Figure 2-7 for an example.

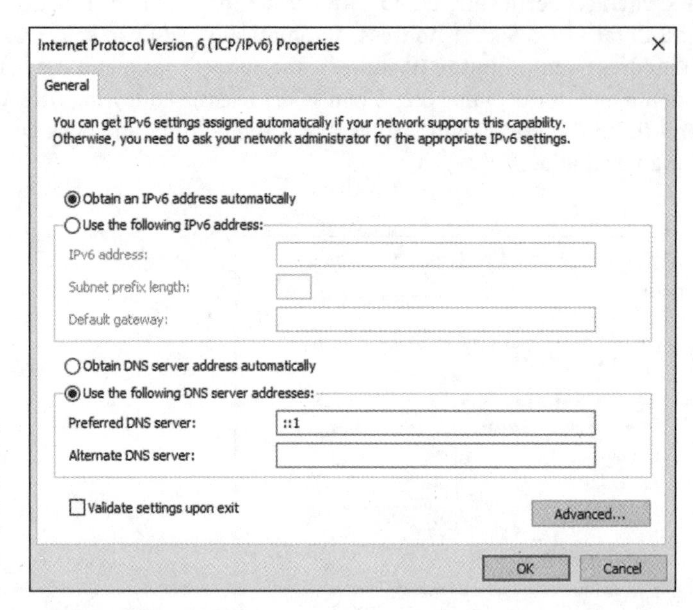

FIGURE 2-7: IPv6 properties are set through the network adapter Properties dialog box.

Installing network features

Installing new network features isn't done as often as it used to be. The option is still available from within the network adapter Properties dialog box. Follow these steps:

1. **From the network adapter's Properties dialog box, click Install.**

   You have a choice of Client, Service, or Protocol.

2. **Select Protocol and click Add.**
3. **Select Reliable Multicast Protocol and click OK.**

The new protocol now shows up in the list of installed protocols, as shown in Figure 2-8.

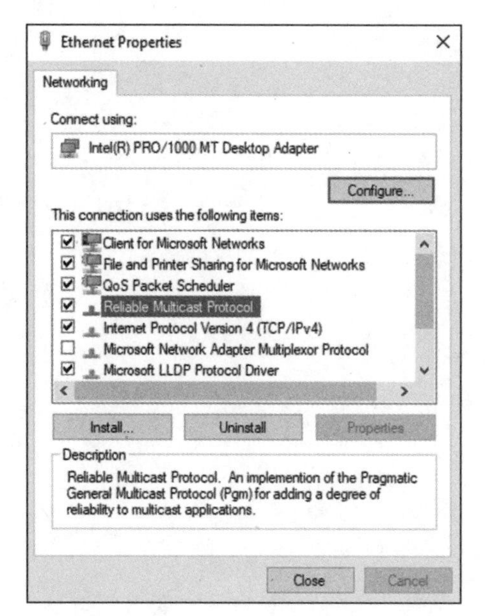

FIGURE 2-8: It's simple to install clients, services, and protocols like the Reliable Multicast Protocol.

Uninstalling network features

Uninstalling network features is similar to adding them. You simply select what you want to uninstall, and then you click the Uninstall button.

WARNING

Be careful uninstalling network features. Uninstalling the wrong thing, like Client for Microsoft Networks, could have very bad outcomes.

You remember how we installed Reliable Multicast Protocol in the preceding section? Let's uninstall that now:

1. **Select Reliable Multicast Protocol.**
2. **Click Uninstall.**

   You get a dialog box asking you to confirm that you want to uninstall it.
3. **Click Yes.**

It really is that simple to uninstall a network feature. Always keep in mind that there is risk when removing features like these on production servers, so you should always use a test environment first to ensure that the change you're making is a safe one.

» Working with Network Policy and Access Services on Windows Server 2022

» Troubleshooting network issues at the command line

Chapter 3 Accomplishing Advanced Network Tasks

Knowing how to set up the basics of Windows Server networking is a must, but knowing how to set up some of the more advanced services can be very important, too. Although the network provides access to resources, sometimes you need to allow remote access or set up access for network devices to leverage your Active Directory (AD) infrastructure for authentication.

In this chapter, I explain how to set up Remote Desktop Services (RDS), and tell you what's required to get it set up properly. I also discuss installing the Network Policy Server (NPS) component of the Remote Access Service (RAS) role, which allows network devices to leverage a protocol called RADIUS, which in turn allows for authentication against AD, even if the device itself is not AD-aware.

Working with Remote Desktop Services

Remote Desktop Services, formerly known as Terminal Services, allows for multiple Remote Desktop Protocol (RDP) connections to the same server. By default, Windows Server 2022 allows for two remote connections. RDP allows you to connect to a remote system and view the desktop, just as if you had the actual console of the server up. To allow Remote Desktop, you need to enable it on the server, and you must allow TCP and UDP 3389 if there is a firewall between you and the server that you want to RDP to.

Using RDS, you can give your users their own virtual desktops to work from. This is great for applications where the install is complicated or costly, and it simplifies upgrades because you only need to upgrade the application on the server, not on multiple PCs. You can also use RDS for RemoteApps, which allows you to run an application on the server but present the application to the user as if it were installed on their desktop.

Installing Remote Desktop Services

Remote Desktop Services is a role, but the installation is a bit different from the past roles that you may have installed. Here's how to install RDS:

1. **Starting from Server Manager, click Manage and then click Add Roles and Features.**

   The Add Roles and Features Wizard opens.

2. **On the Before You Begin screen, click Next.**
3. **On the Select Installation Type screen, select Remote Desktop Services Installation, as shown in Figure 3-1, and click Next.**
4. **On the Select Deployment Type screen, select Quick Start and click Next.**
5. **On the Select Deployment Scenario screen, choose Session-Based Desktop Deployment and click Next.**
6. **On the Select a Server screen, the server you are on is already selected, so just click Next.**
7. **On the Confirm Selections screen, check the Restart the Destination Server Automatically If Required check box, and then click Deploy.**

   You see a progress screen as the role installs (see Figure 3-2). When the role is installed, the server reboots.

8. **After the server reboots, click Close.**

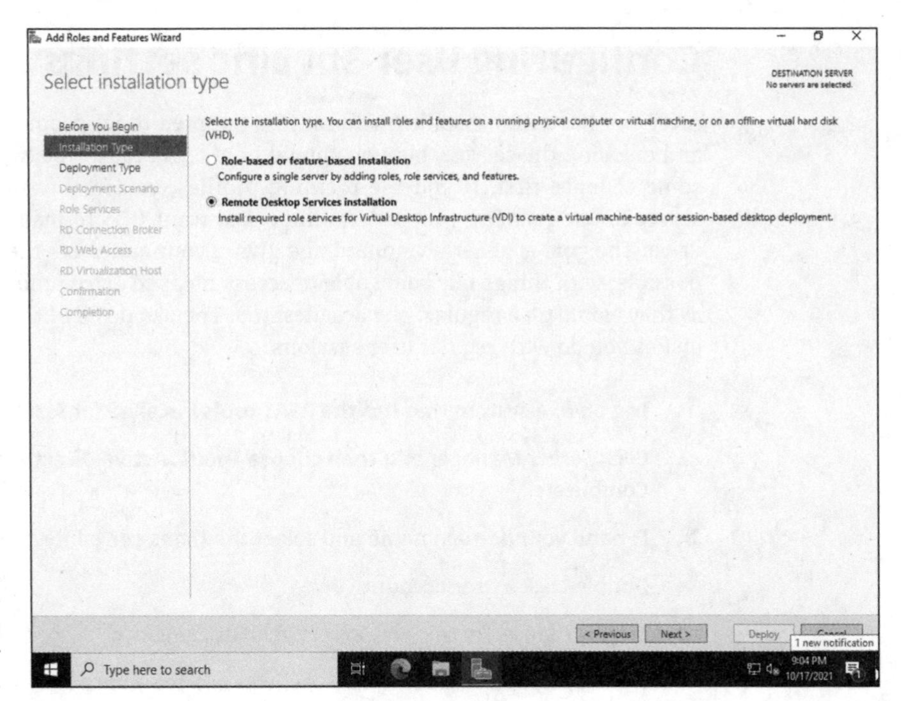

FIGURE 3-1: Selecting Remote Desktop Services is a departure from the usual steps for installing a role.

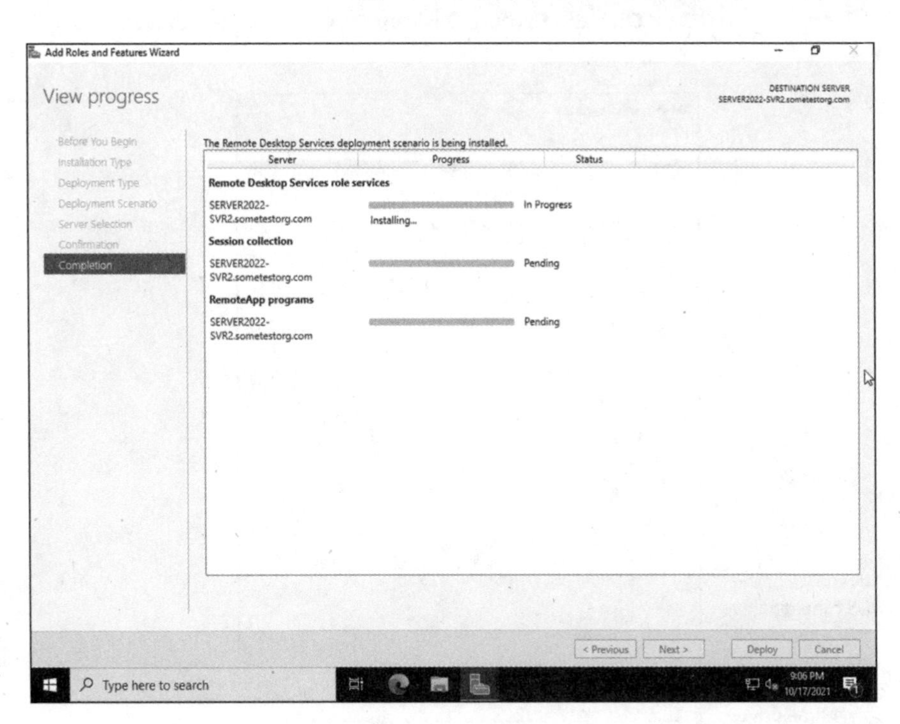

FIGURE 3-2: Installing the RDS role on a single server via the Quick Start installation.

Configuring user-specific settings

After you have RDS installed, you may be tempted to let people start connecting and enjoying the service, but you should hold off because you may want to make some changes first. If you use roaming profiles, you may want to set the RDS server to use roaming profiles. You may also want RDS to map the user's home drive. The goal is generally going to be giving your end users the best experience possible, with things like being able to access mapped drives and applications just as they would on a regular, physical desktop. You can do all of these things in RDS just as you do with regular user sessions.

1. **Log on to a system that has the RSAT tools installed for Active Directory.**
2. **Click Server Manager and then choose Tools⇨Active Directory Users and Computers.**
3. **Expand your domain name and select the Users container.**
4. **Double-click a user account.**

   In this example, I'm using my Karen Smith account.

   The Properties dialog box for the user you selected appears.
5. **Click the Remote Desktop Services Profile tab (see Figure 3-3).**

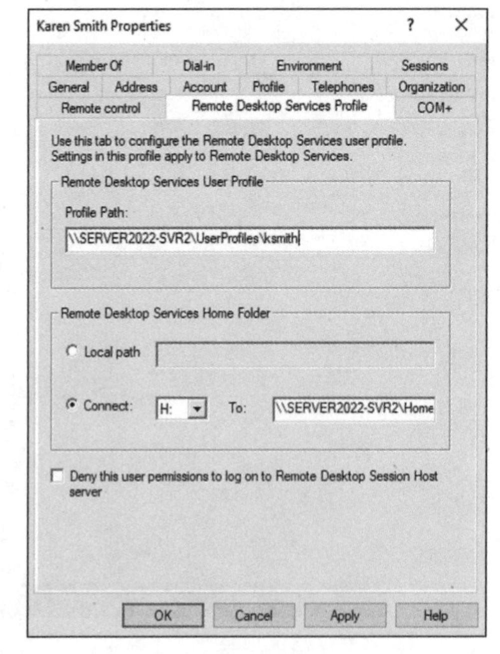

FIGURE 3-3: Setting up a user's profile to take full advantage of the RDS functionality involves a roaming profile and mapping a home drive.

6. **Fill in the information as needed.**

   Here are your options:

   - **Profile Path:** Fill this in if you're using roaming profiles.
   - **Remote Desktop Services Home Folder:** This sets the home drive for the user when they log into a session on RDS.
   - **Deny This User Permissions to Log on to Remote Desktop Session Host Server:** Checking this box blocks the user from connecting through RDS, regardless of whether they're part of a security group that allows it.

Configuring apps

One of the really useful things about RDS is the ability to publish apps. You may be publishing them to a user's desktop, or you may want to make them available through RD Web Access. There are common reasons to share applications through RDS. Applications that rely on old versions of Java or Flash or Internet Explorer, for example, are great candidates for this service. By using RDS to support this legacy application, you only have the old version of the software installed on one system, rather than all the systems where users need to use the legacy application. This reduces the amount of vulnerable software on your network. Windows Server 2022 has three apps already shared by default (assuming you deployed with the Quick Start method — see "Installing Remote Desktop Services," earlier in this chapter): Calculator, Paint, and WordPad. Let's say the business doesn't want users to work with WordPad through RDS; instead, it wants users to work with Notepad++. You can install Notepad++ on the server and configure it to be used over RDS. In that case, you would follow these steps:

1. **Download the Notepad++ executable and install it on the Windows server you installed RDS on earlier.**
2. **From Server Manager, click the side menu where it says Remote Desktop Services.**
3. **Click QuickSessionCollection located under Collections.**
4. **In the RemoteApp Programs window, click Tasks and then click Publish RemoteApp Programs.**
5. **Click Add, locate the executable for Notepad++, make sure it's selected (see Figure 3-4), and then click Next.**
6. **On the Confirmation screen, click Publish.**
7. **On the Completion screen, click Close.**
8. **Right-click the newly added app, and choose Edit Properties.**

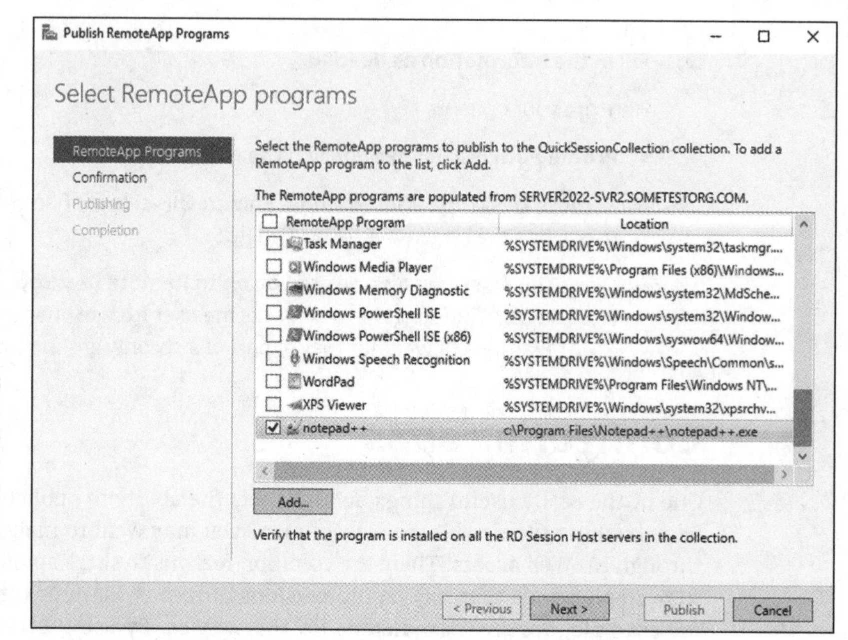

FIGURE 3-4: Adding a non-Windows program to the list of RemoteApp programs available is simple as long as you know where the executable is.

9. **Click User Assignment.**

   By default, everyone has access to the new app. However, you can change that by selecting Only Specified Users and Groups.

10. **Make your changes if desired, and click OK.**

That's all there is to configuring an application and making it available to your users with RDS.

Using RD Web Access

RDS has a great feature known as RD Web Access, which allows you to connect to applications or Remote Desktop sessions via a standard web browser. By default, when you install RDS in Quick Start mode this is enabled, and the address follows this format: `https://`*`serveraddress`*`/RDWeb`.

You can control which applications are visible in RD Web Access from the same screen where you publish new apps. By default, they're published in RD Web Access when you publish them on the RDS Server. If you don't want an app to be in RD Web Access, you can edit the properties of the app. The General tab has an area titled Show the RemoteApp Program in RD Web Access. Simply change the button to No and click OK to remove it.

Before you can access the RD Web Access page, you need to configure a Secure Sockets Layer (SSL) certificate. In previous versions of Windows Server, a self-signed certificate was automatically created and you could just bypass certificate warnings. A self-signed certificate is no longer created by default, however, so let's create one and attach it to the RDWeb service. Follow these steps:

1. **Open Server Manager and select Remote Desktop Services from the menu on the left.**
2. **On the Overview screen, click the Tasks drop-down next to Deployment Overview.**
3. **Select Edit Deployment Properties.**
4. **Select Certificates to expand it.**
5. **Click RD Web Access**
6. **Click Create New Certificate**

   You can alternatively click Select Existing Certificate if you have a certificate issued from your enterprise Public Key Infrastructure (PKI) system.
7. **For certificate name, you need to enter a fully qualified domain name (FQDN).**

   For my example, I've entered server2022-svr2.sometestorg.com.
8. **Enter a password.**
9. **Select the Store This Certificate check box and choose a location to save the certificate.**
10. **Check the Allow the Certificate to Be Added to the Trusted Root Authorities Certificate Store on the Destination Computers check box.**
11. **Click OK.**

    State will say Ready.
12. **Click Apply.**

    After you've clicked Apply, State will say Successful.

Keep in mind that you only need to perform steps 6 through 11 if you don't have a certificate already issued by a trusted certificate authority. If you'd like to know more about deploying your own certificate authority in Windows Server, check out Book 5, Chapter 6.

I assume if you're reading and you've done this setup that you want to play with RD Web Access. Open your web browser (on a system that is not the RD web server) and enter the address for your RD web server. In my case, the address is `https://server2022-svr2.sometestorg.com/RDWeb`.

You may get an error due to an untrusted certificate. This is expected because the certificate is self-signed if you chose to create a certificate in the previous instructions. You can safely disregard this message for now. After you've entered the address, you should see a screen similar to Figure 3-5.

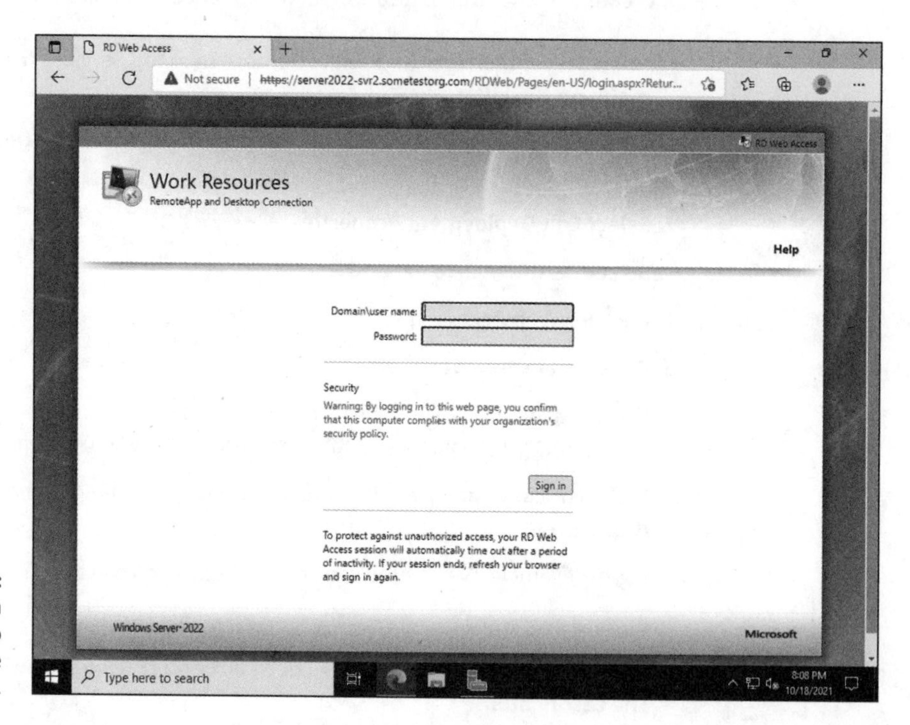

FIGURE 3-5: The logon screen for RD Web Access is simple and elegant.

When you log in to RD Web Access, you see all your published apps that you have permissions to. If you were following along in the previous section and installed the Notepad++ application, you see the icon for Notepad++ in the list. Click that icon to launch Notepad++.

You may get a message saying that a website is trying to run a RemoteApp program. It's complaining because Microsoft doesn't recognize the publisher. Because you know that Notepad++ is safe, go ahead and click Connect. You see a dialog box that says Starting Your App (see Figure 3-6). After the remote connection to the server is established, the application will open just as if it were installed on your system.

Configuring and using RDS licensing

Licensing for RDS is done with client access licenses (CALs). A CAL is essentially a commercial license that allows you to consume a service on a server. The RDS CAL specifically is used to license either a user or device that is connecting to the RDS server.

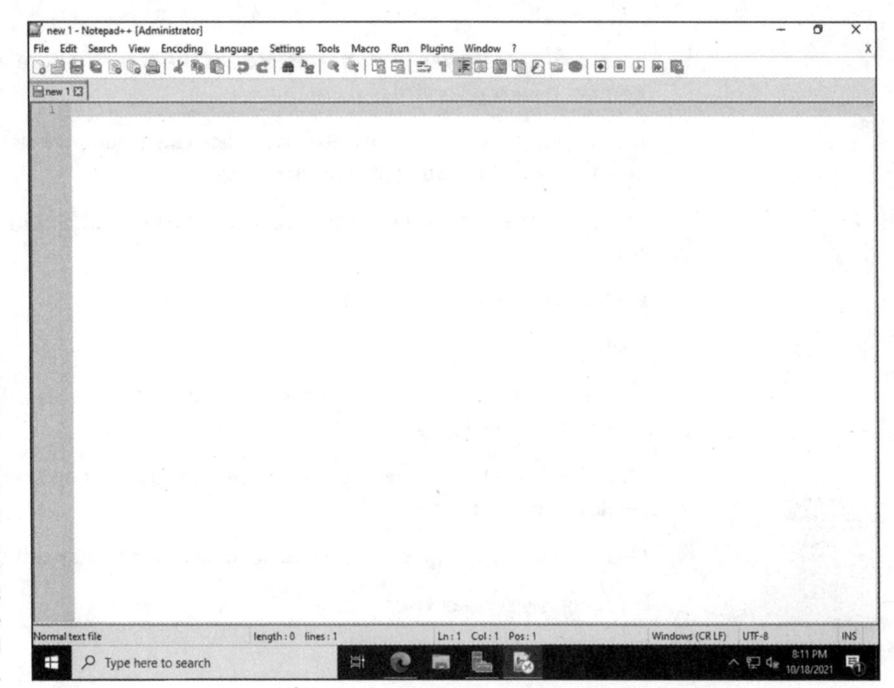

FIGURE 3-6: When you select an application in the RD Web Access console, a remote connection is established, and then the application opens as if it were actually installed on your system.

There are two types of Remote Desktop Services (RDS) licensing:

- **Device-based licensing:** Device-based licensing has the following characteristics:
  - It's physically assigned to a device.
  - It can be tracked regardless of whether a device is in Active Directory or not.
  - Temporary CALs are good for up to 89 days.
  - CALs can't be over-assigned.
- **User-based licensing:**
  - CALs are assigned to users in Active Directory.
  - CALs can't be tracked via workgroup; they can only be tracked via Active Directory.
  - Temporary CALs aren't available.
  - CALs can be over-assigned (which allows you to go over what you're allowed in your licensing agreement).

You need to have an active licensing server to be able to assign licenses. I find it strange that this piece doesn't get installed with the other components for RDS. Here's how to install the licensing role onto your RDS server:

1. **From Server Manager, click the left side of the screen where it says Remote Desktop Services.**
2. **In the Deployment Overview window, click the green plus sign labeled RD Licensing to start adding the RD licensing role.**
3. **Select your server and click the arrow to move your server into the selected box.**
4. **Click Next and then click Add.**
5. **Click Close.**

   Now that RD Licensing is installed, the plus sign has been replaced by a ribbon icon, as shown in Figure 3-7.
6. **Click the Tasks button right above RD Licensing and choose Edit Deployment Properties.**
7. **Click RD Licensing and select which licensing mode you want to use.**

   I'm going to select Per User.
8. **Click OK.**

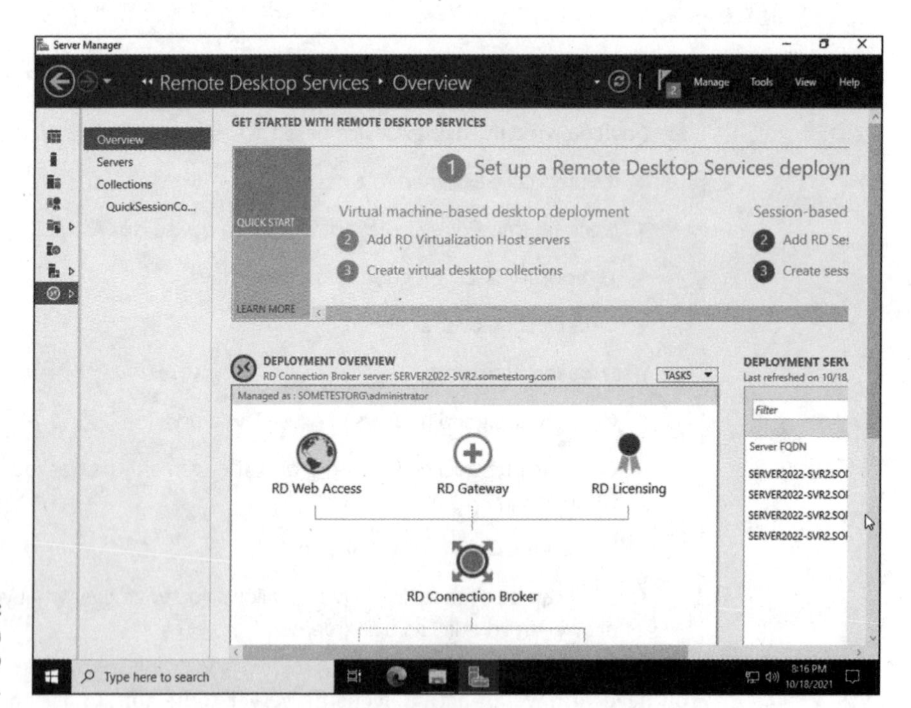

FIGURE 3-7: Click the plus sign to install the RD Licensing role.

Now that the licensing server is installed, let's look at how to add a license to the RDS server:

1. **From Server Manager, choose Tools⇨Remote Desktop Services⇨RD Licensing Manager.**
2. **Right-click the server name and choose Activate Server.**
3. **In the Welcome to the Activate Server Wizard, click Next.**
4. **Under Connection Method, leave it on Automatic Connection and click Next.**
5. **Enter your company's information, and click Next.**
6. **Skip the next screen because it's asking for additional optional information — just click Next.**

   You see a message that says "Activating the License Server."

   If activation was successful, you see the Completing the Activate Server Wizard screen.
7. **Click Finish.**

Now that your RDS server is activated, you can install your CALs.

1. **Right-click the server and choose Install Licenses.**
2. **On the Welcome screen, click Next.**
3. **Select the program that you purchased your licenses through, and click Next.**
4. **Enter the information for your license program (a license code, agreement/authorization code, or something along those lines), and click Next.**
5. **Select your product version, your license type, and the number of licenses you want to install, and click Next.**

   You server will reach out to Microsoft to retrieve the licenses.
6. **After the licenses are retrieved, click Finish.**

Your RDS server is now installed and licensed. Your users are able to connect to remote desktop sessions and do their work as if they were sitting in front of a desktop in the office. Or maybe they can access a legacy application that has historically had difficulty running on their regular desktops. Setting up RDS is not a small or simple task, but it's a very useful service to support your business's needs.

Working with Network Policy and Access Services

It's time to switch gears from Remote Desktop Services to Network Policy and Access Services (NPAS). Network Policy and Access Services is a server role that includes Network Policy Server (NPS).

First, you need to install NPAS. Then I introduce NPS and explain what you can do with that.

1. **From Server Manager, click Manage and then click Add Roles and Features.**
2. **On the Before You Begin screen, click Next.**
3. **On the Select Installation Type screen, click Next.**
4. **On the Select Destination Server screen, click Next.**
5. **On the Select Server Roles screen, scroll down and select Network Policy and Access Services, click Add Features, and then click Next.**
6. **On the Select Features screen, click Next.**
7. **On the Network Policy and Access Services screen, click Next.**
8. **On the Confirm Installation Selections screen, choose Install.**
9. **After the installation has completed, click Close.**

Network Policy Server

Network Policy Server is a component of NPAS that allows you to centrally manage authentication, authorization, and accounting for various devices on the network. It's very commonly used to support authentication on network switches and firewalls. After you've installed NPAS, you can access NPS from Server Manager by choosing Tools➪Network Policy Server.

Registering your RADIUS server

A Remote Authentication Dial-In User Service (RADIUS) server is capable of performing authentication, authorization, and accounting. This is one of the more common reasons why you may install NPS. It was originally used to authenticate remote users connecting through dial-up, but these days it's more commonly used to authenticate systems against Active Directory that don't normally support AD authentication. Frequently, that is network switches, routers, wireless access points, and firewalls. RADIUS servers allow administrators to log in to devices that support the RADIUS protocol with their Active Directory credentials, instead of having to remember a local login on each device.

Before you can use a brand-new RADIUS server to authenticate against Active Directory, you need to register it. This is a very important step that is often missed when setting up a RADIUS server. Here are the steps to register your RADIUS server:

1. **From Server Manager, choose Tools⇨Network Policy Server.**
2. **Right-click NPS (Local) and select Register Server in Active Directory.**

   You're prompted to allow the server to read the user's dial-in properties.
3. **Click OK.**

   NPS confirms that it can now read the dial-in properties.
4. **Click OK.**

That's all there is to it. It's a very simple step, but a crucial one that is often missed.

Understanding RADIUS Proxy

A RADIUS Proxy server forwards both authentication and accounting messages on to a RADIUS server. These can be helpful in providing authentication when domain trusts become involved.

Setting up a RADIUS client

For this scenario, let's say you have a Cisco switch. You've been logging into the switch locally, but you want to configure a central source of authentication so that you don't have to remember a username and password for every switch in your environment. The configuration of the switch to point to the RADIUS server is out of scope for this book, but the configuration of the client is not. Here's how to set up the RADIUS client for a Cisco switch:

1. **From Server Manager, choose Tools⇨Network Policy Server.**
2. **Expand RADIUS Clients and Servers and select RADIUS Clients.**
3. **Right-click RADIUS Clients and choose New.**

   The New RADIUS Client dialog box appears.
4. **Fill in the friendly name, the IP address, and the shared secret.**

   The dialog box should look similar to Figure 3-8.

5. **Click the Advanced tab.**
6. **From the Vendor Name drop-down list, select Cisco.**
7. **Click OK.**

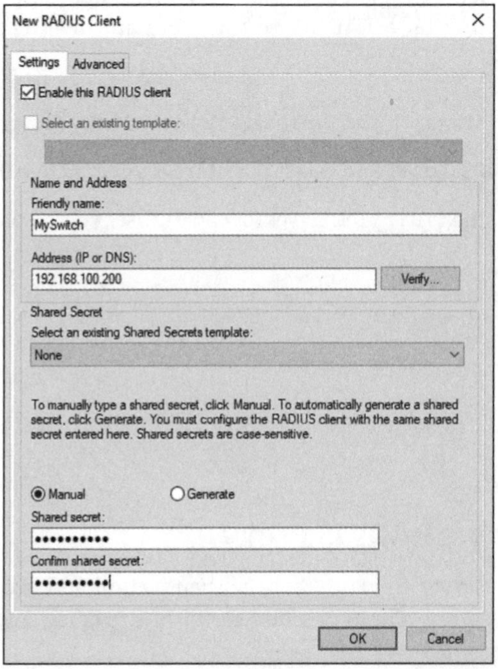

FIGURE 3-8: The Settings tab is where the bulk of the configuration happens for a RADIUS client.

Setting up a network policy

Network policies control what is allowed to authenticate through your RADIUS server. Here's how to configure a basic network policy:

1. **From Server Manager, choose Tools⇨Network Policy Server.**
2. **Expand Policies.**
3. **Right-click Network Policies and select New.**

4. **Give the policy a name and specify the type of network access server.**

   In this case, I've named the policy Domain Admins to Network Switches and I've left the type of network access server on Unspecified.

5. **Click Next.**
6. **On the Specify Conditions screen, specify what conditions must be met for the policy to apply.**

   I followed these steps:

   (a) Select the condition User Groups.

   (b) Click Add.

   (c) Click Add Groups.

   (d) Type **Domain Admins** and then click OK.

   (e) Click OK one more time.

   (f) Click Add.

   (g) Select the condition Client Vendor.

   (h) Click Add.

   (i) Choose Cisco

   (j) Click OK.

7. **Back on the Specify Conditions screen, you can add another condition or click Next.**
8. **On the Specify Access Permission screen, select Access Granted, and click Next.**
9. **On the Configure Authentication Methods screen, click Next.**
10. **On the Configure Constraints screen, click Next.**
11. **On the Configure Settings screen, click Next.**
12. **On the Completing New Network Policy screen, click Finish.**

You can see in Figure 3-9, my new Allow policy on top of the deny policies. If an incoming connection doesn't match my policy, it will automatically be denied.

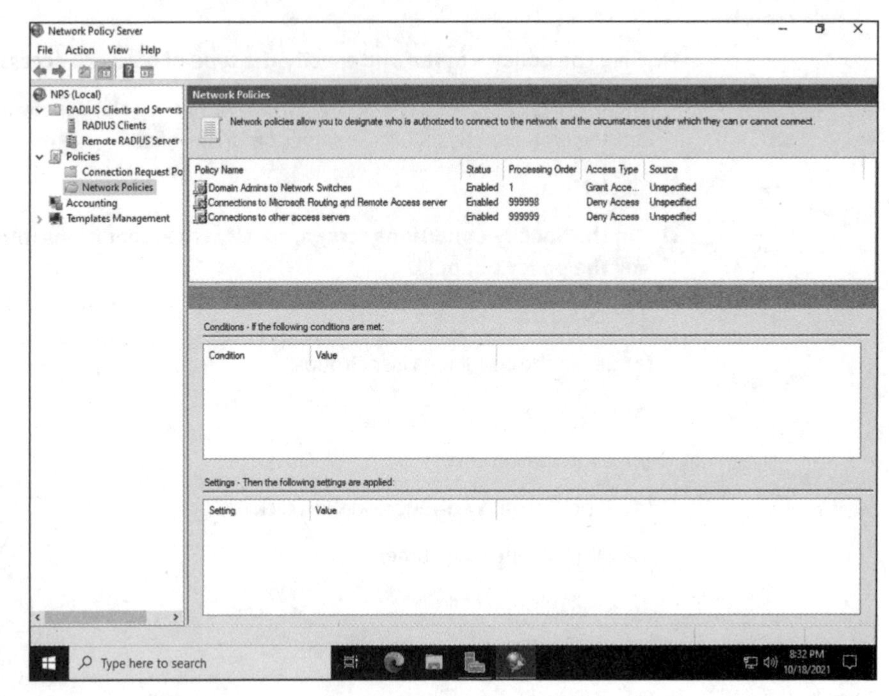

FIGURE 3-9: A network policy allowing traffic needs to have the lower processing number so that it will be processed before the default deny polices.

Troubleshooting at the Command Line

PowerShell is installed on every Windows Server by default. It has many different modules to assist in different tasks. In this section, I walk you through a few of the modules that may assist you in troubleshooting your servers and their network connectivity.

The `Test-NetConnection` cmdlet is a powerful tool in your troubleshooting arsenal. Used by itself, the `Test-NetConnection` cmdlet shows you basic information regarding the network, in addition to a ping. There are some other parameters that really help you dig into specific issues, as shown in Table 3-1.

There are several different combinations that you can use with the parameters. For instance, let's look at a basic check to ensure that you're able to connect to a remote port. Open PowerShell, and type the following command:

```
Test-NetConnection -Port 443 -InformationLevel "Detailed"
```

This command tests to see if you can make a connection over port 443 (HTTPS) and gives detailed output on the results, as shown in Figure 3-10.

TABLE 3-1 **Test-NetConnection Parameters**

Parameter	Description
`-Port`	Used to test remote connectivity to a destination server. You specify a TCP port number and it tests the connection.
`-InformationLevel`	There are two values you can use with this parameter: Detailed returns quite a bit of information, and Quiet returns basic information.
`-DiagnoseRouting`	Runs route diagnostics.
`-TraceRoute`	Runs tracert against the destination system. Tracert displays each hop until it reaches its destination.

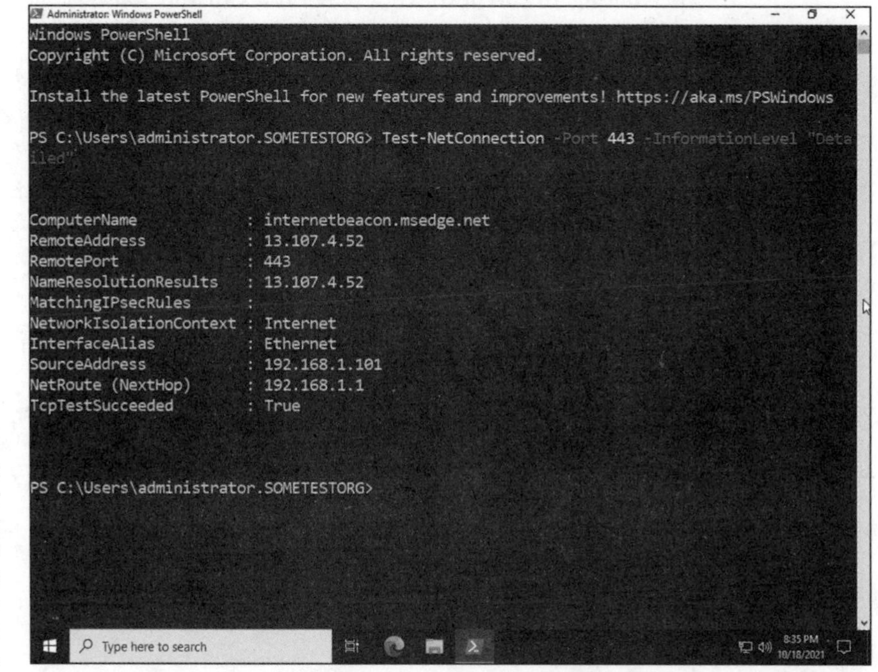

FIGURE 3-10: You can test basic connectivity with just a few parameters added to the `Test-NetConnection` cmdlet.

The prior test checks for basic connectivity. Picture this scenario: Your users have complained that when they're on their remote desktop sessions, they can't reach a certain website. They say that they can reach the website when they use their regular systems. though. You can try the basic connectivity test again, but this time specify a destination. I've used the address `www.dummies.com` for my example.

```
Test-NetConnection -ComputerName www.dummies.com -Port 443
```

If this command succeeds, the TcpTestSucceeded field will say True. If this is the case, the connection to the site is good. If the TcpTestSucceeded field says False, then you may have a routing issue. You can confirm whether this is a routing issue or not with the -TraceRoute parameter.

```
Test-NetConnection -ComputerName www.dummies.com -TraceRoute
```

The preceding command performs a TraceRoute and will show you each hop along the path to the destination. If the TraceRoute does not succeed, there is potentially a routing issue or your Internet service provider (ISP) may be having issues.

As a last step, you can run routing diagnostics to see if there is an issue with the routing to the destination site. You can run routing diagnostics with the -DiagnoseRouting parameter. The command looks like this:

```
Test-NetConnection -ComputerName www.dummies.com
  -DiagnoseRouting -InformationLevel "Detailed"
```

As you can see, this one command can give you a lot of information and can aid in your troubleshooting efforts. You'll be able to find the source of the problem and get network connectivity issues resolved more quickly.

» Repairing individual network connections on Windows Server

» Identifying and fixing common network configuration errors

» Troubleshooting with command-line utilities

» Using third-party troubleshooting tools

Chapter 4
Diagnosing and Repairing Network Connection Problems

It never fails: It's Friday and you're getting ready to head home. Just as you're leaving, you get a call that there is a networking issue on one or more of your servers. Maybe your systems are down. Maybe they're intermittently up and down. Your mission, should you choose to accept it (do you have a choice?), is to find the issue and fix it.

This chapter covers some of the built-in troubleshooting capabilities of the operating system and some of the more common configuration issues you may experience with new systems and older systems.

Using Windows Network Diagnostics

Sometimes the network issue is super obvious, and sometimes it isn't. For instance, your server may indicate that its network cable is unplugged, yet when you look, there is still a cable plugged in. That could mean you have a bad cable. a bad switchport, or a bad network interface card (NIC) on the server. Those are hardware issues, and unfortunately, you're on your own when you're troubleshooting hardware issues. Software issues on the other hand, can be addressed with the Microsoft Windows Network Diagnostics. Follow these steps:

1. **Click the Start menu, and then click the gear icon for the Settings menu.**
2. **Click Network & Internet.**
3. **On the Status page, scroll down and select Network Troubleshooter.**

   The wizard searches for and tries to fix the issue. If it doesn't find anything wrong, you see a screen similar to Figure 4-1.

4. **Click Close to close out of the troubleshooter.**

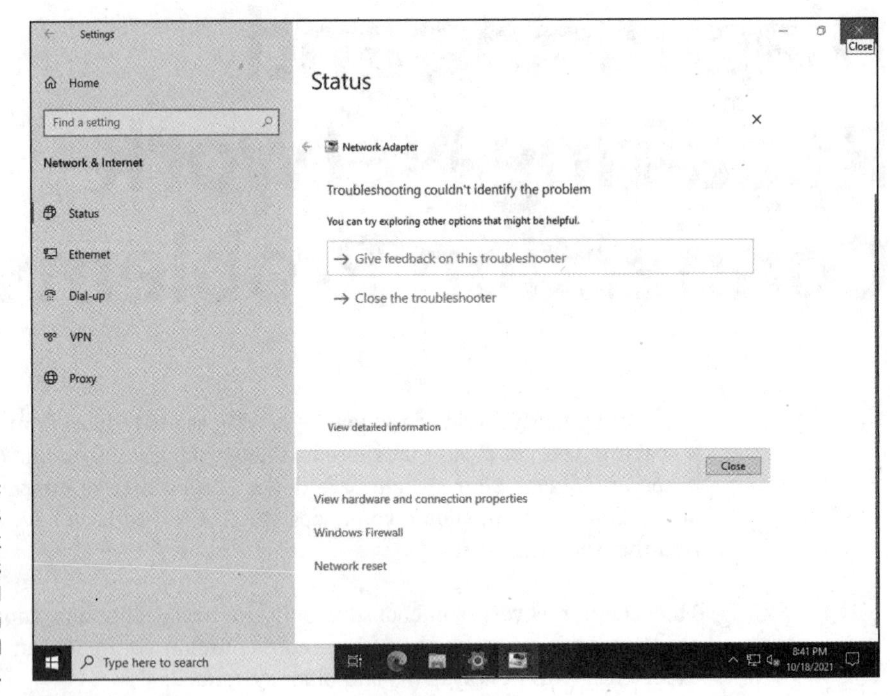

FIGURE 4-1: The Windows Network Diagnostics screen is a wizard that helps you diagnose and repair issues.

There is an additional method to launch an Internet-specific troubleshooting tool. This tool is not focused on internal network issues. It's focused specifically on Internet connectivity issues. To get to the Internet Connections troubleshooting tool, follow these steps:

1. **Click the Start menu, and then click the gear icon for the Settings menu.**
2. **Click Update & Security.**
3. **Select Troubleshoot.**

   The Troubleshoot utility runs and lets you know if it finds anything.

4. **If it doesn't find anything wrong, you can click the Additional Troubleshooters link, where you gain access to a wide array of troubleshooting utilities, shown in Figure 4-2.**
5. **Click Close to close out of the troubleshooter.**

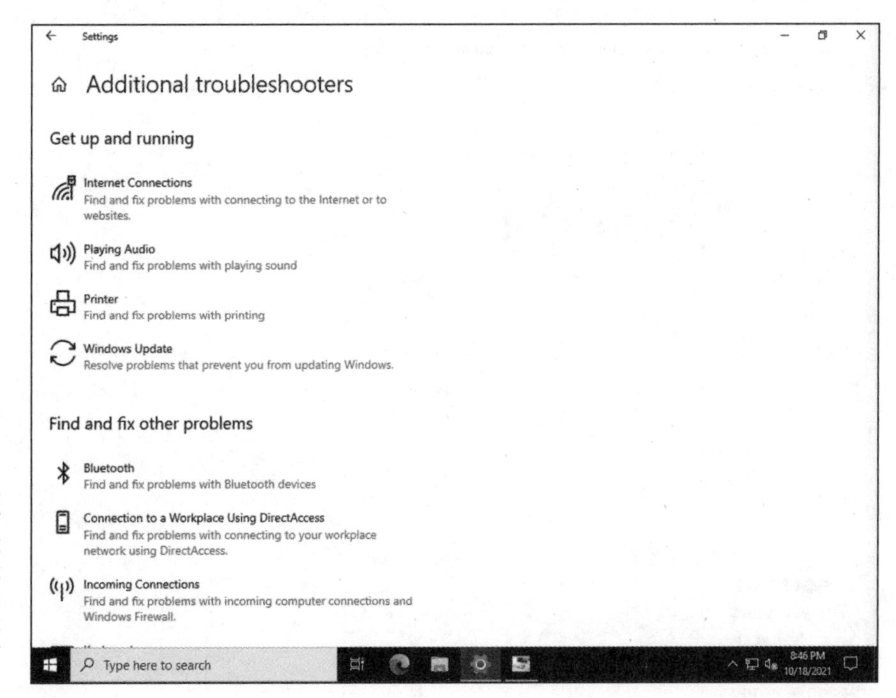

FIGURE 4-2: The Additional Troubleshooters link contains many troubleshooting tools, including one that can help with Internet connectivity issues.

Repairing Individual Connections

You can work directly with the network adapter that is having the issue. This can be very beneficial if a system has multiple network adapters and you need to test them one at a time. The software utility that Microsoft provides is pretty good at finding software issues like disabled adapters and misconfigurations. Follow these steps:

1. **Click the Start menu, and then click the gear icon for the Settings menu.**
2. **Click Network & Internet.**
3. **On the Status page, scroll down and select Change Adapter Options.**
4. **Right-click the network adapter you want to check and select Diagnose.**

   The Windows Network Diagnostics wizard launches and tries to find an error. If it's successful, it will let you know what issue it found. In my example in Figure 4-3, it has found that the network adapter is disabled and has fixed the issue by re-enabling the adapter.

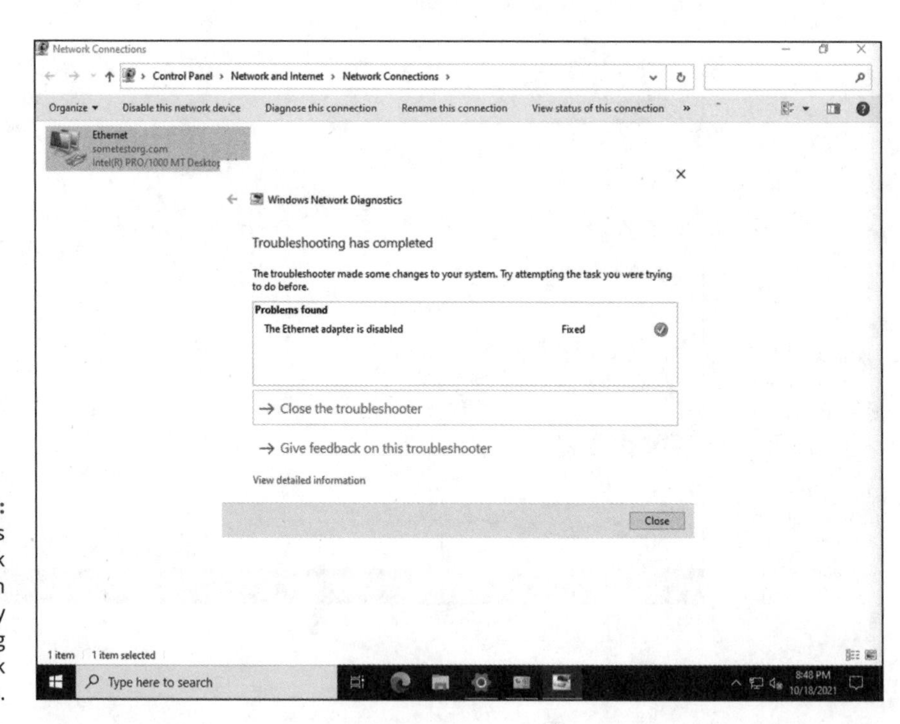

FIGURE 4-3: Windows Network Diagnostics can diagnose many issues, including disabled network adapters.

Of course, just because a system is having a network issue, it doesn't mean that there is a software issue. Hardware issues can be harder to track down because the troubleshooting utility won't really be able to help. Here are some common issues that can affect the network connections on a system:

- **Bad cable:** Ethernet cables have a piece of copper running down the middle of the cable. Fiber-optic cable has glass fibers running down the length of the cable. If the medium used for transmission is damaged, you'll lose your connections. Look out for kinks in the cable and sharp bends. Look for areas where the cable appears to have been stretched. The fix for this issue is to replace the damaged cable.
- **Bad cable connectors:** The connectors on the cables can go bad as well. The connectors used with Ethernet and with fiber-optic cabling are usually made out of plastic. They can crack if too much strain is put on them. The medium inside of the cable may become disconnected if the cable is pulled on rather than the connector. Fixing this issue is simple if you have the right tools. To replace the connector on an Ethernet cable, all you need to do is strip back the UTP cabling and press the eight wires into their appropriate channels. The inner wires are color coded; make sure that you're following the right standard, or the new connector won't work. Replacing the connector on a fiber-optic cable is a little more complicated. You have to cut the fiber-optic cable and polish the glass end. Repairing fiber is easier to leave to the pros when you can.
- **Faulty switchport:** The switchport that the system is connected to may be having issues. Sometimes ports can go bad, or a network administrator may have mistakenly turned off the port. To see if this is the issue, have the network administrator see if the port is up. If it isn't, have the network administrator re-enable it. If the port should be up but it's not working, you can move the cable to a different network port on the switch once your network administrator gets it set up for you.
- **Bad NIC:** If your NIC goes bad, you'll need to replace it. Unfortunately, this means that your system will need to be powered down so that you can replace the card.

Network Troubleshooting at the Command Line

If the Windows Network Diagnostics Wizard doesn't find anything wrong, there are a few more things you can try. Each of the following commands needs to be run in the Command Prompt. Follow these steps:

1. **Click Start and scroll down to Windows System.**
2. **Click Command Prompt and try each of the following commands (in this order):**

   (a) Reset the Transmission Control Protocol/Internet Protocol (TCP/IP) stack.

   netsh winsock reset

   ```
   netsh int ip reset
   ```

   (b) Release your old IP address.

   ```
   ipconfig /release
   ```

   (c) Renew your IP address.

   ```
   ipconfig /renew
   ```

   (d) Flush your Domain Name System (DNS) cache on your system.

   ```
   ipconfig /flushdns
   ```

There are a few additional commands that can be useful when troubleshooting from the Command Prompt.

- `ping`: The `ping` command gives you simple feedback. It lets you know how many packets it sent, how many packets it received, and what the latency was between the sending and receiving. By default, `ping` will send four packets, but you can adjust the number of pings, or make it a continuous ping if needed. (See Figure 4-4 for an example of `ping`.)
- `tracert`: The `tracert` (trace route) utility can help you pinpoint where a problem exists. It reports back along each hop until it gets to the destination that you specified. By default, it will go to a max of 30 hops.
- `pathping`: Works very similar to the `tracert` command and can provide information about network latency and network packet loss.
- `telnet`: Can be used to test if a certain port is open. You need to have the Telnet client installed for this to work. Your security team may not be happy to find Telnet on a system, so make sure that you aren't violating company policy by installing it. Many systems have been configured to not display banners — if you don't get an error, you were probably successful in connecting to whichever resource you wanted to test.

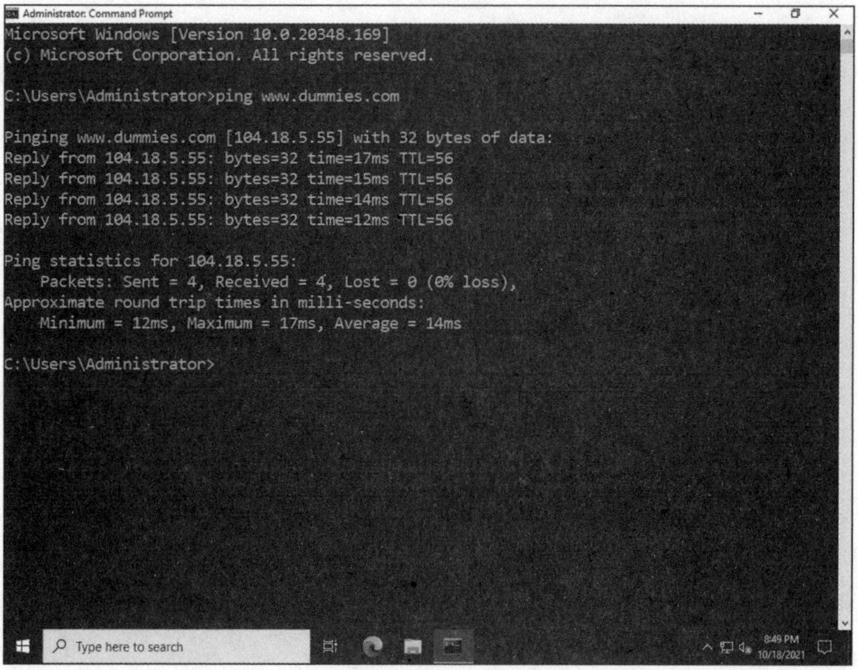

```
Administrator: Command Prompt
Microsoft Windows [Version 10.0.20348.169]
(c) Microsoft Corporation. All rights reserved.

C:\Users\Administrator>ping www.dummies.com

Pinging www.dummies.com [104.18.5.55] with 32 bytes of data:
Reply from 104.18.5.55: bytes=32 time=17ms TTL=56
Reply from 104.18.5.55: bytes=32 time=15ms TTL=56
Reply from 104.18.5.55: bytes=32 time=14ms TTL=56
Reply from 104.18.5.55: bytes=32 time=12ms TTL=56

Ping statistics for 104.18.5.55:
    Packets: Sent = 4, Received = 4, Lost = 0 (0% loss),
Approximate round trip times in milli-seconds:
    Minimum = 12ms, Maximum = 17ms, Average = 14ms

C:\Users\Administrator>
```

FIGURE 4-4: The `ping` utility gives you a simple readout and can point out an issue between a source and the destination.

Working with Windows Firewall

Network issues can be very frustrating to troubleshoot. You may find that everything looks just fine, and you aren't sure what's happening. If you suspect that Windows Firewall is the issue, you can use Event Viewer to see if Windows Firewall has blocked traffic to or from your server. Follow these steps:

1. **From Server Manager, choose Tools⇨Event Viewer.**
2. **Double-click Applications and Services Log to expand it.**
3. **Click Microsoft, Windows, and Windows Firewall with Advanced Security.**
4. **Double-click Firewall.**

   When the Windows Firewall blocks something, you'll see a message similar to Figure 4-5.

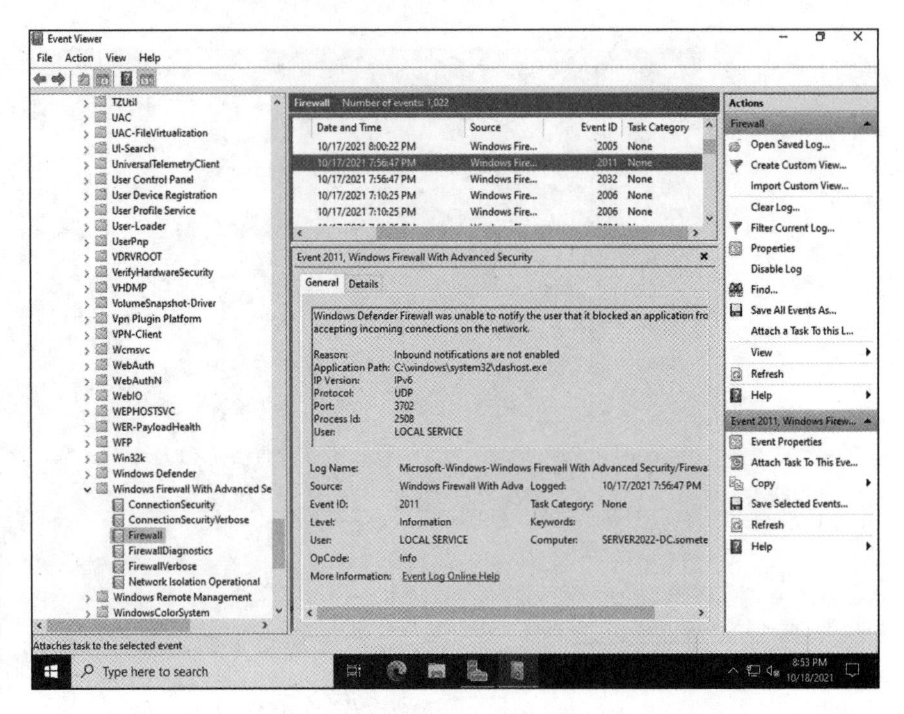

FIGURE 4-5: Event Viewer can show you when Windows Firewall has blocked an incoming packet.

If you can't find anything that was blocked, you can temporarily turn off Windows Firewall to see if that will resolve the issue. Just make sure that doing so isn't against organizational policy before you attempt it.

To turn off the firewall, follow these steps:

1. **Click the Start menu, scroll down to Windows System, and choose Command Prompt.**
2. **Type the following command and press Enter.**

```
netsh advfirewall set allprofiles state off
```

To turn the firewall back on, follow these steps:

1. **Click the Start menu, scroll down to Windows System, and choose Command Prompt.**
2. **Type the following command and press Enter.**

```
netsh advfirewall set allprofiles state on
```

Making Sense of Common Configuration Errors

Server hardware is usually pretty reliable. Some issues may be caused by problematic hardware, but most often the issue stems from a misconfiguration of some kind.

In the following sections, I walk you through some common issues that system administrators have to deal with.

Duplicate IP addresses

Symptom: You get a message stating that there is a duplicate IP address on your network.

Solution: The best solution is to use Dynamic Host Configuration Protocol (DHCP) so that IP addresses are assigned and tracked automatically. If you don't have DHCP in your environment, try to use another IP address.

No gateway address

Symptom: Your system is able to communicate with other systems in the same subnet, but it can't communicate with anything outside of the subnet.

Solution: Set a default gateway address. This will tell the system where to send traffic to if the traffic is not destined for a system on the local network.

No DNS servers set

Symptom: You can't resolve names like `www.dummies.com`, or when you try to join to an Active Directory domain, you get a message that states that the domain name can't be found.

Solution: Set the appropriate DNS servers for your network. This will allow you to do internal name resolution. If your system is going to be joined to an Active Directory domain, it requires a valid entry for a DNS server.

An application is experiencing network issues

Symptom: The system is on the network, and basic functionality like ping and file sharing work. The application on the server is not responding to network requests.

Solution: Check to see if the Windows Defender Firewall is enabled. Verify that there is a rule that allows the traffic that is supposed to be going to the application. It's amazing how often this gets missed when provisioning applications.

Everything should be working, but it's not

Symptom: The hardware looks good, the Windows Network Diagnostics utility says it can't find a problem, but your system is still unable to communicate over the network.

Solution: Check your IPv4 settings. It is very easy to mistype an IP address or a subnet mask. If either of these is incorrect, your system won't function properly.

Working with Other Troubleshooting Tools

Some third-party utilities can be very handy in helping you find network issues. Table 4-1 lists a few of my favorites. Some are free, and some cost money. In general, free products may have no or limited support, so you may want to consider that if you choose to look at third-party tools.

TABLE 4-1 Network Troubleshooting Tools

Tool Name	Cost	Description	Website
Cacti	Free	A network monitoring utility that can be used to create highly customizable graphs.	`www.cacti.net`
Nagios Core	Free	Basic network monitoring with tons of plug-ins and add-ons available to expand its usefulness.	`www.nagios.org/downloads/nagios-core/`
Nagios Network Analyzer	$1,995	Provides network analysis, monitoring, and reports on bandwidth utilization.	`www.nagios.com/products/nagios-network-analyzer/`
Nagios XI	Starts at $1,995	A monitoring solution for applications, services, and networks.	`www.nagios.com/products/nagios-xi/`
SolarWinds ipMonitor	Starts at $1,570	Gives a nice simple up/down console for networks, servers, and applications.	`www.solarwinds.com/ip-monitor`
SolarWinds Netflow Traffic Analyzer	Starts at $1,072	Analyzes netflows for issues and monitors for bandwidth usage.	`www.solarwinds.com/netflow-traffic-analyzer`
SolarWinds Network Performance Monitor	Starts at $1,638	Monitors the performance of the network and alerts you to issues.	`www.solarwinds.com/network-performance-monitor`
Wireshark	Free	A packet sniffer that allows you to filter on the types of traffic you want to see and reconstruct whole TCP/User Datagram Protocol (UDP) streams. It can also show you at a glance if you have lots of retransmissions occurring on the network.	`www.wireshark.org`

5

Managing Security with Windows Server 2022

Contents at a Glance

» Securing files and folders

» Working with the local security policy in Windows Server

» Understanding Windows Security

Chapter 1

Understanding Windows Server 2022 Security

Given the number of security breaches in the news today, it isn't surprising that Microsoft has invested heavily in improving the security of Windows Server. But before you can get into the really cool, new security features of Windows Server 2022, you need a firm grasp on basic security concepts in general and a working knowledge of Windows security in particular.

In this chapter, I cover security basics. Think of this chapter like a security primer, with general security topics first, followed by Windows Server–specific topics (like .NET security, file and folder security, and the Windows Security App).

Understanding Basic Windows Server Security

Securing your server is arguably one of the most important things you need to do in your daily work. After all, you don't want to go through all the effort to build and configure a server just to have it attacked.

This section covers security basics so that you understand the terminology I use throughout this book.

The CIA triad: Confidentiality, integrity, and availability

The CIA triad (shown in Figure 1-1), consisting of confidentiality, integrity, and availability, is one of the most basic concepts in information security. The closer you get to any one point of the triangle, the farther away you are from the others. For example, if you have a system that keeps records completely confidential, that system won't be available to your end users.

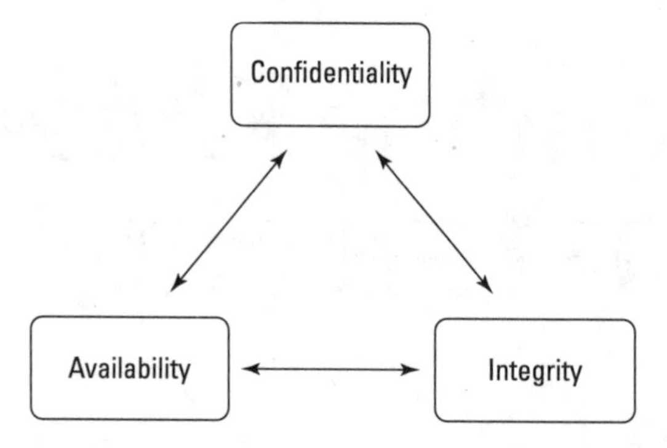

FIGURE 1-1: The CIA triad is one of the most basic concepts in information security.

Here's what each of these terms means for you as a system administrator:

- **Confidentiality:** Confidentiality refers to keeping access to data out of the hands of those who shouldn't have access to it. For instance, when you're using an online banking site or an e-commerce site, the connection should use HTTPS, which means that it's encrypted. Encryption protects sensitive information from being captured by someone eavesdropping on the network.
- **Integrity:** Integrity means that the data has not been changed or tampered with in any way. Data integrity can apply to data at rest or data in transit. Version control can be useful to roll back accidental changes. Potentially malicious changes can be detected by file integrity monitoring software, which creates a hash of a file. If the hash changes, then the file has changed as well.

TECHNICAL STUFF

A *hash* is a mathematical function that is run against a file. It creates a unique "thumbprint" that will change if any modifications are made to the file that it was generated from. By comparing the thumbprints of two files (the original and a copy, for instance), you can tell if the file has been modified.

» **Availability:** Availability is the part of the triad that most people are familiar with. As a system administrator, your goal is to ensure that your systems are up and available to your end users. You may build in redundancy and fault tolerance to make the likelihood of the system going down lower, or you may be responsible for the backups that will allow you to restore data if something does happen.

Authentication, authorization, and accounting

The next set of terms that you need to know are authentication, authorization, and accounting, collectively referred to as triple A:

» **Authentication:** Authentication is how you prove to a system who you are. This may be a username and password, or username and biometrics or a personal identification number (PIN). Think of authentication like showing a security guard your badge and being allowed inside the gate that the guard is responsible for protecting.

» **Authorization:** After you're authenticated, any time you access a resource, the system will check to see if you're authorized to access that resource. Authorization is similar to swiping your badge at the doors inside of a secure building. If you're authorized to enter an area, you'll be allowed through. If you aren't authorized, you won't be allowed through.

» **Accounting:** Accounting, sometimes referred to as auditing, is having a log of when authentication and authorization events occurred. In a Windows system, this can be accomplished with something as simple as Event Viewer.

Access tokens

Access tokens are used by the Windows Server operating system to identify a user who is interacting with an object. The token usually contains the user's security identifier (SID), SIDs for groups that the user is a member of, and the source of the access token. A SID is a unique value that is assigned to an object to identify it. With users, the SID is what allows you to change the name of the user without impacting the user's access. The user's name may change, but the SID will not.

Security descriptors

Security descriptors contain useful information related to the security of an object that is secured or able to be secured. It can include the SID for the owner of the object, access control lists that specify who is allowed to access the object, and which access events should generate audit records.

Access control lists

Access control lists (ACLs) contain access control entries (ACEs) that grant or deny specific access to users or groups. The Windows Server operating system has two types of ACLs that you need to be aware of:

- **Discretionary access control lists (DACLs):** DACLs are what most system administrators think of when asked about ACLs in Windows Server. DACLs are used to grant or deny access based on a user account or group membership. Deny entries always take precedence over allow entries. See Figure 1-2 for an example of a DACL on a folder.
- **System access control lists (SACLs):** SACLs are not as widely known about as DACLs. They can be used to determine what type of event should be audited. They can audit successful access, failed access, or both. In the example in Figure 1-3, I'm auditing both success and failures for any Domain Admin account, but I'm only logging failures for my ksmith account.

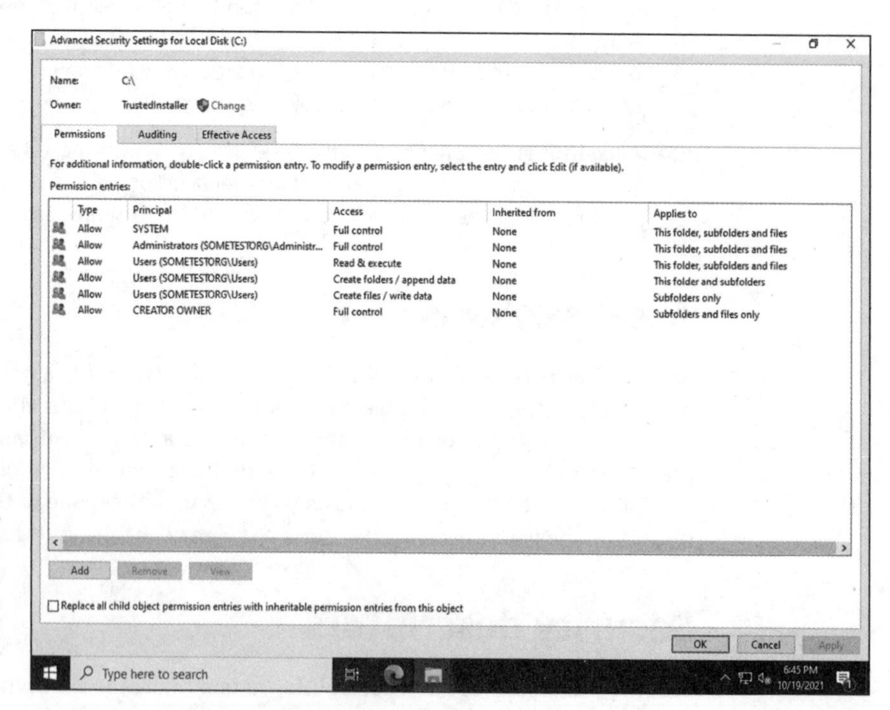

FIGURE 1-2: Discretionary access control lists can be used to determine who should have access to a folder or file.

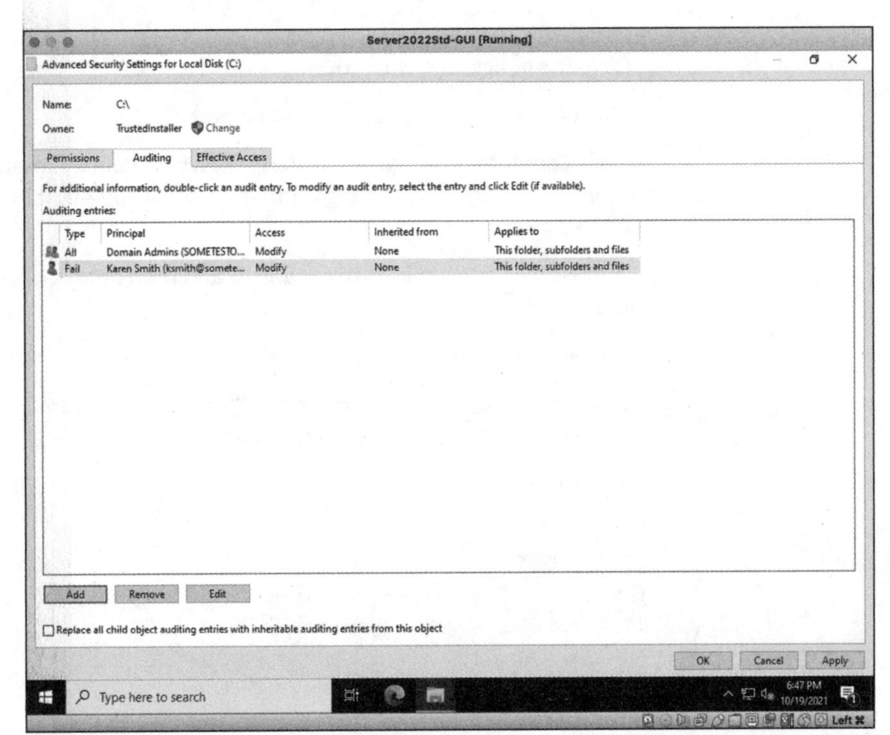

FIGURE 1-3: Using an SACL to audit privileged access to a folder is simple.

SACLs are empty by default, you have to configure what you want them to audit. If you only have one or two servers, configuring this setting manually isn't so bad. If you have numerous servers, though, configuring SACLs quickly becomes unmanageable. Group Policy to the rescue!

Group Policy allows you to make configuration settings in one place and then apply those changes to multiple systems. Here's how to turn on feature through Group Policy.

1. **From Server Manager, choose Tools⇨Group Policy Management.**
2. **Expand a domain by double-clicking the domain.**
3. **Choose the policy you want to set this through.**

   In my case, I'm going to create a new Group Policy Object (GPO) called File Servers.
4. **Right-click the domain name, and select Create a GPO in This Domain, and Link It Here.**

TIP

Creating your GPO this way means that it will apply to all domain systems. Although this is fine in a demo environment, you'll most likely want to link the GPO to an organizational unit (OU) that contains the file servers.

5. **Name the policy, and click OK.**
6. **Right-click your new policy and choose Edit.**
7. **Under Computer Configuration, double-click on Policies, then Windows Settings, then Security Settings, and then Advanced Audit Policy Configuration.**
8. **Double-click Audit Policies to expand it, and then double-click Object Access.**
9. **Double-click Audit File System.**
10. **Check the Configure the Following Audit Events check box, and select Success, Failure, or both (see Figure 1-4).**
11. **Click OK.**

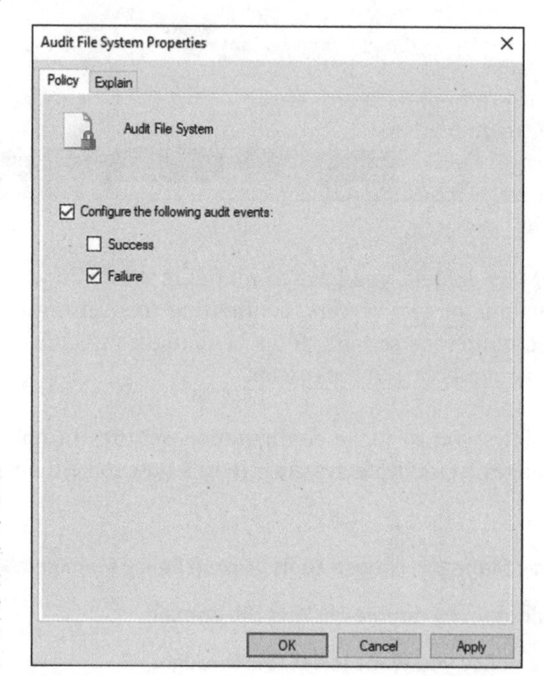

FIGURE 1-4: You can use Group Policy to set your SACLs so that you can apply them across the organization.

That's all there is to setting up file auditing in Group Policy. You also need to enable auditing on the folder that you want to monitor. On the file server where the folder is located, follow these steps:

1. **Right-click the folder that you want to enable auditing on and choose Properties.**
2. **Click the Security tab, and then click the Advanced button.**

3. **Click the Auditing tab, and then click the Add button.**
4. **Click the Select a Principal hyperlink, enter a username or group in the dialog box, and click OK.**
5. **Change the Type drop-down list to All, and click OK.**
6. **Click OK again to exit the Advanced Security Settings for <*folder_name*> dialog box.**
7. **Click OK one more time to close out of the Properties dialog box.**

Working with Files and Folders

Working with files and folders is pretty much the bread and butter of a system administrator's work. Although some folders can be left open to the world, you'll most likely be asked to lock certain folders or shares down so that only certain people can access them. On Windows Server, you can have conflicting NT File System (NTFS) and share permissions, which can make troubleshooting access issues much more difficult. Let's examine the different types of permissions and how we can check effective permissions.

TECHNICAL STUFF

Modern versions of Windows Server support NTFS, which is the standard file system you see on the majority of Windows servers today. NTFS made significant improvements in security and reliability and added support for larger volumes.

Setting file and folder security

With NTFS, we gained the ability to be able to set ACLs, a huge improvement over the FAT32 days. Although file servers do take advantage of NTFS permissions, they can also use shares. The idea behind shares is to make it so that users can access a directory on the server without direct access to the server. That's a great win for security, but the *effective permissions* (the combined permissions of the NTFS permissions and the share permissions) can sometimes cause unexpected access issues.

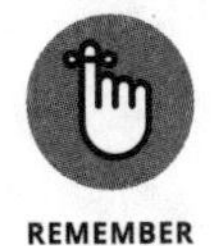

REMEMBER

Share permissions govern what a user can access across the network. NTFS permissions control what a user can do both on the server and across the network.

NTFS permissions

Editing the NTFS permissions on a file or folder is simple: Simply right-click the folder or file that you want to change permissions on and click Properties. Then click the Security tab (see Figure 1-5). There are several different levels of permissions in NTFS file systems:

- **Full Control:** Full Control gives you the ability to read, write, and execute files within a folder, as well as the ability to set permissions. It also allows for the ability to delete files and folders. Full Control is a highly privileged level of permission and should only be granted to those who have administrative access.
- **Modify:** Allows you to change the contents and/or titles of folders and files, and allows for the deletion of files and folders.
- **Read & Execute:** Allows you to open files and folders, and launch programs including scripts.
- **List Folder Contents:** Allows you to view the titles of files and folders, but does not allow you to open the files.
- **Read:** Allows you to open the files to read them, but does not allow you to modify them.
- **Write:** Allows you to add a file or subfolder and make changes to a file.
- **Special Permissions:** These are special sets of permissions that are set through the Advanced dialog box.

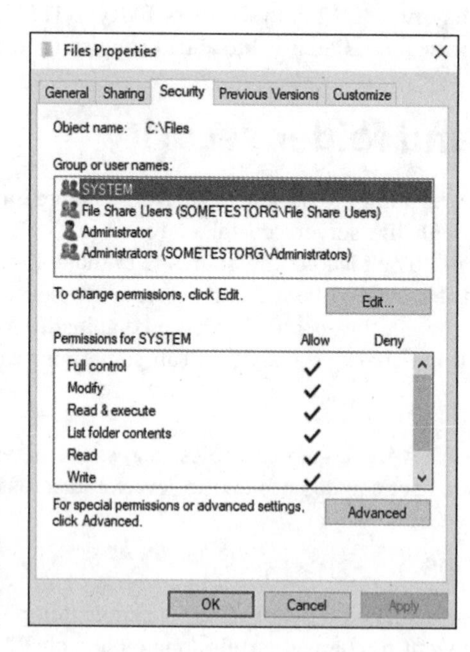

FIGURE 1-5: You can set permissions very granularly with the Security tab in the File Properties dialog box.

Share permissions

Share permissions are set using the Sharing tab is the File Properties dialog box (refer to Figure 1-5) — just click the Advanced Sharing button and then click Permissions. Share permissions are much simpler than NTFS permissions. You see something similar to Figure 1-6.

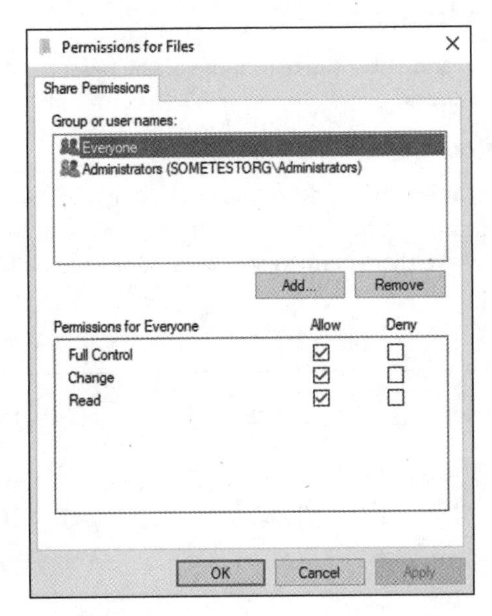

FIGURE 1-6: Share permissions allow you to grant Full Control, Change, or Read access to users or groups.

You have three simple levels of permissions that you can set on a share:

- **Full Control:** Allows you to read, modify, and delete items within the share. You can also change permissions and take over ownership of the files.
- **Change:** Allows you to do everything that Full Control can with the exception of setting permissions.
- **Read:** Allows you to view files and folders, but not edit in anyway.

Effective permissions

Effective permissions are the ones that can cause problems. You may get your permissions set up right, but then a user calls in and says they can't access their files. In Windows Server 2016, Microsoft introduced an Effective Permissions tab that allows you to check what permissions a user will actually have based on the combination of NTFS permissions and share permissions.

To get to the Effective Permissions tab, follow these steps:

1. **Right-click the folder that you want to check and select Properties.**
2. **Click the Security tab.**
3. **Click Advanced.**
4. **Click the Effective Access tab.**
5. **Click Select a User, and enter the username of the person whose permissions you want to check.**
6. **Click View Effective Access.**

   You see a screen similar to Figure 1-7. It shows you that my user Karen Smith doesn't have permissions to the folder. In reality, she has no NTFS permissions, although she does have share permissions. Because she has no NTFS permissions, she is denied.

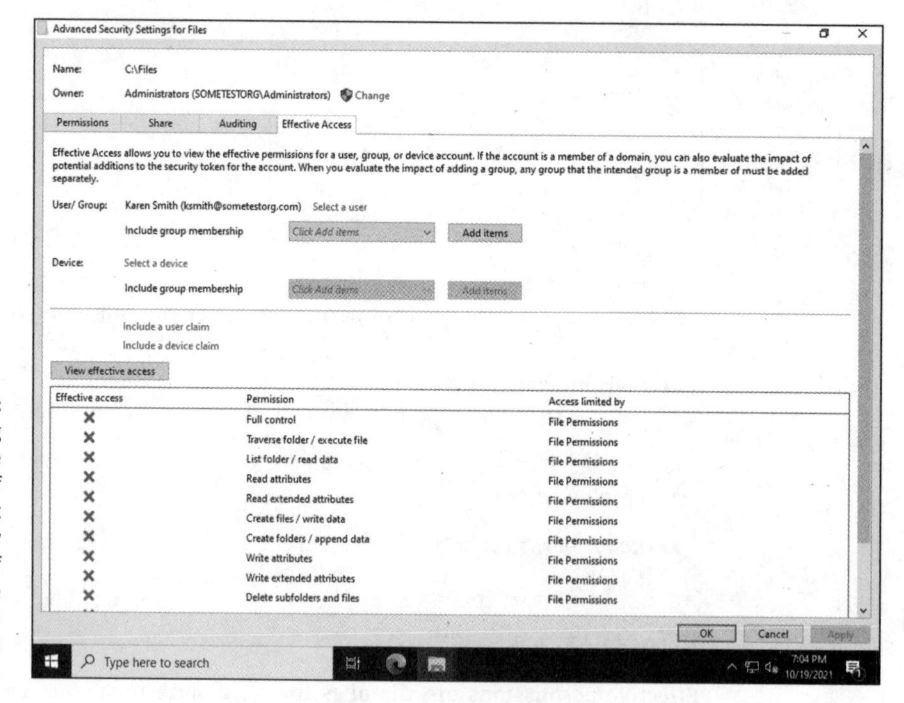

FIGURE 1-7: Checking the effective permissions of a user account is a great way to validate if they have the permissions you expect them to have.

Creating a Local Security Policy

Local security policies allow you to define quite a few settings related to the computer that you're on. They can be very useful if you need a setting to be applied on a specific server that is not currently being applied by Group Policy. Local policy applies to the server first, followed by AD Group Policy. If there is a conflict between the settings of the Local Security Policy and the AD Group Policy, the settings in the AD Group Policy will apply because it's applied to the system after the Local Security Policy. To access the Local Security Policy, follow these steps:

1. **Click the Start menu.**
2. **Type** secpol.msc **and press Enter.**

   The Local Security Policy window opens.

From this screen, you can change quite a few settings related to the security of your system. If file server auditing is not enabled, for instance, you can turn it on for this individual server. See an example of the Local Security Policy in Figure 1-8.

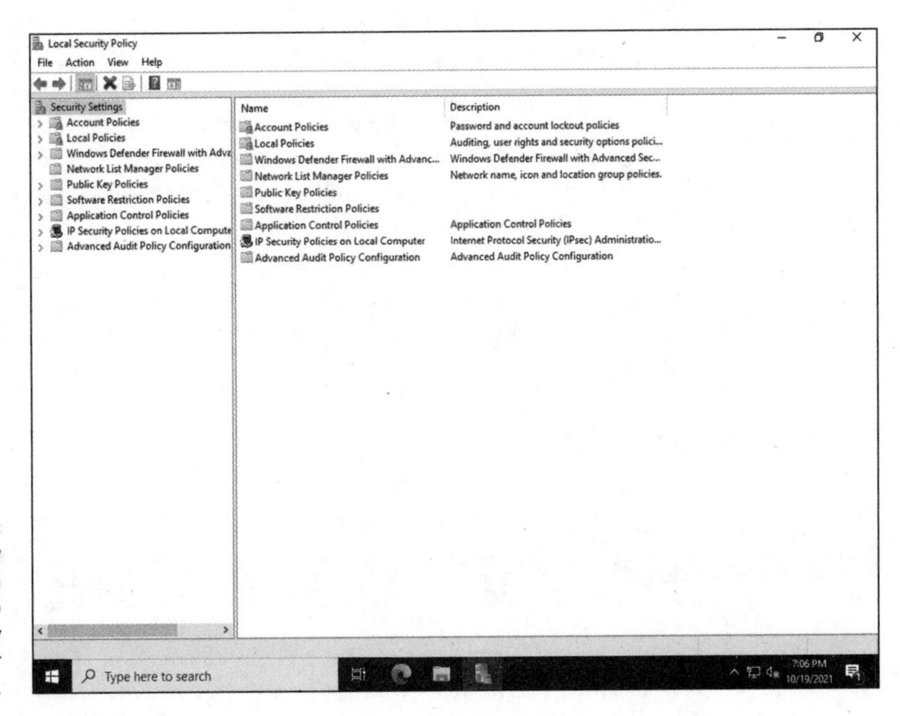

FIGURE 1-8: The Local Security Policy screen allows you to set local security settings on your system.

REMEMBER

Group Policy will overwrite these settings. If you set a super strict Local Security Policy, but you have lax password requirements being pushed through Group Policy, your system will inherit the lax password requirements. Some key areas of interest for any System Administrator would be:

- **Account Policies:** Account Policies contains the Password Policy, Account Lockout Policy and Kerberos Policy. The Password Policy allows you to set password requirements. The Account Lockout Policy defines when an account will be locked out and for how long it will be locked. The Kerberos Policy allows you to define the lifetime for the different ticket types.
- **Audit Policy:** The Audit Policy is located under Local Policies. It allows you to specify which types of events you want to audit. These can be set to Success, Failure, or both.
- **User Rights Assignment:** User Rights Assignment is the second section under Local Policies. This allows you to assign accounts to various things on the system. For instance, if you have a service account that is used to launch non-interactive scripts, you need to add the user account to the logon as a batch job section.
- **Security Options:** Security Options is also located under Local Policies. This is a long list of security settings that can be defined for your system. They're organized into like groups of configuration options like Accounts, Audit, DCOM, Devices, Domain Controller, Interactive Logon, Microsoft Network Client, Microsoft Network Server, Network Access, Network Security, Recovery Console, Shutdown, System Cryptography, System Objects, System Settings, and User Account Control.
- **Advanced Audit Policy Configuration Settings:** This area is all about auditing and the types of things that you can audit. It allows you to set up auditing for all different categories of events including: Account Logon, Account Management, Detailed Tracking, DS Access, Logon/Logoff, Object Access, Policy Change, Privilege Use, System, and Global Object Access Auditing.

Paying Attention to Windows Security

The Windows Security app provides a central location to view the security health of your system. You can view the status of your antivirus software under Virus & Threat Protection, check your firewall settings in Firewall & Network Protection, look at whether an app is trusted with App & Browser Control, and gain additional protections with Device Security.

Virus & Threat Protection

Microsoft Defender Antivirus is a full-featured anti-malware solution. It's able to check for regular updates and does scheduled and real-time scanning. It even provides ransomware protection capabilities. Check out Figure 1-9 for an idea of what the Virus & Threat Protection dashboard looks like. From this dashboard, you can access previous scans and check settings for Microsoft Defender Antivirus.

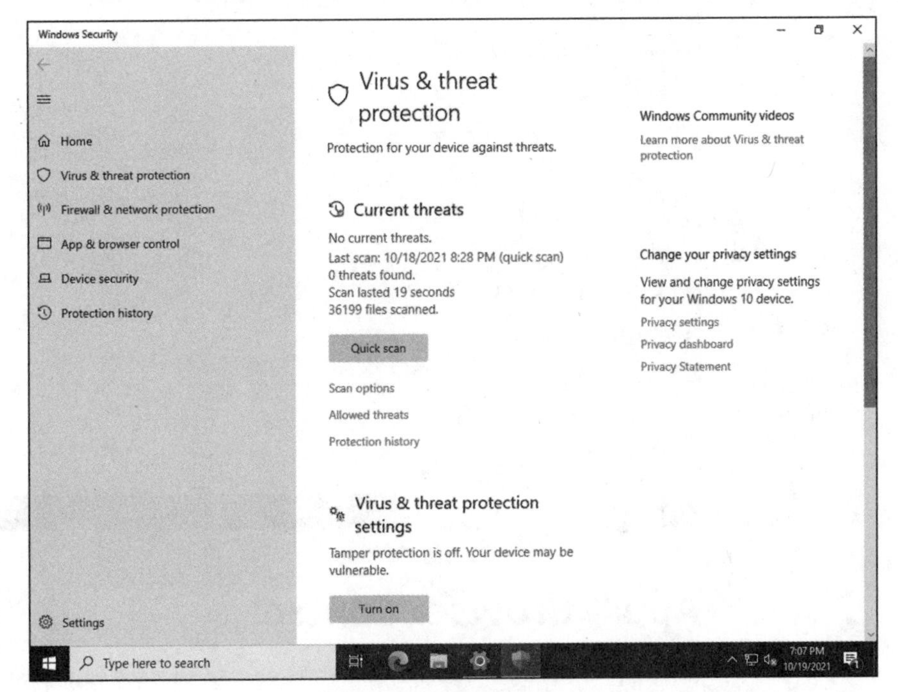

FIGURE 1-9: The Virus & Threat Protection dashboard offers a full-featured anti-malware solution.

You can choose the settings for the Virus & Threat Protection screen across the organization using Group Policy. Simply make the changes that you want to push out to the organization by opening the Group Policy Management Editor. Open the GPO that you want to edit, and then double-click on Computer Configuration, followed by Policies, then Administrative Templates, then Windows Components, and then Microsoft Defender Antivirus.

Firewall & Network Protection

The Firewall & Network Protection area allows you to work with the various profiles in your Windows Firewall. You have a private profile and a public profile, and if your system is domain-joined, you have a domain profile as well. From here, you can add exceptions for applications so that they're allowed through the firewall and you can adjust the notification settings of the Windows Firewall. See Figure 1-10 for an idea of what the Firewall & Network Protection area looks like.

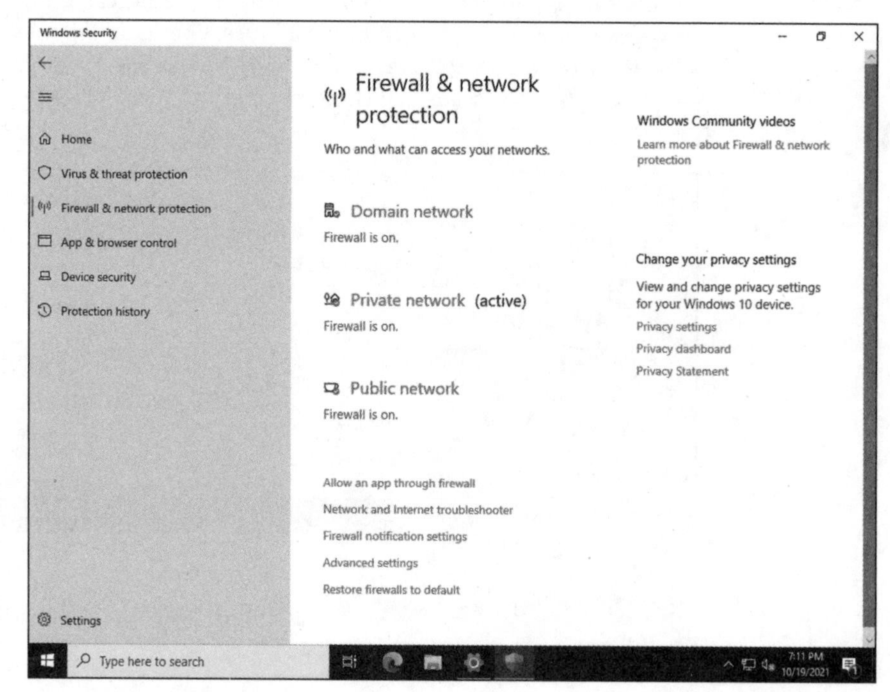

FIGURE 1-10: Changing the firewall settings from Firewall & Network Protection is simple with the links provided.

App & Browser Control

The App & Browser Control section is able to protect your system by checking applications and files that are downloaded from the web for threats. You can set it to block or warn when it comes across these files, or you can turn it off. The default is set to warn. In addition you can tweak the exploit protection settings built into Windows Server 2022 by clicking Exploit Protection Settings. App & Browser Control is shown in Figure 1-11.

The Exploit Protection screen is worth a little more scrutiny (see Figure 1-12). It can protect you against multiple types of exploits and is on by default. It features Control Flow Guard, which helps to ensure the integrity of indirect calls made to the system, and Data Execution Prevention, which prevents code from being run in memory pages that are reserved for data. Plus, it offers Address Space Layout Randomization (ASLR), which provides randomization for the locations where executables are stored in your server's memory. This provides protection against buffer overflow attacks.

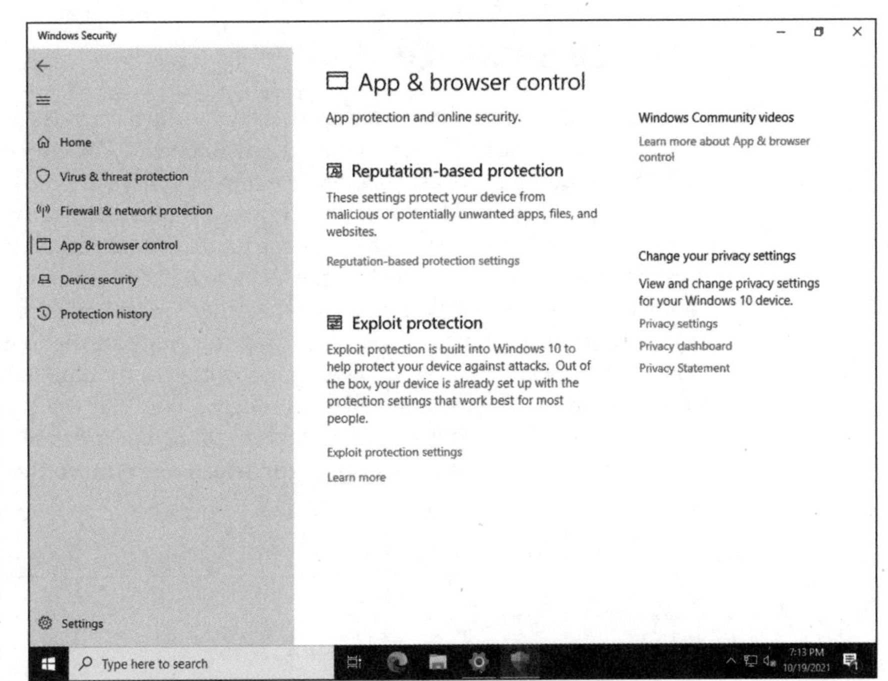

FIGURE 1-11: App & Browser Control gives you the ability to protect yourself against downloaded executables and files, as well as configure Exploit Protection.

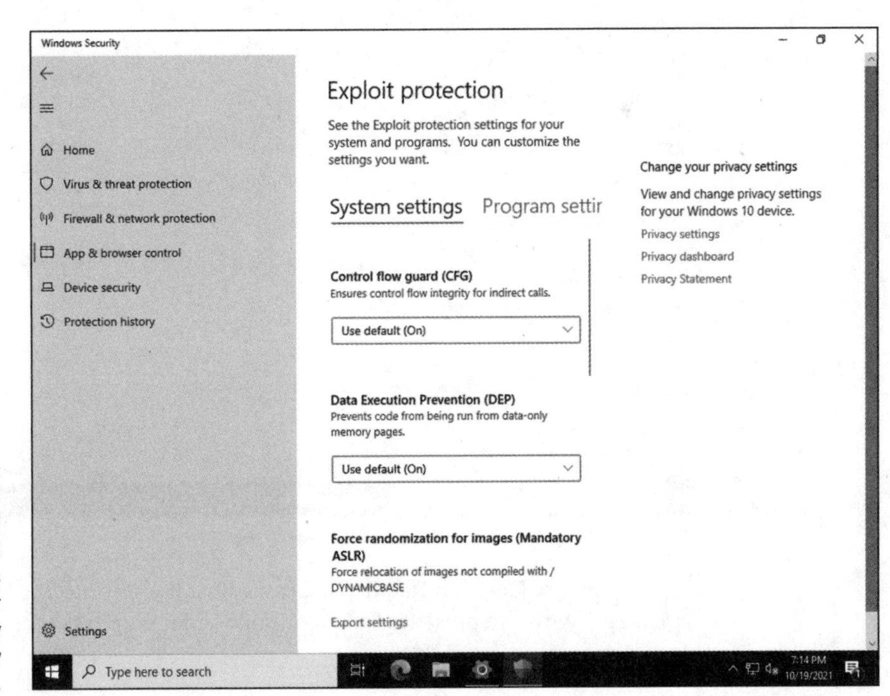

FIGURE 1-12: Exploit Protection provides several more advanced mechanisms to protect your system, already enabled by default.

Device Security

Last but not least is the Device Security section, which provides utilities that allow you to interact with your Trusted Platform Module (TPM) chip (if you have one) and virtualization controls. The TPM is a chip on your motherboard that generates cryptographic keys and keeps half of the key; the other half of the key is stored on disk. This prevents a thief from stealing a hard drive and decrypting it on another system. There is a button to clear the TPM, and users can receive recommendations to update their TPM firmware when there is an update available. There is also a Hypervisor Control Integrity setting that you can use to enable or disable this functionality. It's used to determine if software running in kernel mode like drivers is safe software. You can see in Figure 1-13 that the Hypervisor Control Integrity piece is running on my virtual machine, but it has no TPM exposed to it (the TPM doesn't get passed through VirtualBox v6.x), so the TPM options do not exist.

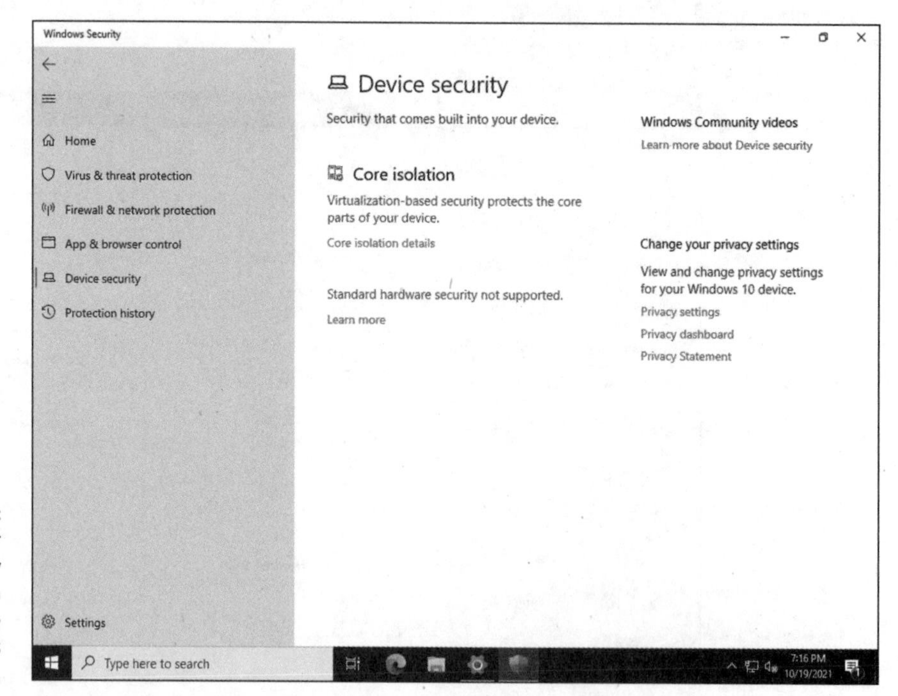

FIGURE 1-13: The Hypervisor Control Integrity feature is shown under Core Isolation on this virtual machine.

If you click Core Isolation Details, you can adjust the settings for Hypervisor Control Integrity. You're presented with a simple slider switch that allows you to enable or disable memory integrity.

with file system security

» Sharing resources on Windows Server 2022

» Setting up federated access with Active Directory Federation Services

» Understanding and deploying Active Directory Rights Management Services

Chapter 2

Configuring Shared Resources

One of the most common tasks that a system administrator faces when getting systems on the network is making sure that the resources on those systems are available.

In this chapter, I explain shared folder permissions, printer sharing, and how to share other items of interest from your server. I also explain how to set up Active Directory Federation Services (AD FS), which allows you to use the same authentication source against other entities and applications, as well as configure Active Directory Rights Management Services (AD RMS) to protect documents that you're sharing internally and externally.

Comparing Share Security with File System Security

Making resources available to your end users without giving them access to the servers the resources are on is a must in organizations. By sharing a folder, you can map a drive to the shared folder for your end users automatically. The end user sees another drive (think home drives or department drives), when in reality it's a folder that resides on a server. One of the most confusing things for new system administrators is how to share a folder. There are two tabs: Security and Sharing. For an explanation on discretionary access control lists (DACLs) and the different permission levels available on the Security tab and the Share tab, check out Book 5, Chapter 1.

As a best practice, you'll typically set more open access on the shares, and then restrict access with the New Technology File System (NTFS) permissions on the Security tab. This simplifies administration greatly, and with the Effective Permissions tab, you can check to ensure that users are getting the permissions to the share and its contents that they're supposed to have.

TECHNICAL STUFF

You may be wondering, "Why do I have set permissions in two places? That's just silly!" I agree. The reason predates the NTFS file system. Before NTFS, there was FAT16 and FAT32. These files systems didn't allow you to set access control the way you can with NTFS. So, to properly secure shares, the Sharing tab was introduced, which provided the basic three settings: Read, Change, and Full Control. When the NTFS file system was introduced, all of a sudden you had access to much more granular access controls. And that brings us to where we are today. Many system administrators set the same permissions on both NTFS and the shares. This does work, but it adds a great deal of complexity to the management of the shares. What Microsoft and the majority of Microsoft trainers recommend is to set pretty open permissions on the share, and then restrict access through NTFS using the Security tab.

Shared folder permissions

When you create a share, you create a Universal Naming Convention (UNC) path (`\\servername\sharename`) that your users can then use to connect to that folder. The great thing about setting up folder shares is that you can use shares to give your users access to the intended folder, without needing to give them any access to log in to the server.

Security settings on shares tend to have more open permissions, while restrictions are now done on the Security tab and are enforced by the file system.

In Figure 2-1, you can see the share permissions for the Software folder. Notice that everyone has full control.

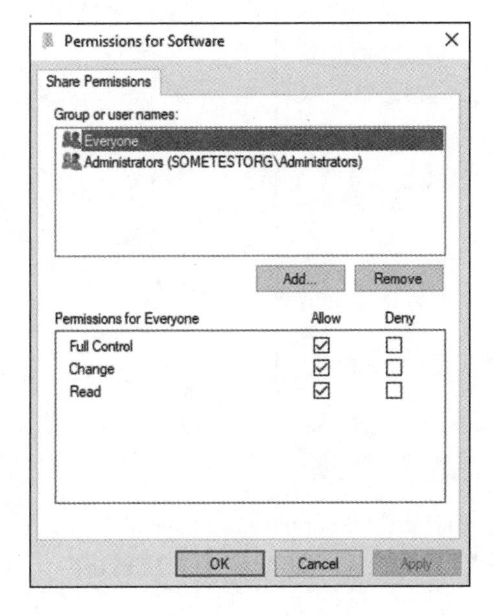

FIGURE 2-1: Share permissions are pretty open in this example. Everyone has full control.

File system security

NTFS was first introduced in 1993, but it didn't gain popularity until quite a bit later. NTFS was the first file system supported by Windows that offered more granular access control. Rather than the three basic permissions provided by shares, it has six permissions that can be assigned: Full Control, Modify, Read & Execute, List Folder Contents, Read, and Write.

The current best practice is to have more open permissions on shares, while using the NTFS file system to restrict access to the folder. The reasoning behind this best practice is that it would be difficult to set both share and NTFS permissions and keep them both in sync. To avoid this issue, you would want to use one over the other. Because NTFS permissions offer more granularity, they're the logical best choice.

In Figure 2-2, you can see that a user named Karen Smith has Read & Execute, List Folder Contents, and Read for permissions. In the next section, I explain why this is important.

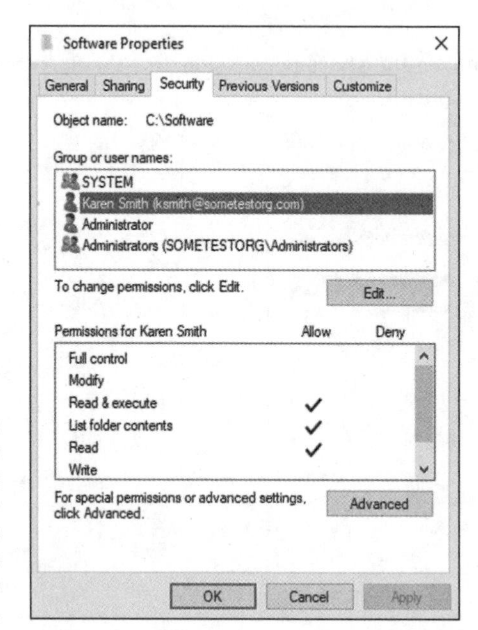

FIGURE 2-2: Here, NTFS permissions set on the Security tab are more restrictive than the share permissions.

Effective permissions validation

Effective permissions refers to the permissions that a user has actually been granted as a combination between file system permissions and share permissions. Securing your shares with NTFS is preferred due to its granularity and the fact that it applies to both local and network access. In either case, the more restrictive permission will apply. If a user does not explicitly have access granted, he'll be denied access by default. See Table 2-1 for examples of effective permissions.

TABLE 2-1 **Effective Permissions**

Share Permissions	NTFS Permissions	Effective Permissions
Read	Full Control	Read
Full Control	Read	Read

Now that you have an idea of how effective permissions work, let's go back to the example that I show you in the previous sections. The share permissions on my server had everyone set to Full Control. The NTFS permissions gave the Karen Smith user much more limited permissions. Check the Effective Permissions tab to see what she actually has:

1. **Right-click the folder you want to check permissions on and choose Properties.**
2. **Click the Security tab.**
3. **Click Advanced.**
4. **Click the Effective Access tab.**
5. **Fill in the user or group you want to check and then click View Effective Access.**

You can see the results for the Karen Smith user in Figure 2-3.

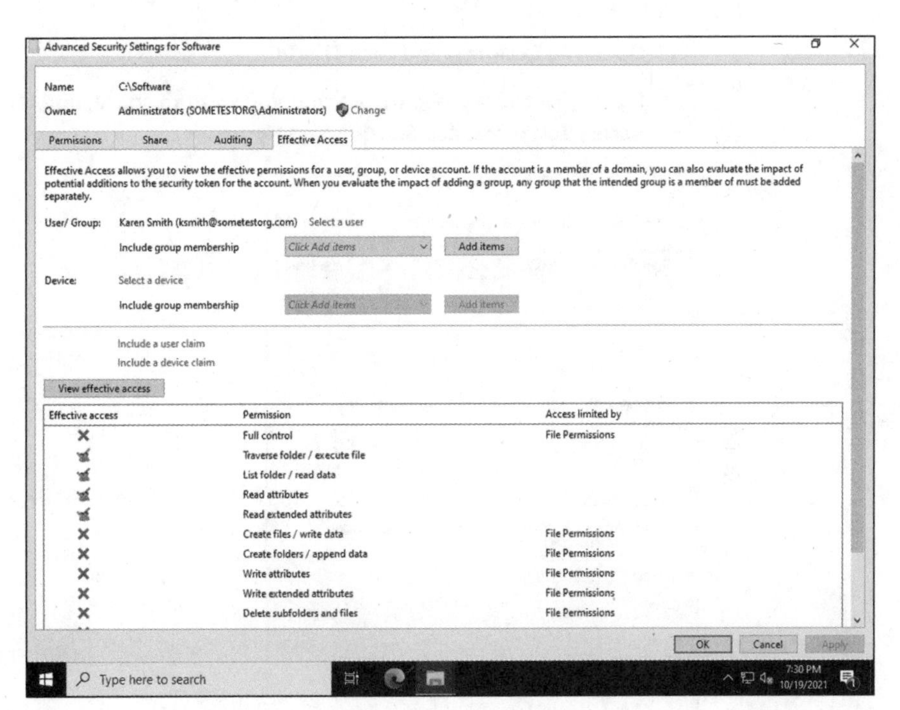

FIGURE 2-3: The Effective Access tab shows you what a user or group actually has for permissions.

The check-marked items are the access that was given to the Karen Smith user on the Security tab. Also, notice that each of the items with a red X says File Permissions under the Access Limited By column.

Sharing Resources

You may want to share other items beyond the typical files and folders. For example, you may want to share device drives or other hardware. In this section, I walk you through a few of the things that you may want to share and explain how to do so.

Storage media

You may need to share access to a DVD drive or an external storage device that is attached to your machine. This is a very simple thing to do. Just follow these steps:

1. **Open File Explorer, and click This PC.**
2. **Right-click the storage media that you want to share, and choose Give Access To⇨Advanced Sharing (see Figure 2-4).**

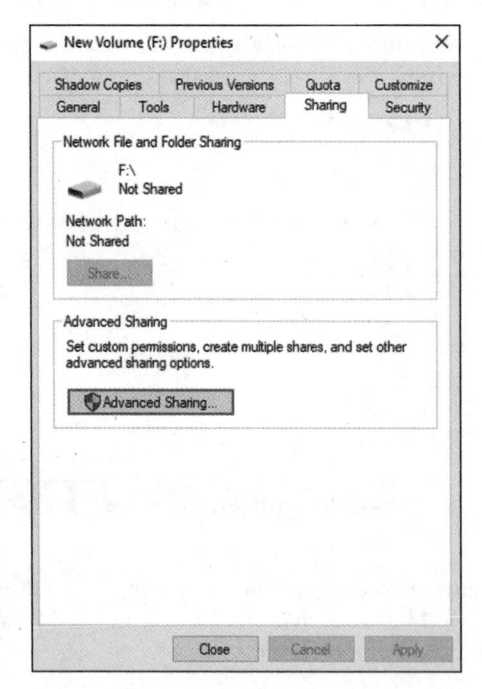

FIGURE 2-4: Sharing a storage device follows similar steps to sharing a folder.

3. **On the Sharing tab, click the Advanced Sharing button.**
4. **Select the Share This Folder check box.**

5. **Create a more descriptive drive name.**

   By default, the share name is the drive letter of the volume/media you're sharing.

6. **Click Permissions, add whoever you want to have access to the drive, and click OK.**

7. **Click OK again.**

   Your drive will be shared.

You can tell the storage drive is shared because an icon with two people will show up underneath it after you've shared it.

Printers

Sharing a printer is relatively straightforward as well, though the steps are a little different than they are for sharing storage devices. In more than a few organizations, I've seen a local printer shared where there was an enterprise-wide print server. You might ask why they were sharing a printer from a workstation rather than use it through the print server. The answer was usually along the lines of a one-off printer.

Follow these steps to share a printer:

1. **Click Settings and then click Devices.**
2. **Click Printers & Scanners.**
3. **Select the printer you want to share, and click Manage.**
4. **Click Printer Properties.**

   The Printer Properties dialog box appears.

5. **Click the Sharing tab.**
6. **Click Change Sharing Options.**
7. **Select the Share This Printer check box and choose any other options that you want.**

   TIP

   I recommend keeping Render Print Jobs on Client Computers checked because this will reduce the load on your system when people print.

8. **Click OK.**

Other resources

How you share other resources will depend on what you want to share. In a Windows Server environment, the options to share should be very similar to what was already covered, so if you understand how to share folders and printers, you should be able to handle any sharing requests that come your way.

Configuring Access with Federated Rights Management

Sharing isn't always limited to things like folders or hardware. Sometimes you may want to "share" credentials with another entity like using your Active Directory on-premises to authenticate to the Microsoft Azure portal. This isn't a true sharing, of course. By setting up Active Directory Federation Services (AD FS), you aren't giving Microsoft your credentials; you're simply federating a trust.

You may also find that you want to have greater control over your files when they leave your organization. Maybe you want to password encrypt them or force them to expire. You can do these things with Active Directory Rights Management Services (AD RMS).

In this section, I dig into AD FS and AD RMS, explaining what each of these does and how to configure them.

Working with Active Directory Federation Services

If you've ever worked with Active Directory before, you probably liked the fact that so many of your internal applications could take advantage of Active Directory, too. Your users were able to use just one username and password to access all the things that they needed to do their jobs. But what about these third-party cloud services that are popping up everywhere? Wouldn't it be great of your users could use the same usernames and passwords that they know to access these third-party solutions? Great news! They can use their usernames and passwords, in the third-party-supported AD FS.

So, how does AD FS work? Let's look at the authentication process. Site 1 is your local network where your domain controllers and your AD FS and AD FS Proxy servers live. Site 2 hosts the third-party website you want to authenticate to using your Active Directory credentials. There is a trust set up between Site 1 where you are and Site 2 where you want to access resources. The AD FS server in Site 1 is referred to as the *identity provider*, because that's where the authentication takes place. Site 2 is referred to as the *trusting party*; it relies on the token from AD FS to determine if you're an authenticated user or not. It works like this:

1. You open your browser and type in the URL for Site 2.
2. Site 2 redirects the request to the Site 1 AD FS Proxy server, which asks for your username and password.
3. After the request is authenticated, the Site 1 Proxy returns you to Site 2 with a token that contains the claims about your user account including your identity.
4. You're logged into Site 2.

REMEMBER

AD FS was designed with a very specific goal in mind: to provide a single sign-on (SSO) experience to your end users to web applications. It works really well for that purpose, and it's what Microsoft recommends in most cases for authentication with Office365. AD FS was not designed to help authenticate other things that require the more traditional Windows NT token like access to file shares, Exchange Server (email), RDP, or older web applications that don't understand claims.

At this point, you may be wondering how AD FS fits into a chapter on sharing resources. You aren't sharing Active Directory after all. When you federate a trust with another organization, you can access that other organization's resources. In essence, the other organization can share its resources (like a web application) with you without needing you to authenticate separately to its systems.

Now that you know what AD FS is, and the basics of how it works, let's actually set it up. You need a Secure Sockets Layer (SSL) certificate to finish the configuration. For the purposes of these instructions, I'll create a self-signed certificate, but you'll want a certificate from a public certificate authority for an actual production AD FS server.

WARNING

Do not install AD FS on to a domain controller. AD FS proxy servers need to be public facing if you're using them to authenticate against external resources like Software as a Service (SaaS) providers, and you certainly don't want a domain controller exposed to the Internet.

To install, configure, and test an AD FS server, follow these steps:

1. **From Server Manager, click Manage and then click Add Roles and Features.**
2. **On the Before You Begin screen, click Next.**

3. **On the Select Installation Type screen, click Next.**
4. **On the Select Destination Server screen, click Next.**
5. **On the Select Server Roles screen, select Active Directory Federation Services, and then click Next.**
6. **On the Select Features screen, click Next.**
7. **On the Active Directory Federation Services (AD FS) screen, click Next.**
8. **On the Confirm Installation Selections screen, click Install.**
9. **Click Close.**

The first step is complete: You've installed AD FS. To make it useful, though, you need to configure it. Follow these steps:

1. **Click the flag in Server Manager, and then click Configure the Federation Service on This Server (see Figure 2-5).**

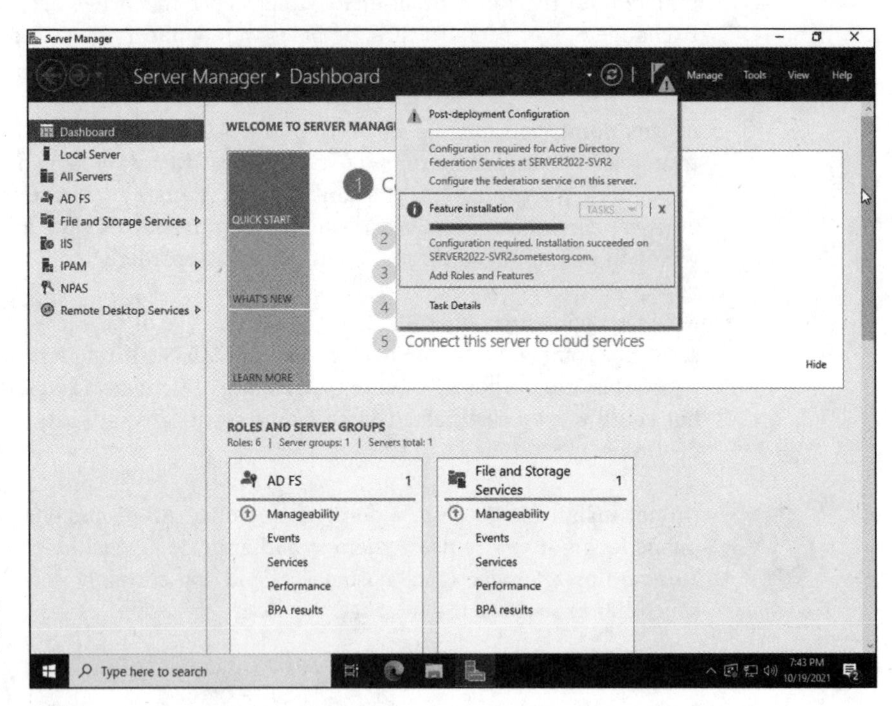

FIGURE 2-5: Configuring the AD FS role after installation.

2. **On the Welcome screen, select Create the First Federation Server in a Federation Server Farm, and click Next.**

3. **On the Connect to Active Directory Domain Services screen, enter the name of a Domain Administrator, and click Next.**

4. **On the Specify Service Properties screen, select the SSL certificate you want to use.**

   The Federation Service Name will fill in based on the common name on the certificate.

5. **Fill in the Federation Service Display Name, and click Next.**

   On the next screen, you need to create a group managed service account, but this option will most likely be grayed out because the KDS Root Key has not been created.

6. **Right-click Start, and select Windows PowerShell (Admin).**

7. **Type in the following command, and then press Enter:**

```
Add-KdsRootKey -EffectiveTime (Get-Date).AddHours(-10)
```

   The KDS Root Key is created.

TECHNICAL STUFF

   Key Distribution Service (KDS) keys are required for you to be able to generate passwords for Group Managed Service Accounts (gMSAs). A gMSA is a type of managed service account that is able to synchronize its password to multiple systems automatically. The service using the gMSA needs to support gMSAs for this to work.

8. **Return to the Active Directory Federation Services Wizard and click OK on the error box.**

   You may need to click Previous and then click Next to get the option for Create a Group Managed Service Account to be available.

9. **Select Create a Group Managed Service Account, type the name you want to use, and click Next.**

   I'll use gmsa_adfs.

10. **On the Specify Configuration Database screen, select Create a Database on This Server Using Windows Internal Database and click Next.**

TIP

   In a production environment, you'll most likely want to select an actual SQL Server to host the database for AD FS. The Windows Internal Database is limited and not very scalable.

11. **On the Review Options screen, click Next.**

   Your review screen should look similar to Figure 2-6.

Configuring Shared Resources

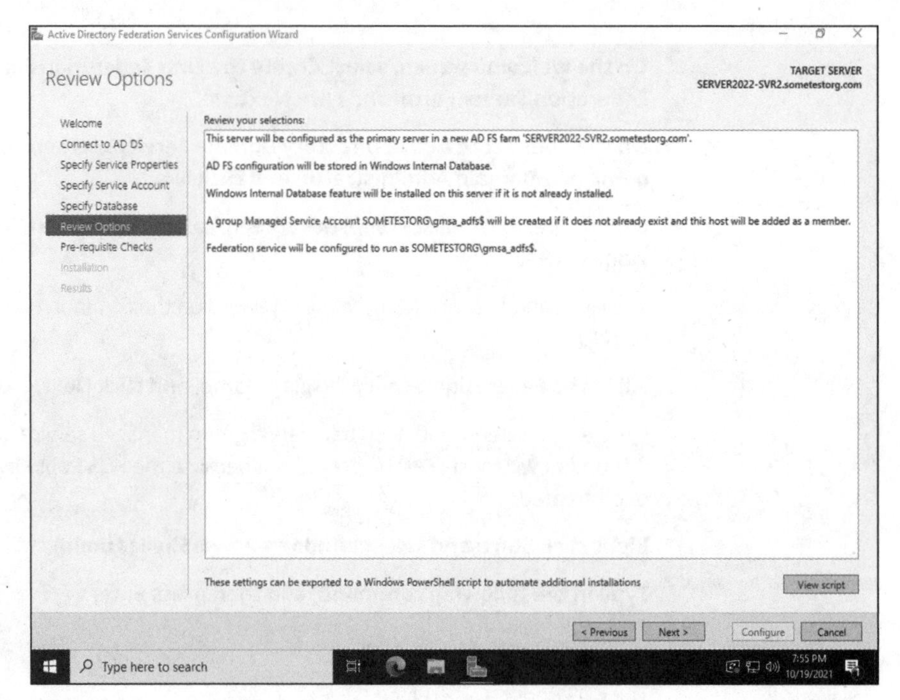

FIGURE 2-6: Review your AD FS settings before completing the configuration.

12. **On the Pre-requisite Checks screen, click Configure.**

    The installation screen will show you the progress and any errors it ran into. When it's complete, you'll see the Results screen, with the message "This server was successfully configured."

13. **Click Close.**

14. **Restart the server for the AD FS role to finish its configuration.**

To make sure that AD FS is up and running, you need to enable the page you're going to use for testing. Starting in Windows Server 2016, the idpinitiatedsignon.aspx page was disabled by default. Follow these steps:

1. **Right-click the Start Menu and choose Windows PowerShell (Admin).**
2. **Type in the following command, and press Enter:**

```
Set-AdfsProperties -EnableIdPInitiatedSignonPage $true
```

3. **Open Internet Explorer and navigate to** `https://ADFS_FQDN/adfs/ls/idpinitiatedSignOn.aspx`.
4. **Click the Sign In button.**
5. **Enter an Active Directory username and password, and click OK.**

   You're presented with a screen that says "You are signed in."

After you've confirmed that AD FS is working, you can disable the page you used to test it with by entering the following command:

```
Set-AdfsProperties -EnableIdPInitiatedSignonPage $false
```

Working with Active Directory Rights Management Services

AD RMS allows you to provide data protection on documents made in RMS-aware applications like Microsoft Office. This protection can follow the document off-premises and can be applied even when there is no network connection. In short, the data is protected regardless of where it goes. You can set encryption and access expiration requirements. And you can even prevent emails or documents from being copied and pasted or printed.

AD RMS is installed through the already familiar Server Manager interface. Configuration isn't exceptionally complex either. Follow these steps to get AD RMS up and running in your environment:

1. **From Server Manager, click Manage and then click Add Roles and Features.**
2. **On the Before You Begin screen, click Next.**
3. **On the Select Installation Type screen, click Next.**
4. **On the Select Destination Server screen, click Next.**
5. **On the Select Server Roles screen, select Active Directory Rights Management Services, click Add Features, and then click Next.**
6. **On the Select Features screen, click Next.**
7. **On the Active Directory Rights Management Services screen, click Next.**
8. **On the Select Role Services screen, accept the default with just AD RMS selected, and click Next.**
9. **On the Web Server Role (IIS) screen, click Next.**
10. **On the Select Role Services screen, accept the defaults and click Next.**
11. **On the Confirm Installation Selections screen, click Install.**
12. **Click Close.**

Now that you have AD RMS installed, there is some post-install configuration that needs to be done. Follow these steps:

1. **Click the flag in Server Manager and click the Perform Additional Configuration link shown in Figure 2-7.**

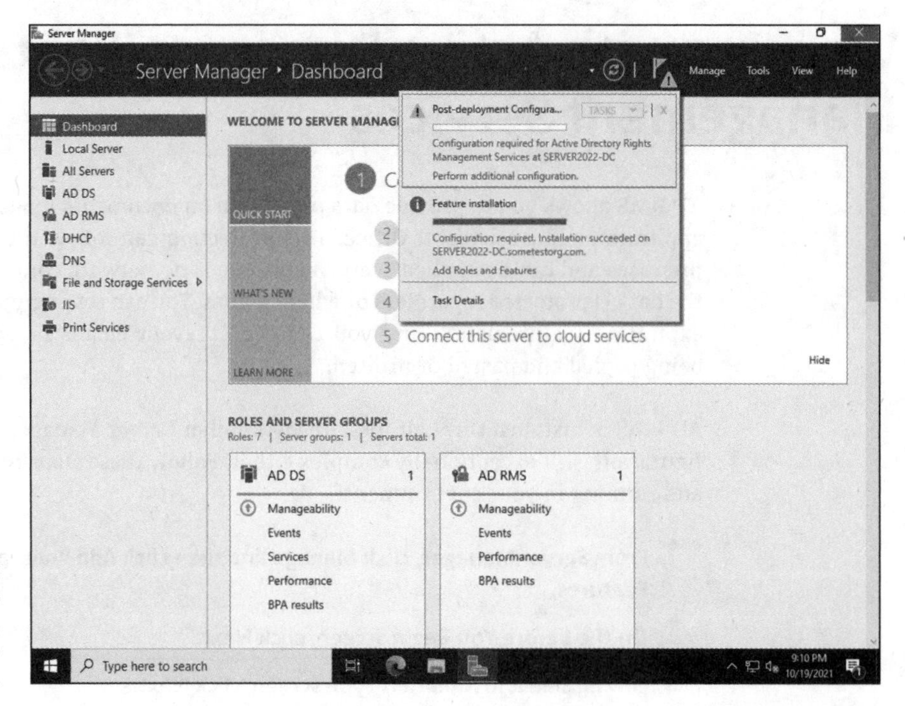

FIGURE 2-7: After AD RMS is installed, you need to configure it.

2. **On the Active Directory Rights Management Services screen, click Next.**
3. **On the Create or Join an AD RMS Cluster screen, choose Create a New AD RMS Cluster, and click Next.**
4. **On the Select Configuration Database screen, enter the name of your SQL server and click Next.**

   For lab environments, it's okay to choose Windows Internal Database.

5. **On the Specify Service Account screen, click Specify, enter the username and password of the service account for AD RMS, click OK, and then click Next.**

TECHNICAL STUFF

A quick note here to preserve your sanity: In a production environment, the service account for AD RMS should be a member of Domain Users and should have no additional privileges assigned to it. If you're in your lab environment, and you've installed AD RMS on a domain controller, you'll need to place the service account into the Domain Admins group. If you don't, you'll get an error along the lines of `Invalid credentials were presented. Verify the correctness of the provided password.`

6. **On the Specify Cryptographic Mode screen, select Cryptographic Mode 2 and click Next.**

TIP

Unless there is a need for legacy support, you should always choose Cryptographic Mode 2. It has longer private keys (2048 bits) and uses a stronger hashing algorithm. This makes it more difficult for an attacker to break through your encryption because it becomes more time intensive with larger key sizes.

7. **On the Specify AD RMS Cluster Key Storage screen, select Use AD RMS Centrally Managed Key Storage and click Next.**
8. **On the next screen, set a password to protect the cluster key, and then click Next.**
9. **Select the website that you want to use for AD RMS, and then click Next.**

   In this case, I'll just use the Default Web Site.

10. **On the Specify Cluster Address screen, set the Connection Type to Use an SSL-Encrypted Connection, give the cluster a name (this must be a fully qualified domain name), and then click Next.**
11. **On the Choose a Server Authentication Certificate screen, choose an existing certificate and click Next.**

    Note that you can create a self-signed certificate or choose a certificate later. If you don't have a certificate available, select Self-Signed for now and replace it with a valid certificate as soon as possible.

12. **On the Server Licensor Certificate screen, click Next.**
13. **On the next screen, leave the default Register the SCP Now selected and click Next.**

    You're finally on the Confirmation page!

14. **If everything looks correct, click Install.**
15. **After installation has finished, click Close.**

Now you have AD RMS installed and configured. I'm sure you want to jump right in and start playing around. If you try to open the AD RMS console right now, though, you get an error that says "The request failed with HTTP status 401: Unauthorized." This happens because a group was created, called AD RMS Enterprise Admins. This group is created locally on the server, and though your account is added to it automatically, you don't get the benefits of the new permissions until you log out and then log back in. Go ahead and do that now. I'll wait.

After you've logged out and logged back in, Server Manager will launch automatically. Choose Tools⇨Active Directory Rights Management Services. You see a console similar to Figure 2-8.

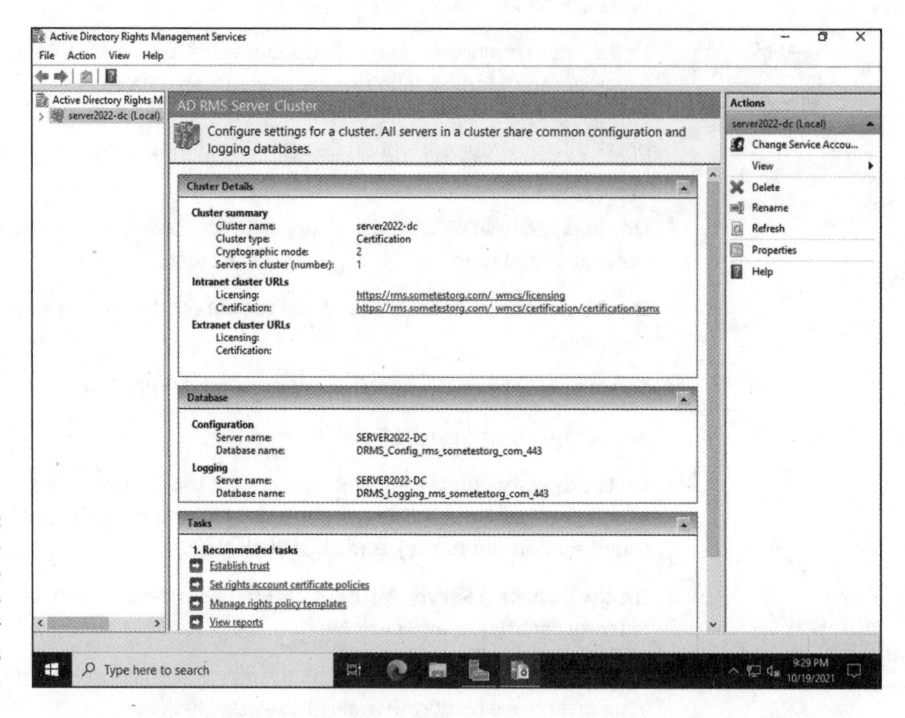

FIGURE 2-8: You can manage AD RMS from the console available through Server Manager's Tools menu.

Now you need to set up a *rights policy template*, which is what helps to define any rules and/or conditions that you want to apply to data that is protected by the template. Follow these steps:

1. **In the Active Directory Rights Management Services console, click Rights Policy Templates in the menu on the left.**
2. **Click Create Distributed Rights Policy Template from the menu on the right.**

   The screen shown in Figure 2-9 appears.
3. **Click Add.**
4. **Fill in the Template Identification Information: Language, Name and Description, click Add, and then click Next.**
5. **Select the users who can access the protected data and what rights they'll have.**

   In my case, I've given my user Karen Smith only view permissions, as shown in Figure 2-10.

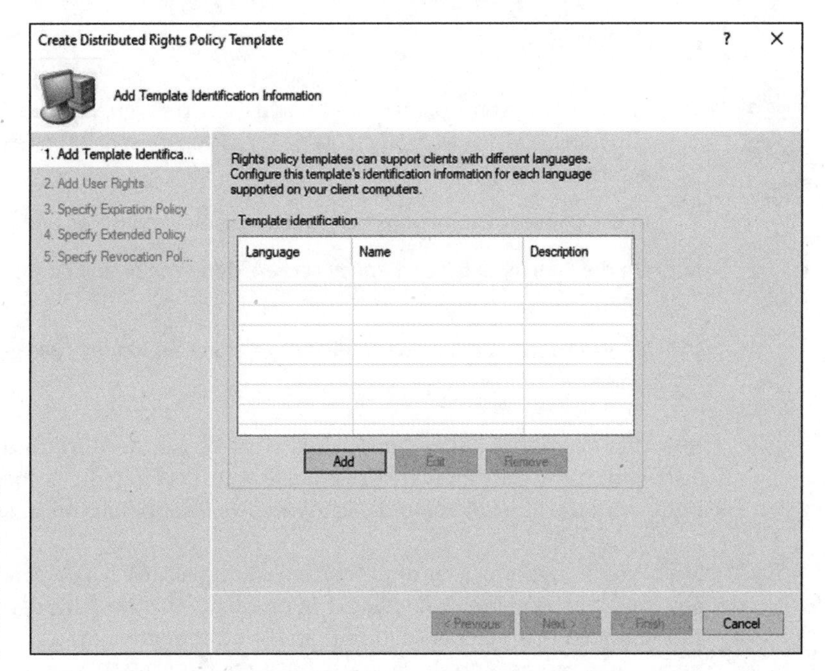

FIGURE 2-9: Creating the Rights Policy Template defines what you want to apply to protected data.

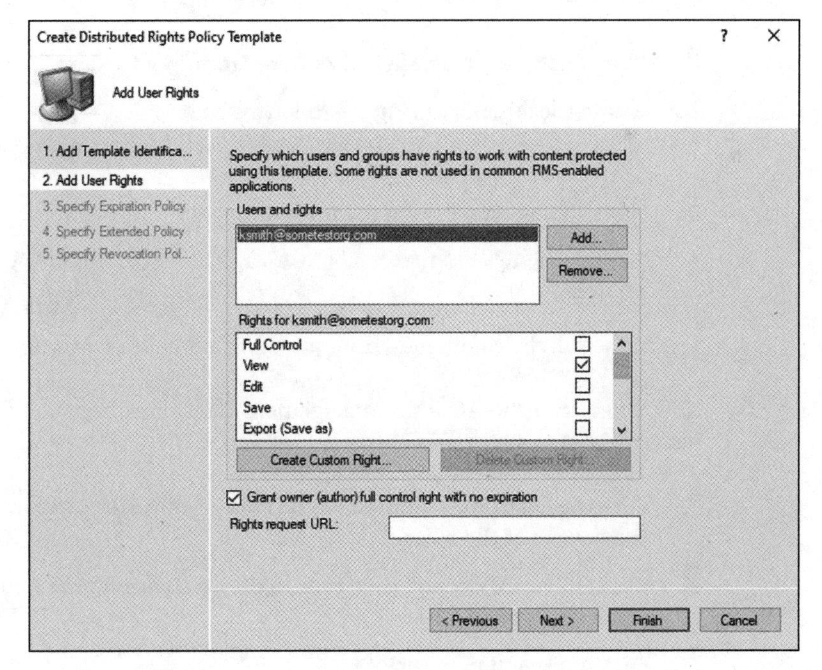

FIGURE 2-10: Ensuring that a user can view but do nothing else is simple with AD RMS.

6. **Click Next.**

In a production environment, you would want to assign rights via groups rather than individually named users. The group must have an email address associated with it.

7. **In the Content Expiration section, choose Never Expires and click Next.**
8. **On the Specify Extended Policy screen, don't make any changes. Just click Next.**
9. **On the Specify Revocation Policy page, don't make any changes. Just click Finish.**

Now that the rights policy template is created, you need to configure the actual distribution for the template. You do this in PowerShell. To open PowerShell, right-click the Start Menu and click Windows PowerShell (Admin).

First, you'll create a new directory on a storage drive. You can place this new directory wherever you would like. I'm placing it on `C:` for this demonstration, but I recommend that you don't store it on the system drive for a production deployment. Then you need to share the directory to the AD RMS service account. You do this for the AD RMS templates. Follow these steps:

1. **Open Windows PowerShell as Administrator.**
2. **Create the directory using the following code:**

```
New-Item C:\ADRMS_Templates -ItemType Directory
```

3. **Share the directory using the following code:**

```
New-SmbShare -Name ADRMSTemplates -Path C:\ADRMS_Templates
   -FullAccess sometestorg\ad_rms
```

You can see all the commands from my environment in PowerShell in Figure 2-11.

Now go back to the AD RMS Management Console, where you'll set up the file location for it to save templates:

1. **Click the Change Distributed Rights Policy Templates File Location link at the bottom of the screen.**
2. **Click Enable Export and then type in the UNC path for the folder share, as shown in Figure 2-12.**
3. **Click Apply and then click OK.**

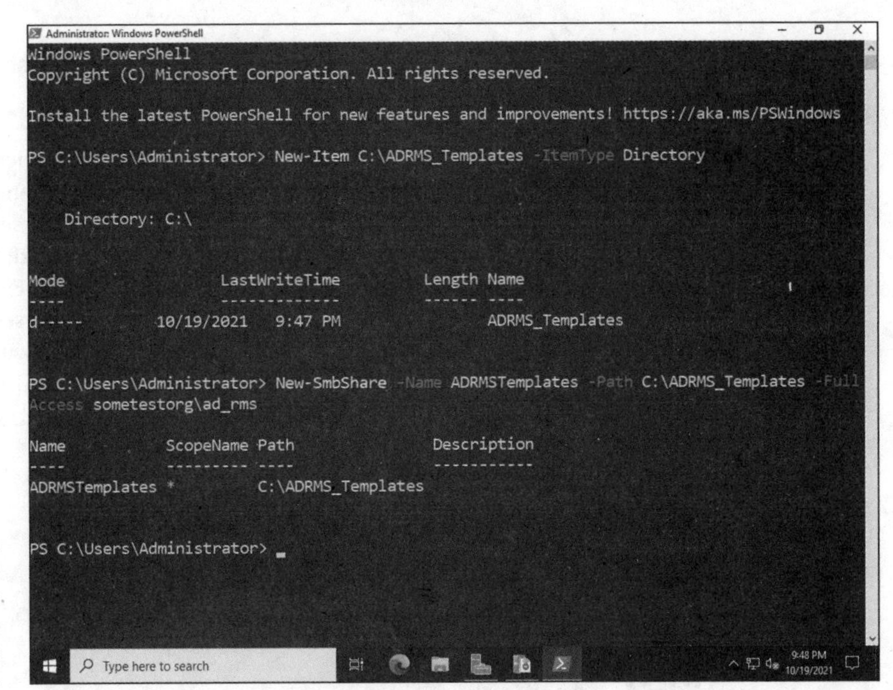

FIGURE 2-11: Creating the folders and the shares is simple through Windows PowerShell.

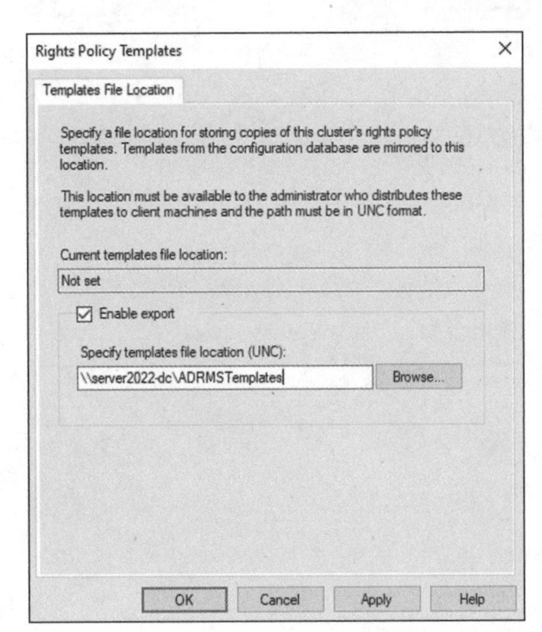

FIGURE 2-12: When the share is created, you need to point AD RMS to where you want the templates exported to.

To verify that this is working, go to your template location. You should see an XML document with the name that you gave it earlier.

Now that the server is set up, your users will be able to restrict access to their documents with the templates you've given them access to. When your users want to protect a document in Microsoft Word, for example, they can select the File tab, click Info from the menu on the left, click Restrict Access, and then click Connect to Rights Management Servers and Get Templates, as shown in Figure 2-13. Templates you've defined on the RMS server will display, and your user can choose the appropriate template based on the kind of data that is in the document.

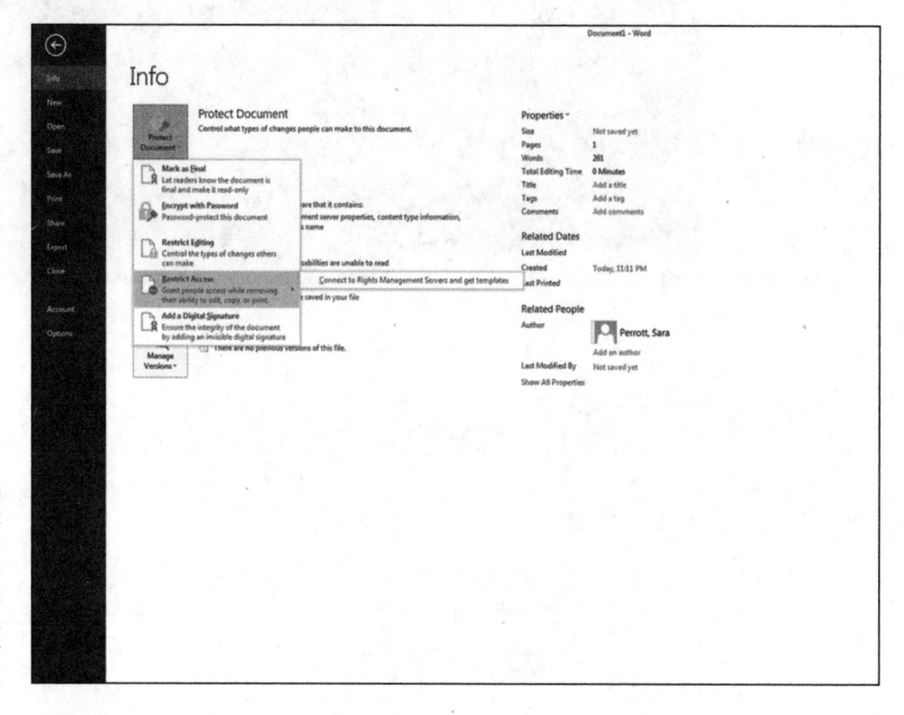

FIGURE 2-13: The Protect Document button allows you to select an RMS template that is defined on your AD RMS server.

» Administering user passwords in Windows Server 2022

» Understanding and configuring Credential Guard

» Configuring the startup and recovery options in Windows Server

» Protecting your server

Chapter 3
Configuring Operating System Security

Windows Server 2022 provides multiple built-in security mechanisms that allow you to better secure your system for no extra cost past that of the server OS license.

In this chapter, you learn about some of these built-in mechanisms — how they work and how to take advantage of them.

Understanding and Using User Account Control

You can't have a discussion about Windows internal security without a discussion of User Account Control (UAC). Love it or hate it, it serves a purpose.

Using User Account Control to protect the server

If you've worked with Windows operating systems at all, you've no doubt been prompted that something is trying to install. You've most likely received a screen similar to Figure 3-1, rolled your eyes, and allowed it. This interruption is annoying when you're trying to install something, but imagine if it prompted you to install something when you weren't attempting an install. This is what UAC is designed to protect against.

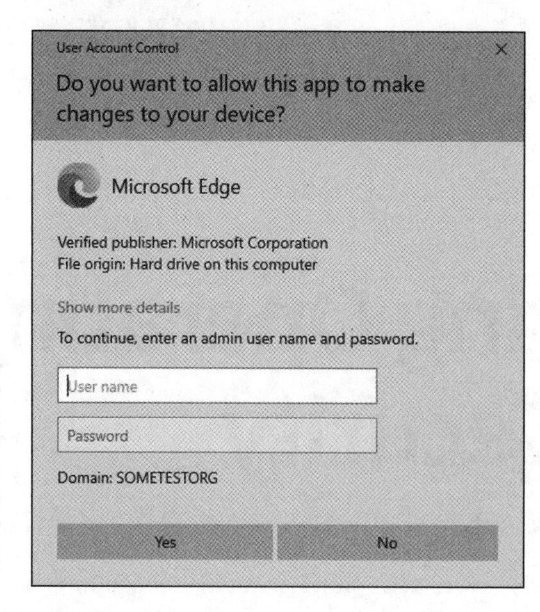

FIGURE 3-1: The familiar User Account Control window protects your system from unauthorized software or unauthorized users.

In the past, if you accidentally went to a website or opened an email that had malware, that malware could potentially run with administrator permissions without interference, and you wouldn't even know that it had run. UAC shored up this weakness. Now, if you download malicious code and it attempts to elevate its privileges, you're prompted with the familiar box asking if you want to allow it to run as administrator. This gives you the ability to stop it from even getting the chance to run in the first place.

WARNING

It's not a good idea to access the Internet on a server. It may even be against your organization's policies. If you must download software for your server from the Internet, download it to your workstation first, make sure that it's scanned by your antivirus software, and then copy it to the server.

Running tasks as administrator

You may be wondering how you can tell if an application will give you a UAC prompt when you run it. If the application's icon has a small shield in its lower-right corner, you'll get a UAC response from it.

TIP

UAC is great for interactive sessions, but what if you've scheduled tasks that you need to run. These tasks are run in a non-interactive session, so how do you get them to run as administrator without UAC interfering? When you're creating the task in Task Scheduler, make sure that you select the Run Whether a User Is Logged On or Not radio button, and select the Run with Highest Privileges check box. With those items selected, your scheduled tasks will be able to run as administrator with no interference from UAC.

Watching out for automatic privilege elevation

This feature allows administrative accounts to go about their business without receiving any of the UAC prompts. Sounds great, right? The downside is that this is very similar to what you would get if you turned off UAC completely.

Maybe you understand the risks inherent with doing this, and you still want to have automatic privilege elevation. If so, see the "Using the Local Security Policy to Control User Account Control," later in this chapter. There, I go over the necessary steps to allow automatic privilege elevation.

Overriding User Account Control settings

There are several ways to override or permanently disable UAC. Some methods provide more granularity than others. I never recommend permanently disabling UAC. I do, however, recommend tuning it so that it works more seamlessly with your users and administrators. UAC is protecting your system, after all, so it's best to allow it to do exactly that.

Working with User Accounts in Control Panel

One of the common methods for dealing with UAC is to just turn it off when it annoys you. I don't recommend this course of action, but it can be accomplished quite easily with the User Accounts applet:

1. **Open the Control Panel, and click User Accounts.**
2. **Click User Accounts again.**
3. **Click Change User Account Control settings, as shown in Figure 3-2.**

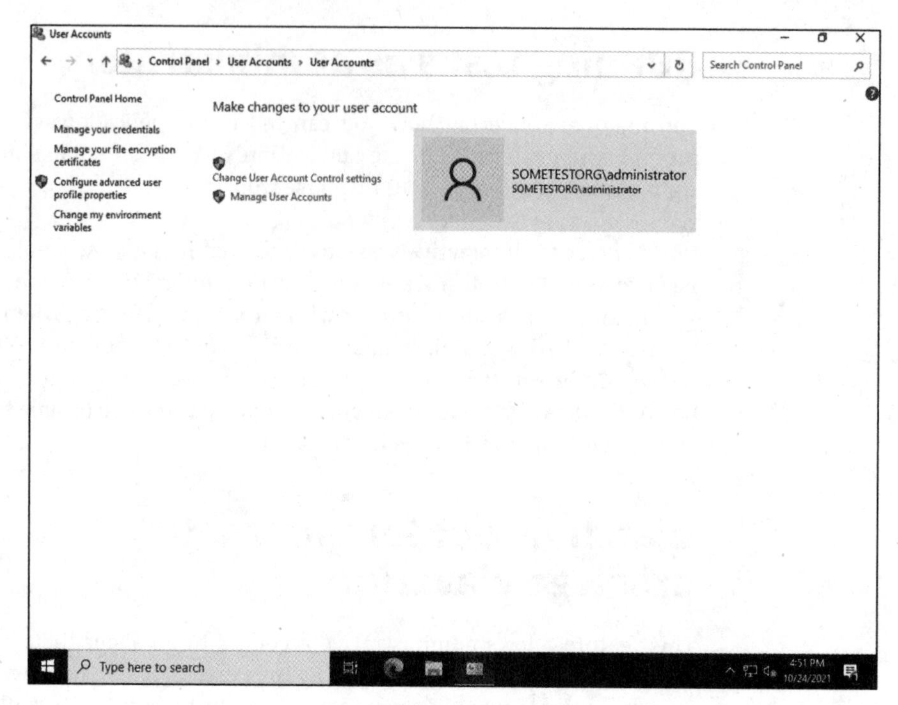

FIGURE 3-2: Turning off User Account Control in User Accounts is simple, but it should be done with caution.

4. **In the User Account Control Settings box, choose Never Notify (see Figure 3-3).**

   Your four options are as follows:

   - Never Notify
   - Notify When Apps Try to Make Changes (Don't Dim the Desktop)
   - Notify When Apps Try to Make Changes (Default)
   - Always Notify When Apps Try to Make Changes or You Try to Make Changes

Using the Local Security Policy to control User Account Control

In most cases, you're going to want more granularity than you have through User Accounts in the Control Panel. This is where the Local Security Policy is really handy.

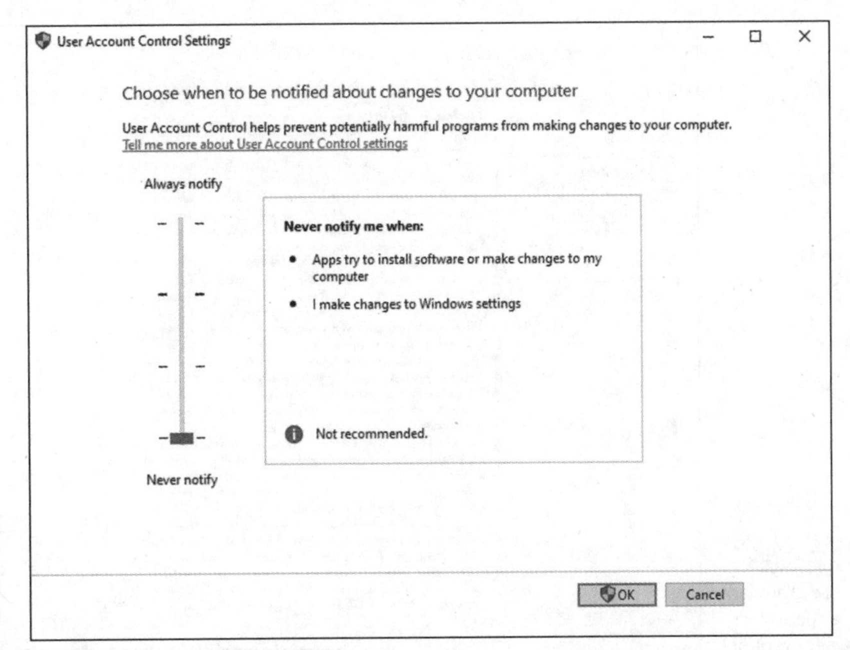
User Account Control Settings
Choose when to be notified about changes to your computer
User Account Control helps prevent potentially harmful programs from making changes to your computer.
Tell me more about User Account Control settings
Always notify
Never notify me when:
Apps try to install software or make changes to my computer
I make changes to Windows settings
Not recommended.
Never notify
OK
Cancel

FIGURE 3-3: Choosing Never Notify disables UAC for the logged-in user.

To bring up the Local Security Policy box, click the Start Menu and then type **secpol.msc** and press Enter. Select Local Policies and then Security Options and scroll all the way down. You see multiple options for UAC configuration toward the bottom, all sorted by User Account Control, as shown in Figure 3-4. If you want to change these settings across your organization, make the changes discussed here in Group Policy instead of the Local Security Policy.

I promised you earlier in this chapter that I would show you how to automatically elevate administrative permissions without the UAC prompts. Here's how:

1. **Double-click User Account Control: Behavior of the Elevation Prompt for Administrators in Admin Approval Mode.**
2. **Select Elevate without Prompting and click OK.**

This automatically elevates your permissions as needed.

Several other settings are also behaviors you can choose from for Admin Approval Mode:

- **Elevate without prompting:** Allows a privileged account to run executables and make changes to the system with no prompts.

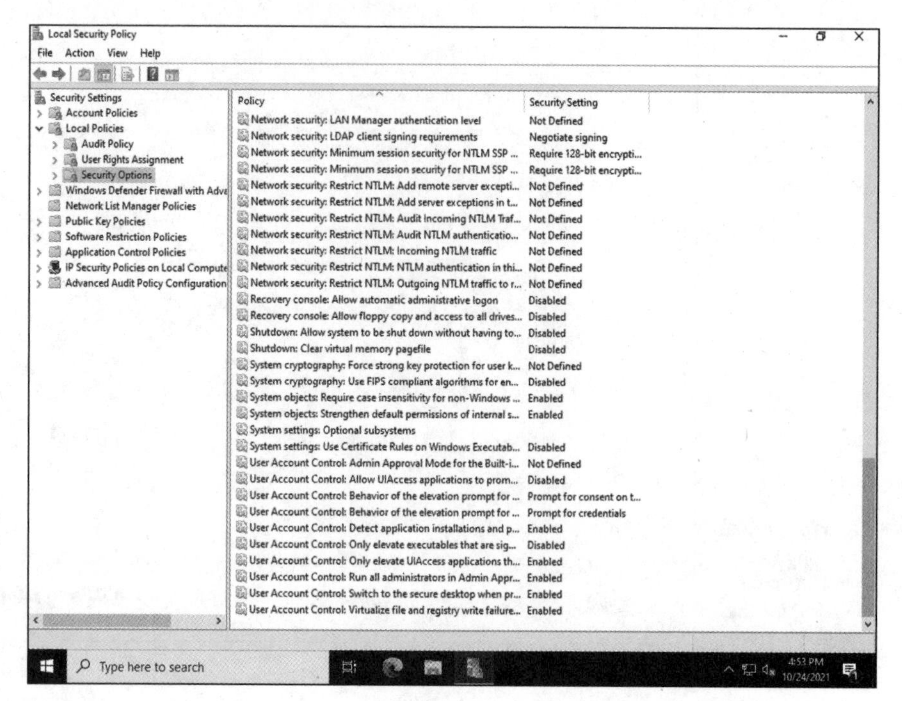

FIGURE 3-4: The Local Security Policy allows for more granular configuration of User Account Control.

- **Prompt for credentials on the secure desktop:** The user's screen will be dimmed, and the user will be asked for credentials. The user won't be able to interact with the desktop until they either enter credentials or cancel the operation they were attempting.
- **Prompt for consent on the secure desktop:** The user's screen will be dimmed, and the user will be asked to permit or deny the action. The user won't be able to interact with the desktop until they make a choice.
- **Prompt for credentials:** When something requires administrative privileges, the user is prompted for administrative credentials.
- **Prompt for consent:** When something requires administrative privileges, the user is asked to permit or deny the action. If the user permits it, it will run with the user's highest level of permissions. If the user doesn't have high enough permissions, then the action will not occur.
- **Prompt for consent for non-Windows binaries (default):** When something that is not a Windows program requires administrative privileges, the user is asked to permit or deny the action. If the user permits it, it will run with their highest level of permissions. If the user doesn't have high enough permissions, the action won't occur.

Managing User Passwords

With most systems today, passwords are managed by Active Directory — the exception to that being in a workgroup environment, where password management is done locally on each system. Understandably a lot of people don't have the best understanding of how to manage local passwords. Luckily, Windows Server 2022 happens to have a tool that simplifies working with passwords locally and on the network.

Regardless of whether you're using a local sign-on or a domain account to sign on, Windows Server 2022 (and Windows 10!) uses a product called Credential Manager to save your web and Windows credentials. You can access Credential Manager in the Control Panel by clicking User Accounts and then clicking Credential Manager.

The great thing about Credential Manager is that it gives you the ability to save credentials so that you can have an almost single sign-on type of experience. You can back up and restore credentials, as well as add Windows credentials, certificate-based credentials, and generic credentials all from the same pane of glass, shown in Figure 3-5.

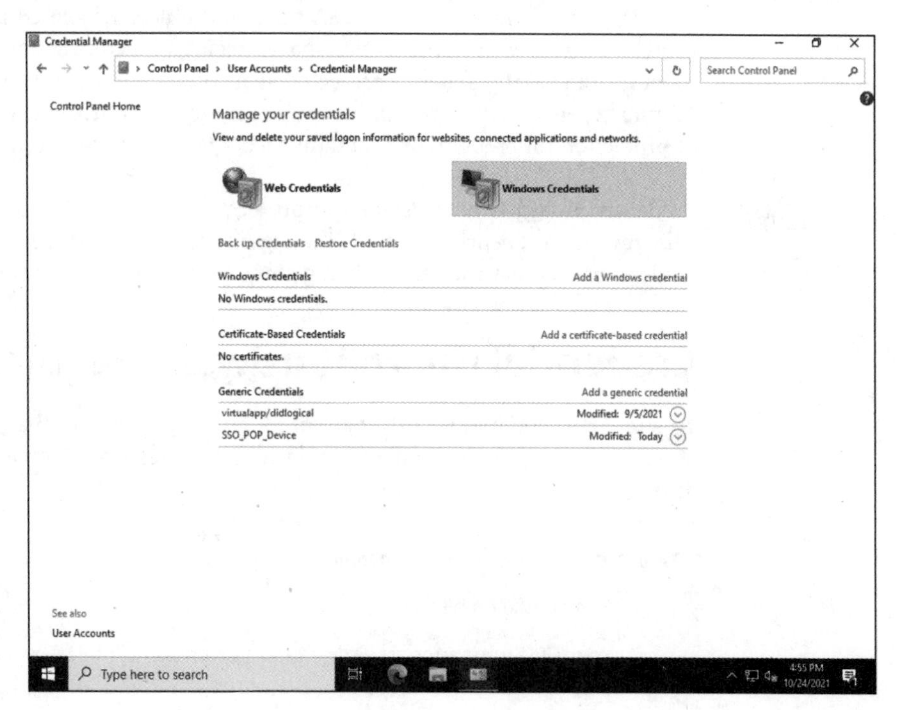

FIGURE 3-5: Credential Manager allows you to manage your network passwords from one single location.

Configuring Operating System Security

Understanding Credential Guard

Credential Guard was introduced in Windows Server 2016 as a way to mitigate some of the password attacks that were becoming more common. It protects just about anything that could be considered a credential on a Windows server including NT LAN Manager (NTLM) password hashes, Kerberos Ticket Granting Tickets (TGT), and credentials that are stored by applications as domain credentials. NTLM was a protocol that was developed by Microsoft to facilitate authentication. It stores hashed password values on the server or on the domain controller. Kerberos provides stronger encryption capabilities and uses ticket-based authentication. The ticket granting ticket (TGT) is used to request access to resources. Credential Guard is available on Windows Server 2016/Windows 10 and newer.

WARNING

Do not enable Credential Guard on Domain Controllers. It can cause crashes.

How Credential Guard works

Traditionally, Windows stored secrets in the Local Security Authority (LSA). These secrets were stored in memory, which made them vulnerable to credential theft. With Credential Guard enabled, LSA talks to the newer isolated LSA process that protects secrets with virtualization-based security. Secrets are no longer stored in memory and are much better protected against credential theft. Hacking tools like Mimikatz can't extract credentials from memory anymore, which provides a layer of protection for credentials that didn't exist before Credential Guard.

REMEMBER

Credential Guard is not designed to protect credentials that are stored in Active Directory or the Security Accounts Manager (SAM). It's designed to protect secrets while in use so that they aren't being stored in memory where they can be stolen.

Credential Guard Hardware Requirements

To support Credential Guard, there are a few requirements that need to be met. Most modern systems should be able to meet these requirements without any difficulty:

- Hardware must support virtualization-based security.
  - You need a 64-bit CPU.
  - Your CPU must support virtualization extensions (Intel VT-x or AMD-V) and extended page tables (SLAT).
  - Your system needs to be able to run the Windows hypervisor, Hyper-V.

- Hardware must support secure boot.
- You must have TPM version 1.2 or 2.0.
- UEFI lock is preferred, though not a requirement. It prevents an attacker from disabling Credential Guard with a registry change.
- The UEFI firmware on your server must support secure boot and be on firmware version 2.3.1.c or higher.

How to enable Credential Guard

You have a few options for enabling Credential Guard. In this section, I cover the most common methods you would use in an Enterprise setting. Group Policy is one of the more common methods because it will apply to any domain-joined system. The registry can also be a very useful method, especially if you have systems that aren't domain-joined.

Group Policy

Enabling Credential Guard by Group Policy is by far the simplest method because it has the least amount of steps and room for error. Follow these steps:

1. **From Server Manager, click Tools and then click Group Policy Management.**
2. **Expand Forest and Domains, and then expand the domain you want to apply Credential Guard to.**
3. **Right-click Group Policy Objects and select New.**
4. **Name your policy and click OK.**

   I named mine Credential Guard.
5. **Expand Group Policy Objects if you haven't already and click your new policy to select it.**
6. **Right-click your policy and choose Edit.**
7. **Navigate through Computer Configuration, then Policies, then Administrative Templates, then System, and then Device Guard.**

   Device Guard is used to prevent malicious code from running, while Credential Guard is focused on protecting credentials from compromise. They're different features but complement each other very well.
8. **Double-click Turn On Virtualization Based Security and select Enabled.**

9. **Under Select Platform Security Level, choose Secure Boot.**

TECHNICAL STUFF

Secure boot prevents unauthorized software and/or drivers from loading as the system starts.

10. **In the Credential Guard Configuration box, select Enabled with UEFI Lock.**

Your settings should look like Figure 3-6.

11. **If your settings match Figure 3-6, click OK.**

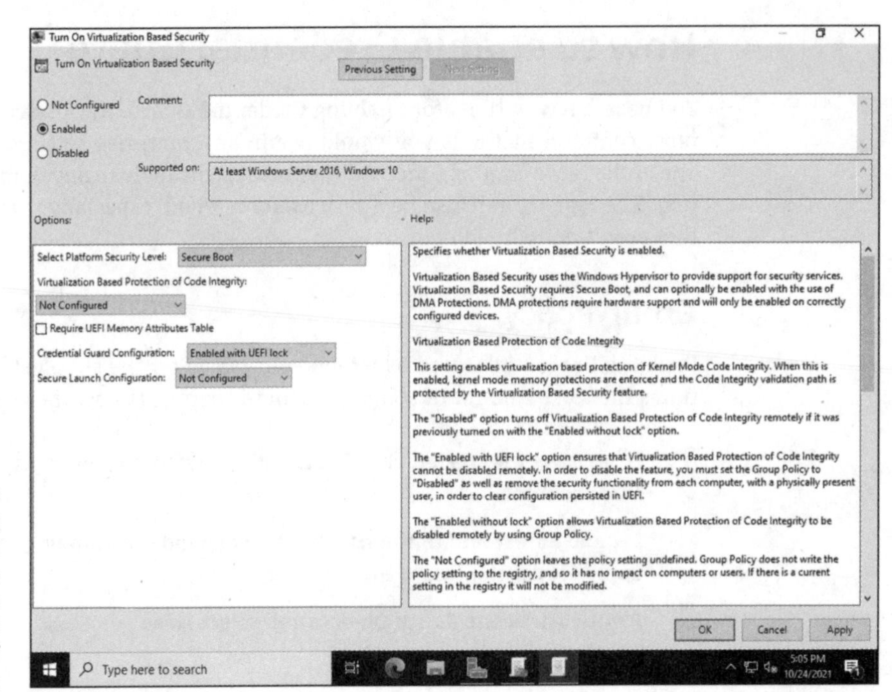

FIGURE 3-6: Enabling Credential Guard with Group Policy is the simplest and fastest method of deployment.

Registry

On some systems, you can't apply Group Policy and you need another means of enabling Credential Guard. There are a few steps that you will need to perform to enable it through the registry. This section walks you through each of these steps.

ENABLE VIRTUALIZATION-BASED SECURITY

This first step only needs to be performed if the system on which you want to turn Credential Guard on is a Windows 10 build prior to build number 1607. Windows Server 2016 and newer does not require this step.

1. **Open the Registry Editor by clicking the Start Menu and typing** regedit.exe **and then pressing Enter.**
2. **Navigate to HKEY_LOCAL_MACHINE\System\CurrentControlSet\Control\DeviceGuard.**
3. **Add the following DWORDs to the DeviceGuard key by right-clicking DeviceGuard and choosing New, and then choosing DWORD (32-bit) Value.**
   - EnableVirtualizationBasedSecurity: Set the value to 1 to enable.
   - RequirePlatformSecurityFeatures: Set the value to 1 for Secure Boot.

You can see the newly created DWORDs in Figure 3-7. When these DWORDs are created, you can turn on Credential Guard.

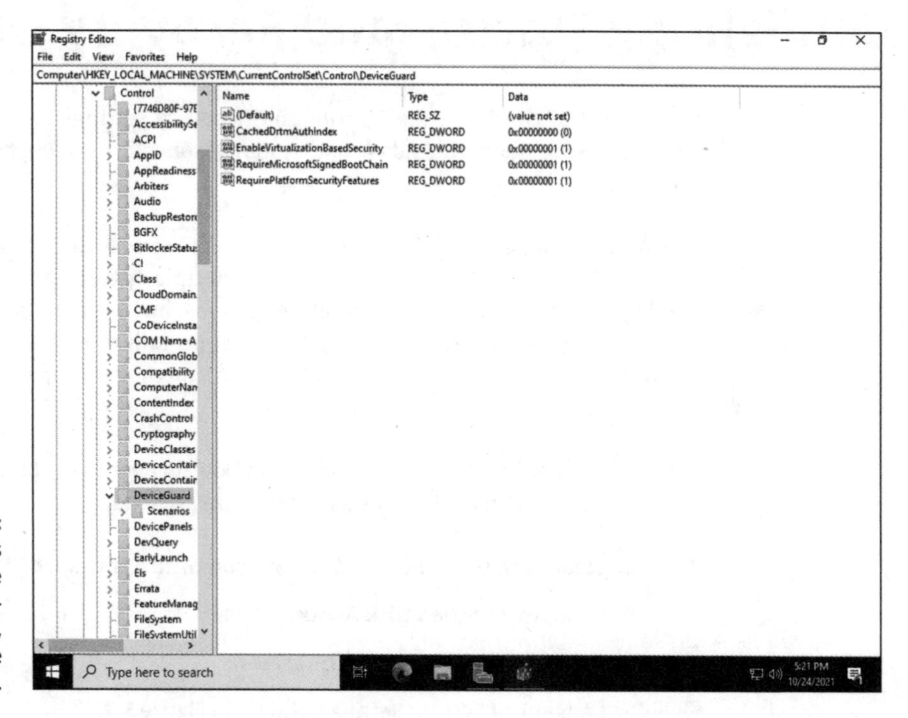

FIGURE 3-7: With the two keys set that enable virtualization-based security, you can enable Credential Guard.

ENABLE CREDENTIAL GUARD

The second step in this adventure is to actually enable Credential Guard. If you're on Windows Server 2016 or newer, or Windows 10 1607 or newer, you may have skipped right to this step. If you aren't sure how to access the Registry Editor, check Step 1 in the preceding section.

1. **With Registry Editor open, navigate to HKEY_LOCAL_MACHINE\System\CurrentControlSet\Control\Lsa.**
2. **Add a new DWORD named LsaCfgFlags and set the value to 1 to enable Credential Guard with UEFI lock.**
3. **Close the Registry Editor.**

TIP

These settings will take effect after a reboot, because they were made in the HKEY_LOCAL_MACHINE (HKLM) hive.

Configuring Startup and Recovery Options

If you need to configure the Startup and Recovery settings for your server, you can do this through the Advanced System Properties dialog box in Windows Server 2022.

The Startup and Recovery dialog box allows you to choose things like which OS you want the system to start with (if you have multiple operating systems installed), the time to show the list of the operating systems, and the amount of time to show recovery options. You can also configure System failure to automatically restart the systems and whether you want it to create a memory dump or not. You can see these options in Figure 3-8.

You may be thinking, "Well, that's cool, but how do I get to the Startup and Recovery dialog box?" I'm glad you asked. Follow these steps:

1. **Click Start and then click the Settings icon, which is shaped like a gear.**
2. **Click System and then click About.**
3. **Scroll down to the bottom of the screen.**
4. **Click Advanced System Settings, shown in Figure 3-9.**
5. **Click the Advanced tab, and then click the Settings button in the Startup and Recovery section.**

Startup and Recovery

System startup

Default operating system:

Windows Server

Time to display list of operating systems: 30 seconds

Time to display recovery options when needed: 30 seconds

System failure

Write an event to the system log

Automatically restart

Write debugging information

Automatic memory dump

Dump file:

%SystemRoot%\MEMORY.DMP

Overwrite any existing file

Disable automatic deletion of memory dumps when disk space is low

OK Cancel

FIGURE 3-8: The Startup and Recovery dialog box allows you to set the startup and recovery functionality that you want your server to use.

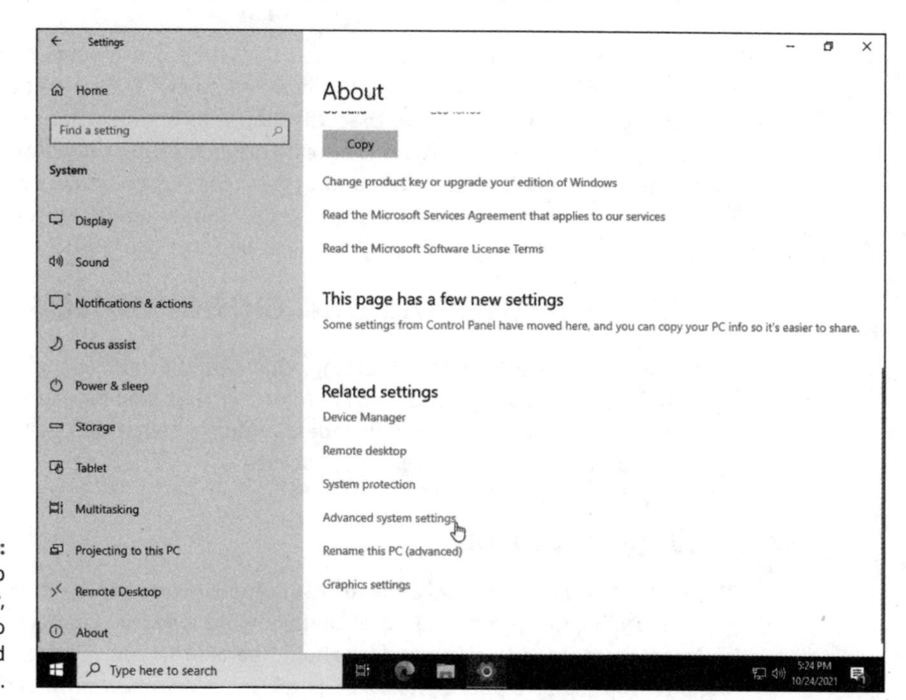

FIGURE 3-9: To access Startup and Recovery, you need to select Advanced System Settings.

In the Startup and Recovery dialog box, you can set the default operating system that you want to boot to, as well as how long you want the list of available operating systems to show up. This setting is helpful if you have multiple operating systems installed on the same server, and you want to specify which operating system should be considered the default. You can also configure what you want the system to do in the event of a failure, and whether you want a memory dump to occur, as well as where you want it stored. By default, if it encounters a failure, Windows Server 2022 automatically restarts and creates a memory dump. The memory dump is helpful later in diagnosing the issue that caused the system to fail.

And there you have it. That's how you change the startup and recovery options in Windows Server 2022.

Hardening Your Server

Windows Server 2022 made some improvements with the default strength of the cryptographic protocols and cipher suites available. The newer Transport Layer Security (TLS) protocol, TLS 1.3 is available and enabled by default.

When a system requests a secure connection, the highest supported TLS version is the one that will be chosen. Historically that has been TLS 1.2, but with the addition of TLS 1.3 in the mix, I believe that will change as organizations choose to support TLS 1.3. You can view the default priority order for the protocols and cipher suites on the Microsoft Documentation site at: `https://docs.microsoft.com/en-us/windows/win32/secauthn/tls-cipher-suites-in-windows-server-2022`.

Cipher protocols and cipher suites

At this point you might be wondering what cipher protocols and cipher suites are. So let's take a quick dive into these topics so that you can come away from this section of the book with a better understanding of what they are and how you can use them to make your system more secure.

Cipher protocols

A cipher protocol is what we use to ensure that we have a secure connection for our data. The cipher protocol is what defines what a secure connection should have. As an example, different vendors should be able to communicate securely using non-proprietary cipher suites. TLS is the most common nowadays; the acronym stands for *Transport Layer Security.*

Cipher suites

A cipher suite is a set of instructions that define how data being transmitted should be encrypted. Modern cipher suites utilize TLS, while older cipher suites used Secure Socket Layer (SSL).

Cipher suites appear like this:

```
TLS_ECDHE_ECDSA_WITH_AES_128_GCM_SHA256
```

Let's examine some of the part of the cipher suite string.

TECHNICAL STUFF

» **ECDHE:** This is the key exchange that's used to protect the information needed to create shared keys.

Shared keys are used in symmetric encryption. Symmetric encryption is very fast, but because the keys are the same for both sender and receiver, they need to be protected. Most often, they're protected with asymmetric encryption, where the sender uses a private key to encrypt the shared key, and the receiver uses a public key to decrypt the shared key.

» **ECDSA:** This is the signature, and it's used to digitally sign things like email messages.

» **AES_128_GCM:** This is the bulk encryption algorithm that's used to encrypt/decrypt messages between servers and clients.

» **SHA256:** This is the hashing algorithm used to validate the integrity of a file or message.

Changing the default priority order of cipher suites

There may come a time when you need to change the default priority of the cipher suites on your system. I have had to do this in the past to ensure that older cipher suites weren't listed, or in some cases to move a less favorable cipher suite to first in line because an application required it. Changing the default order is relatively simple:

1. **Open the Local Group Policy editor by clicking Start, typing** gpedit.msc, **and pressing Enter.**
2. **Navigate through Computer Configuration, Administrative Templates, and Network, and then double-click SSL Configuration Settings.**

   You should see a screen similar to Figure 3-10.

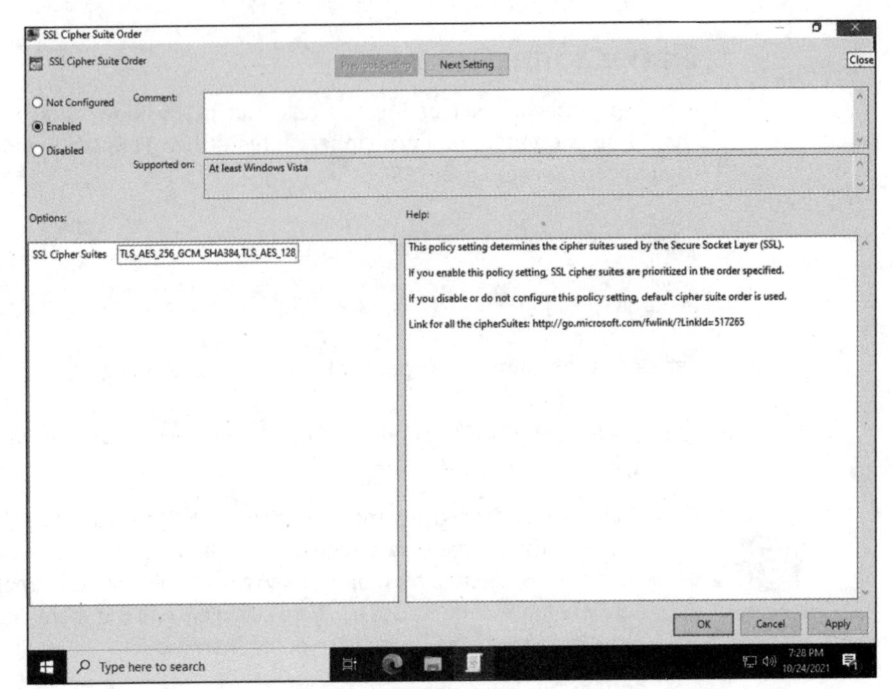

FIGURE 3-10: Modifying cipher suites is relatively simple with the Local Group Policy editor.

3. **Right-click the Cipher Suites box and choose Select All.**
4. **Choose Copy.**
5. **Paste the text into Notepad and make your changes.**

   I recommend keeping a copy of the original to refer back to.

6. **Copy the updated cipher suite list and paste it into the SSL Cipher Suites window in the Local Group Policy editor window.**

Disabling older protocols

Last but not least on the system hardening agenda is to disable older protocols like TLS 1.0 and TLS 1.1. By disabling these protocols, you also disable some of the older cipher suites that go along with them. The security protocols are located in the Registry:

```
HKEY_LOCAL_MACHINE\SYSTEM\CurrentControlSet\Control\
  SecurityProviders\SCHANNEL
```

Navigate to the `Protocols` folder, and you'll see the protocols listed there if a change has been made to them before. There is a folder for the Server component and a folder for the Client component of each. If this is a fresh system, no folders will appear under Protocols and you'll need to create them.

I use TLS 1.1 in my example (see Figure 3-11). I've disabled the TLS 1.1 server component by adding the DWORD *Enabled* and setting it to 0.

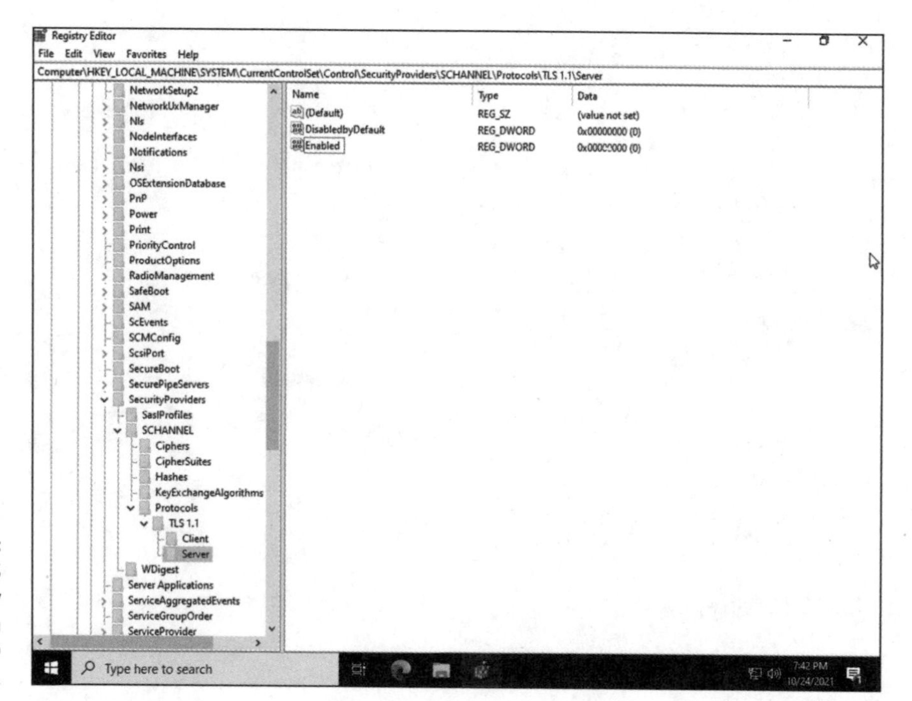

FIGURE 3-11: Disabling TLS 1.1 in the Registry can be done with a new DWORD value.

Supported protocols are either Enabled, Disabled by Default, or Disabled. Each of these states has a different meaning:

- **Enabled:** This protocol can be used unless the other system negotiating communications has it disabled.
- **Disabled by Default:** This protocol won't be used unless it's specifically requested.
- **Disabled:** This protocol will not be used.

If a protocol is enabled, the Enabled DWORD will have a value of 1 and DisabledByDefault will be set to 0. If the protocol is disabled, the Enabled DWORD will be set to a 0 and the DisabledbyDefault can be either a 1 or a 0.

» Windows Defender Firewall

» Configuring and Managing the Windows Defender Firewall in Windows Server 2022

Chapter 4

Working with the Internet

There are few things more basic to security than the principle of keeping bad things out and letting good things in. You may choose to block or restrict inbound traffic from risky protocols or port numbers. You might even lock down outbound traffic to only approved protocols and port numbers. Defining the acceptable forms of inbound and outbound traffic is typically accomplished with the Windows Defender Firewall.

In this chapter, I introduce you to the Windows Defender Firewall, including the configuration and usage tasks that every System Administrator should know.

Firewall Basics

The Windows Defender Firewall is a *stateful firewall.* This means that you can create a rule to allow inbound traffic, and established traffic will automatically be let back out. If you create an outbound rule, traffic going out will automatically be allowed back in. It can inspect all traffic passing through it and track the state of the connection. This is a great improvement over the older, traditional firewalls, referred to as *stateless firewalls,* for which you had to create a rule to allow traffic in both directions in an access control list. Stateless firewalls do not inspect traffic; they only allow or block based on source and destination IP addresses or ports.

The Windows Defender Firewall, like most of the firewalls out there, operates on a default deny for inbound connections. Essentially, if there is not a rule allowing traffic in, then it will be blocked. Outbound connections are typically allowed by default.

TIP

You may see the firewall referred to as *Microsoft Defender Firewall* in some Microsoft documentation and *Windows Defender Firewall* in other documentation. The product in the operating system is still called Windows Defender Firewall, so this is the terminology I'm sticking with. Just remember that these two terms apply to the same product to avoid confusion.

Getting acquainted with the Windows Defender Firewall profiles

The Windows Defender Firewall uses profiles to define trust levels of network traffic. The profiles can be assigned to specific network adapters, though by default all the profiles are enabled for each network adapter. As an example, the domain profile is used when a system is connected to a domain and will typically be more permissive than the public profile, which is designed to be used when a network adapter is connected to an untrusted network like the Internet.

The Windows Defender Firewall has three profiles:

- **Domain:** This profile is available only if the system is joined to a domain. It's the least restrictive of the profiles.
- **Private:** This profile is used if the system is sitting on a network that it has no association with; for instance, a system on a network that doesn't have a domain would use the Private profile. The Private profile should be more restrictive than the Domain profile because there is less trust on a Private network than on a Domain network.
- **Public:** This profile is used when a connection is made to a public network like a hotel, restaurant, or coffee shop. It should be the most restrictive profile because it's connecting with the least amount of trust to the network.

Enabling and disabling the Windows Defender Firewall

When you're troubleshooting connectivity issues, the typical request is, "Can we disable the firewall?" I can't tell you how often I was asked to do this by a vendor when troubleshooting connectivity issues with their applications. Technically, it is possible to disable the Windows Defender Firewall, but you should check your organization's policy as to whether you're allowed to do so.

WARNING

Disabling the Windows Firewall isn't a good idea. It's a layer of protection for your server.

With that disclaimer out of the way, let's look at how to disable and enable the Windows Defender Firewall. There are three ways to enable and disable the firewall. You can do it through the graphical user interface (GUI), through PowerShell, or through the command line.

Disabling/enabling through the graphical user interface

Disabling the firewall through the GUI is definitely the longest process of the three. You can't simply disable all at once; instead, you have to turn it off for each individual profile.

Here are the steps involved with turning off the firewall for the Domain profile (the steps are the same for the Public and Private profiles, just substitute the desired profile in Step 4):

» **Right-click the Start menu and click the gear icon to go into the Settings menu.**

» **Click Update & Security and select Windows Security.**

» **Under Protection areas, click Firewall & Network Protection (shown in Figure 4-1).**

» **Click Domain Network (Active).**

TIP

If you don't see a domain profile, it's because the system is not connected to a domain. This profile does not show up unless the system is domain joined.

» **To disable the Windows Defender Firewall, click each profile you want to disable, and then click the Windows Defender Firewall toggle switch, currently in the on position in Figure 4-2, to slide it to off.**

» **To re-enable the firewall, simply click the toggle switch again to slide it back to the on position.**

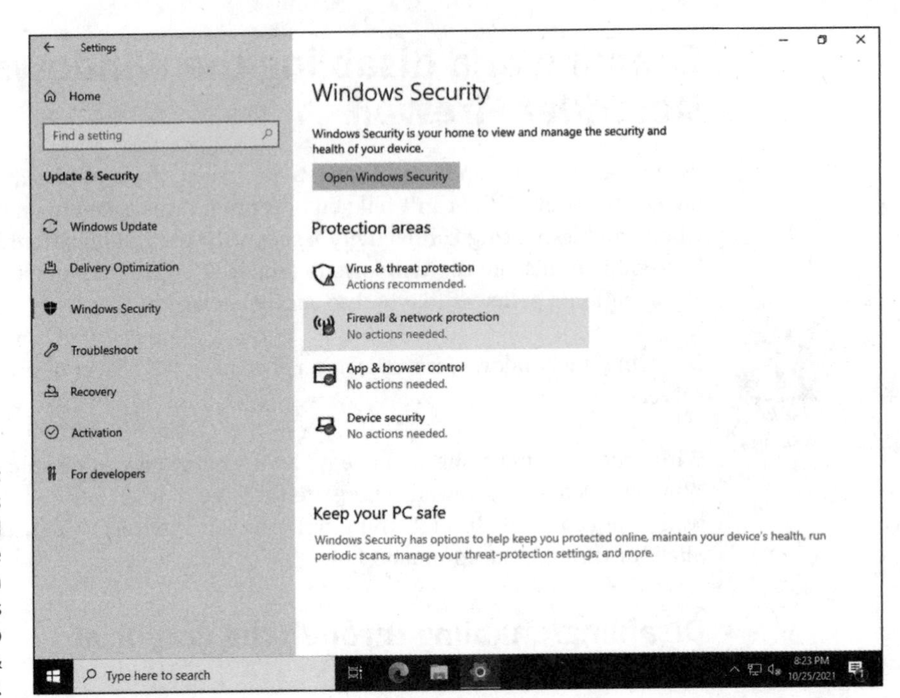

FIGURE 4-1: Windows Defender Firewall settings are available through the Windows Security app in Update & Security.

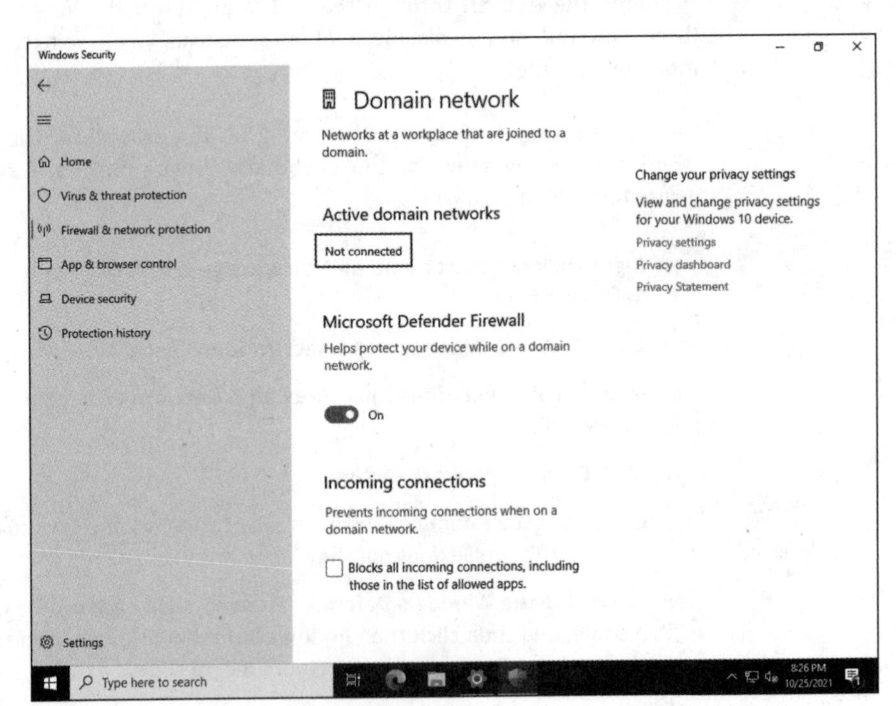

FIGURE 4-2: Each Windows Defender Firewall profile can be disabled individually with the toggle switch.

Disabling/enabling through PowerShell

One of the methods to disable or enable the firewall is with PowerShell. It's a simple one-line command to turn it off for all the Windows Defender Firewall profiles or to turn it off for a specific profile. Here's the command to disable the firewall for all profiles:

```
Set-NetFirewallProfile -Profile Domain,Public,Private -Enabled
   False
```

To re-enable the firewall for all profiles, use the following:

```
Set-NetFirewallProfile -Profile Domain,Public,Private
   -Enabled True
```

If you only want to disable/enable one of the profiles, you would use just that profile name instead of `Domain,Public,Private`.

Disabling/enabling through the Command Prompt

Disabling and enabling via the Command Prompt is a one-line command that can be used to disable or enable all or specific profiles. Here's how to disable the firewall for all profiles:

```
netsh advfirewall set allprofiles state off
```

To re-enable the firewall for all profiles, use the following:

```
netsh advfirewall set allprofiles state on
```

If you only want to disable/enable one of the profiles, you would use just that profile name (`domain`, `public`, or `private`) instead of `allprofiles`.

Configuring Windows Defender Firewall with Advanced Security

To configure Windows Defender Firewall, you need to get into the Advanced Settings. Follow these steps:

1. **Right-click the Start menu and click the gear icon to go into the Settings menu.**
2. **Click Update & Security and select Windows Security.**

3. **Under Protection Areas, click Firewall & Network Protection.**

4. **Scroll down and click Advanced Settings.**

The Advanced Settings screen looks familiar if you've worked with Windows Defender Firewall in the past. It shows you, at a glance, which profiles are enabled, as well as a basic overview of what kind of blocking state the firewall is in (see Figure 4-3).

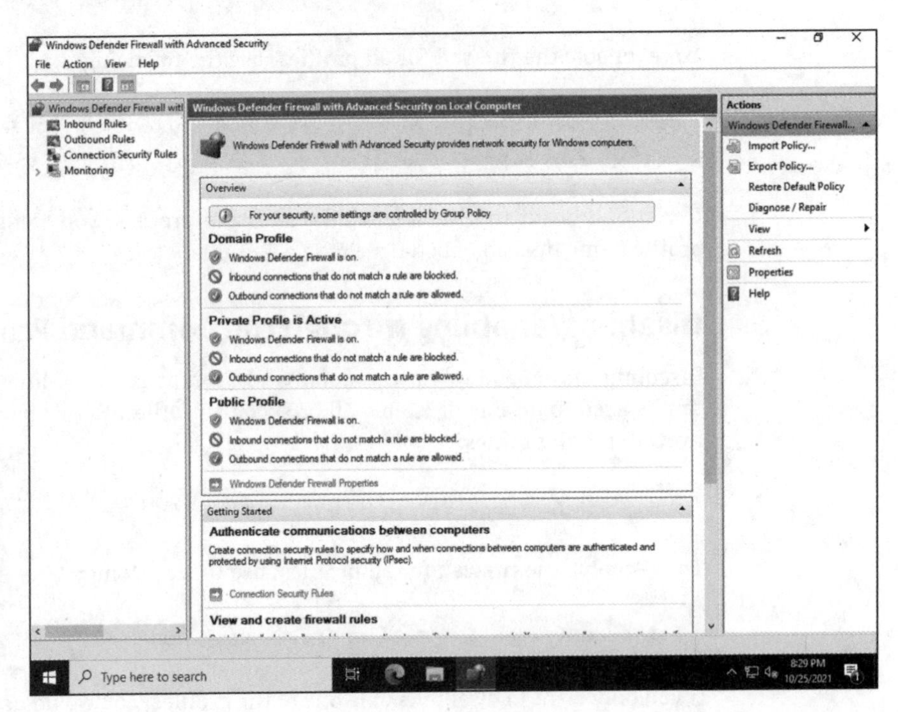

FIGURE 4-3: The Advanced Settings screen of the Windows Defender Firewall allows you to make granular changes to the way the firewall behaves.

The left side of the screen give you several options:

- **Inbound Rules:** If an inbound connection doesn't match a rule in the Inbound Rules area, then by default it will be blocked. Clicking this area gives you the ability to enable built-in rules or create custom rules.
- **Outbound Rules:** If an outbound connection doesn't match a rule in the Outbound Rules area, then by default it will be allowed. Clicking this area gives you the ability to enabled built-in rules or create custom rules.

- **Connection Security Rules:** This area allows you to configure your system to use IPSec to protect communications between endpoints.
- **Monitoring:** This area lets you see at a quick glance which profile rules are enabled and if there are any connection security rules configured. It also lets you see if there are any Security Associations set up currently. See the "Understanding IPSec" section, later in this chapter, if you aren't sure what an SA is.

Working with profile settings

For most people, the default profile settings work well. But what if you work for an employer who wants to be very strict about outbound connections not allowing traffic out unless it's explicitly allowed? You can change the behavior of the profile.

From inside the Advanced Settings screen, right-click Windows Defender Firewall with Advanced Security and select Properties. You see tabs for each of the profiles and a tab for IPSec. Start with the tab for the Public Profile, shown in Figure 4-4.

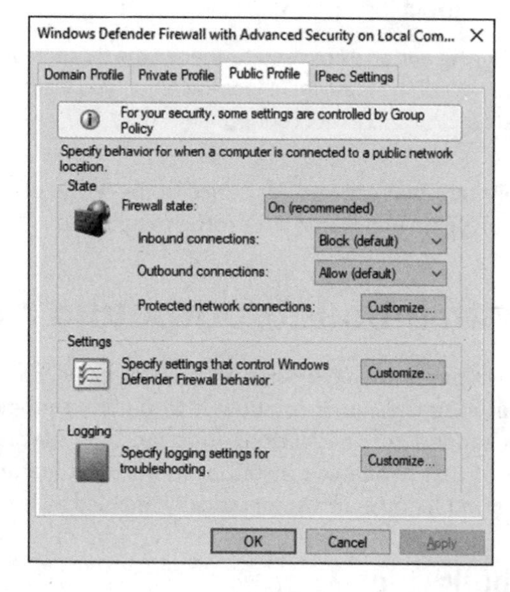

FIGURE 4-4: Profiles can be changed from the default behavior in the Properties screen for Windows Defender Firewall.

There are a few settings that I want to call your attention to:

- **Firewall State:** Here you have the choice of on or off. This can be set to disable or enable the firewall for the specific profile you're on.

- **Inbound Connections:** You have three options here:
  - **Block:** Blocks anything that is not allowed by a rule.
  - **Block All:** Blocks everything regardless of whether there is a rule allowing it or not.
  - **Allow:** Allows traffic regardless of whether the traffic is allowed by a rule.

  The default setting for this is Block.
- **Outbound Connections:** You have two options:
  - **Block:** Blocks anything that is not allowed by a rule.
  - **Allow:** Allows traffic regardless of whether the traffic is allowed by a rule.

  The default setting for this is Allow.
- **Protected Network Connections:** Clicking this button will allow you to select which network adapters you want the profile to be applied to.
- **Settings:** Clicking the Settings button lets you customize the settings of the profile that you're on. This includes the settings regarding notifications, allowing *unicast* (one-to-one transmission), and whether you want to merge rules that are pushed out through Group Policy.
- **Logging:** The Logging button allows you to change the location of the firewall logs, set the size limit of the logs, and choose whether you want to log dropped packets and/or successful connections.

As you can probably tell, you can get very specific in terms of how you want a profile to behave and where you want that profile to apply to.

Working with inbound/outbound rules

Windows Server 2022 has quite a few firewall rules already created for you. The ones that are essential to the server to allow it to function properly are already enabled. Rules that support core networking functions and file and print sharing are great examples of that. When you install new roles and features, the firewall rules for these roles and features are automatically enabled as well.

Enabling prebuilt rules

Enabling the prebuilt rules is very simple: Simply right-click the rule that you want to enable and choose Enable Rule, as shown in Figure 4-5.

To disable the rule, simply right-click it and choose Disable Rule.

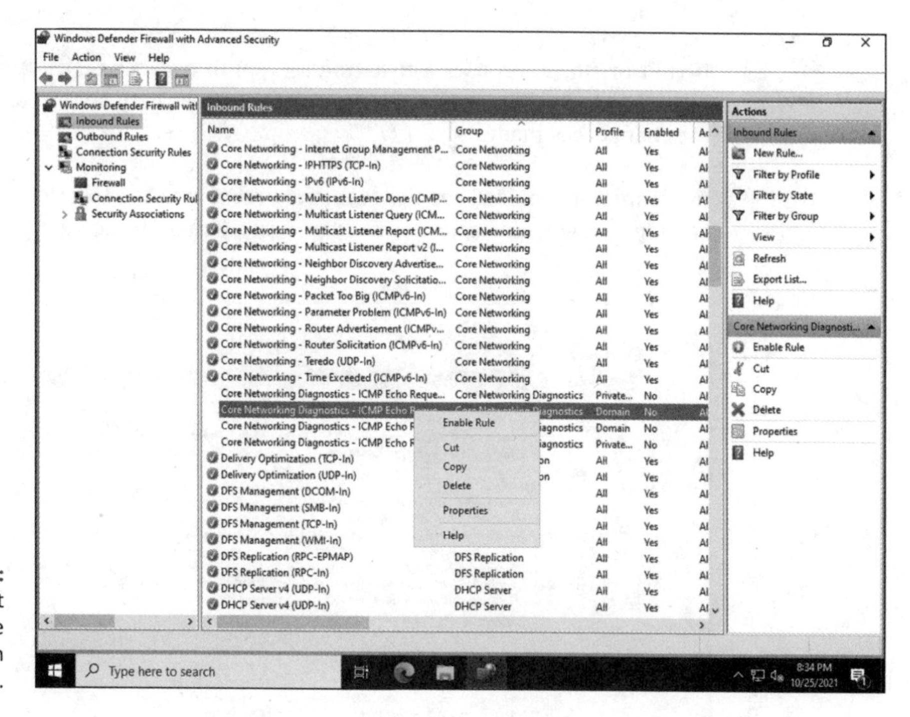

FIGURE 4-5: Enabling prebuilt rules can be done from the main screen.

Creating a custom rule

The prebuilt rules are convenient, but they tend to be very specific to Microsoft services. What if you need to install a vendor product with specific port needs?

Let's create a rule from a use case: You're a system administrator, and you've been asked to allow inbound connections to a domain-connected system hosting a MySQL database. You know that MySQL needs TCP port 3306, so let's create a rule to allow this traffic:

1. **Select Inbound Rules and then click New Rule (located in the menu on the right side of the screen).**
2. **On the Rule Type screen, select Port and click Next.**
3. **On the Protocols and Ports screen, leave the TCP selected and with Specific Local Ports selected, type** 3306 **into the text box, and then click Next.**
4. **On the Action screen, select Allow the Connection, and then click Next.**
5. **On the Profile screen, leave Domain checked, but uncheck Private and Public, and then click Next.**

   By doing this, you're only allowing domain traffic to reach the MySQL database.
6. **On the Name screen, give it a meaningful name, and then click Finish.**

TIP

If your organization uses a ticketing system to track changes, putting the ticket number into the Description field of the rule can be very helpful to document why the change was made.

In Figure 4-6, you can see the rule that I created at the top. You can tell that it's already enabled because it has the green check mark beside it.

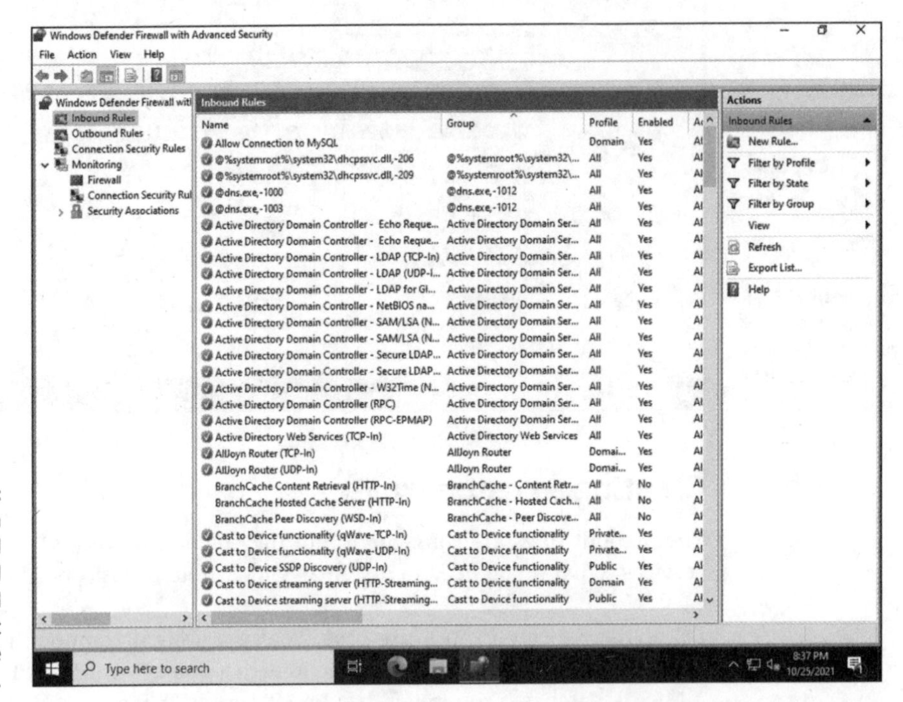

FIGURE 4-6: The custom rule created for MySQL will allow inbound TCP/3306 traffic coming from the domain.

I can hear the voice inside your head saying, "That was easy, but where is the granularity?" You can get exceptionally granular on the rule. Let's take a peek at the settings. Double-click a rule. I'll use the MySQL rule that I just created.

There are eight tabs in all. Each tab allows for you to make a change to the rule. Some of the settings are the basic stuff that you set when you initially created the rule; others allow for more granularity than what was inside the rule wizard:

- **General:** This tab allows you to adjust the name of the rule, whether it's enabled, and whether you want to allow or block the traffic.
- **Programs and Services:** This tab allows you to specify an application or service that should be allowed through the firewall.

- **Remote Computers:** On the Remote Computers tab, you can choose to allow connections from only specific computers or to skip the rule for specific computers.
- **Protocols and Ports:** The Protocols and Ports tab allows more depth than the wizard does when Port is selected. When I created the rule, I had the choice between TCP and UDP because I chose a Port rule. On this screen, however, I have many more options, shown in Figure 4-7. I could get these through the wizard if I chose Custom instead of Port.

  I can change the local and remote ports on this tab as well. Currently, in my rule, I have TCP 3306 allowed.

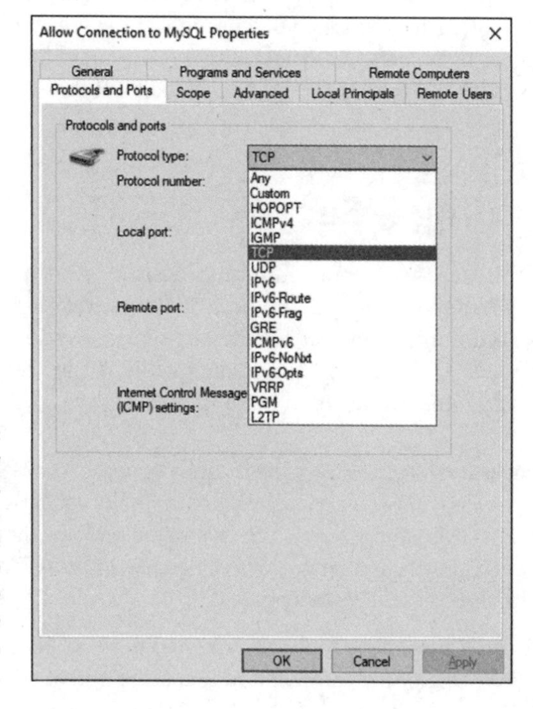

FIGURE 4-7: There are many more protocols available through this tab than there were through the wizard because I created it as a Port rule.

- **Scope:** The Scope tab allows you to determine who should be allowed to connect. Local IP, for example, may be set if your system has multiple IP addresses but you only want to allow the connection from the rule through one of the IP addresses. By using the Remote IP, you're setting which IPs the application on the server can reach out to. This can increase security posture greatly; for instance, with my MySQL server, maybe I only want to allow it to talk to the application server, but not the web server. I can set that with Scope.

- **Advanced:** Advanced allows me to set which profiles to apply this rule to, which network adapter to apply it to, and whether you want to allow edge traversal. Normally, you want this to be set to Block Edge Traversal, which is the default setting.
- **Local Principals:** The Local Principals tab allows you to set specific local users who are allowed to connect through the rule, or specify which users should be able to skip the rule.
- **Remote Users:** The Remote Users tab allows you to set specific remote users who are allowed to connect through the rule, or which users should be able to skip the rule.

You could have made all these settings through the wizard if you had selected Custom instead of Port. I like to take you through the exercise of looking at the individual tabs, though, because I've met quite a few system administrators who don't understand what the tabs mean or how to edit the rule properly after it's created.

Understanding IPSec

IP Security (IPSec) is used to secure communications over an IP-based network. It's typically set up to support and secure network-to-network, host-to-host, or host-to-network communication. For businesses that deal with sensitive information, IPSec provides a method to encrypt data while it's in transit. There are a few terms you should understand when talking about IPSec:

- **Security Association (SA):** The SA is the most basic part of the IPSec connection. It's an agreement between endpoints on how they'll establish secure communication — everything from which cryptographical algorithm to use, which key to encrypt with, and other relevant network information. The key exchange used for IPSec follows two phases:
  - **Main Mode or Aggressive Mode:** Also referred to as IKE Phase 1. This phase securely creates the communication channel and handles the key exchange. This negotiation sets up the SA.
  - **Quick Mode:** Also referred to as IKE Phase 2. All subsequent key exchanges are done through Quick Mode because it's less resource intensive on the system.
- **Internet Key Exchange (IKE):** IKE is used to handle negotiations and authentication for the creation of IPSec SAs.

Configuring the IPSec settings

The IPSec tunnel can be configured through the Connection Security Rules section in Windows Defender Firewall. For this section, I'm going to create a server-to-server tunnel.

1. **With the Advanced Settings screen open, right-click Connection Security Rules and select New Rule.**
2. **On the Rule Type screen, select Server-to-Server and click Next.**
3. **On the Endpoints screen, define which endpoints will meet the criteria for the tunnel, and click Next.**

   I'm going to leave both of these on Any IP Address. In a production environment, you would want this to be more specific.
4. **On the Requirements screen, you can select whether you want to request or require authentication, and click Next.**

   I'll leave it on Request Authentication for Inbound and Outbound Connections.
5. **On the Authentication Method screen, you can select Computer Certificate if you have an internal public key infrastructure (PKI) that can support this use.**

   I'll click Advanced, and then click Customize.
6. **For this demo, I've elected to use NTLMv2 as my primary authentication method, with Kerberosv5 as a secondary authentication option, shown in Figure 4-8.**
7. **Click OK.**
8. **Click Next.**
9. **On the Profile tab, keep Domain selected but uncheck Private and Public, and then click Next.**
10. **On the Name page, give your IPSec tunnel a meaningful name, and then click Finish.**

Now you have the connection security rule finished. This needs to be set up on any system on which you want to use IPSec. As an example, I set it up on one of my other systems and the SA came up for it right away, as shown in Figure 4-9.

TECHNICAL STUFF

I set up the connections using NTLM and Kerberos mainly because it was the simplest way to do it for the demonstration in the book. Although this does work, certificate authentication is preferred because it proves the identity of the sender given that the sending system/user should be the only entity with the private key. With the Customize button, you have the ability to add a pre-shared key if you would like, but this is not recommended because it's stored in plaintext.

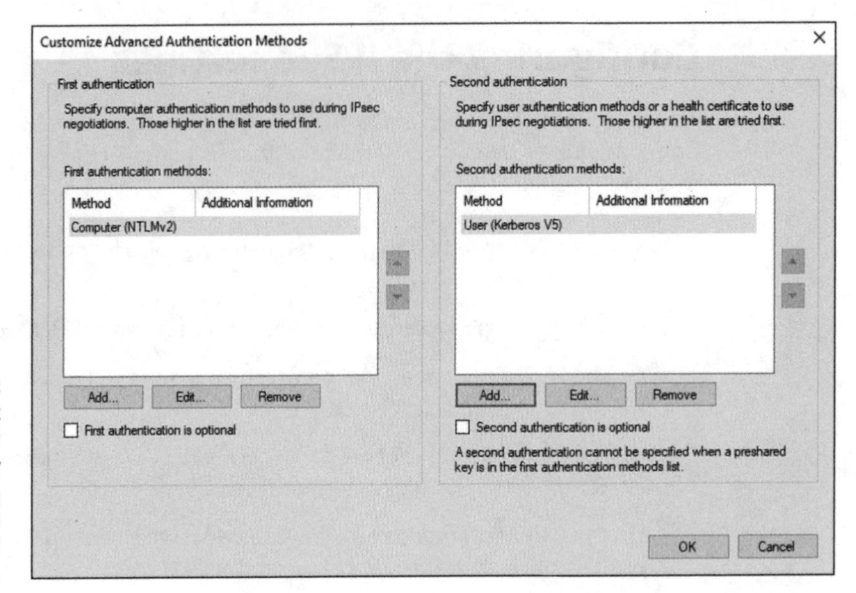

FIGURE 4-8: You can set primary and secondary authentication options for both computers and users.

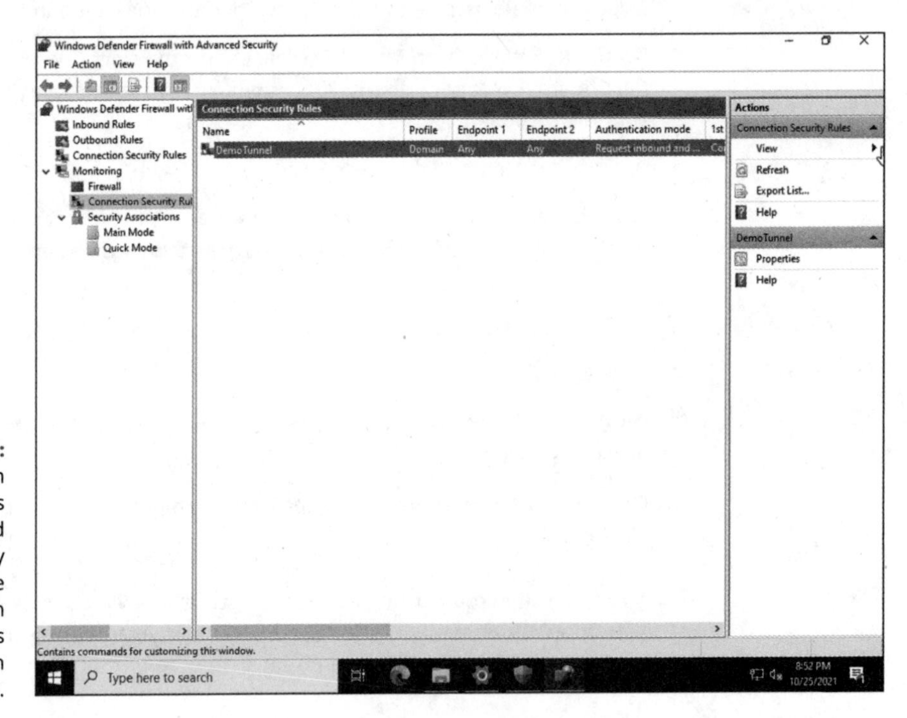

FIGURE 4-9: The SA between my two systems connected automatically after the connection security rules were set up on both systems.

» Examining certificate types in Active Directory Certificate Services

» Selecting the appropriate user certificate

» Choosing the right computer certificate

Chapter 5
Understanding Digital Certificates

Everyone knows that certificates are a good thing. Far fewer people understand what a certificate is and how it works.

When you go to a website, you might check that the site is using HTTPS. But what does the HTTPS actually mean? Its short for HyperText Transfer Protocol over SSL (which stands for Secure Sockets Layer). A certificate is used to secure the communication channel.

In this chapter, I explain certificates in general and tell you what type of certificates can be issued with Active Directory Certificate Services (AD CS). The certificates I discuss in this chapter are definitely not an exhaustive list, but they are some of the more commonly used certificates.

Certificates in Windows Server 2022

Windows Server 2022 provides AD CS, which is the focus of the next chapter in this minibook. AD CS allows you to stand up your own public key infrastructure (PKI), which allows you to issue certificates for users and internal systems that are trusted.

You can still install certificates from third-party certificate authorities, like GoDaddy and DigiCert; in fact, this is a must if the certificate is securing a resource that people outside your organization will access. However, if a resource will only be accessed by people within your organization, then it's a prime candidate for an internal certificate issued by your organization's PKI. This saves you the expense of the external certificate as well.

Cryptography 101

Cryptography is used to secure data in transit and at rest. Cryptography uses mathematical algorithms to generate "keys," which are used to encrypt data. A single key may be used that can encrypt and decrypt the data (see "Symmetric cryptography"), or you may have a private key and a public key, which are mathematically linked. One key encrypts and the other decrypts (see "Asymmetric cryptography").

TECHNICAL STUFF

A key is used by a cryptographic algorithm to change plain text into encrypted text, which is referred to as *cipher text.* It is also used to change the cipher text back into plain text. The key itself consists of a string of bits, the length of which is determined by what the algorithm you're using supports, and the length that you specify.

There are two types of cryptography: symmetric and asymmetric.

Symmetric cryptography

Symmetric cryptography is the ability to encrypt and decrypt with the same key. When you're using a shared secret or a password, you can think of that as symmetric cryptography. Because it only uses one key, it's faster, but it's also less secure because an attacker would only need to find out what the key is to decrypt the data. You can see a diagram of how symmetric cryptography works in Figure 5-1.

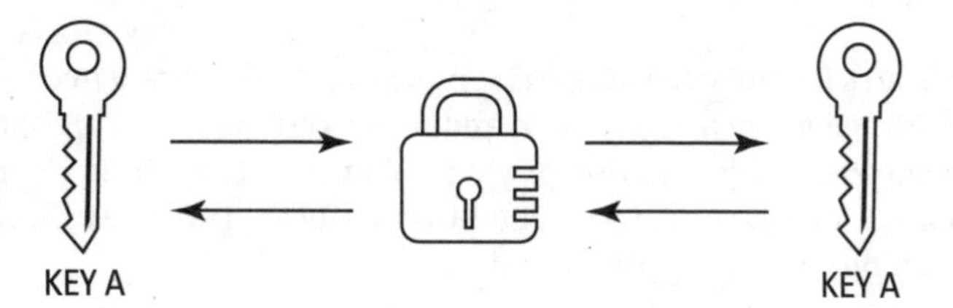

FIGURE 5-1: Symmetric cryptography uses one key to encrypt and decrypt data.

TECHNICAL STUFF

A *cipher* is the algorithm that is being used to perform the encryption and decryption.

Some common symmetric ciphers include

- 3DES
- AES
- TwoFish
- BlowFish
- IDEA

Asymmetric cryptography

Asymmetric cryptography (or *public key cryptography*) is an encryption scheme where data is encrypted by one key and decrypted by another key. The keys are mathematically linked and will only work with each other. The private key is kept safe and offers *non-repudiation* (meaning that it can prove identity) because it isn't distributed. The public key is given to others and is used to decrypt data that was encrypted by the private key. Asymmetric cryptography is what is used in public key infrastructure (PKI). A common example of asymmetric cryptography is when you use HTTPS. Say you go to your bank's website. Your browser is presented with the public certificate, which it uses to encrypt and decrypt traffic with the bank's web servers. Your bank's web servers have the private key, which proves that the bank is who it says it is, and allows the bank to decrypt your traffic. You can see a diagram of how asymmetric cryptography works at a high level in Figure 5-2.

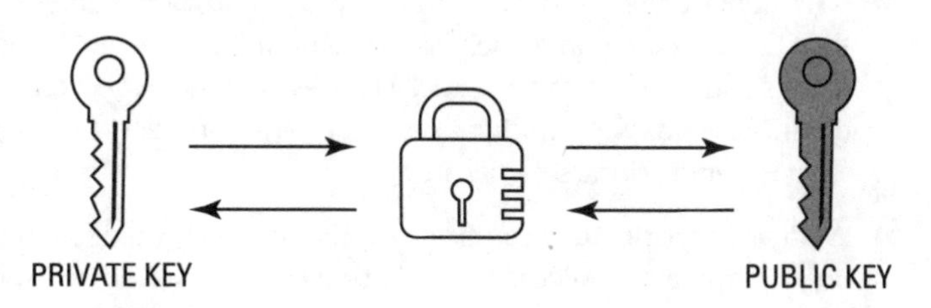

FIGURE 5-2: Asymmetric cryptography uses two mathematically linked keys to encrypt and decrypt data.

Certificates fall under asymmetric cryptography. When you need to create a certificate, you generate a private key, and then a *certificate signing request* (CSR), which is an encoded representation of your public key. The CSR is given to a certificate authority, internal or external, and the certificate is created from there. The private key never leaves your possession.

Common asymmetric ciphers include

- Diffie-Hellman
- DSS\DSA
- RSA
- ECDH
- ECDSA

Certificate-specific concepts

No introduction to certificates or cryptography would be complete without clearing up a few terms. If you understand these terms, then this chapter and the next will be a lot simpler:

- **CRL:** The Certificate Revocation List keeps track of certificates that have been revoked, which makes them no longer valid. By default, the base CRL is updated every seven days, and the delta CRL is updated once a day.
- **OCSP:** The Online Certificate Status Protocol gives near real-time revocation information on certificates. This is an improvement over using strictly CRLs because CRLs don't update as frequently.
- **FQDN:** A Fully Qualified Domain Name is the hostname and domain name. For instance, `Server2.sometestorg.com` would be an FQDN.
- **CN:** The common name on the certificate is typically going to be the same as the FQDN.
- **SAN:** Subject Alternative Names allow you to add more names to a certificate than just the common name. This is useful when you need to support short names and IP addresses, especially for development work. You can add whole other FQDNs as well, which makes these certificates very useful when you want to avoid wildcard certificates.
- **Wildcard certificate:** A certificate that is essentially valid for any host on your domain. A wildcard for `sometestorg.com` would be expressed as `*.sometestorg.com`. Wildcard certificates can be used to save money to secure multiple websites and/or servers.

Types of Certificates in Active Directory Certificate Services

There are many types of certificates in AD CS. The standard non-domain-joined server has a set of certificate templates out of the box, but a domain-joined Enterprise certificate authority (CA) has even more certificate templates to choose from.

In the following section, I discuss the more common types of user and computer certificates that you encounter, and what their uses are.

User certificates

User certificates are all about — you guessed it — the users. These certificates are typically used to establish the identity of a user. Here are some of the more common User certificate types that you may run into:

- **User:** This certificate template is used for the traditional authentication-style certificate. It's most commonly used in two-factor authentication (2FA) solutions as the second factor of authentication, after username and password. This is especially popular with virtual private network (VPN) solutions. You can always look at the template on the issuing CA, and it will tell you what purposes it's approved for. In Figure 5-3, you can see the template properties for the User template.

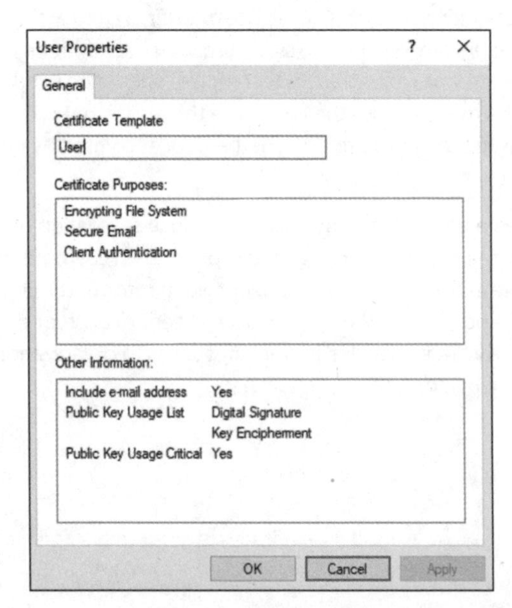

FIGURE 5-3: The User certificate template is good for EFS, secure email, and client authentication.

- **Code Signing:** When you need to run internal PowerShell scripts or executables and you want to ensure that they haven't been changed or altered in any way, you want a code signing certificate. The Code Signing certificate validates that the code has not been changed or altered in any way since it was last signed. If you want to run remote PowerShell scripts, this is a great way to ensure that only good, tested scripts are run from authorized users because you can set the permissions on who can enroll with the template, as well as the execution policy of PowerShell scripts in your environment.

Encrypting File System (EFS) can be used to encrypt files or folders and can only be decrypted by the user who encrypted them or an authorized recovery agent (EFS Recovery Agent). Please note that this is different from BitLocker, which offers full disk encryption, not file/folder-level encryption.

- **Basic EFS:** If you're using EFS, your systems will automatically request a Basic EFS certificate from one of your certificate authorities the first time one of your users tried to encrypt a file, assuming they don't already have a User certificate already. The Basic EFS certificate is used for EFS operations exclusively.
- **EFS Recovery Agent:** This certificate template is also used in conjunction with EFS, but it's used to decrypt data that has been encrypted by EFS. This may be due to someone leaving the company or having been terminated. It may even be an accidental deletion. With the EFS Recovery Agent certificate, the data can be decrypted. By default, all members of the Domain Admins and Enterprise Admins groups can enroll in this certificate.
- **Key Recovery Agent:** The Key Recovery Agent certificate is used by an authorized administrator to decrypt private keys. It can be used to recover private keys assuming that the CA has been configured to archive and allow recovery of the private key that is associated with the public key it was given when a certificate was requested. This template should be used very sparingly because it gives the user with the certificate the ability to recover private keys and, by extension, the ability to decrypt the data encrypted by the certificates to which the private key belongs.

It is considered a best practice by Microsoft to use separation of duties if you want to utilize the Key Recovery Agent template. The recommendation is to allow someone in the Certificate Manager role to retrieve the private key but not decrypt it, and to allow the Key Recovery Agent to decrypt the private key but not retrieve it. This provides better safeguards to organizational data because no one person can decrypt all data.

Computer

Computer certificates are similar to user certificates in that they're verifying identity. The main difference is that they're verifying the identity of a machine rather than the identity of a user. Here are some of the more common Computer certificate templates and their uses:

- **Computer:** The Computer template can be used for both workstations and servers. This is often used for VPNs to determine whether a system is authorized, but it can also be used for encryption. By default, the name of the system is pulled from Active Directory, though it can be made a manual process. You can always look at the template on the issuing CA and it will tell you what purposes it's approved for. In Figure 5-4, you can see the template properties for the Computer template.
- **Domain Controller:** The Domain Controller template is good for both client and server authentication, as well as the use of smart card logon support. The biggest difference between it and the Computer template is that the Domain Controller template is designed to help facilitate secure replication between domain controllers. I always recommend making sure that your domain controllers have a certificate from this template because it allows them to support Lightweight Directory Access Protocol (LDAP) over SSL (LDAPS), which encrypts authentication traffic across the wire. By default, the name of the system is pulled from Active Directory, though it can be made a manual process.
- **Web Server:** The Web Server template is used for supporting HTTPS on internal websites. This template is typically one that you would want to set up to ask for the common name and any storage area networks (SANs) that you want to use so that users don't get certificate untrusted messages. For internal certs, I often use the FQDN for the CN, and will add the short name and IP address as SANs.
- **Subordinate Certification Authority:** This is the template used by a root or issuing certificate authority to issue certificates to subordinate certificate authorities. This deals a little more with the hierarchy of certificate authorities, which is covered in the next chapter. Simply put, this template is used on any system that sits in the certificate issuing chain between the root CA and the end user or system's certificate. It's one of the few certificate templates that does not have an intended purpose added into the description of the template. In fact, if you look at its intended purpose, it simply says, "All."

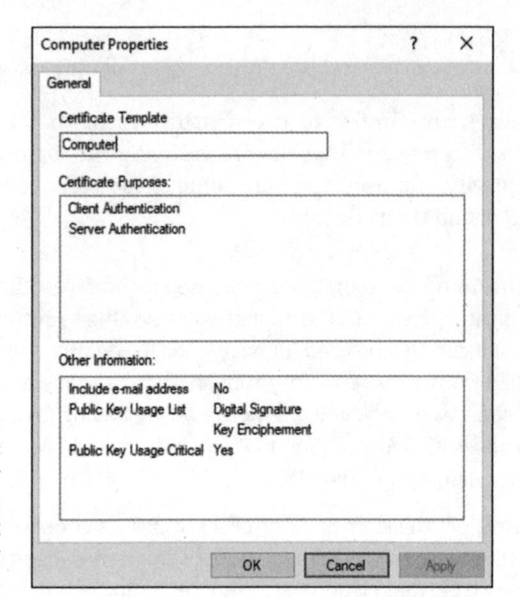

FIGURE 5-4: The Computer template provides client and server authentication.

Hierarchy

» Installing a Windows Server 2022 certificate authority

» Configuring certificate auto-enrollment with Active Directory Certificate Services

Chapter 6
Installing and Configuring AD CS

At some point in your career, you will most likely need to work with certificates. If you're managing your own Windows Public Key Infrastructure (PKI), you'll most likely be using Active Directory Certificate Services (AD CS).

If you're not sure what a certificate is or why you might want one, check out Chapter 5 of this minibook. There, I cover what certificates are and some of the more common templates that are used to issue certificates. As a brief recap, certificates are used to prove identity and/or encrypt data. Certificates are issued by servers in a public key infrastructure (PKI). Because certificates purchased from third-party certificate authorities can become cost-prohibitive for organizations, many organizations install their own internal PKIs to support internal systems or applications.

In this chapter, I fill you in on AD CS specifically. I explain certificate authority (CA) architecture and how to install and configure a Windows Server 2022 CA.

Introducing Certificate Authority Architecture

Before you build out your PKI, you need to plan out how you're going to architect it. The PKI that you architect will support the certificate needs of your organization — everything from allowing encryption of credentials to secure replication to code signing and more. There are important decisions that need to be made, such as whether your root CA will be an offline CA or an enterprise CA, and how many issuing CAs you need, as well as whether you should have a separate policy CA. Don't worry if you didn't understand that last sentence. In the following sections, I define each of these CA roles and why you might need or want them.

Root certificate authorities

The root CA is the first level of trust for all certificates. It has the highest level in the certificate trust chain when it comes to validating that the certificate is good. It's the only CA that will have a self-signed certificate. As such, it must be protected properly, or you could have an attacker issuing certificates from your certificate authority that are trusted by everything on your network.

A root CA should never be used to issue day-to-day certificates. The best practice is to have one root CA and one issuing CA at a minimum. The root CA issues a subordinate CA certificate to the issuing CA, and that should be it. Technically, you can combine the two roles and have the root CA also be an issuing CA, but that approach is not recommended.

If the root CA is offline, it can't be attacked. This is the most secure type of root CA; it's known as an offline root CA. With this architecture, the root CA is brought online to issue certificates based off of the subordinate CA template for the issuing CAs. After it has issued the certificate and the certificate revocation lists (CRLs) have been updated, it's turned off again.

A CRL is used to identify which certificates have been revoked, meaning that they're no longer valid certificates.

Offline root certificate authorities are the most secure option and should be the preferred choice. The downside to offline root CAs, however, is that you have to distribute its certificate through Group Policy and manually publish the CRLs, which can be time consuming. For a small IT department, the skill set may not be there to support this kind of maintenance. If your environment doesn't have strict regulatory requirements regarding the safeguarding of your PKI, an enterprise root CA is a good solution. The enterprise root CA is attached to Active Directory, so it can publish its own certificate and CRLs automatically. This is the simplest method of deployment, too, because there is very little manual work. If you choose to go this route, make sure that the operating system is hardened, that access is limited to the bare essential users, and that it has good anti-malware software running on it. You need to take some extra precautions with an enterprise root CA because it's online all the time.

Issuing certificate authorities

Issuing CAs are the workhorses of the PKI world. When a user or a system requests a certificate, that request is routed to your issuing CA, which will fulfill the request and return the certificate to you. You can have standalone issuing CAs, but the most common configuration is an enterprise issuing CA. This means that there is integration between the certificate authority and Active Directory. The issuing CA is able to communicate with Active Directory, including the publishing of CRLs.

Policy certificate authorities

The policy CA is a special use case and is only traditionally seen in very large and heavily secured enterprises. If a policy CA is in place, it will issue certificates to an issuing CA. Policy CAs are used to create and enforce the policies and procedures regarding the validation of identity as it pertains to certificate holders and to secure the CAs in the CA architecture.

In most common smaller organizations, the issuing CA and the policy CA are one and the same and the policy CA enforces its own policies and procedures that are set.

Installing a Certificate Authority

So, you've gotten to the point where you've figured out what your PKI architecture is going to look like and you're ready to build it out. I'm going to assume that your Windows Server 2022 servers are already built, and that they're ready to go. In the following sections, I cover the steps involved.

Creating the CAPolicy.inf file

For any certificate authority that you build, you should use a `CAPolicy.inf` file. This sets a lot of the basic parameters including the renewal date and CRL validity length, and it can specify a policy if you're using the policies in Certificate

Services. This is only used to control the CA renewal, not the renewal of other certificates.

Creating this file is a simple process. There is a lot of configuration that can be done with the CAPolicy.inf file, so I recommend doing your research to determine what you want the settings in the file to be. Pay close attention to the validity periods that you define. You don't want your certificate authority certificates to be extremely long lived, but too short of a lifespan will impact issued certificates and their validity periods. Issued certificates can't have longer validity periods than the CA they were issued on.

```
[Version]
Signature="$Windows NT$"
[Certsrv_Server]
RenewalKeyLength=2048
RenewalValidityPeriod=Years
RenewalValidityPeriodUnits=5
AlternateSignatureAlgorithm=0
```

You can type this into Notepad and save it as CAPolicy.inf into the C:\Windows directory. The name and location are important. If these are not correct, then the settings will not be applied. The [Version] section is the only required section in the file, and it always has to be at the beginning.

You may be wondering what the other settings are for. This is what each of them is doing:

- RenewalKeyLength: This sets the size of the key when a certificate is renewed. When the certificate is first created, it uses whatever the CA is set to use for certificates.
- RenewalValidityPeriod: This specifies what type of timeframe you want a renewed certificate to be good for. You can choose from hours, days, weeks, months or years.
- RenewalValidityPeriodUnits: This is where you specify the actual number you want the certificate to be good for. For instance, in the example file above, I have selected 5 which ends up being a validity period of 5 years.
- AlternateSignatureAlgorithm: When set to 0, this is enabled and will create a certificate request that includes the PKCS #1 (RSA) signature format.

TECHNICAL STUFF

If you're going to install an offline CA, it's important that you add a CRL Period to the CAPolicy.inf file. If you'll only bring the root CA up once a year, then set it to once a year, as in the following example. You need to add the following two lines into the [Certsrv_Server] section:

```
CRLPeriod=years
CRLPeriodUnits=1
```

After you've created your `CAPolicy.inf` file and placed it in `C:\Windows`, you're ready to build out your new certificate authority. Let's look at root CAs first.

Installing the root certificate authority

Installing the root CA is one of the very first steps that you must take to establish your PKI. On Windows Server, the Active Directory Certificate Services (AD CS) role will allow you to do that. The installation process between the offline root and enterprise root CAs are very similar. In the following sections, I walk you through the installation process for both of them.

Offline root certificate authority

I'm going to start your journey into certificate authorities with the installation of an offline root CA. As you might recall from earlier in this chapter, this is the most secure form of root CA, but it also requires some manual work. Follow these steps:

1. **From Server Manager, click Manage and then click Add Roles and Features.**
2. **On the Before You Begin screen, click Next.**
3. **On the Select Installation Type screen, click Next.**
4. **On the Select Destination Server screen, click Next.**
5. **On the Select Server Roles screen, select Active Directory Certificate Services, click Add Features, and then click Next.**
6. **On the Select Features screen, click Next.**
7. **On the Active Directory Certificate Services screen, click Next.**
8. **On the Select Role Services screen, select Certificate Authority, and leave everything else unchecked; then click Next.**
9. **On the Confirm Installation Selections screen, click Install.**
10. **After installation has completed, click Close.**
11. **Click the flag at the top of Server Manager and click Configure Active Directory Certificate Services on the destination service, shown in Figure 6-1.**

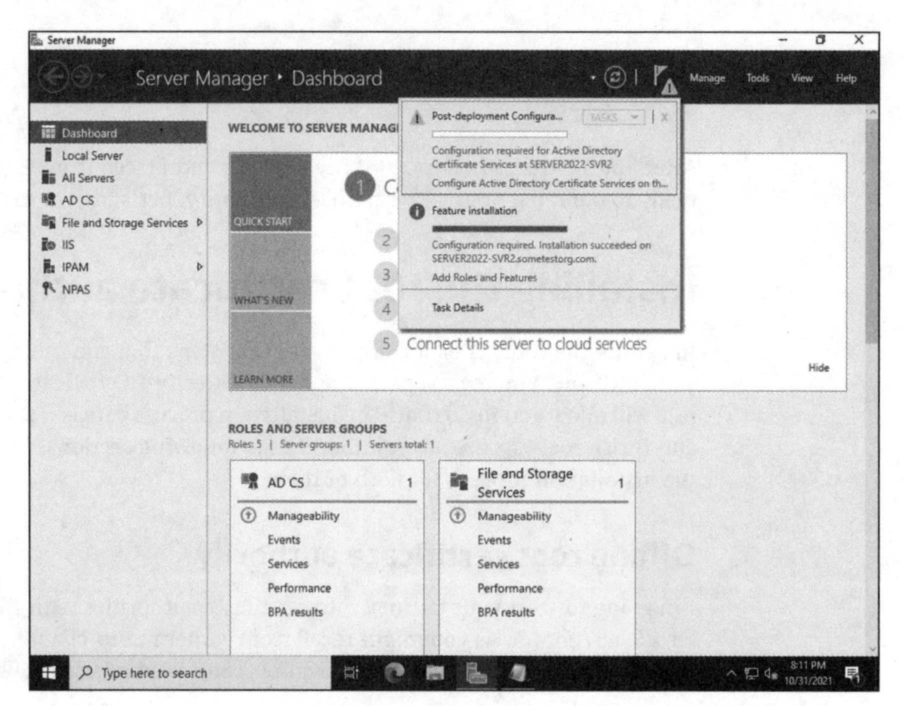

FIGURE 6-1: After AD CS has been installed, it needs to be configured.

12. **On the Credentials screen, enter an account that is in the Local Administrators group on the server, and click Next.**
13. **On the Role Services screen, select Certification Authority, and click Next.**
14. **On the Setup Type screen, select Standalone CA (as shown in Figure 6-2), and click Next.**
15. **On the CA Type screen, select Root CA, and click Next.**
16. **On the Private Key screen, select Create a New Private Key, and click Next.**
17. **On the Cryptography for CA screen, ensure your key length is a minimum of 2048, and select SHA256 for certificate signing (see Figure 6-3); then click Next.**
18. **On the CA Name screen, change the Common Name to whatever you like and then click Next.**

    I changed mine to ROOTCA.
19. **For the Validity Period screen, accept the default of five years and click Next.**
20. **On the CA Database screen, click Next.**

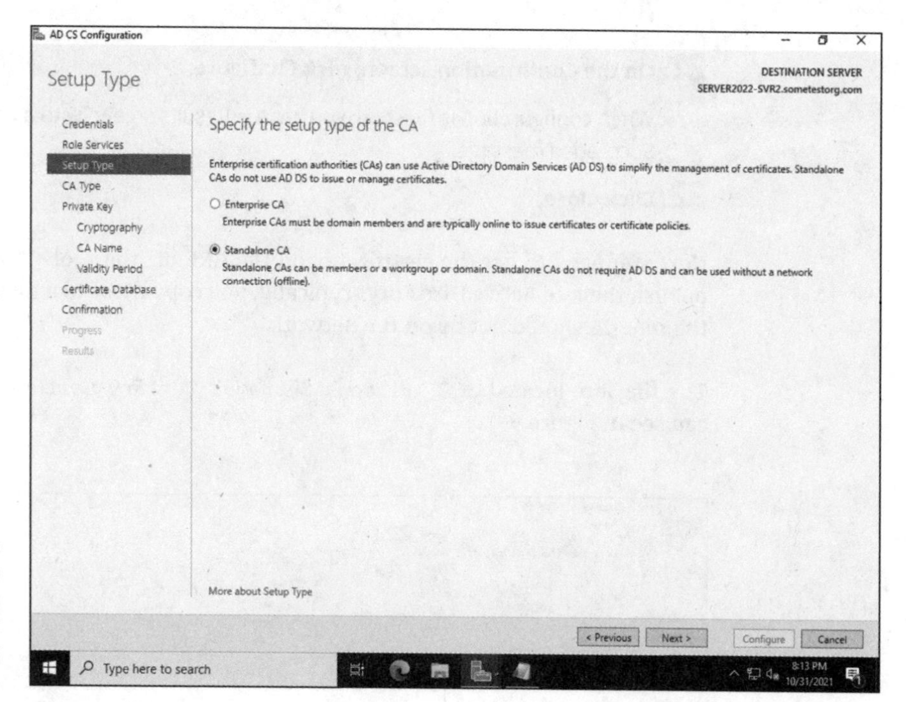

FIGURE 6-2: You can't change this selection later, without reinstalling AD CS.

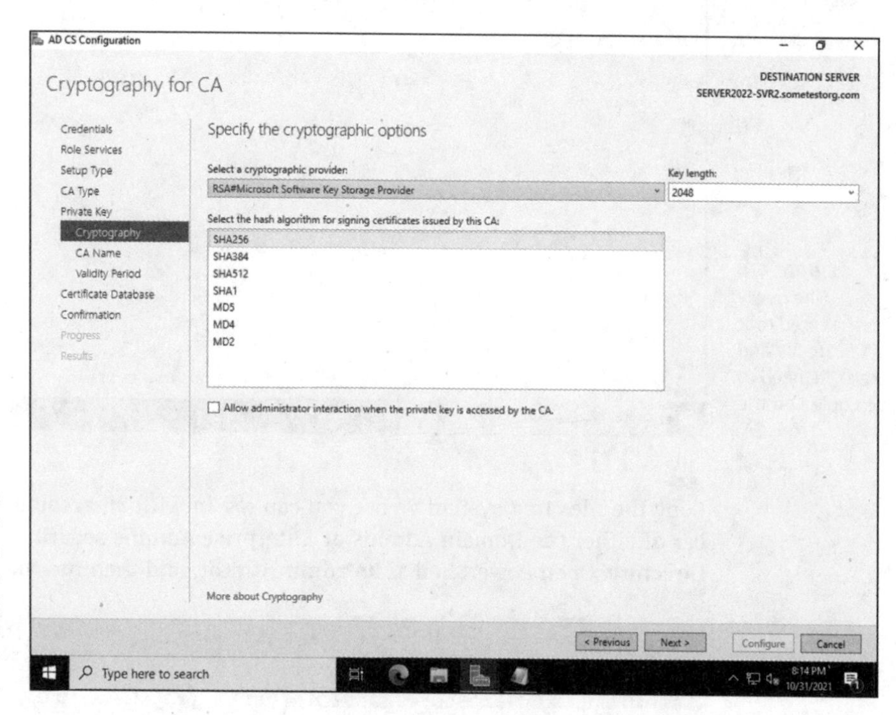

FIGURE 6-3: The Cryptography for CA screen allows you to set important security parameters for your CA.

21. **On the Confirmation screen, click Configure.**

When configuration is finished, you get the Results screen with a Configuration Succeeded message.

22. **Click Close.**

Next you need to get the certificate and the CRL off the root CA so that you can publish them to Active Directory. Typically, you copy these to a flash drive because the root CA should not be on the network.

The files are located at `C:\Windows\System32\CertSrv\CertEnroll`, which you can see in Figure 6-4.

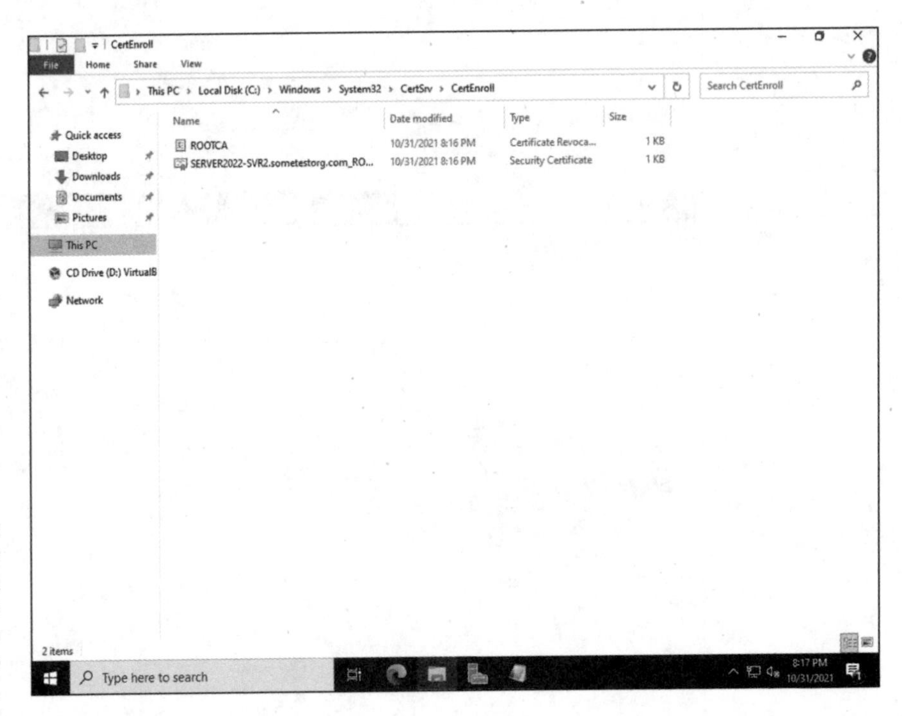

FIGURE 6-4: The newly created root certificate file and the CRL need to be copied off the root CA.

Copy the files to a system where you can log in with an account that is a member of either the Domain Admins or Enterprise Admins security groups in Active Directory. Open PowerShell as an administrator, and then run these commands:

```
certutil.exe -dspublish -f "certificatename.crt" RootCA
certutil -f -dspublish "CRL_Name.crl"
certutil.exe -addstore -f root "certificatename.crt"
```

The first command publishes the certificate to Active Directory where it will be replicated out to any AD-joined systems within approximately eight hours. The second command publishes the new CRL, and the third command ensures that the root CA cert gets added to the relevant certificate stores on any subordinate CAs.

Enterprise root certificate authority

The Enterprise root certificate authority is integrated into the Windows domain and is left powered on. It automatically publishes CRLs through Active Directory.

1. **Follow steps 1 through 13 from the preceding section.**
2. **On the Setup Type screen, select Enterprise CA and click Next.**
3. **On the CA Type screen, select Root CA and click Next.**
4. **On the Private Key screen, select Create a New Private Key and click Next.**
5. **On the Cryptography for CA screen, ensure your key length is a minimum of 2048, and select SHA256 for certificate signing (refer to Figure 6-3); then click Next.**
6. **On the CA Name screen, change the Common Name to whatever you like and then click Next.**

   I changed mine to ROOTCA.
7. **On the Validity Period screen, accept the default of five years and click Next.**
8. **On the CA Database screen, click Next.**
9. **On the Confirmation screen, click Configure.**

   When configuration is finished, you get the Results screen with a Configuration Succeeded message.
10. **Click Close.**

After the configuration is complete, you don't have any manual tasks to complete as you did with the offline root. The certificates and CRLs are published directly into Active Directory.

Installing the issuing certificate authority

When you have your root CA built, you have a great foundation. Now you need an issuing CA that will actually issue the certificates for you. The following steps are built off the assumption that you have built an enterprise root CA.

1. **Follow steps 1 through 13 from the "Offline root certificate authority" section.**
2. **On the Setup Type screen, select Enterprise CA and click Next.**

3. **On the CA Type screen, select Subordinate CA and click Next.**
4. **On the Private Key screen, select Create a New Private Key and click Next.**
5. **On the Cryptography for CA screen, ensure your key length is a minimum of 2048, select SHA256 for certificate signing (refer to Figure 6-3), and click Next.**
6. **On the CA Name screen, select a common name for your certificate authority and click Next.**

   I chose ISSUECA.
7. **On the Certificate Request screen, select Send a Certificate Request to a Parent CA and click the Select button, click the root CA, and then click OK.**

   You should see the previously created enterprise root CA in the list, as shown in Figure 6-5. Note that if you created the offline root CA, you can select the other bubble on this screen to generate a certificate signing request (CSR).

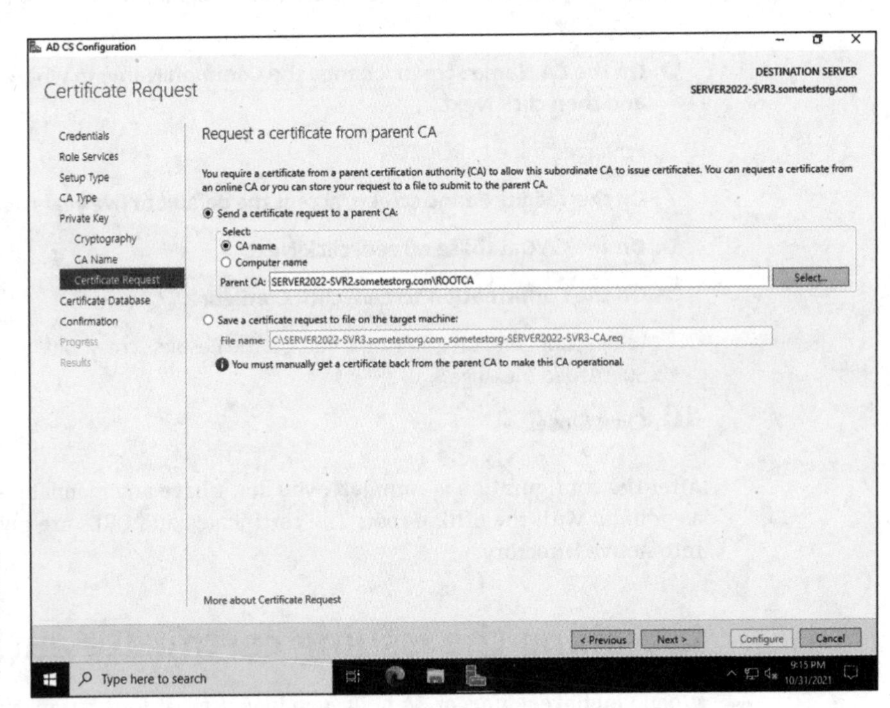

FIGURE 6-5: Selecting the root CA, assuming the root CA is an enterprise CA, to issue the issuing CA's certificate.

8. **On the Certificate Request screen, click Next.**
9. **On the CA Database screen, click Next.**

10. **On the Confirmation screen, click Configure.**

    When configuration is finished, you get the Results screen with a Configuration Succeeded message.

11. **Click Close.**

Enrolling for certificates

Now you have certificate authorities set up. What should you do first? I like to issue certificates to my domain controllers so that I can start using LDAP over SSL (LDAPS). This will encrypt replication traffic between domain controllers and directory queries. I have logged on to my domain controller. Here are the steps to enroll in a certificate from the machine where you want the certificate to be:

1. **Click Start, type** mmc.exe, **and press Enter.**
2. **Choose File⇨Add/Remove Snap-in.**
3. **Choose Certificates, and then click Add (as shown in Figure 6-6).**

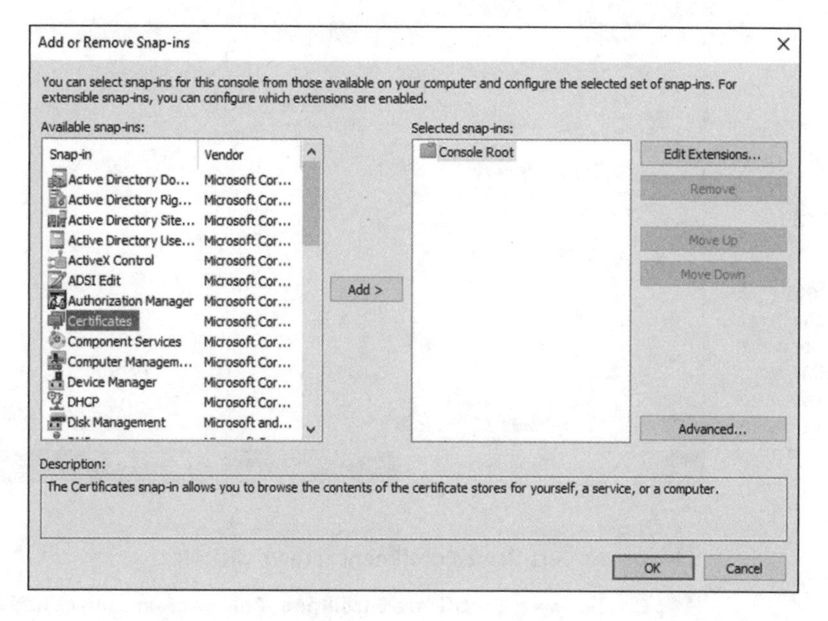

FIGURE 6-6: The Microsoft Management Console (MMC) gives you lots of configuration options for your system, including certificates.

4. **On the Certificates Snap-in screen, choose Computer Account, and click Next.**
5. **On the Select Computer screen, leave Local Computer selected and click Finish.**
6. **Click OK.**
7. **Expand Certificates and then expand Personal.**
8. **If there is a Certificates subfolder select that; if there isn't do the next step on the Personal folder.**
9. **With the Certificates folder selected, right-click the white space next to it, choose All Tasks and then Request New Certificate (see Figure 6-7).**

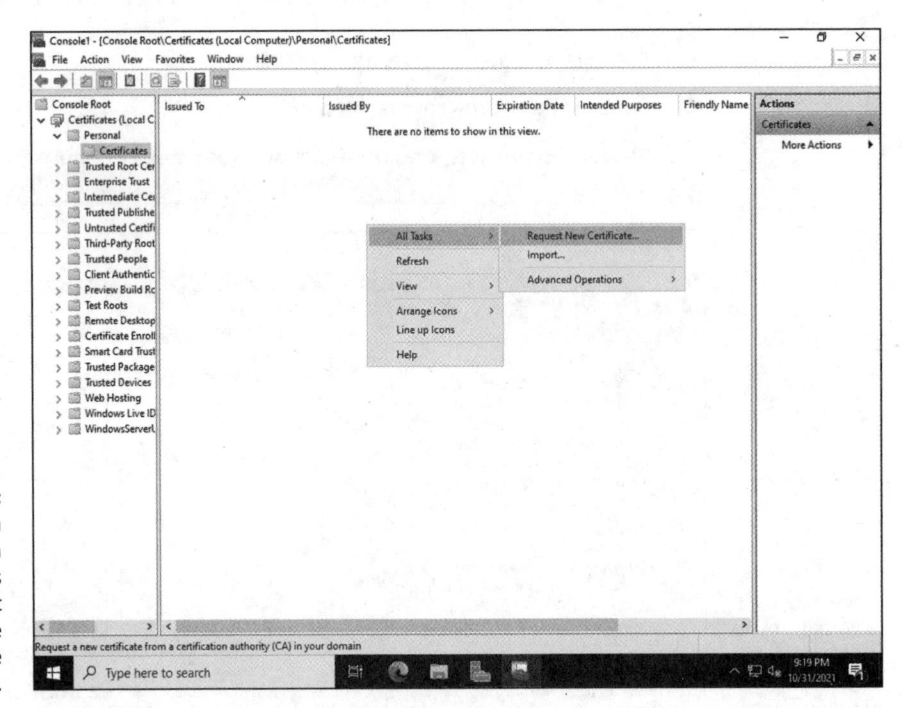

FIGURE 6-7: Requesting a certificate within the MMC starts when you select the certificate store and make the request.

10. **On the Certificate Enrollment screen, click Next.**
11. **On the Select Certificate Enrollment Policy screen, select Active Directory Enrollment Policy and click Next.**

    You see all the available certificate templates.

12. **Choose Domain Controller, as shown in Figure 6-8, and then click Enroll.**

    The server will make the request and return with a Success message after the certificate has been issued.

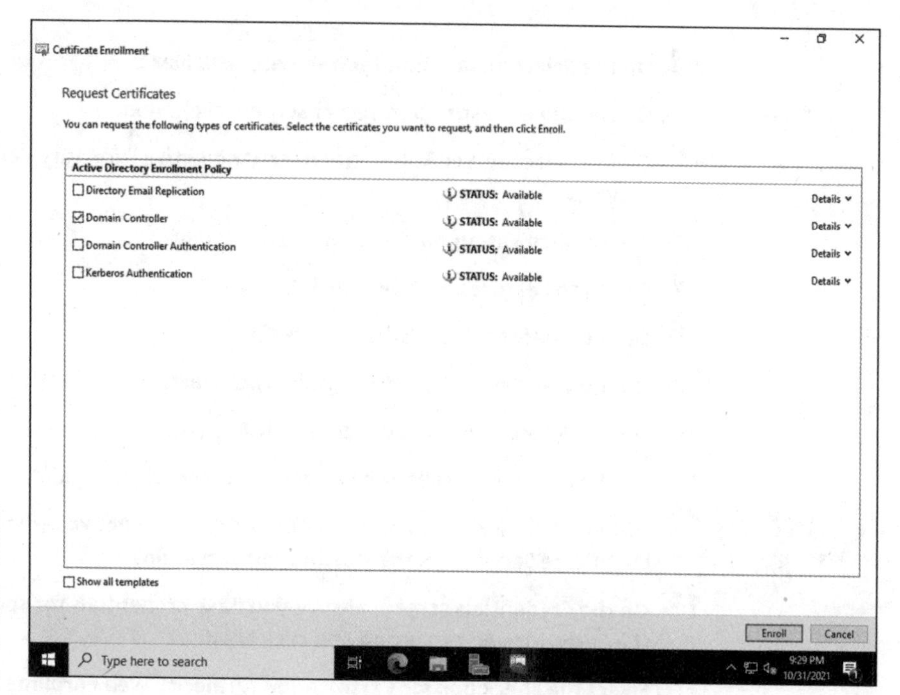

FIGURE 6-8: You're presented with the certificate templates that you have permissions to use.

That's all you have to do to enroll in a certificate from an enterprise issuing CA on an AD-joined system.

Setting up web enrollment

With enterprise CAs, you can enroll in certificates from whatever machine you're on. You may have some devices that are not connected to Active Directory. Are those systems just out of luck? Of course not! The web enrollment interface allows you to issue certificates with a certificate signing request (CSR) that you've generated from just about any device.

TECHNICAL STUFF

A CSR is generated when you want to create a certificate. It contains the public key that you want to include in the certificate, as well as the information you want to include on the certificate like the name, organization, and so on. The name is referred to as a common name and consists of the fully qualified domain name of the system (for example, `www.dummies.com`).

Let's return to the enterprise issuing CA you installed earlier and add the web enrollment piece:

1. **From Server Manager, click Manage, and then click Add Roles and Features.**
2. **On the Before You Begin screen, click Next.**

3. On the Select Installation Type screen, click Next.
4. On the Select Destination Server screen, click Next.
5. On the Select Server Roles screen, expand Active Directory Certificate Services.
6. Select Certification Authority Web Enrollment.
7. Click Add Features and then click Next.
8. On the Select Features screen, click Next.
9. On the Web Server Role (IIS) screen, click Next.
10. On the Select Role Services screen, click Next.
11. On the Confirm Installation Selections screen, click Install.
12. After installation is complete, click the Configure Active Directory Certificate Services on the Destination Server link.
13. On the Credentials screen, ensure you have credentials for someone in the Enterprise Admin group and click Next.
14. Select the check box for Certification Authority Web Enrollment and click Next.
15. On the Confirmation screen, click Configure.
16. On the Results screen, click Close.

You may be asking what all that work was for. You've given your users a web page where they can request and receive certificates on their own with no need to ask you for them. This is a fantastic way to enable your users to secure their internal applications without having to contact someone else to create the certificate for them. Your users can access the web interface (shown in Figure 6-9) at `http://servername/CertSrv/Default.asp`.

Installing Online Certificate Status Protocol

Historically, revocation information for certificates was received through certificate revocation lists (CRLs). The base CRL could get quite large, so delta CRLs were introduced to work with a smaller time frame. By default, the base CRL was updated once a week, while the delta CRL was updated once a day. This meant that if you revoked a certificate it may not be in the delta for a day. A better way was needed to get more timely revocation information. Online Certificate Status Protocol (OCSP) was created for that purpose. It provides near real-time revocation information. Here's how to install OCSP:

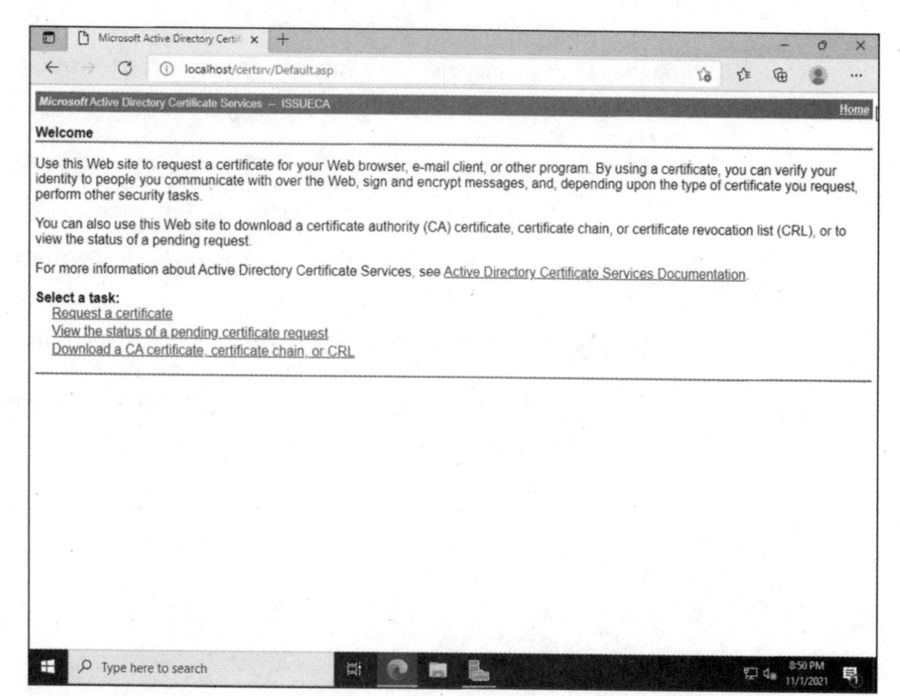

FIGURE 6-9: The Web Enrollment page for certificates is simple and easy to use.

1. **From Server Manager, click Manage, and then click Add Roles and Features.**
2. **On the Before You Begin screen, click Next.**
3. **On the Select Installation Type screen, click Next.**
4. **On the Select Destination Server screen, click Next.**
5. **On the Select Server Roles screen, expand Active Directory Certificate Services, select Online Responder, click Add Features, and then click Next.**
6. **On the Select Features screen, click Next.**
7. **On the Confirm Installation Selections screen, click Install.**
8. **After installation is complete, click the Configure Active Directory Certificate Services on the Destination Server link.**
9. **On the Credentials screen, ensure you have credentials for someone in the Enterprise Admin group and click Next.**
10. **Select the check box for Online Responder and click Next.**
11. **On the Confirmation screen, click Configure.**
12. **On the Results screen, click Close.**

Now that OCSP is installed, you need to enable its certificate template for when you configure it. Follow these steps:

1. **From Server Manager, choose Tools➪Certification Authority.**
2. **Expand the CA name and select Certificate Templates.**
3. **Right-click Certificate Templates, and choose New➪Certificate Template to Issue.**
4. **Scroll down and select OCSP Response Signing and then click OK.**

You should now see the OCSP Response Signing template listed in Certificate Templates, as shown in Figure 6-10.

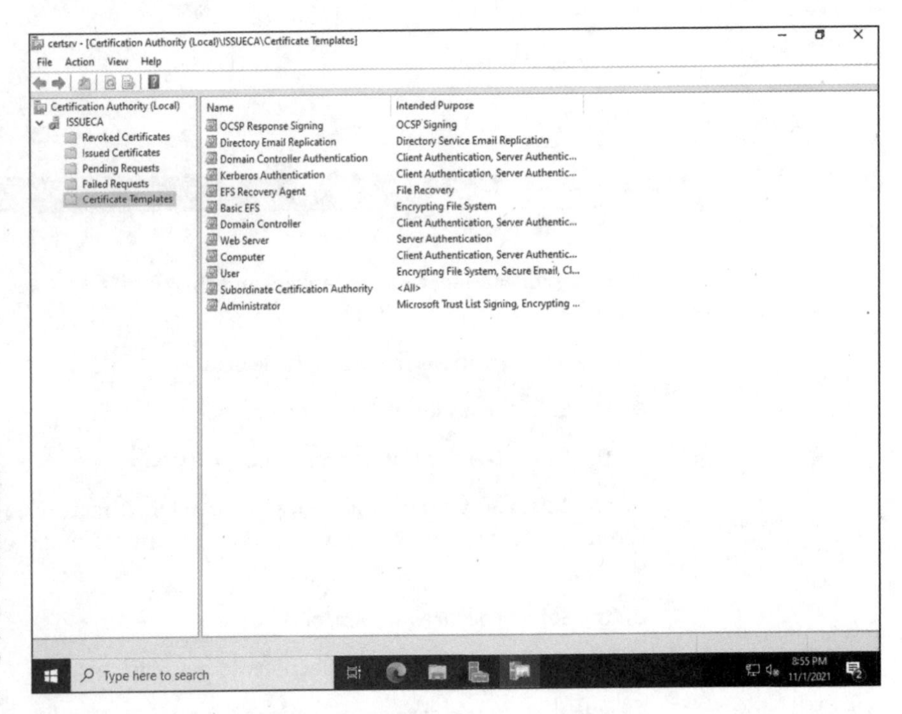

FIGURE 6-10: The OCSP Response Signing template is needed to set up the revocation information as part of the OCSP configuration.

You've installed the Online Responder, and you've enabled its certificate template. Now you're ready to configure it!

1. **From Server Manager, choose Tools➪Online Responder Management.**
2. **Right-click n Revocation Configuration and choose Add Revocation Configuration.**
3. **On the first screen of the Add Revocation Configuration Wizard, click Next.**

4. **Give your configuration a name and click Next.**

   The name can be anything you like. I'll name mine ISSUECAREV.

5. **On the Select CA Certificate Location screen, leave the default Select a Certificate for an Existing Enterprise CA selected and click Next.**

6. **On the Choose CA Certificate screen, click the Browse button next to Browse CA Certificates Published in Active Directory.**

7. **Choose the certificate that belongs to your issuing CA and then click OK.**

8. **Click Next.**

9. **On the Select Signing Certificate screen, select Automatically Select a Signing Certificate and check the Auto-Enroll for an OCSP signing certificate check box. Then click Browse.**

10. **Select your issuing CA and click OK.**

    It will automatically select the OCSP Response Signing template, shown in Figure 6-11.

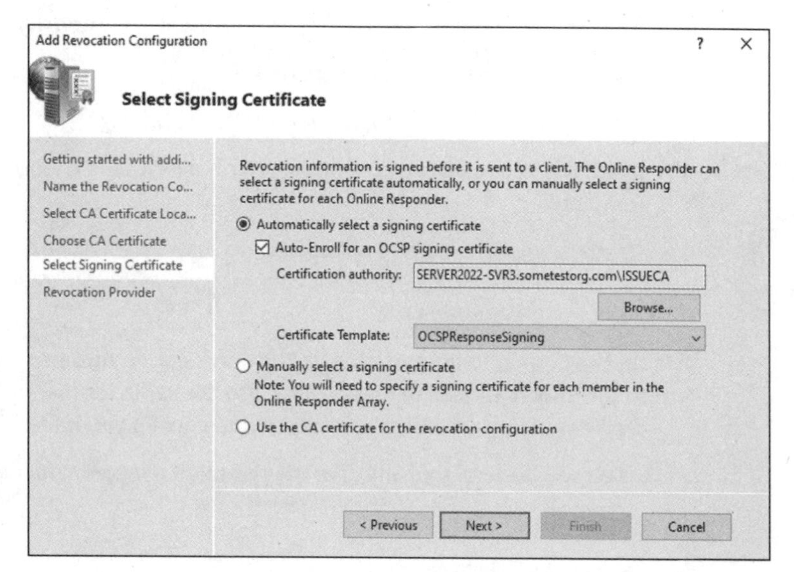

FIGURE 6-11: Setting the OCSP Revocation Configuration requires the OCSP Response Signing certificate template.

11. **Click Next.**

    The revocation provider is initialized.

12. **Click Finish.**

Your OCSP responder is now up and running! Now when a certificate is presented, the OCSP service can respond with a real-time response that indicates if the certificate is still valid or if it has been revoked.

Configuring Certificate Auto-Enrollment

You may want to have auto-enrollment occur on some certificates. A common use case would be for user certificates if your users need them for virtual private network (VPN) connections or for two-factor authentication (2FA). In this section, you set up the User certificate template to auto-enroll.

Configuring the template

The first step to setting up auto-enrollment is to set it up in the certificate template itself. So, let's set up the template to auto-enroll the Domain Users group.

1. **From Server Manager, choose Tools⇨Certification Authority.**
2. **Double-click the CA name to expand it.**
3. **Right-click Certificate Templates, and choose Manage.**
4. **Select the User certificate template, right-click it, and choose Duplicate Template.**

   WARNING

   Always choose Duplicate Template when modifying certificate templates. Custom templates could get overwritten with updates or upgrades if you use the out-of-the-box template.
5. **On the Compatibility tab, change the Certification Authority drop-down to the lowest version of CA you have. Do the same for the Recipient drop-down for the lowest server/desktop version you have.**
6. **Select the General tab, and give the template a meaningful name.**

   I'll call mine VPN User Cert.
7. **Select the Security tab.**
8. **Select Domain Users, and check Read and Autoenroll.**

   Enroll will already be selected.

9. **Click OK.**
10. **Close out of the Certificate Templates Console by clicking the X in the upper-right corner.**
11. **Right-click Certificate Templates again, but this time choose New⇨Certificate Template to Issue.**
12. **Select your new VPN User Cert (or whatever you named it in Step 6) and click OK.**

Now the CA is set to issue the new user certificate. You're ready to set up Group Policy so that domain-joined users will automatically get their certificates when they log in.

Configuring Group Policy

You need to move to a system that has the Group Policy Management tools installed. I've moved back over to my domain controller. Follow these steps:

1. **From Server Manager, choose Tools⇨Group Policy Management.**
2. **Right-click the Default Domain Policy (or whichever policy you want to place this in) and choose Edit.**
3. **Double-click User Configuration, then Policies, then Windows Settings, then Security Settings, and then Public Key Policies.**
4. **Double-click Certificate Services Client Auto-Enrollment.**
5. **Change the configuration model to Enabled.**
6. **Check both of the check boxes below Configuration Model, as shown in Figure 6-12.**
7. **Click OK.**

Your users will now receive the auto-enrolled certificate that you set up on the CA the next time they log in. If you want to verify that their system received the certificate, you can check the Issuing CA to check for the presence of their certificate, or you can open the MMC as discussed earlier and check the Personal certificate store to ensure that the certificate is there.

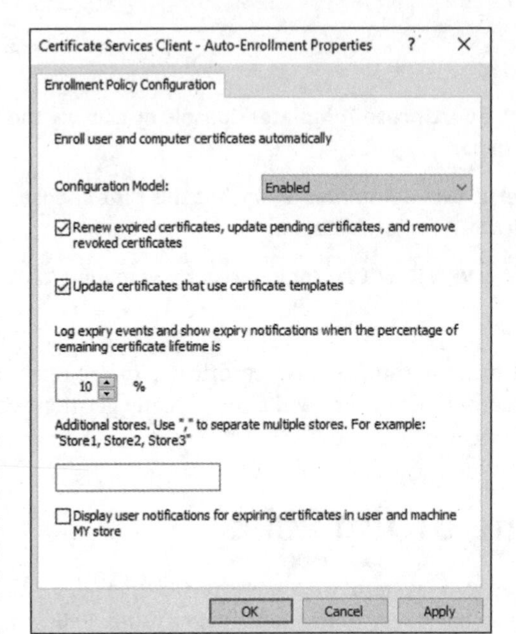

FIGURE 6-12: The last part of setting up auto-enrollment is to set Group Policy to auto-enroll user certificates.

» Examining DANE and how it complements DNSSEC

» Securing DNS with HTTPS

Chapter 7
Securing Your DNS Infrastructure

Domain Name System (DNS) is a requirement of Active Directory and is what makes networks easier to work with. With DNS, you don't have to remember IP addresses — you can just remember simple names instead. Imagine, though, that a criminal was able to make your systems believe that their DNS server was your DNS server. You and your users could be redirected to a malicious site.

In this chapter, you learn how to secure your DNS infrastructure using DNS Security Extensions (DNSSEC) and DNS-based Authentication of Named Entities (DANE).

Understanding DNSSEC

DNSSEC was designed to prevent attackers from hijacking the DNS lookup process and protect users from being given addresses to malicious servers. DNSSEC signs zones and records, which allows the endpoint that made the query to validate that a DNS record is a valid record, or if it's redirecting to an invalid and potentially malicious location instead (DNS cache poisoning).

By digitally signing the root zone in your DNS infrastructure, you can give your users assurances that their systems are getting responses from valid DNS servers. Digital signatures create a hash — think of it as a unique thumbprint. If anything about a record has changed, the hash won't match and the record will be invalid. It's important to remember that DNSSEC does not encrypt data at all; it's only validating the identity of the DNS server doing the DNS lookup.

The basics of DNSSEC

With DNSSEC, you have two keys to be aware of:

- **Key Signing Key (KSK):** The KSK is a long-term key used to sign ZSKs and validate DNSKEY records.
- **Zone Signing Key (ZSK):** The ZSK is a short-term key used to sign the actual DNS records.

To secure a zone with DNSSEC, one of the first things that occurs is the grouping of like record types into a resource record set (RRset). The RRset is then digitally signed, which provides protection for all the individual records within the resource record set.

The private key of the ZSK is used to digitally sign each RRset, and the public key of the ZSK is used to verify the signature. The digital signature of each RRset is saved as an RRSet Signature (RRSIG) record.

Records used for DNSSEC

DNSSEC introduced several new record types to work with and contains the new cryptographic features that DNSSEC added. Here are the record types that you need to remember:

- **RRSIG:** Contains the digital signature of an RRset that has been signed by the ZSK, or a DNSKEY record that has been signed by the KSK.
- **DNSKEY:** DNSKEY records can contain the public key of the ZSK or the public key of the KSK.
- **DS:** The Delegation Signer (DS) record allows for the transfer of trust from the parent zone to a child zone, which in turn allows the child zone to be DNSSEC-enabled. The DS record contains a hashed copy of the parent zone's DNSKEY.

- **NSEC:** Traditional DNS returns an empty response if there is no match. The issue with that is that it isn't providing an authenticated response. The NSEC records return an authenticated response with the next available secure record. For example, say you request `totallybogus.sometestorg.com`. There is no record for totallybogus, so the DNS server may just return the record for www.
- **NSEC3:** NSEC3 does the same thing as NSEC, but the owner name in the record is stored in a hashed state. By storing the owner's name in hash, you're protecting against zone enumeration, which is a form of reconnaissance that allows an attacker to construct the DNS zone based on redirects to other records.
- **NSECPARAM/NSEC3PARAM:** The NSECPARAM and/or NSEC3PARAM resource records select the NSEC or NSEC3 records to include in a negative response.

Configuring DNSSEC

Now you have an understanding of how DNSSEC works at a high level. Let's actually configure it on our Windows server.

Enabling DNSSEC

When you've decided that you want to use DNSSEC, the first thing that you have to do is actually go in and sign the DNS zone. You need a system that has the DNS Manager administrative tool installed. Follow these steps:

1. **From Server Manager, choose Tools⇨DNS.**
2. **Expand your server and Forward Lookup Zones, and then select the domain on which you want to enable DNSSEC.**
3. **Right-click the domain and choose DNSSEC⇨Sign the Zone, as shown in Figure 7-1.**

   The DNS Security Extensions (DNSSEC) Wizard appears.
4. **Click Next.**
5. **On the Signing Options screen, select Customize Zone Signing Parameters and click Next.**
6. **On the Key Master screen, select The DNS Server <*yourservername*> Is the Key Master and click Next.**
7. **On the Key Signing Key screen, click Next.**

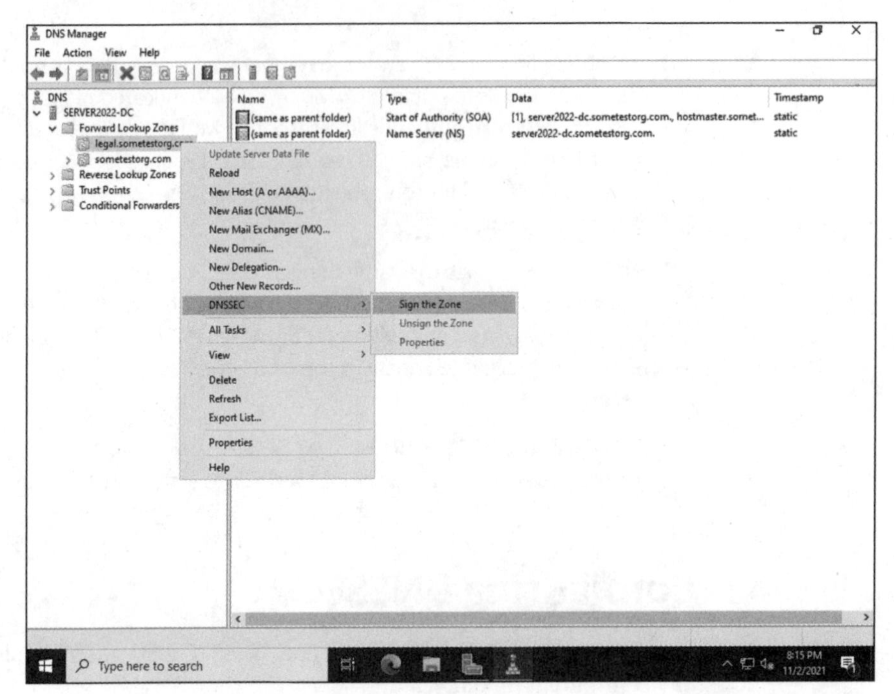

FIGURE 7-1: You can enable DNSSEC on a zone by selecting Sign the Zone located under DNSSEC in the menu.

8. **On the second Key Signing Key screen, click Add.**
9. **Accept the default KSK settings, shown in Figure 7-2, and click OK.**
10. **Click Next.**
11. **On the Zone Signing Key screen, click Next.**
12. **On the second Zone Signing Key screen, click Add.**
13. **Change the key length field to 2048, and then click OK.**
14. **Click Next.**
15. **On the Next Secure (NSEC) screen, choose Use NSEC3 and click Next.**
16. **On the Trust Anchors screen, select the Enable the Distribution of Trust Anchors for This Zone check box, leave the other check box checked, and click Next.**
17. **On the Signing and Polling Parameters screen, change the DS record generation algorithm to just SHA-256, as shown in Figure 7-3.**
18. **Click Next.**
19. **On the final screen, click Next.**

    The zone will be signed.
20. **After it's signed, click Finish.**

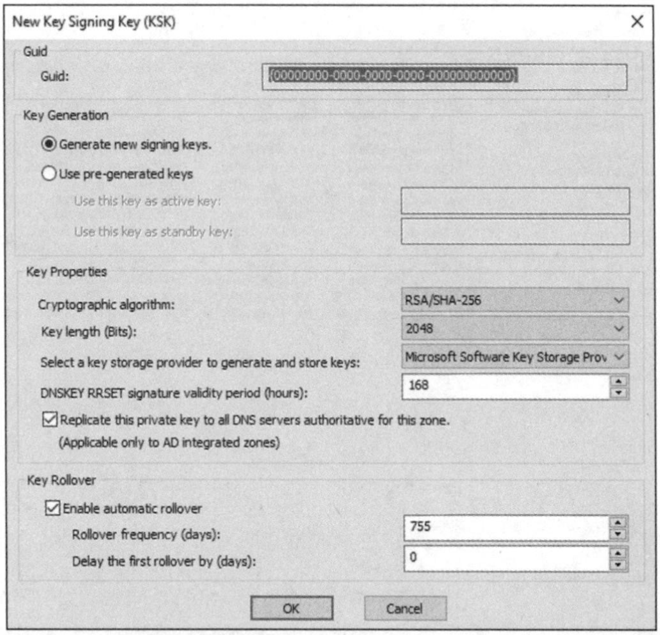

FIGURE 7-2: Creating the KSK is simple — you can safely accept the defaults.

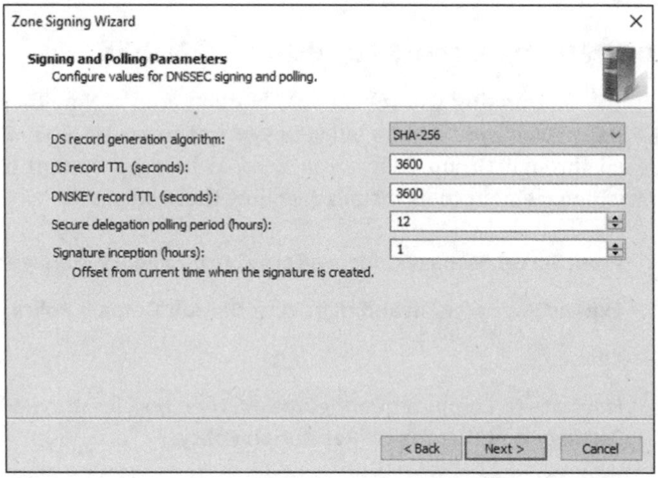

FIGURE 7-3: Considering that the DS record is responsible for the transfer of trust, you want to select only secure algorithms, such as SHA-256.

If you want to verify that DNSSEC is enabled, you may need to refresh, but you'll see all the special record types I mention earlier. My example is shown in Figure 7-4.

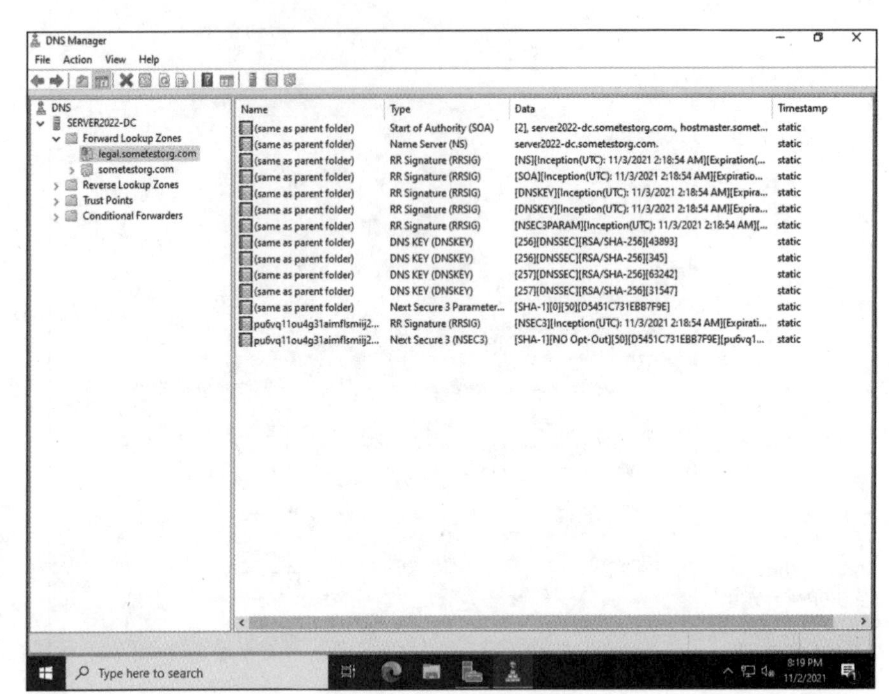

FIGURE 7-4: After the zone is signed, the special DNSSEC record types will show up in the signed zone.

Configuring clients to require DNSSEC

You may be thinking you're all done, but you aren't. Now that your DNS zone is DNSSEC enabled, you have to tell your systems to use DNSSEC. This is best accomplished through Group Policy. You need to be on a system that has the Group Policy Management tools installed. Follow these steps:

1. **From Server Manager, choose Tools⇨Group Policy Management.**
2. **Expand your domain and right-click Default Domain Policy.**
3. **Click Edit.**
4. **Navigate to Computer Configuration, then Policies, then Windows Settings, and then Name Resolution Policy.**
5. **In the Create Rules section, ensure that Suffix is selected in the drop-down box, and then fill in the domain name.**
6. **On the DNSSEC tab, check Enable DNSSEC in This Rule and Require DNS Clients to Check That Name and Address Data Has Been Validated by the DNS Server (as shown in Figure 7-5).**
7. **Click Create.**

That's all there is to it. With your Name Resolution Policy in place, domain-joined systems will use DNSSEC.

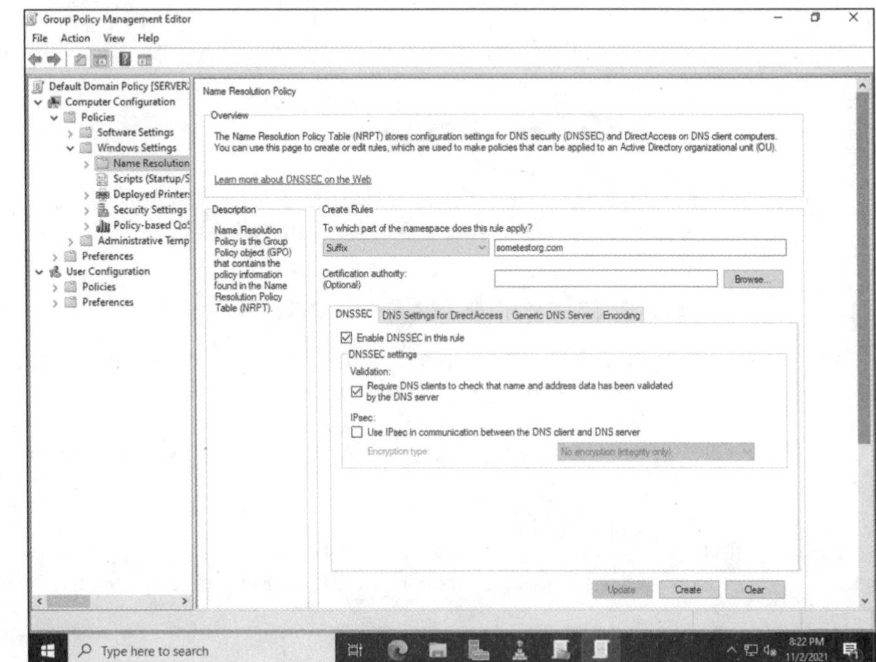

FIGURE 7-5: Creating a Name Resolution Policy from within Group Policy is the simplest way to guarantee that domain-joined systems will use DNSSEC.

Understanding DANE

DANE is built on DNSSEC. Although DNSSEC tries to protect against DNS cache poisoning and spoofing, DANE adds an additional layer of protection by allowing the administrator of a domain name to specify which certificate authority is allowed to issue certificates for their organization's domain, as well as provide a way to authenticate client and server certificates with a certificate authority. This prevents an attacker from issuing a certificate from a certificate authority and trying to pass it off as your own. It requires DNSSEC to work its magic.

The basics of DANE

DNS-based Authentication of Named Entities (DANE) uses a newer record type known as a Transport Layer Security Authentication (TLSA) record. The TLSA record is used to associate a domain name with a TLS certificate. Not all DNS hosting providers support DANE, but if your DNS servers are running Windows Server 2016 or newer, you can use DANE because Windows Server 2016 supports unknown record types.

DANE will allow you to bind a certificate to a specific DNS name using DNSSEC.

Configuring DANE

Configuring DANE consists of two parts: First, you must generate the TLSA record, and then you can install the TLSA record with PowerShell.

Generating the TLSA record

To generate the TLSA record, you need a copy of the certificate that you're wanting to protect. Follow these steps:

1. **Open a browser and go to the site that you want to protect.**

   I'm using my own `https://www.sometestorg.com` site for this demonstration.

2. **Click the padlock icon in the URL bar and click View Certificates.**

   This may vary depending on browser. I'm using the default browser, which is Edge.

3. **When the certificate properties screen comes up, click the Details tab.**
4. **Click the Copy to File button.**
5. **On the Welcome to the Certificate Export Wizard screen, click Next.**
6. **On the Export File Format screen, select Base-64 encoded, and click Next.**
7. **On the File to Export screen, select Browse and choose a name and location for the file and click OK.**

   I'm saving it to my Desktop for now.

8. **Click Next.**
9. **On the Completing the Certificate Export Wizard screen, click Finish.**
10. **Open a browser and go to the following address:**

    ```
    https://www.huque.com/bin/gen_tlsa
    ```

11. **Accept the default selections in the first three fields.**
12. **Click the Start menu, click Windows Accessories, and then click Notepad.**
13. **Open the certificate that you exported with Notepad by choosing File ⇨ Open, and then selecting the exported certificate file.**

14. **Copy the text from the file starting with the:**

```
-----BEGIN CERTIFICATE-----
```

and ending with the:

```
-----END CERTIFICATE-----
```

15. **Go back to the TLSA Generator page and paste the text into the Enter/ Paste PEM Format X.509 Certificate Here box.**

16. **For port number, put in whatever applies.**

    In my case, 443 for HTTPS traffic.

17. **Enter the transport protocol and the domain name.**

    In my example, the transport protocol is `tcp` and the domain name is `sometestorg.com`.

    Your screen should look similar to Figure 7-6.

18. **Click Generate.**

    You receive an output similar to Figure 7-7. When you have this in hand, you can install the TLSA record on the server.

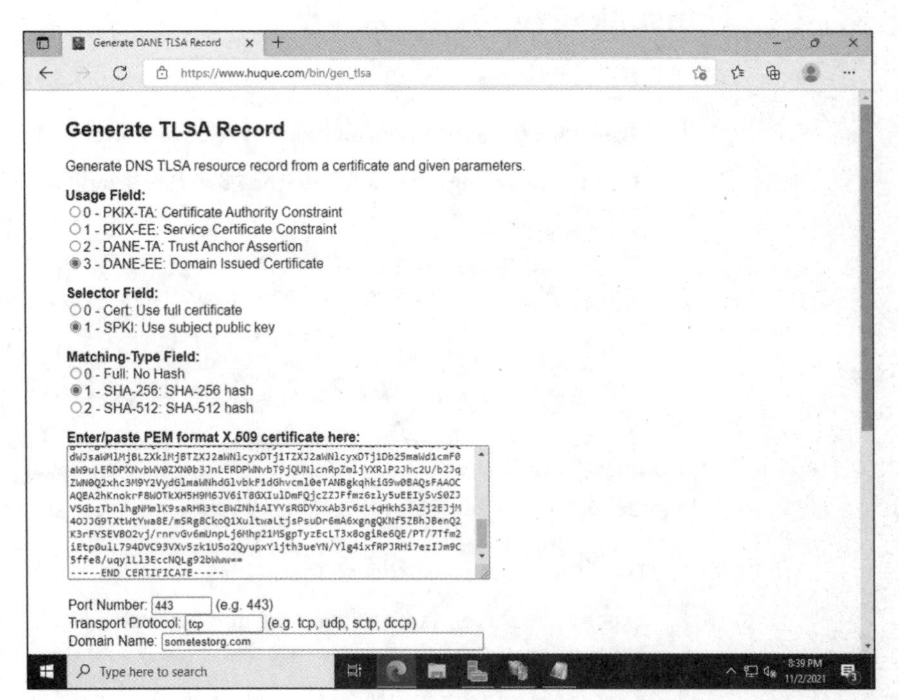

FIGURE 7-6: You need to generate the TLSA record before you can use it.

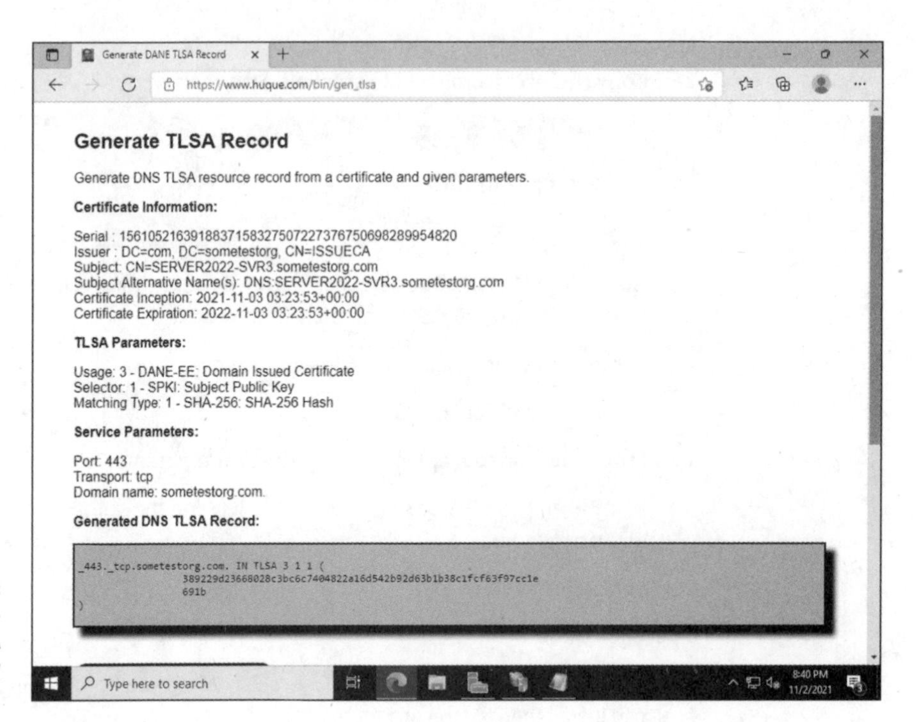

FIGURE 7-7: The page generated the TLSA record and now you can install it.

Installing the TLSA record

To install the TLSA record, follow these steps:

1. **Right-click the Start menu and select Windows PowerShell (Admin).**
2. **Type the following command into the PowerShell window:**

```
Add-DnsServerResourceRecord -TLSA -CertificateAssociation
   Data "9933c1848f2f492f4715abff9e79f74025fdd219a2f77a
   34d3cc9a00f36c8a0b" -CertificateUsage DomainIssued
   Certificate -MatchingType Sha256Hash -Selector
   FullCertificate -ZoneName sometestorg.com -name _
   443._tcp.www.sometestorg.com
```

The TLSA record is now installed. The `-CertificateAssociationData` parameter uses the long numerical string at the end of the generated record. You can verify that the record was successfully created by following these steps:

1. **From Server Manager, choose Tools⇨DNS.**
2. **Expand you server name, expand the domain name, and in my case I expand `com`, then `sometestorg`, then `www`, then `_tcp`.**

   You see a record that says Unknown under Data.

3. **Double-click that Unknown record.**

   If you see a Type of 52, it was recognized as a valid TLSA record. An example is shown in Figure 7-8.

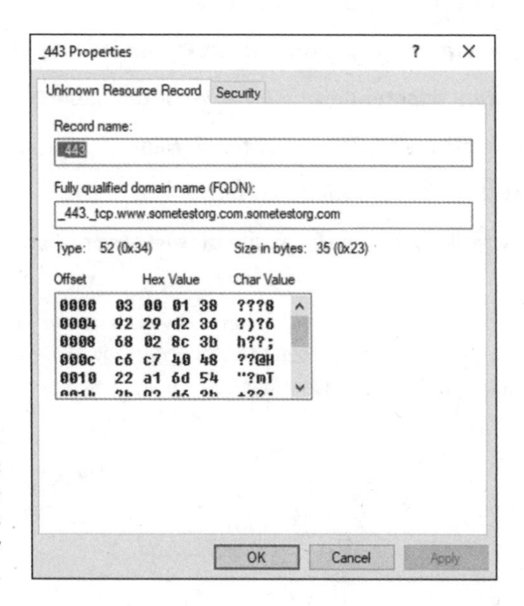

FIGURE 7-8: Verifying that the TLSA record was created properly is important.

Protecting DNS Traffic with DNS-over-HTTPS

Every time I read about DNS-over-HTTPS (DoH), I hear "Doh!" in Homer Simpson's voice. In this case, however, rather than an expression of frustration, DoH is an awesome way to protect your critical DNS traffic with a commonly used secure protocol, HTTPS. It's also very simple to configure.

DNS-over-HTTPS allows DNS traffic to be encrypted, which provides protection and confidentiality and also effectively masks your DNS traffic as it travels over the HTTPS port 443, rather than the traditional DNS port 53.

This sounds great doesn't it? You may be telling yourself that this is too good to be true. There couldn't *possibly* be many DNSs that support DoH could there? In fact, there are several, and I suspect more will be added in time. If you'd like to see the current list of available systems, you can run the following command:

```
Get-DNSClientDohServerAddress
```

You'll see a list of known providers that support DoH.

Enabling DoH in Server 2022

To enable DoH in Windows Server 2022, follow these steps:

1. **Click the Start menu, and then click Settings.**
2. **Select Network & Internet, and then select Ethernet.**
3. **Select the network adapter on which you want to enable DoH.**

   In my case, I'll choose my only adapter, which is Network 2.
4. **Scroll down to DNS Settings and click Edit.**
5. **Under DNS Encryption, choose from one of the three options shown in Figure 7-9.**

   I'll choose Encrypted Preferred, Unencrypted Allowed because this is my lab environment. You could choose to set it to Unencrypted Only, which is the default, or to Encrypted Only, which will enforce DoH.
6. **Click Save.**

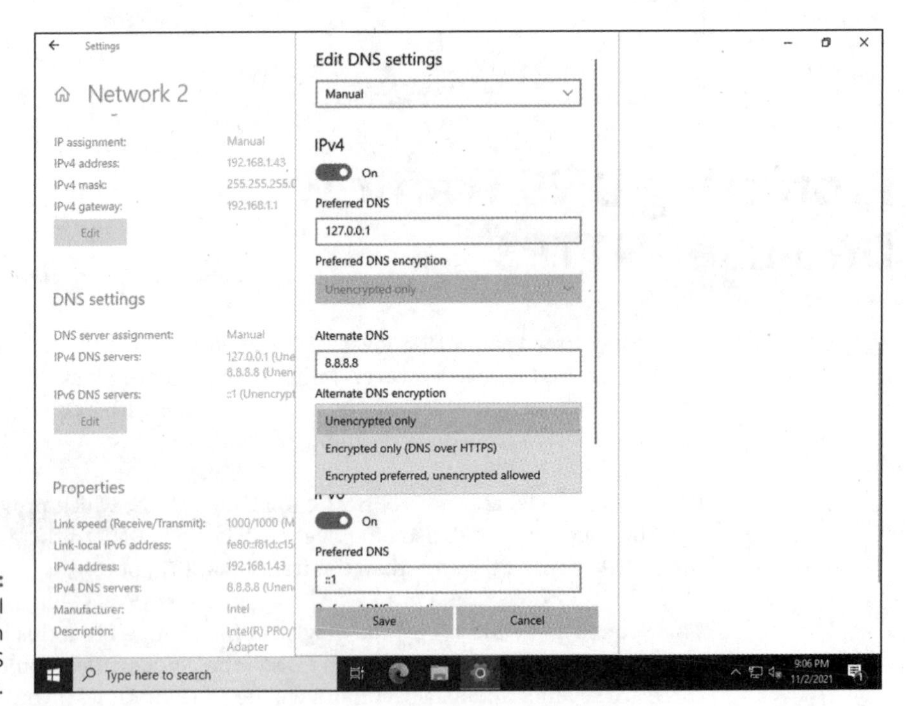

FIGURE 7-9: You have several encryption options for DNS traffic.

Using Group Policy to enable DoH

Being able to enable DoH is great, but let's face it: You don't want to enable it individually on systems across the enterprise. Not surprisingly, Microsoft gives you the ability to enable DoH through Group Policy. Follow these steps:

1. **Open Server Manager.**
2. **Click Tools and then click Group Policy Management.**
3. **Right-click Default Domain Policy and choose Edit.**
4. **Navigate to** `Computer Configuration\Policies\Administrative Templates\Network\DNS Client` **and double-click Configure DNS-over-HTTPS (DoH) Name Resolution, as shown in Figure 7-10.**

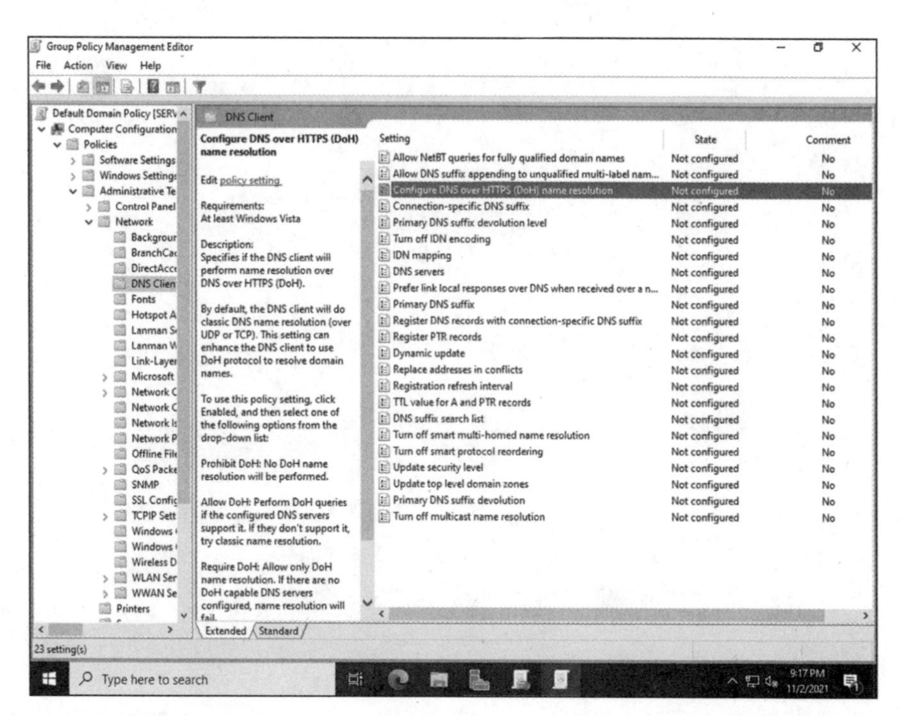

FIGURE 7-10: The new DoH feature can be configured via Group Policy.

5. **Choose Enabled, and then choose which option you want to use for DoH.**

   You have the option to prohibit, allow, or require DNS-over-HTTPS. I'll choose Allow DoH for now.

6. **Click OK.**

That's all there is to it! If you've followed along in this chapter, you've come a long way in improving the security of your DNS infrastructure.

6 Working with Windows PowerShell

Contents at a Glance

» Using PowerShell day to day

» Using PowerShell to remotely administer systems

» Getting help in PowerShell

» Addressing security issues with PowerShell

Chapter 1

Introducing PowerShell

PowerShell is the wave of the future. As server automation becomes more commonplace, there will be a much higher demand for a system administrator who has PowerShell skills. Additionally, remote administration with PowerShell reduces the need to interact with the graphical user interface (GUI) and can allow you to make changes to one or many systems from your workstation.

This chapter covers the basics of PowerShell, from the beginning terminology to using PowerShell remoting.

Understanding the Basics of PowerShell

Before you start using a scripting language, you need to learn it. In this section, I fill you in on the basics of PowerShell so the rest of this chapter makes sense.

Objects

In PowerShell, an *object* is a single instance of something, like a service or a process. If you run `Get-Process` or `Get-Service` inside of the PowerShell, each row is an object (assuming you're using the default formatting). In Figure 1-1, there is a service on each line, and each service is an object.

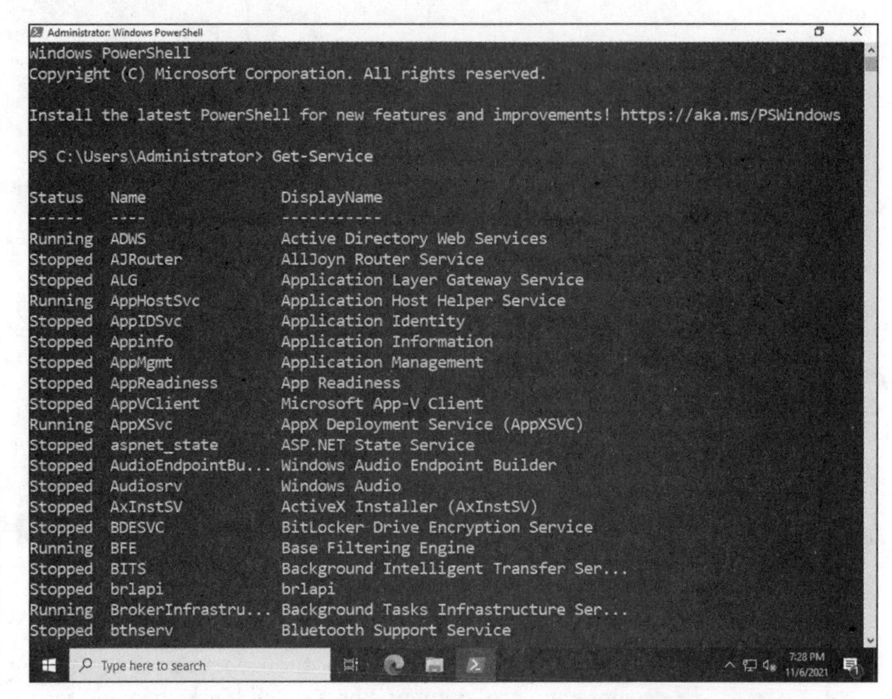

FIGURE 1-1: Each row returned in the `Get-Service` command is an object.

Pipeline

The *pipeline* allows you to take the output of one command and send it (pipe it) to the next command. For instance, if you're trying to figure out what methods and properties are available for a cmdlet, you can use the output of the cmdlet and pipe it to `Get-Member`. In Figure 1-2, I've entered in `Get-Service` with no filters, and then added `Get-Member`. In this case, because there are no filters, `Get-Member` is able to return all methods and properties associated with `Get-Service`.

You can do lots of things after the pipeline. Usually, you see things like formatting and filters, sometimes even export commands.

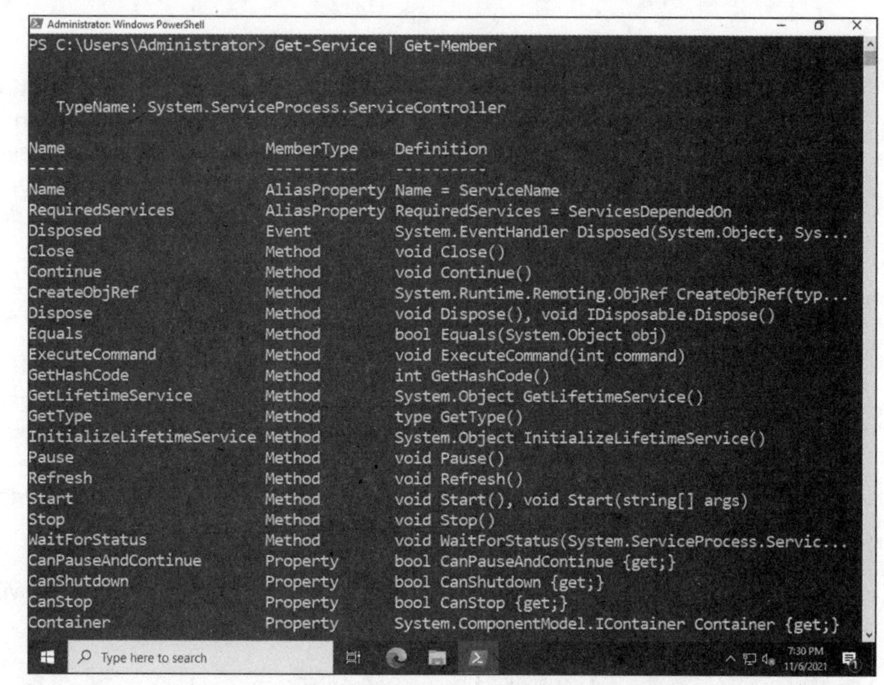

FIGURE 1-2: Using the pipeline allows you to send the output of one command to the input of another.

Providers

PowerShell *providers* allow you to access various data sources as if they were a regular data drive on your system. The providers that are built into PowerShell are listed in Table 1-1. You can view a list of the built-in providers on your system with the following command:

```
Get-PSProvider
```

TABLE 1-1 **PowerShell Built-in Providers**

Provider	Drive Example
Registry	HKLM:
Alias	Alias:
Environment	Env:
FileSystem	C:
Function	Function:
Variable	Variable:

PowerShell providers allow you to interact with your system in different ways. For instance, the FileSystem provider allows you to work with the file system on your server using the PSDrive cmdlets. `Get-PSDrive` will return a list of all the drives that are available from within the current PowerShell session, including your system drives, as well as the drives that can be used to work with the other providers. You can also create and remove a PSDrive with the `New-PSDrive` and `Remove-PSDrive` cmdlets.

Variables

Think of a *variable* as a container in which you can store something. You can create a variable at any time. Variables are not case-sensitive, and they can use spaces and special characters. As a best practice, Microsoft avoids using spaces and special characters in variable names. It recommends sticking with alphanumeric characters (A–Z, 0–9) and the underscore character (_).

Variables are declared with the dollar sign ($). For instance, if I wanted to create a variable to store my first name, I could type the following:

```
$FirstName = "Sara"
```

REMEMBER

Variables are not case-sensitive. I capitalize the first letter of each word because it makes it easier for me to read it, but you don't have to.

When you want to display the value of a variable, all you need to do is type it with the preceding dollar sign. In Figure 1-3, you can see that I set my variable `FirstName` to my first name `Sara`. Then I typed in the name of the variable and it showed me the value of the variable.

Sessions

Sessions are used to connect to PowerShell on a remote system. This can be done to run commands or to interact directly with the PowerShell on the remote system.

For example, `New-PSSession` will create a new PowerShell session. This can be done on a local system or a remote system. Note that `New-PSSession` creates a persistent connection, whereas `Enter-PSSession` connects to a remote system but only creates a temporary connection that exists for as long as the command or commands are running.

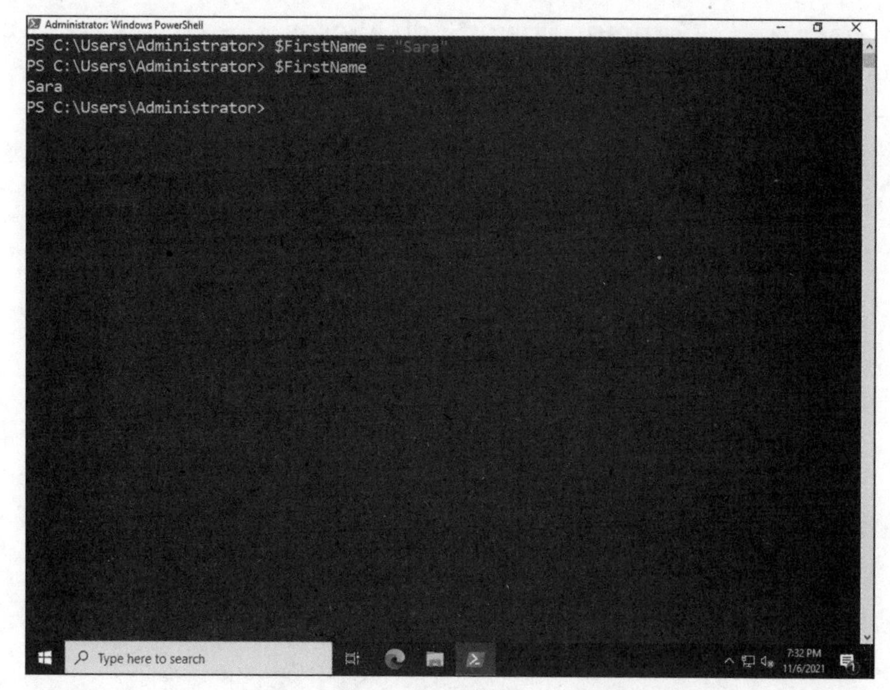

FIGURE 1-3: Using the dollar sign tells the system you're working with a variable. You can create and display variables by adding the dollar sign in front of the variable name.

Comments

Whenever you write PowerShell scripts, you should always add *comments*, which help others understand how the code works and what it's doing. Comments can also help you if you ever need to change the code, or if you don't run it very often. Comments start with the hash symbol (`#`). Everything that comes after the hash symbol on that line is a part of the comment. See the following example:

```
#This is a comment...
Write-Host "This is not a comment"
```

Aliases

Aliases are shortcuts for full commands. There are far too many to list all of the PowerShell aliases, but Table 1-2 lists some of the more common ones.

Cmdlets

A *cmdlet* is a piece of code that consists of a verb and a noun. Common verbs are `Get`, `Set`, `New`, `Install`, and so on. With the cmdlet `Get-Command`, `Get` is the verb and `Command` is the noun.

TABLE 1-2 Common PowerShell Aliases

Alias	Full Command
gcm	Get-Command
sort	Sort-Object
gi	Get-Item
cp	Copy-Item
fl	Format-List
ft	Format-Table
pwd	Get-Location
cls	Clear-Host
ni	New-Item
sleep	Start-Sleep
write	Write-Output
where	Where-Object

Running Get-Command returns all cmdlets, aliases, and functions. You can find cmdlets, aliases, and functions that you're looking for with the Get-Command cmdlet. Look at the following:

```
Get-Command -Noun *network*
```

When you run this command, PowerShell returns a list of all commands where the noun includes *network*. The asterisks are wildcards and basically tell PowerShell that you don't care if there is text before or after the noun you're looking for. An example of this command is shown in Figure 1-4.

Parameters allow you to refine what you're interested in the command doing. For instance, using the Get-Command cmdlet by default will return cmdlets, aliases, and functions. By using the -All parameter, you can get the Get-Command cmdlet to return cmdlets, aliases, functions, filters, scripts, and applications. You can find out which parameters are available for a given command by checking the help documentation for the command. I cover using help within PowerShell later in this chapter.

FIGURE 1-4: You can use the cmdlet `Get-Command` to find other commands even if you don't know the whole noun's name.

Using PowerShell

In this section, I show you the typical day-to-day usage of PowerShell and how to accomplish the things that will make your life as a system administrator so much better.

Writing PowerShell commands and scripts

You may end up spending a significant amount of time typing out one-liners in PowerShell, or you may end up writing whole scripts. Several tools work well for writing PowerShell. I cover them in the following sections.

PowerShell Integrated Scripting Environment

You can always type directly into PowerShell, however if you're newer to PowerShell, I recommend using the PowerShell Integrated Scripting Environment (ISE). It makes suggestions based on what you're typing and is very handy if you aren't sure. As you can see in Figure 1-5, PowerShell ISE is correctly suggesting `Get-Member` based on what I started typing. Plus, you can look up commands in the right-side pane. PowerShell ISE is being deprecated; it still exists in PowerShell v5, but it will be removed in PowerShell v6. At the time of this writing, it is still included in Windows Server 2022.

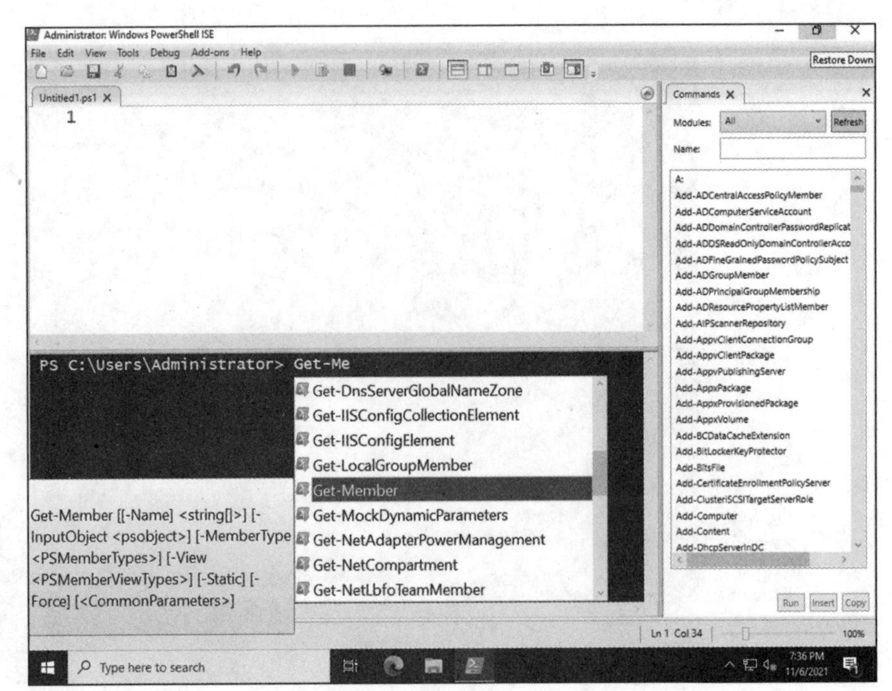

FIGURE 1-5: The PowerShell ISE is a very powerful tool for writing PowerShell scripts.

Text editors

For some system administrators, the simplicity of a simple text editor can be really tempting. Notepad is available by default in the Windows Server operating system. Personally, I love Notepad ++ because it's a simple interface, but it still provides color coding and, with the installation of an extension, you can check the differences between two files. Notepad ++ is an open-source project, and you do have to download and install it (go to `https://notepad-plus-plus.org`).

Visual Studio Code

Visual Studio Code is a code editor that is designed to be a light version of the traditional and more complex products. It's optimized for quick code development and available on Windows, Linux, and macOS, free of charge. It features many of the useful features that you're used to if you've used PowerShell ISE in the past, but it adds some new features that make it a better organizational tool. It can complete commands with IntelliSense and has a much more user-friendly method of browsing PowerShell commands when you install the PowerShell extension.

For system administrators who have some familiarity with Visual Studio, or who want integration with GitHub or other code repositories, Visual Studio Code is a great option. It brings support for Git commands out of the box and can be customized with other extensions past the core PowerShell extension. Going forward, it's what Microsoft recommends using for work in PowerShell, given that the PowerShell ISE is being deprecated.

You can download Visual Studio Code at `https://code.visualstudio.com`. Installation and configuration are relatively simple:

1. **Download the Visual Studio Code installer from** `https://code.visualstudio.com`.
2. **Navigate to your Downloads folder and double-click the VS Code executable to begin installation.**
3. **On the Setup – Visual Studio Code screen, click Next.**
4. **On the License Agreement screen, select the I Accept the Agreement radio button, and then click Next.**
5. **On the Select Destination Location screen, select where you want the software to install to, and then click Next.**
6. **On the Select Start Menu Folder screen, choose a folder in the Start menu if you want the application to be stored somewhere other than the default, and click Next.**

   You can also opt to not create a Start menu folder. I'll accept the defaults.
7. **On the Select Additional Tasks screen, select the Create a Desktop Icon check box, the Add "Open with Code" Action to Windows Explorer File Context Menu check box, and the Register Code as an Editor for Supported File Types check box; leave the Add to PATH check box selected (see Figure 1-6).**
8. **On the Ready to Install screen, click Install.**
9. **On the Completing the Visual Studio Code Setup Wizard screen, leave Launch Visual Studio Code check box selected, and click Finish.**
10. **After VS Code has launched, click Extensions on the left side menu.**

    It looks like a small box being added to three other boxes.
11. **Type** PowerShell **into the search box, and click Install on the extension for PowerShell, as shown in Figure 1-7.**

    In the figure, you can see PowerShell is the first in the list. Make sure you select the one developed by Microsoft.

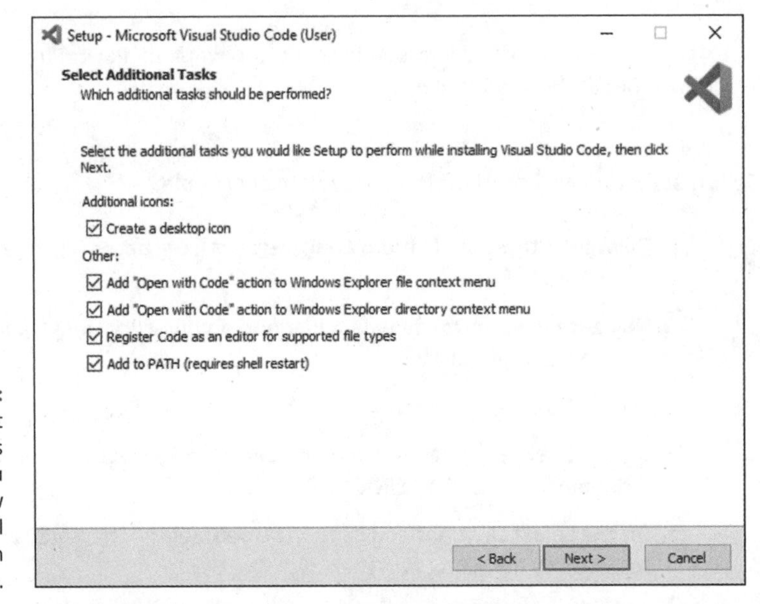

FIGURE 1-6: The Select Additional Tasks screen allows you to customize how and when you'll interact with VS Code.

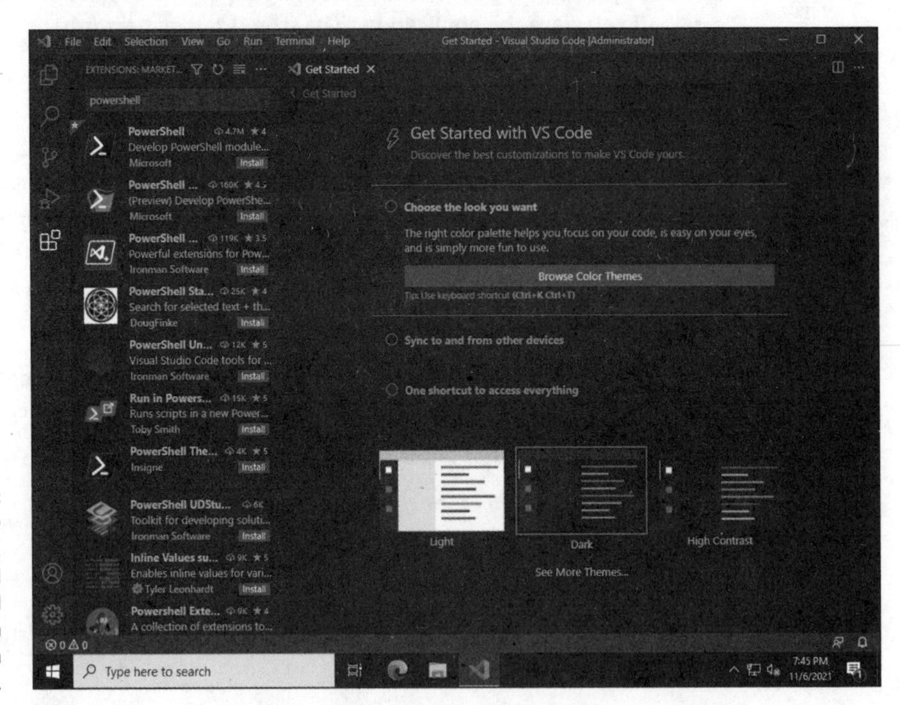

FIGURE 1-7: After VS Code is installed, you can install the PowerShell extension from the Extension Marketplace.

With the PowerShell extension installed in VS Code, you can open the PowerShell Command Explorer (see Figure 1-8). This is helpful when looking up commands. To open it, click Help and then click Show All Commands. Type **PowerShell** into the search box and select PowerShell Command Explorer. If you click the question mark next to a cmdlet, your browser opens to the online help page for that cmdlet. Clicking the pencil icon inserts the cmdlet into the coding window.

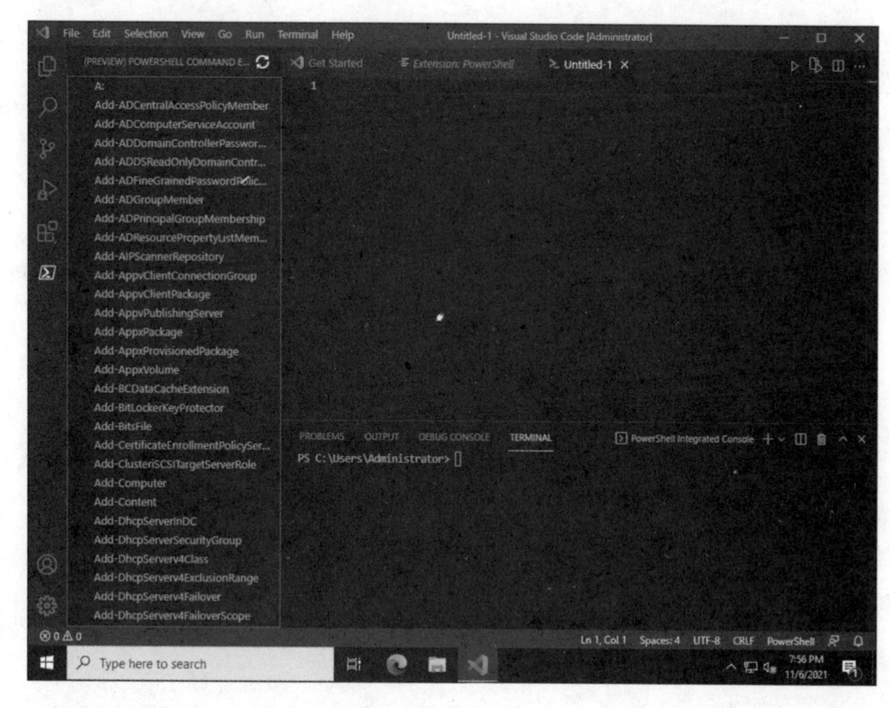

FIGURE 1-8: The PowerShell Command Explorer windows shows you PowerShell cmdlets and gives easy access to help files.

Working with objects

One of the most common things you do in PowerShell is work with objects. In fact, it would be extremely difficult not to! In the following sections, I cover some of the more common objects you use in PowerShell, and some examples of how to work with them.

Properties and methods

Properties and methods are used along with cmdlets to refine what you want the cmdlet to do. Properties are used to view the data that applies to an object, such as checking to see if an object is read-only, or to verify that data regarding an object exists. Properties are prefixed with a dash like this:

- `Get-Command -version`: Returns the version numbers on everything returned by the `Get-Command` cmdlet
- `Get-Command -verb Get`: Returns aliases, cmdlets, and functions that use the verb `Get`

Methods are different from properties. You call them by putting a period (.) before the method name. Methods are typically used to specify some kind of action you want to take on an object. Consider the `Replace` method in the following example:

```
'This is a great For Dummies book!'.Replace('great','super')
```

Here, you're using the `Replace` method to change the word *great* to the word *super*.

Variables

As I mention earlier, variables are used to store values. They can store commands, values, and strings. You'll use variables extensively if you do any amount of scripting because they make things simpler. You can call a variable instead of having to type a long command, for example.

Arrays

An *array* is a type of variable. When you create the array variable, you assign multiple values to the same variable. For instance:

```
$Alphabet = "A,B,C,D,E,F"
```

This code creates an array variable named `$alphabet`, which contains the letters *A*, *B*, *C*, *D*, *E*, and *F*.

To read back the contents of the array, you do the same thing that you would do to display the contents of a variable. You call the variable in the PowerShell window — in this case, `$Alphabet`. You get a display similar to Figure 1-9 if you use my example.

Arrays are a very useful type of variable for system administration work. For example, when you want to store a list of usernames or computer names that you exported from Active Directory so that you can run commands against each of them, an array is a perfect fit. I haven't covered loops yet, but you can run a command against each entry in an array with a loop, instead of having to run a command against an individual system.

Working with the pipeline

A *pipeline* is essentially a group of commands that are connected with a pipe to form a pipeline. The first command sends its output to the input of the second command, and the second command sends its output to the input of the third command.

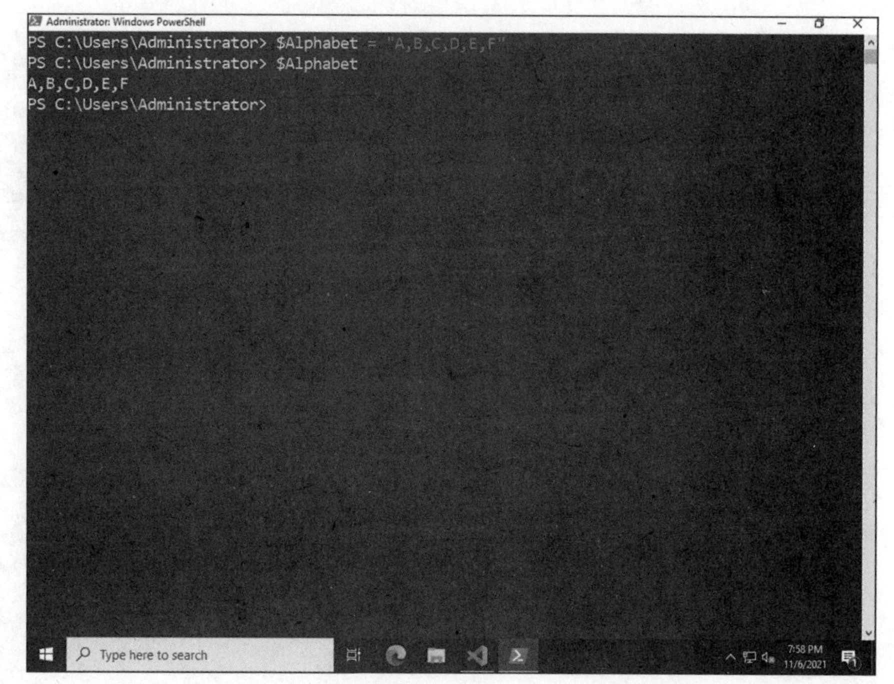

FIGURE 1-9: Creating an array variable and displaying an array variable are very similar to what you used earlier to work with variables.

The pipeline can be very handy for passing commands and/or data to another command so that you can act on it. Say, for instance, the calculator is running and you hate that pesky calculator. You can run the following command to end it.

```
Get-Process win32calc | Stop-Process
```

Sure you could've hit the red X to close the calculator, but what fun would that be? In the real world, you would use this command to stop a process that isn't responding. You could use a similar version with `Start-Process` or other useful cmdlets, and run them against one or more servers to simplify remote administration tasks.

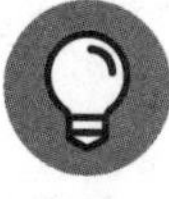

TIP

The $_ variable represents whatever objects happens to be in the pipeline at that point in time. If you need to filter on two or more properties, this represents a really nice shorthand method to do this. Consider the following example:

```
Get-Process | Where {$_.CPU -gt 10}
Get-Process | Where {$_.CPU -gt 10 -AND $_.Handles -gt 700}
```

In this case, the $_ is passing the output of the `Get-Process` cmdlet and is allowing you to access the properties of the `Get-Process` command to run the comparison operators against. You can see the actual output of these two commands in Figure 1-10.

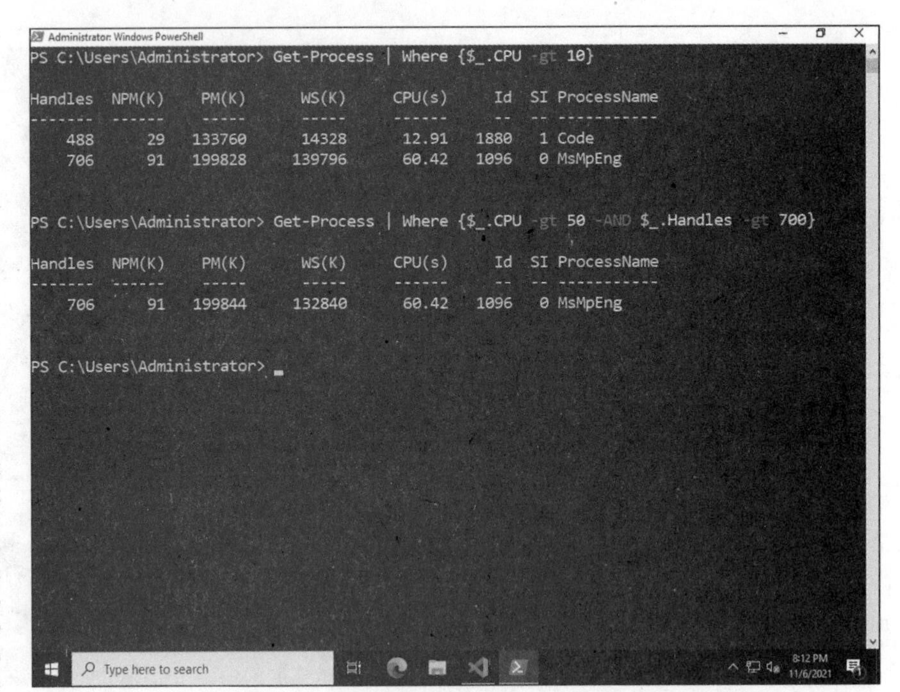

FIGURE 1-10: Using a pipeline with $_ to pass on the output of the first cmdlet, which allows you to use the cmdlet's properties to filter using the comparison operators.

TIP

So, when should you use the pipeline? I believe that you should use the pipeline anytime you need the output of one command to be passed to the next command. This can result in some very long pipelines, but using the pipeline is simpler than breaking each cmdlet out on its own.

Working with modules

PowerShell is very powerful out of the box, but you'll run into scenarios when it can't accomplish what you want it to do. For example, PowerShell out of the box does not know how to work with Active Directory objects. To tell it how to interact with Active Directory, you import the ActiveDirectory module. At its simplest, a module is just a package that contains cmdlets, providers, variables, and functions.

Browsing available modules

Modules are pretty cool, right? Your next question may be: How do I find out what modules are available to me? I'm glad you asked. The command is simple:

```
Get-Module -ListAvailable
```

If you run the command by itself, you'll get a list of the modules that are available currently within your session. By adding the `-ListAvailable` parameter, you can get a list of modules that are installed on your computer and available for use. This command will take some time to return results, but when it does, you'll have a nice list of all the PowerShell modules that are available to you on your system. I've included an example of the output in Figure 1-11. Please note that your output may look different because you'll have additional modules listed based on the roles and features installed on your system.

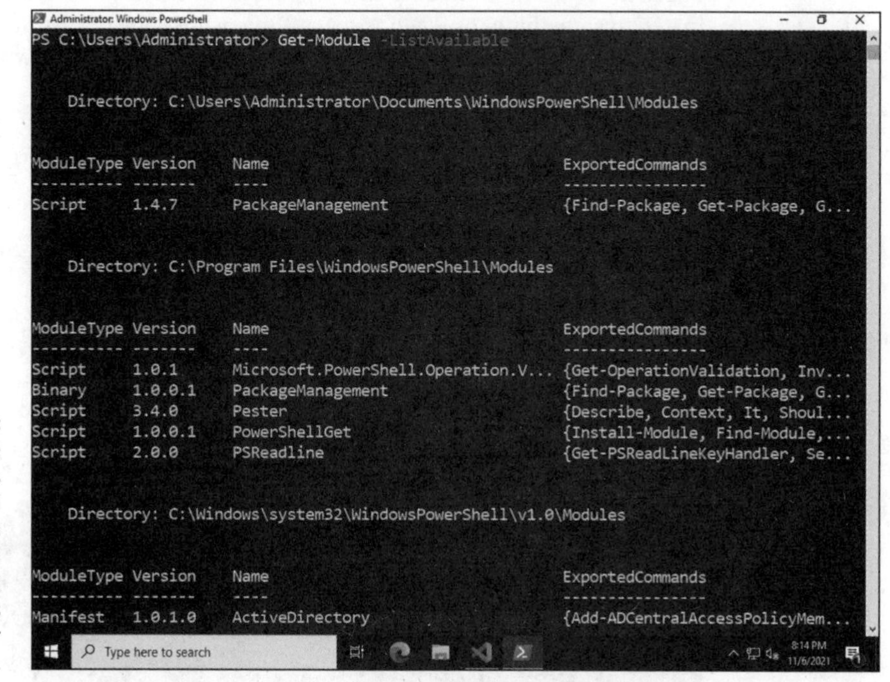

FIGURE 1-11: Viewing the available modules in PowerShell gives you an idea of how powerful it can be, and how many tools you have at your disposal.

Browsing the properties of a module

`Get-Member` is the simplest method to find out what properties are available to you with modules. The following line of code will give you a printout of all the properties associated with `Get-Module`. The last part of the command cuts the results down to just the name column because you aren't interested in the other columns for this purpose.

```
Get-Module | Get-Member -MemberType Property | Format-Table Name
```

Working with comparison operators

Comparison operators are very useful when you need to see if two objects match or don't match. For instance, you may use a comparison operator to find all the services on your system that are disabled. In this section, I cover the most common comparison operators.

-eq and -ne

Equal to (-eq) and not equal to (-ne) are used when you need to find an exact match to something, or when you want to ensure your results do not match. The response you get back is a true/false response. Consider the following code sample:

```
$num = 2
$othernum = 3
$num -eq $othernum
```

When the last line runs, it will return a false because 2 does not equal 3. Let's run not equal to (-ne) now, which returns a true:

```
$num -ne $othernum
```

You can see these little snippets in Figure 1-12 with their outputs.

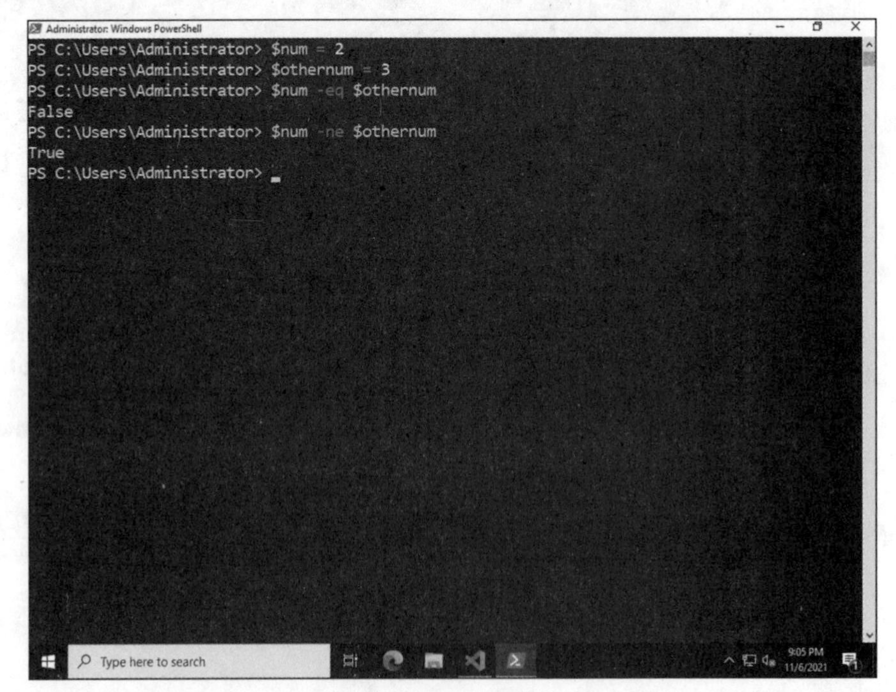

FIGURE 1-12: If you're trying to determine whether one object matches another, a simple equal to (-eq) may be your best bet.

-gt and -lt

Greater than (-gt) and less than (-lt) are also comparison operators that return a true/false response. You can play with a really simple version of this by setting a variable to a value and then testing it. For example:

```
$x = 4
$x -gt 8
```

The preceding code will check to see if the value of x (in this case 4) is greater than 8. It is not, so the response will be false.

-and and -or

-and and -or also return true/false responses based on the conditions that they're given. -and, for example, returns a true if both statements it is fed are true. -or returns a true if one of the statements it is given is true.

For example, the following statement will be true if $a is less than $b, and $b is less than 50:

```
($a -lt $b) -and ($b -lt 50)
```

This equation is similar but will be true if $a is less than $b, or if $b is less than 50:

```
($a -lt $b) -or ($b -lt 50)
```

Getting information out of PowerShell

It's all fun and games to write equations and have PowerShell put the answer right on the screen, but the truth of the matter is that you're usually going to want PowerShell to output the information in a different format. A common use case is exporting information on systems out of Active Directory to a CSV, so that you can filter on properties of each system like the operating system version, and whether service packs are installed. Or you may want it to write the result on the screen, but with some kind of text to give the information context. In the following sections, I examine a few of the ways to get PowerShell to output text.

Write-Host

Write-Host is used to write things onscreen. This command is very helpful when you want to give the result of something some context.

For example, you can set the value of a variable, in this case, $x to 2:

```
$x = 2
```

Then you can use Write-Host to print a sentence and the output of the variable:

```
Write-Host "The value of x is:" $x
```

This will end up printing out, "The value of x is: 2."

Write-Output

Write-Output prints the value of a variable to a screen, much like simply typing the variable does. The main reason to use this particular command is that you can tell it not to enumerate data. If you're working with arrays, this can be helpful because the array will be passed down the pipeline as a single object rather than multiple objects.

An example from Microsoft's web documentation for PowerShell showcases this perfectly. In both instances, you create an array with three values inside of it. If you measure the array, you get a count returned of 3, which makes sense. When you add -NoEnumerate, you get a count of 1. The three values are still there, but the array is simply being treated as one object rather than a collection of three, shown in Figure 1-13.

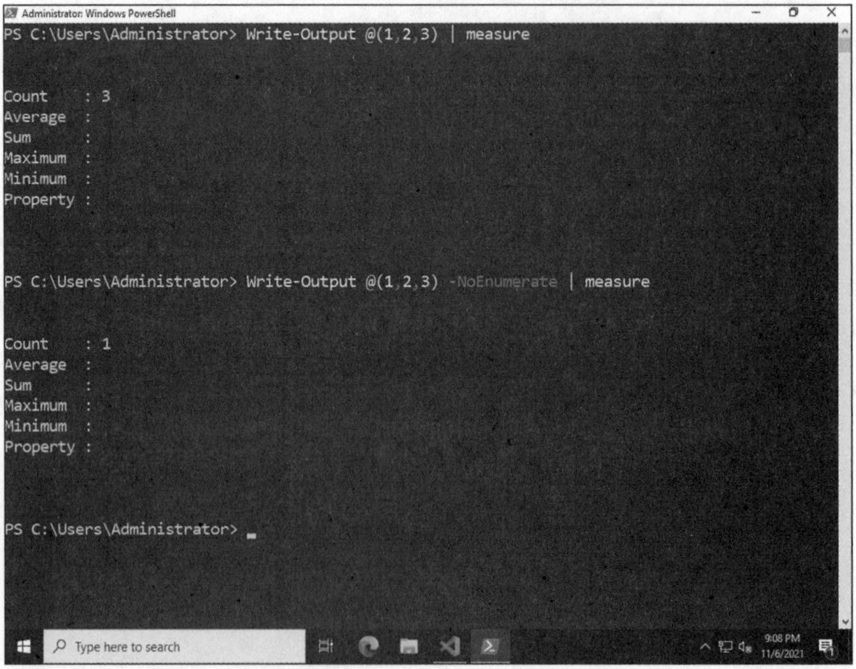

Administrator: Windows PowerShell
PS C:\Users\Administrator> Write-Output @(1,2,3) | measure
Count : 3
Average :
Sum :
Maximum :
Minimum :
Property :
PS C:\Users\Administrator> Write-Output @(1,2,3) -NoEnumerate | measure
Count : 1
Average :
Sum :
Maximum :
Minimum :
Property :
PS C:\Users\Administrator>
Type here to search
9:08 PM
11/6/2021

FIGURE 1-13: Adding -NoEnumerate tells PowerShell to treat arrays differently than normal.

Out-File

`Out-File` is another method to export data from PowerShell into a file outside of PowerShell. This is especially helpful when you need to compile the data from a script, for instance, for analysis later on.

To get a list of processes running on a system so that you can dump them to file, you could do something like this:

```
Get-Process | Out-File -filepath C:\PSTemp\processes.txt
```

Scripting logic

Loops can make a PowerShell script even more powerful. For instance, say you want to run a command against an exported list of computers from Active Directory. You can use a loop to enumerate through the imported CSV file and run the command on each individual entry.

If

The `If` statement tests a condition to see if it's true. If it is true, it will execute the code in the block. If it isn't true, it will check the next condition or execute the instructions in the final block. Here is a silly example:

```
$server = 'Windows'
If ($server -eq 'Linux') {
Write-Host 'This is a Linux server.'
}
ElseIf ($server -eq 'Solaris') {
Write-Host 'This is a Solaris system.'
}
ElseIf ($server -eq 'Windows') {
Write-Host 'This is a Windows system.'
}
Else {
Write-Host "Don't know what kind of system this is."
}
```

In this instance, because `$server` was initialized as Windows, the first two blocks will be skipped, but the third block will execute. The last block is ignored because it was not reached.

ForEach-Object

Using `ForEach-Object` allows you to enumerate multiple objects that have been passed through the pipeline or even imported from arrays and CSV files, just to name a few. Let's say you want to gather the names of all the processes running on your system, but you don't want any of the other information that goes along with it. You could use `ForEach-Object` to accomplish this:

```
Get-Process | ForEach-Object {$_.ProcessName}
```

While

`While` loops are known as pretest loops because the code is not executed if the condition set for the loop is not true. Basically, the code will be executed until the expression it's evaluating becomes false. These can be very useful when you need to increment or decrement a counter. Look at the following example:

```
$myint = 1
DO
{
"Starting loop number $myint"
$myint
$myint++
"Now my integer is $myint"
} While ($myint -le 5)
```

You may be thinking, "Okay, it will count from 1 to 5." But that's not quite true. On the fifth loop, the `While` loop sees that 5 is less than or equal to 5 so it will allow it to continue. Only when it's presented with 6 will it stop execution because 6 is not less than or equal to 5. So it will count from 1 through 6. Give it a try!

Other cool tricks

There are some other neat things you can do with PowerShell, and I would be remiss in my duties as an author if I didn't add them to this chapter. Read on for more.

Exporting and importing CSV files

Exporting a CSV is very handy when you have output that contains multiple columns. This is my go-to command when I'm doing an export from Active Directory because it keeps everything neat and organized.

Here's how to export the processes on a system, similar to what you did with `Out-File` to get to a text file:

```
Get-Process | Export-Csv -Path C:\PSTemp\processes.csv
```

Importing a CSV can be the answer to a system administrator's prayers when you need to work with a lot of data and you don't want to type it in manually. Expanding on the example earlier where you exported a process list, you can import that same list and then format it nicely. Formatting is discussed in "Formatting your output" later in this chapter.

```
$Procs = Import-Csv -Path c:\PSTemp\processes.csv
```

TIP

You only need to specify `-Path` if you want to save output or import output from somewhere other than the directory you're currently in.

Exporting HTML/XML

This one isn't exactly a true export. You're actually converting the output of a command to HTML and then using `Out-File` to write it to that file. Consider the following example:

```
Get-Process | ConvertTo-Html | Out-File c:\PSTemp\processes.html
```

You can see this example in Figure 1-14, where I've used the `Invoke-Item` command to actually open my created HTML file that contains the output of the `Get-Process` command.

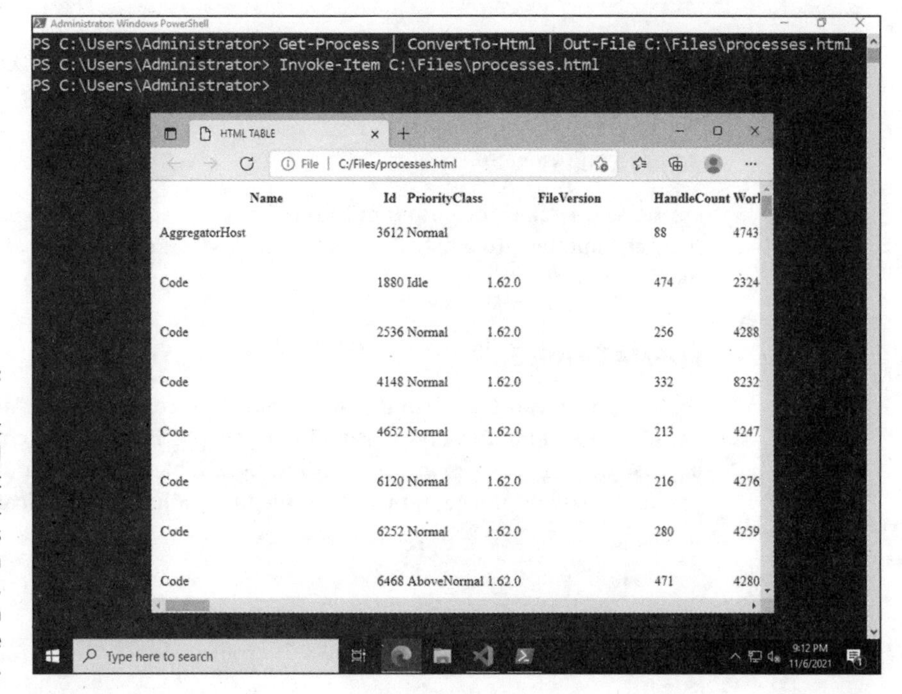

FIGURE 1-14: The `ConvertTo-Html` cmdlet can be used to convert the output of a previous command to an HTML format, which can then be written to file with `Out-File`.

Sorting through objects

Telling PowerShell how you want it to sort objects can be very useful. For example, you may want to sort on memory usage so you can see which processes are using the most resources. The `Sort-Object` cmdlet will sort processes in the order of least CPU-intensive to most CPU-intensive:

```
Get-Process | Sort-Object -Property CPU
```

Filtering through objects

Examples like sorting can be very useful, but it's highly unlikely that you would want to look through all the results for CPU. You probably just want to know what the most resource-intensive processes are. The `Select-Object` cmdlet will return the last five results from that CPU listing, which is a much more manageable and useful list if you're troubleshooting issues.

```
Get-Process | Sort-Object -Property CPU | Select-Object -Last 5
```

Formatting your output

You've probably noticed from playing in PowerShell that sometimes the output isn't as readable as it could be. So let's look at some of the ways to format output to be a little prettier.

Run the `Get-Process` cmdlet to get a baseline for appearance. Figure 1-15 gives you an idea of what the cmdlet's output looks like.

FORMAT-LIST

`Format-List` takes the output of the cmdlet or whatever code is before it and formats the output into a list. The output of `Get-Process | Format-List` will look like Figure 1-16.

FORMAT-TABLE

`Format-Table` can take data in a list format and convert it into a table. With the `Get-Process` cmdlet I've been using in this chapter, it wouldn't change the output because `Get-Process` is already outputting a table. You can add the `-AutoSize` parameter so that the column sizes in the table adjust automatically.

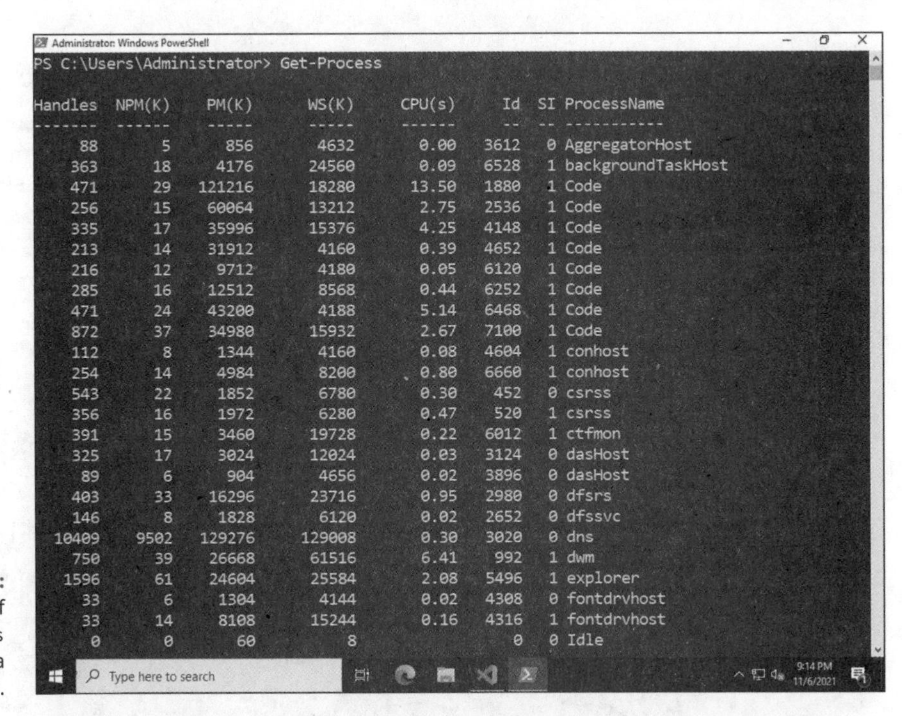

FIGURE 1-15: The output of Get-Process is normally in a table format.

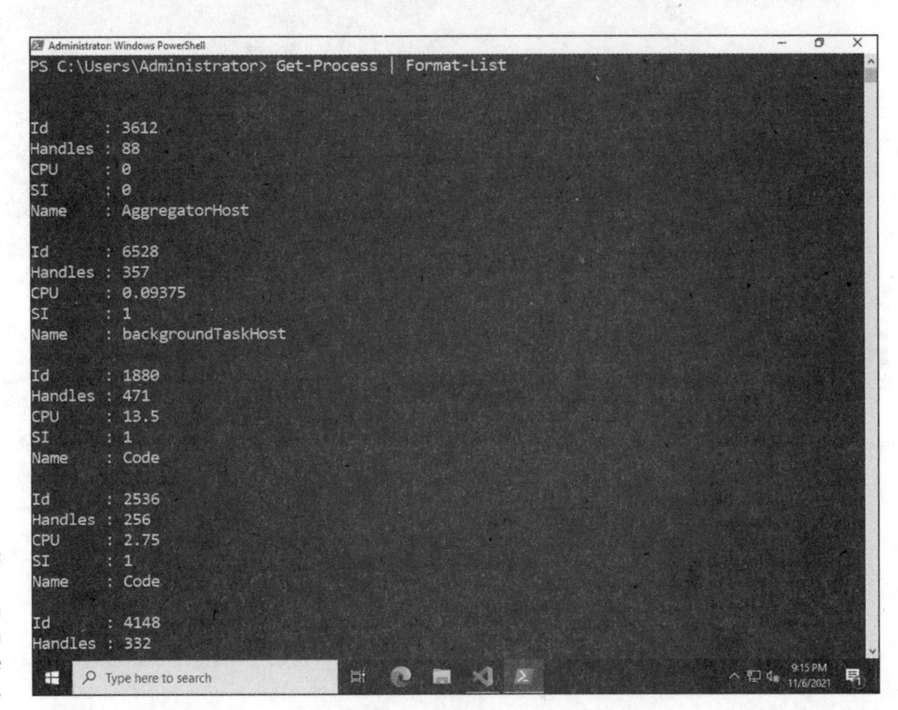

FIGURE 1-16: Get-Process formatted into a list rather than the usual table format.

FORMAT-WIDE

`Format-Wide` will display data in a table format, but it will only show one property of the data that it's presented. You can specify how many columns, and you can specify which property you want it to display. The example in Figure 1-17 is the output of `Get-Process` split into three columns. I haven't specified a property name, but you can see that it's using the Process Name as the property.

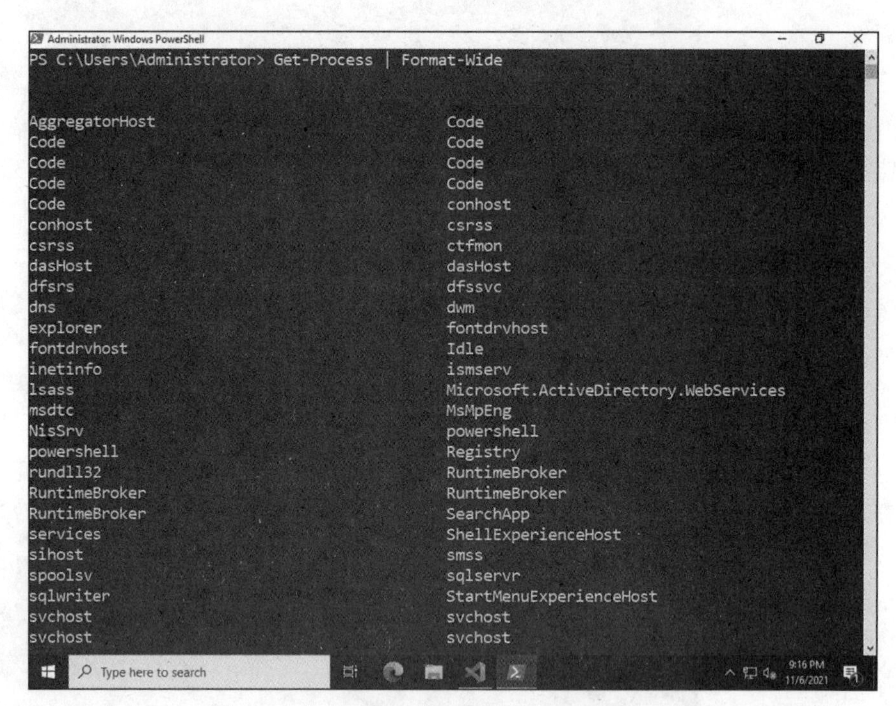

FIGURE 1-17: You can use `Format-Wide` to nicely format and present data in table formats.

Running PowerShell Remotely

Running PowerShell locally can be great for automating work and simplifying administrative work. But the true strength in PowerShell comes from the fact that you can also run it remotely against other systems on your network.

Invoke-Command

`Invoke-Command` can be used locally or remotely. Because the focus of this section is on the remote usage of the cmdlet, that's where I'll focus. Say I wanted to run my favorite Get-Process on a system named Server3. The entire thing would look something like this:

```
Invoke-Command -ComputerName Server3 -Credential domain\username
  -ScriptBlock {Get-Process}
```

New-PSSession

Running New-PSSession allows you to establish a lasting, persistent connection to a remote system on your network. If you want to run command through your PSSession, you use the Invoke-Command cmdlet (see the preceding section). To open the new connection, type the following:

```
New-PSSession -ComputerName Server3
```

Enter-PSSession

Enter-PSSession allows you open an interactive session to a remote computer. The prompt changes to indicate that you're connected to the remote system. In Figure 1-18, you can see the changed prompt, as well as the commands that I ran against the remote server.

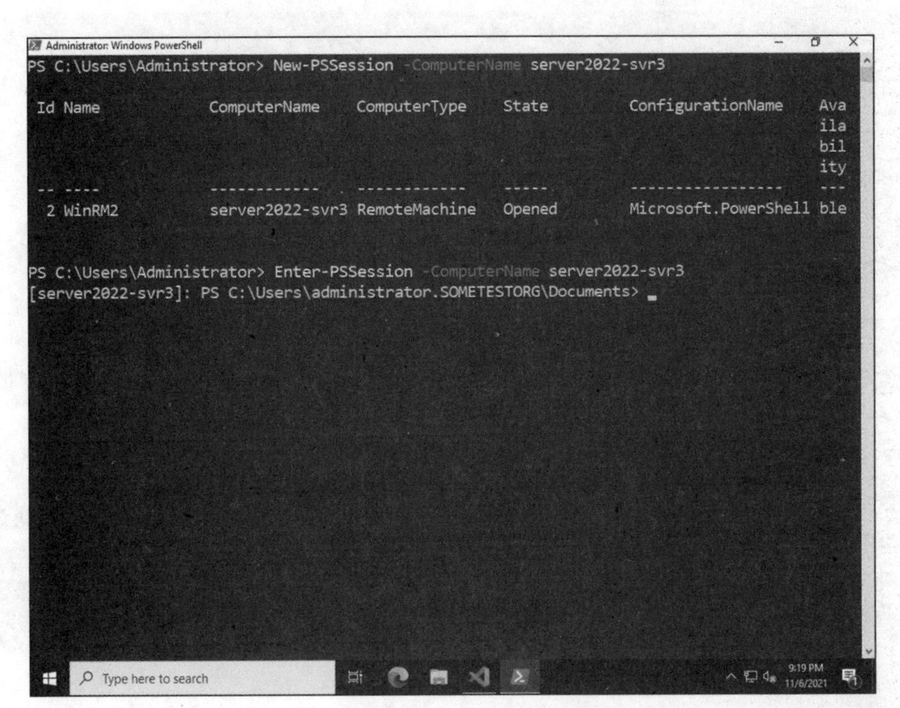

FIGURE 1-18: Interacting with a remote server is intuitive after connecting to it with Enter-PSSession.

This is extremely useful if you're running headless systems like Server Core in your environment. To exit the interactive session, you can type **Exit-PSSession** or simply type **exit**.

Getting Help in PowerShell

You can do so many things with PowerShell that it would be impossible to memorize all of them. This is where the built-in help comes in very useful. The help pages give you a description of what a command can do, along with examples and additional parameters you can use with the command.

Update-Help

One of the first things I like to do with a new system is run the `Update-Help` cmdlet. This cmdlet pulls down the help articles available at the moment you issue the command. It can take a bit of time, so run it when you aren't needing to use it right away. In Figure 1-19, you can see what the update process looks like. Each module displays its progress. This process does require an Internet connection.

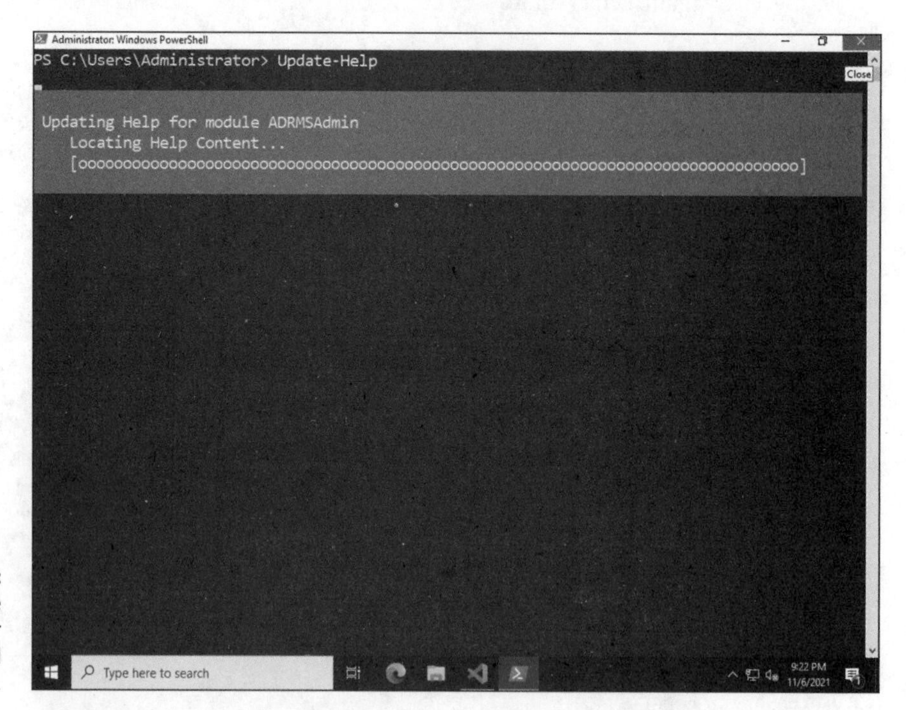

FIGURE 1-19: Updating the help pages for your PowerShell modules.

REMEMBER

The most recent versions of the help articles are going to be on Microsoft's PowerShell Reference pages (`https://docs.microsoft.com/en-us/powershell/scripting`), not by download on your system.

Get-Help

The `Get-Help` cmdlet is like the Swiss Army knife of cmdlets. If you don't know what you're looking for, it can be very helpful. Here are some options:

- `Get-Help Get-Process`: Running this command will return information about the `Get-Process` command. This is most useful if you need to learn more about a cmdlet, or if you need to learn more about how to interact with a cmdlet.
- `Get-Help process`: This command is nice when you don't know the exact name of the cmdlet you're looking for. It will search for help topics that contain the word process and will display them to you afterward.

-Detailed and -Full

Help articles have different levels of detail that they can go into. Normally, the help article will show basic syntax to give you an idea of how to use the cmdlet. To look at more levels of detail, you can use the following:

- `-Detailed`: The `-Detailed` parameter will display descriptions of parameters and examples of how to use them with the cmdlet.
- `-Full`: The `-Full` parameter will truly give you all the information available regarding a cmdlet, including descriptions of parameters, examples of how to use the cmdlet, type of input/out objects, and any additional notes that are in the help file.

For example, you could type `Get-Help Get-Process -Full` to get all f the details available from the help files for the `Get-Process` cmdlet.

Identifying Security Issues with PowerShell

Given the power of PowerShell, you need to be able to secure it properly. There are several things you can do to ensure that only proper and authorized PowerShell scripts are able to run on your network.

Execution Policy

The Execution Policy allows you to define what kind of scripts are allowed to run within your network. You can set the Execution Policy through Group Policy or through the following PowerShell cmdlet:

```
Set-ExecutionPolicy -ExecutionPolicy <policy>
```

There are several policy types that can be put in place of `<policy>` in the preceding example:

- `Restricted`: This is the default policy if no other policy is specified. It prevents PowerShell scripts from running and will not load configuration files.
- `AllSigned`: For a script or configuration file to run, it must be signed by a trusted certificate. I cover how to do this in the next section.
- `RemoteSigned`: This requires that any script that is downloaded from the Internet be signed by a trusted certificate. Scripts created locally do not have to be signed to run.
- `Unrestricted`: This allows you to run all scripts and load all configuration files. You're prompted for permission before a script is run. I don't recommend using this setting.

WARNING

- `Bypass`: This is very similar to Unrestricted, except it doesn't even prompt for permission to run a script. I caution you to never use this setting.
- `Undefined`: This removes whatever Execution Policy is currently set, unless that Execution Policy is being set through Group Policy.

Code signing

To use an Execution Policy that is more secure, a PowerShell script needs to be signed. By signing a PowerShell script, you're validating that it came from a trusted source and that it has not been altered since it was released. If you're using `RemoteSigned`, then you only need to worry about signatures on scripts downloaded from the Internet. However, if your security people have gone wild and it's set to `AllSigned`, you need to sign your PowerShell scripts before you can run them. In the following sections, I show you the steps involved in doing that.

Creating a Code Signing Certificate

To be able to sign a PowerShell script, you need a Code Signing Certificate (CSC). If you're publishing certificates for use on the Internet, you can purchase CSCs from some of the large public certificate authorities like GoDaddy and DigiCert.

If you're creating certificates for internal usage, then you can create a CSC using your internal certificate authority. That's the workflow you can walk through here:

1. **Click Start and then type** CertMgr.msc **and press Enter.**
2. **Right-click Personal, then select All Tasks, then Request New Certificate.**
3. **On the Before You Begin screen, click Next.**
4. **On the Select Certificate Enrollment Policy screen, choose Active Directory Enrollment Policy, and then click Next.**
5. **Select the CSC template, and then click Enroll.**

   On the Certificate Installation Results screen, you should see Succeeded.
6. **Click Finish.**

Importing the certificate into the Trusted Publishers Certificate Store

If you're only going to run scripts on your local system, you can get away with manually exporting the certificate from your Personal Store and adding it to your Trusted Publishers Certificate Store. In an Enterprise situation, you'll want to use Group Policy to push your certificate to the Trusted Publishers Certificate Store on any system you'll run the script from. Follow these steps to do add your code signing certificate to your Trusted Publishers Certificate Store:

1. **Click Start, type** CertMgr.msc**, and press Enter.**
2. **Select Personal, right-click your CSC, and choose All Tasks ⇨ Export.**
3. **On the Welcome to the Certificate Export Wizard, click Next.**
4. **On the Export Private Key screen, leave the selection on No, and click Next.**
5. **Leave the file format on the default .DER and click Next.**
6. **On the File Export screen, click Browse.**
7. **Select a location, name your cert, and click Next.**

   I'll name mine Demo CSC and save it to my Desktop.
8. **On the Completing the Certificate Export Wizard, click Finish.**

   You get a pop-up that says "The export was successful."
9. **Click OK.**
10. **Navigate to where you saved the certificate, and double-click it.**

11. **Click the Install Certificate button.**

12. **On the Welcome to the Certificate Import Wizard screen, select Local Machine and click Next.**

13. **On the Certificate Store screen, select Place All Certificates in the Following Store, and click Browse.**

14. **Choose Trusted Publishers and click OK; then click Next.**

15. **On the Completing the Certificate Import Wizard screen, click Finish.**

    You'll get a pop-up that says "The import was successful."

16. **Click OK.**

Signing your script

After you've created the certificate, and you've imported it into the Trusted Publishers Certificate Store, you can sign your certificate, and your system will trust it. So let's sign the simple `Do While` script that I created earlier.

```
Set-AuthenticodeSignature c:\DoWhile.ps1 @(Get-ChildItem cert:
   \CurrentUser\My -codesign)[0]
```

This signs the script with the certificate that is in my Personal Store. After the script is signed, I can run it. I'm not prompted and it runs without issue. If you look at Figure 1-20, you can see what the script looks like with the signature added to it.

In the example in Figure 1-21, I show you the whole thing from start to finish. I set the Execution Policy to `AllSigned`. Then I try to run an unsigned script. You can see I get an ugly error message saying that the script is not digitally signed. I run the PowerShell cmdlet to sign my script, and then I run the script again and it executes successfully. Pretty cool, right?

Firewall requirements for PowerShell remoting

PowerShell remoting relies on the Windows Remote Management (WinRM) service. WinRM creates two listeners, one for HTTP and one for HTTPS. To allow remote PowerShell commands to work, you need to have ports 5985 and 5986 open. Port 5985 provides HTTP support, and Port 5986 provides HTTPS support.

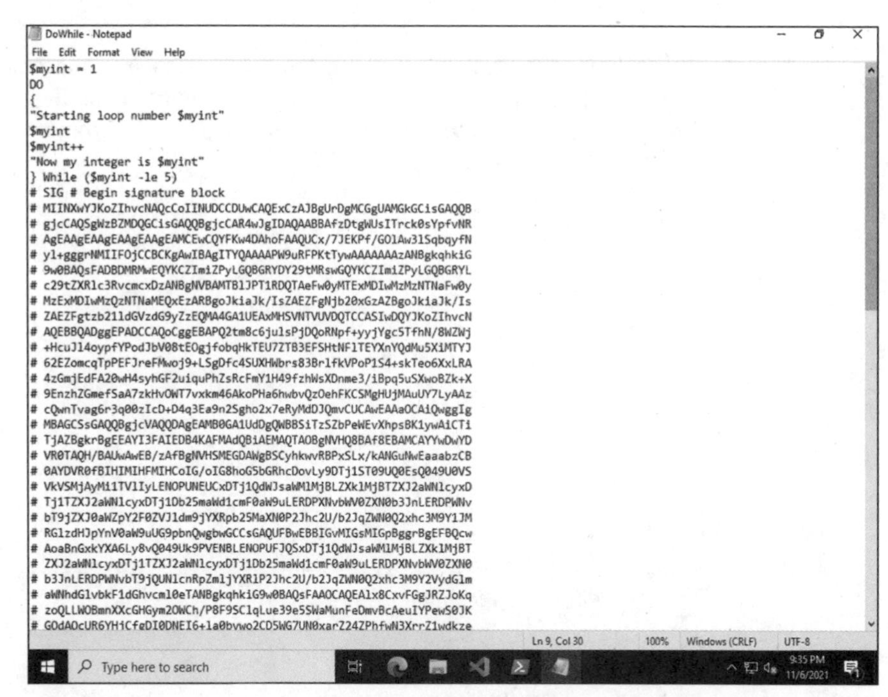

FIGURE 1-20: Signing a script allows it to run even in a restrictive environment that requires all PowerShell scripts to be signed.

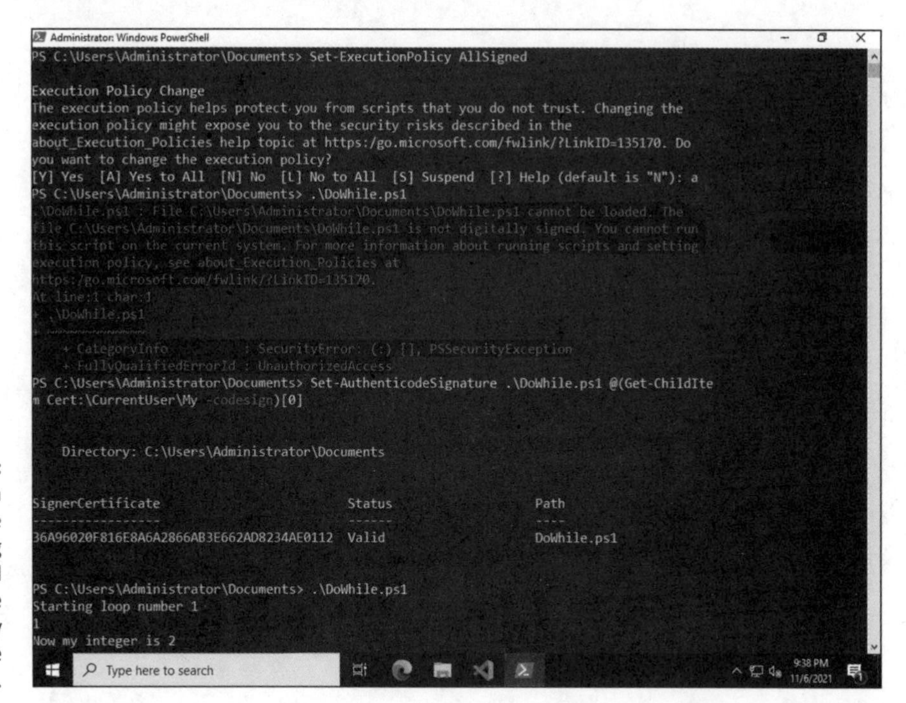

FIGURE 1-21: A demonstration of what you see when running an unsigned script, signing the script, and finally executing the signed script.

On an individual system, running the `Enable-PSRemoting` cmdlet does all the work needed to allow for remote PowerShell to work, including enabling the local firewall rules necessary. This cmdlet needs to be issued from an elevated (use Run as Administrator) PowerShell window.

Framework versions

» Paying attention to what's new in .NET 4.8

» Viewing the contents of the Global Assembly Cache

» Examining .NET Standard and .NET Core

Chapter 2 Understanding the .NET Framework

Many people wonder why they need to understand the .NET Framework and, more important, why it's bundled with PowerShell. Using .NET, not only can you do things in PowerShell that you wouldn't be able to do natively, but you also can create functions and modules with the .NET code that you can reuse.

In this chapter, I explain the basics of .NET and how to interact with it. I don't cover .NET programming — that's a whole other book's worth of material.

Introducing the Various Versions of .NET Framework

Before I jump into .NET versions, I want to make sure that you understand what the .NET framework actually is. A framework allows a programmer to call code instead of having to write the code each time the programmer wants the functionality. The .NET framework gives developers the code they need to write .NET applications without having to custom develop every single little piece of code themselves. .NET is integrated with PowerShell, so you can call the same snippets of .NET code that developers can from within PowerShell, either in the console or in a script. For instance, Figure 2-1 shows a piece of .NET code called from PowerShell that displays processes on the system. This piece of .NET code is actually used for the PowerShell Get-Process cmdlet.

```
Administrator: Windows PowerShell
PS C:\Users\Administrator> [diagnostics.process]::GetProcesses()

Handles  NPM(K)    PM(K)      WS(K)     CPU(s)     Id  SI ProcessName
-------  ------    -----      -----     ------     --  -- -----------
    823      36    25020      60236       0.38    984   1 dwm
    371      14     3188      14876       0.14   2952   1 ctfmon
    385      16    11216      21024       4.70   2304   0 svchost
    129       8     1480       6664       0.02   1960   0 svchost
    217      16     9948      20000       0.08   2944   0 inetinfo
    577      27    13592      54260       0.36   5700   1 StartMenuExperienceHost
    265      14     2492       8164       0.05   2008   0 svchost
     89       6      908       4648       0.00   3924   0 dasHost
    208      16     2448      11196       0.05   3808   0 vds
    227      11     2140      12584       0.14   2936   1 RuntimeBroker
    916      83    40992      57376      14.16   4708   0 svchost
    298      16     3860      16356       0.05   5280   0 svchost
    235      11     2236      11364       0.08    568   1 winlogon
    146       8     1840       6104       0.00   3076   0 dfssvc
    433      32     8084      17532       0.50   1156   0 svchost
  10411   10150   129472     129260       0.41   2928   0 dns
    881      91   216096     153432      56.63   2336   0 MsMpEng
    540      23     9768      42600       0.13   5740   1 TextInputHost
     33       6     1536       4728       0.05   4260   1 fontdrvhost
    303      16     5492      21576       0.30   6072   1 RuntimeBroker
    116       8     1248       7032       0.02   4504   0 svchost
    397      18     5040      15568       0.42   1536   0 svchost
    187      12     2080      12780       0.02   1732   0 svchost
    138       7     1328       6072       0.05   2800   0 svchost
    126       8     1312       6464       0.02   1336   0 svchost
Type here to search                                    10:31 AM 11/7/2021
```

FIGURE 2-1: The .NET Framework expands the functionality of PowerShell greatly with code that can be called on the console or via script.

Each new version of the .NET Framework adds new functionality and fixes old problems. The .NET Framework follows a similar cadence to most products in that it has major and minor versions. The major releases tend to focus heavily on new features, while the minor versions add features and fix issues found in the previous releases.

As of the time of this writing, version 4.8 is the current major version and current minor version. The 4.8 version is what is installed by default on Windows Server 2022. It's very common to have multiple versions of the .NET Framework installed on the same system; there is usually no issue with them co-existing.

TIP

The version of .NET you're currently on is stored in the Windows Registry. To locate the release number, you can use the following command:

```
(Get-ItemProperty 'HKLM:\SOFTWARE\Microsoft\NET Framework Setup\
   NDP\v4\Full' -Name Release).Release
```

This command returns a number. Table 2-1 lists the minimum number for each major version. When I run this command on my system, for example, I get 528449. According to Table 2-1, that means that I have version 4.8, which is correct because the minimum value for 4.8 is 528040.

TABLE 2-1 .NET Versions with Release Values

Version	Minimum Value
.NET Framework 4.5	378389
.NET Framework 4.5.1	378675
.NET Framework 4.5.2	379893
.NET Framework 4.6	393295
.NET Framework 4.6.1	394254
.NET Framework 4.6.2	394802
.NET Framework 4.7	460798
.NET Framework 4.7.1	461308
.NET Framework 4.7.2	461808
.NET Framework 4.8	528040

Source: https://docs.microsoft.com/en-us/dotnet/framework/migration-guide/how-to-determine-which-versions-are-installed#ps_a

TIP

Most of the time, you have one version of .NET installed, and that's it. Sometimes, though, you may have a legacy application that needs an older version of .NET, and a newer application that requires a newer version of .NET. The great thing about .NET is that you can run more than one version side by side. For instance, your legacy application may need .NET 3.5, but your newer application needs .NET 4.5. You can install both versions of .NET on your system, and each application will be able to run with the version of .NET that it needs.

.NET is available in the server operating system as a feature. To install .NET, follow these steps:

1. **In Server Manager, choose Manage ➪ Add Roles and Features.**
2. **On the Before You Begin screen, click Next.**

3. **On the Select Installation Type screen, choose Role-Based or Feature-Based Installation, and then click Next.**
4. **On the Select Destination Server screen, click Next.**
5. **On the Select Server Roles screen, click Next.**
6. **On the Select Features screen, select either .NET Framework 3.5 Features (which installs .NET Framework 2.0, 3.0, and 3.5) or .NET Framework 4.8, and then click Next.**
7. **On the Confirm Installation screen, click Install.**

Focusing on New Features in .NET 4.8

.NET version 4.8 introduced new features in a few key areas:

- **Base Classes:** Cryptography improvements were made when using FIPS mode. You would previously get an error if you used one of the managed versions of a cryptographic provider class that had not been certified for FIPS 140-2.

  Zlib, which is used for compression, has been updated as well to include some issue fixes.

- **Windows Communication Foundation (WCF):** A health endpoint can be enabled to allow access to service health information via WCF. This is referred to as ServiceHealthBehavior, and it can be enabled through code or through a configuration file.
- **Windows Presentation Foundation (WPF):** WPF is a UI-based framework that aids in the creation of desktop applications. In this updated version of WPF, support has been added for high dots per inch (DPI) configurations. This allows applications that need high DPI to be scaled properly so that they're crisp and clear to the end user.
- **Common Language Runtime:** Updates were made to the Just-In-Time (JIT) compiler. It's now based on the .NET Core 2.1 version of the JIT compiler.

  If you're wondering why they're using something so old, well, they aren't. .NET has historically been specific to Windows, and .NET Core is an open-source software framework that allows you to create .NET applications on Windows, Linux, and macOS.

  Some improvements were also made to how memory is managed with the Native Image Generator (NGEN) and malware scanning capabilities, which ensure that all assemblies are scanned, not just those loaded from disk.

Viewing the Global Assembly Cache

Before I dive into the Global Assembly Cache (GAC), you may be wondering what it is. The GAC is responsible for storing assemblies that are shared by multiple applications on a computer. The *assembly,* at its most basic definition, is an executable of some kind. It contains all the code that will be run and serves as the boundary for the application. Many assemblies are installed when .NET is installed on your system.

.NET Framework versions 4 and up store their assemblies in `%windir%\Microsoft.NET\assembly`. The `%windir%` is a placeholder for the Windows directory, which is typically located at `C:\Windows`.

Viewing the assemblies in the GAC is done using a tool called `gacutil.exe`. This tool is a part of the Developer Command Prompt for Visual Studio so you need to install Visual Studio on your system if you want to play with the GAC on your system. Visual Studio IDE Community is the free version and does include the Developer Command Prompt. Figure 2-2 gives you an idea of what the GAC looks like when you view the assemblies. I've typed the following command:

```
gacutil.exe -l
```

This command lists all the assemblies within the GAC.

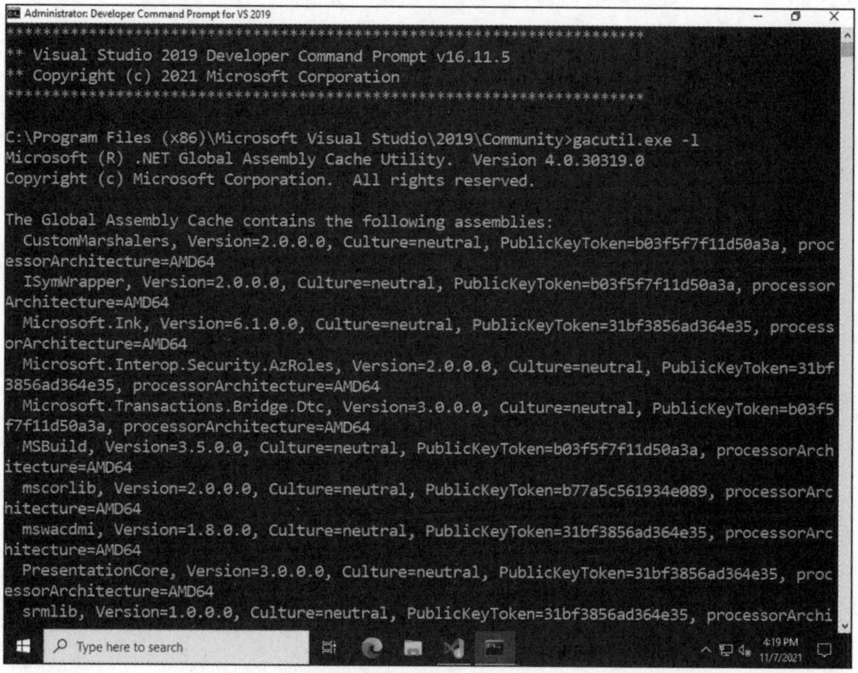

```
Administrator: Developer Command Prompt for VS 2019
**********************************************************************
** Visual Studio 2019 Developer Command Prompt v16.11.5
** Copyright (c) 2021 Microsoft Corporation
**********************************************************************

C:\Program Files (x86)\Microsoft Visual Studio\2019\Community>gacutil.exe -l
Microsoft (R) .NET Global Assembly Cache Utility.  Version 4.0.30319.0
Copyright (c) Microsoft Corporation.  All rights reserved.

The Global Assembly Cache contains the following assemblies:
  CustomMarshalers, Version=2.0.0.0, Culture=neutral, PublicKeyToken=b03f5f7f11d50a3a, proc
essorArchitecture=AMD64
  ISymWrapper, Version=2.0.0.0, Culture=neutral, PublicKeyToken=b03f5f7f11d50a3a, processor
Architecture=AMD64
  Microsoft.Ink, Version=6.1.0.0, Culture=neutral, PublicKeyToken=31bf3856ad364e35, process
orArchitecture=AMD64
  Microsoft.Interop.Security.AzRoles, Version=2.0.0.0, Culture=neutral, PublicKeyToken=31bf
3856ad364e35, processorArchitecture=AMD64
  Microsoft.Transactions.Bridge.Dtc, Version=3.0.0.0, Culture=neutral, PublicKeyToken=b03f5
f7f11d50a3a, processorArchitecture=AMD64
  MSBuild, Version=3.5.0.0, Culture=neutral, PublicKeyToken=b03f5f7f11d50a3a, processorArch
itecture=AMD64
  mscorlib, Version=2.0.0.0, Culture=neutral, PublicKeyToken=b77a5c561934e089, processorArc
hitecture=AMD64
  mswacdmi, Version=1.8.0.0, Culture=neutral, PublicKeyToken=31bf3856ad364e35, processorArc
hitecture=AMD64
  PresentationCore, Version=3.0.0.0, Culture=neutral, PublicKeyToken=31bf3856ad364e35, proc
essorArchitecture=AMD64
  srmlib, Version=1.0.0.0, Culture=neutral, PublicKeyToken=31bf3856ad364e35, processorArchi
Type here to search                                              4:19 PM 11/7/2021
```

FIGURE 2-2: You can view the contents of the GAC with the Developer Command Prompt.

Understanding assembly security

Because the GAC lives in the `Windows` folder, it inherits the permissions of the `Windows` folder. In many cases, you may want to tighten the permissions on the GAC directories so that only administrators can delete assemblies. If someone deletes an assembly that the system or an application relies on to function properly, that application will no longer work.

Identifying the two types of assembly privacy

Two types of assemblies make up the .NET presence on your system:

- **Private:** Private assemblies are deployed with an application and can only be used by that application. Think of them like the children on the playground who won't share.
- **Shared:** Shared assemblies are available to be used by multiple applications on your system. They're stored in the `WinSXS` folder and are installed via Windows Update and Windows Installer packages.

Viewing assembly properties

In older versions of the Windows Server operating system, you could simply right-click an assembly to get all the properties of the file. That function was removed several operating system versions ago. Now if you need to get information on the assembly file, your best bet will be to go through PowerShell. Say that I want to view the version information on the `accessibility.dll` that is in use. This is the command that I would need to run:

```
[Reflection.AssemblyName]::GetAssemblyName('C:\Windows\
  Microsoft.NET\Framework\v4.0.30319\accessibility.dll').Version
```

After I've run this command I'm presented with the major, minor, build, and revision numbers, as shown in Figure 2-3.

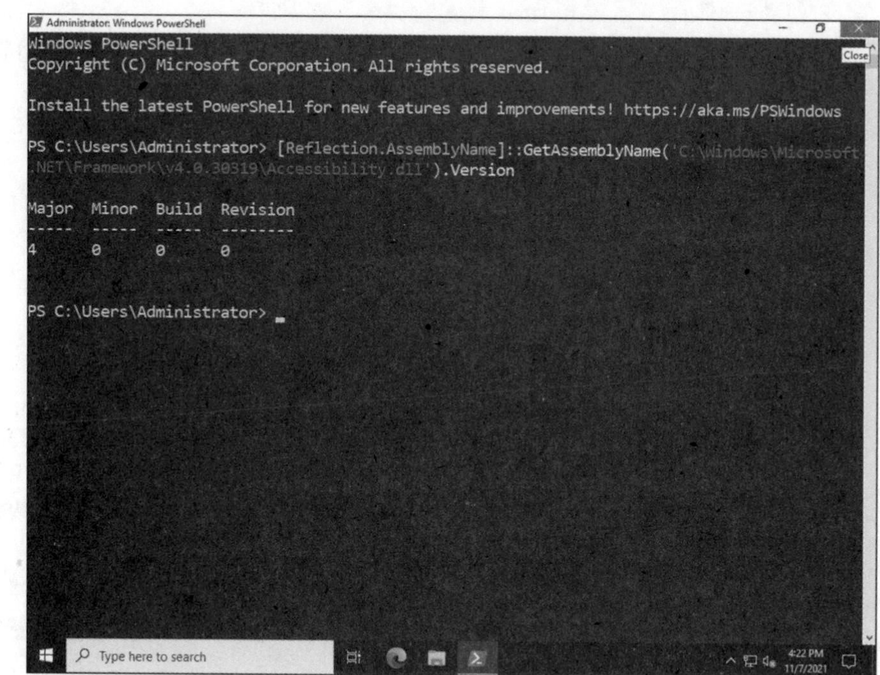

FIGURE 2-3: You need to use PowerShell to view the assembly properties in Windows Server 2022.

Understanding .NET Standard and .NET Core

The .NET Framework has been a staple for many years, but newer frameworks are gaining in popularity.

.NET Core

.NET Core is one of the newest members of the .NET Framework family. It's open source and it can be run on Windows, Linux, and macOS. With .NET Core, you can build applications that are cross-platform. If your application is developed with .NET Core, then only .NET Core applications will be compatible — you won't be able to support Xamarin or the classic .NET Framework. .NET Core is an implementation of the specifications that are set in the .NET Standard.

You may wonder why you would use .NET Core if it isn't compatible with the other runtimes. There are a couple good reasons:

- You can develop .NET Core on Windows, Linux, or macOS.
- If you're coding for a mobile application, .NET Core is optimized for mobile work.

.NET Standard

.NET Standard is a set of APIs that all the .NET frameworks must support. This includes .NET Core, Xamarin, and the classic .NET Framework. It's important to note that .NET Standard is a specification, not a framework. It's used to build libraries that can be used across all your .NET implementations, including the traditional .NET Framework, the newer .NET Core, and Xamarin.

Tying it all together: .NET and PowerShell

PowerShell Core 6.0 uses the newer .NET Core as its runtime. This means that you can now run PowerShell on Windows, Linux, and macOS. PowerShell Core also enables you to take advantage of all the awesome .NET Core APIs in your commands and scripts, which really extends the utility and capabilities of your scripts. You can start working with PowerShell Core without impacting your current installation of PowerShell because both PowerShell and PowerShell Core can be run side-by-side. PowerShell Core is available for download from the PowerShell repo on GitHub at `https://github.com/PowerShell/PowerShell/releases`.

» Running PowerShell scripts or cmdlets

» Working remotely with PowerShell

» Performing administrative tasks with PowerShell scripts

Chapter 3
Working with Scripts and Cmdlets

In Chapter 1, I explain what PowerShell is and fill you in on the basics of working with it. In Chapter 2, I tell you a little bit about .NET and its interaction with PowerShell. This chapter is all about putting that information and your skills to use by starting to build out the scripts that you'll use on a daily basis as a system administrator.

Introducing Common Scripts and Cmdlets

Writing single lines of PowerShell is pretty common when you have administrative tasks that you're trying to perform, but most of the really useful and powerful PowerShell comes from the ability to put multiple cmdlets into a script.

The Microsoft website has some great reference material for PowerShell. What I find the most beneficial is the Reference link off the main PowerShell site (`https://docs.microsoft.com/en-us/powershell/scripting/`). When you click the Reference link, you get all the Help pages that you can request from the PowerShell console but in a nice graphical interface. You can find syntax help and examples of how to use the cmdlets. The default selection is the latest stable version of PowerShell. At the time of this writing, that's version 7.1. However, Windows Server ships with an older version. You can either update to the new version or click the Version drop-down box and choose the older version.

Each cmdlet is grouped by type, and then listed for you to look through or select. Figure 3-1 shows you an example of what the Connect-PSSession page looks like.

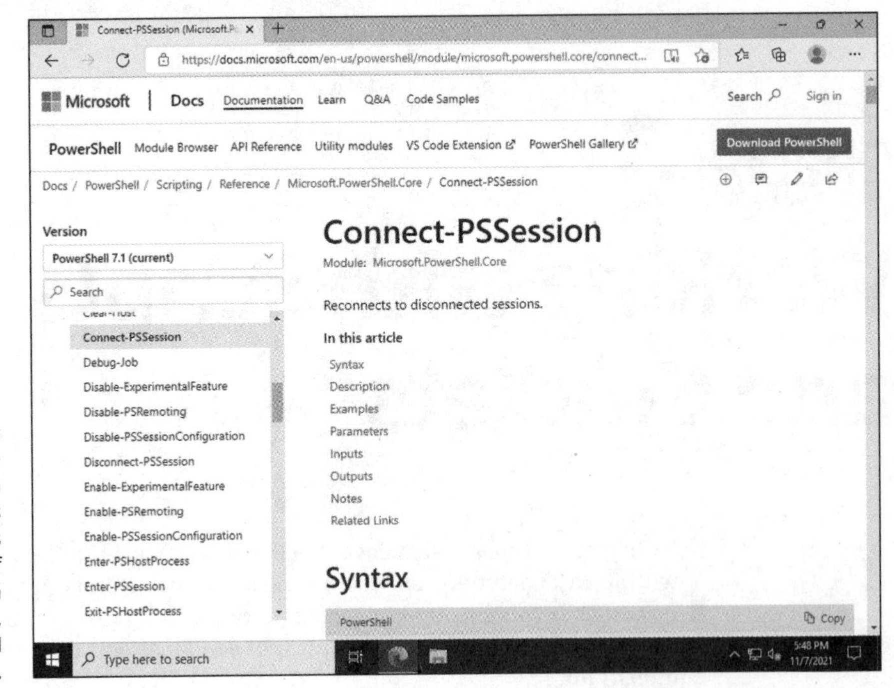

FIGURE 3-1: The Reference page from Microsoft's website contains hundreds of cmdlets with descriptions, syntax, and examples.

If you want to get more examples to work with, the PowerShell Gallery has thousands of example scripts that you can download and start using. What I love about PowerShell Gallery is that you can search for what you want to download. For instance, I've downloaded PowerShell modules written for enterprise-grade firewalls. The vendors only published an API; the community came together and wrote a PowerShell module that allows me to use familiar PowerShell syntax rather than learn the vendor's API.

Executing Scripts or Cmdlets

For the most part, executing a PowerShell script is not that much different from using a Command Prompt. You can right-click it and choose Run with PowerShell, or you can call it from the PowerShell console. Don't forget about the execution policy that was discussed in earlier chapters. If your execution policy is not set to where it will allow the desired script to run, then PowerShell will give you an error when you try to execute the script.

When calling it from the console, you can give it the location to the script:

```
C:\PSTemp\dowhile.ps1
```

Or you can use the `.\` to indicate that you're already in the correct directory:

```
.\dowhile.ps1
```

Working with COM objects

You'll more commonly work with the .NET framework when writing scripts in PowerShell, but you can also work with COM objects. The commands are very similar, but there are syntax differences in how you identify COM objects. This example creates a COM object that represents Internet Explorer, and sets the visible property to `$True` so that you can see the Internet Explorer window. It sets the URL to the `www.dummies.com` website.

```
$IE = New-Object -COMObject InternetExplorer.Application
   -Property @{Navigate2="www.dummies.com"; Visible = $True}
```

If you're wondering what properties you have to work with, you can find them as you would with a regular cmdlet. The previous example, for instance, could be typed as follows, to find out the properties that are available:

```
$IE = New-Object -COMObject InternetExplorer.Application
$IE | Get-Member
```

Figure 3-2 shows you an example of what the display from `Get-Member` would be. You can see `Navigate2` shows up as one of the methods, and if you scrolled down you would find `Visible` toward the bottom of the list of properties.

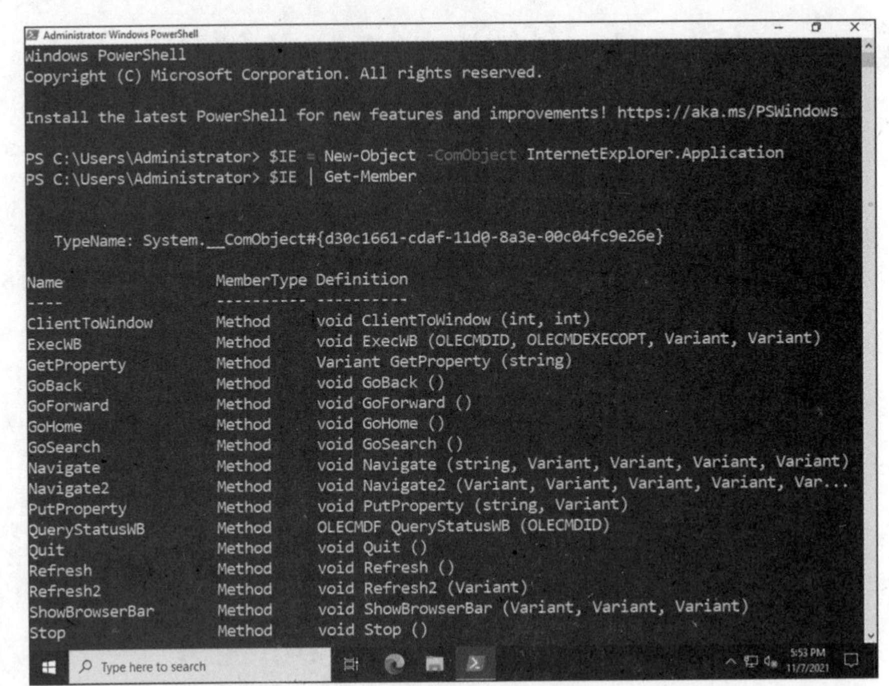

FIGURE 3-2: Checking the properties of COM objects is similar to checking the properties of cmdlets, using `Get-Member`.

Combining multiple cmdlets

Using one cmdlet at a time is not an efficient use of your time, and let's face it, every system administrator wants to be able to save some time.

You've learned about the pipeline and about variables. All these things will enable you to be a PowerShell ninja, one idea at a time.

For instance, you could write this line of code in this manner:

```
$cmd = Get-Command
$cmd | Get-Member
```

Or you could simply combine the two from the beginning like this:

```
Get-Command | Get Member
```

Combining multiple cmdlets instead of running them one at a time, and storing their values into variables, works much better and far more efficiently. In Chapter 4 of this minibook, I show you how to take this to the next level and save your commands into a file that you can write anytime. PowerShell scripts are absolutely wonderful. They're huge time savers — and, if designed correctly, you can use them for years to automate processes and batch jobs.

Working from Another Location

You may need to work with data remotely from time to time without the requirement of logging in to the remote system to do that. There are several methods to accomplish this task. You may also want to be able to get information through the remote console.

To list the items in a share, similar to running the `dir` or `ls` commands:

```
Get-ChildItem \\servername\sharename
```

To check the health of your file shares:

```
Get-FileShare -FileServer (Get-StorageFileServer -FriendlyName
   "servername")
```

Working with Server Message Block (SMB), which is a file-sharing protocol for use over a network, is simple as well. Let's look at a few commands that will allow you to work with SMB shares on Windows file servers. These run locally so you will need to create a session with the system you want to run this against.

`Get-SmbShare` returns a list of all the SMB shares on the local system. This can be helpful when you need to go a step further and list the properties of a share. For example, say I want to see the properties of a share named `MyData`. I can type something like the following:

```
Get-SmbShare -Name "MyData" | Format-List -Property *
```

As you can see in Figure 3-3, you can get some valuable information on your share with that simple command. You can find out the physical path, whether the data is encrypted, whether shadow copies are turned on, and so on.

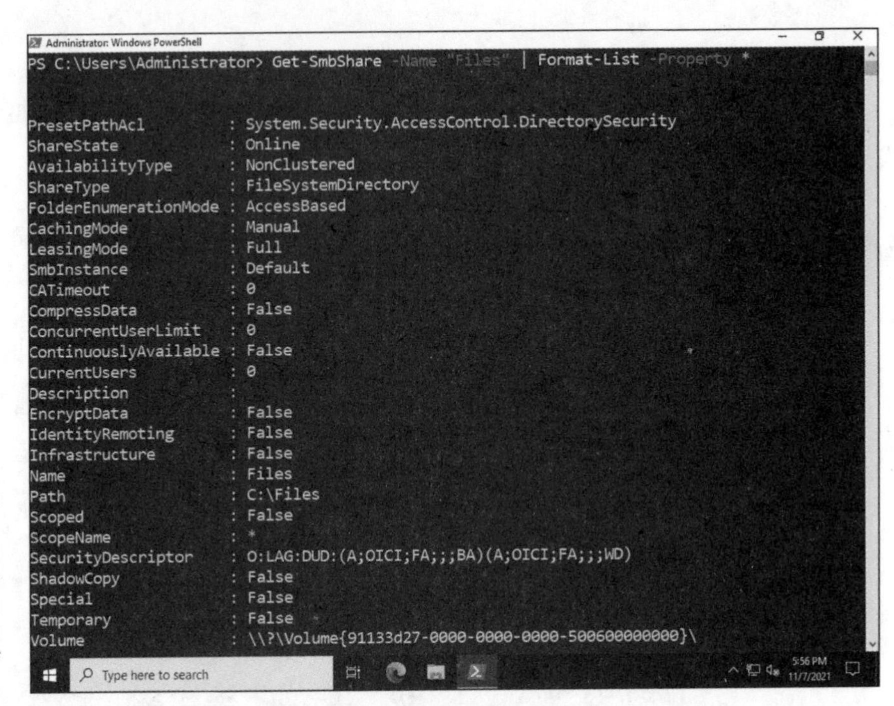

FIGURE 3-3: `Get-SmbShare` is a very powerful cmdlet that gives you the ability to collect a great deal of information.

Performing Simple Administrative Tasks with PowerShell Scripts

In this section, I show you some cool things that you can do with PowerShell scripts that might help you right now.

REMEMBER

Keep in mind that for some of these scripts you may need to install a module to get it to work properly. If you run the Active Directory scripts on your domain controller, you won't need to add the AD module. If you're running from a system without Active Directory (like your desktop, for example), you need to import the Active Directory module with the command `Import-Module ActiveDirectory`. If you have the Remote Server Administration Tools (RSAT) installed on your desktop, the modules will be able to automatically load when needed.

Adding users in Active Directory

Adding users in Active Directory is a pretty common task. You can do this from the graphical user interface (GUI), but this method can be much faster. This example creates a user named George Smith, adds him to the Sales OU, prompts for the password (which is stored securely), enables the account, and ensures that the user will need to change his password after he has logged in.

```
New-ADUser -Name " George Smith" -GivenName "George" -Surname
   "Smith" -SamAccountName "gsmith" -UserPrincipalName "gsmith@
   sometestorg.com" -Path "OU=Sales,DC=sometestorg,DC=com"
   -AccountPassword(Read-Host -AsSecureString "Input Password")
   -Enabled $true -ChangePasswordAtLogon $true
```

As you can see in Figure 3-4, the user account is created and placed into the OU that I specified. The account is enabled and will be ready for the user on his first day.

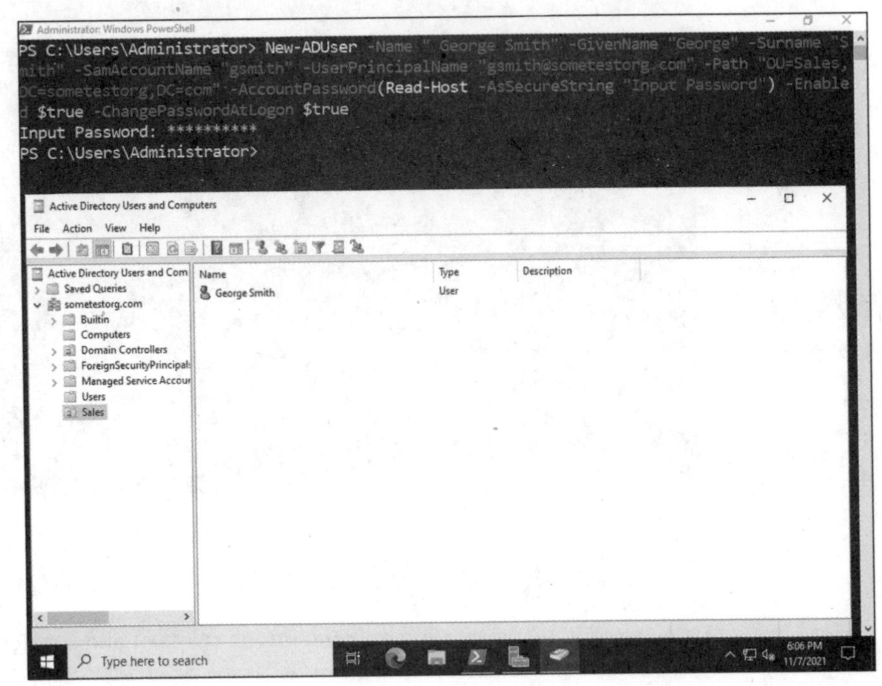

FIGURE 3-4: Creating a user in PowerShell is quick, and the user shows up almost instantaneously, as shown in the Active Directory Users and Computers window.

Creating a CSV file and populating it with data from Active Directory

Pulling data from Active Directory is an important skill for a system administrator to have. The following code queries Active Directory for Server operating systems. You can further define your filter to return specific versions of the server OS as well. Notice in Figure 3-5 that the information from onscreen appears in the CSV. There are no service packs out yet, so the OperatingSystemServicePack field is empty.

```
Get-ADComputer -Filter "OperatingSystem -like '*Server*'"
   -Properties OperatingSystem,OperatingSystemServicePack |
   Select Name,Op* | Export-CSV c:\PSTemp\ServerOSList.csv
```

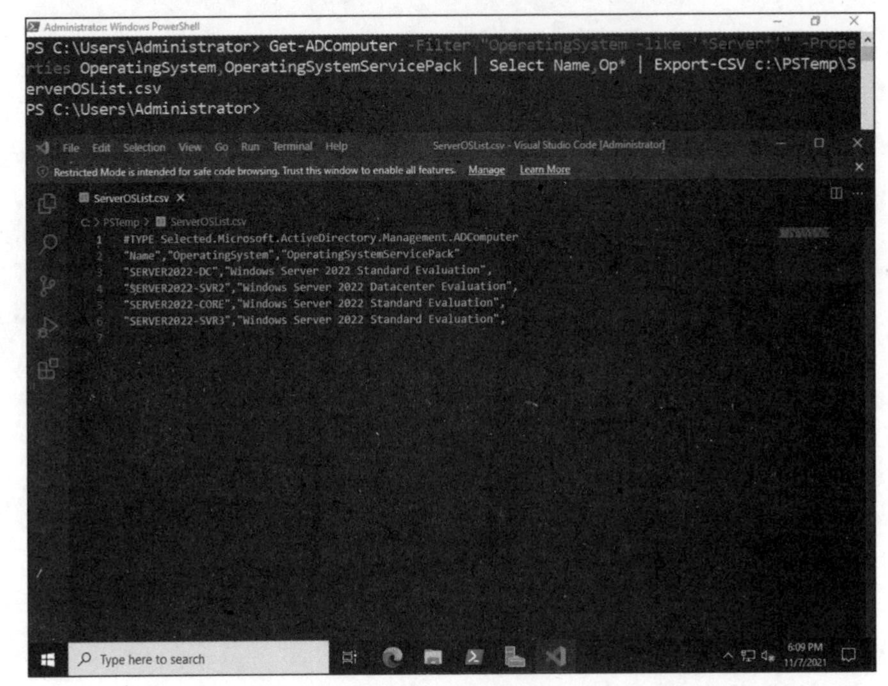

FIGURE 3-5: Running your query and having the information on the screen can be nice, but it's even better when it's in a file you can work with later.

Checking to see if a patch is installed

It never fails — when a big vulnerability is released and a patch is announced, your boss is almost always going to ask, "Is the patch installed?" If you work for a larger organization, you may have purchased a tool that can manage this for you. If you work for a smaller org, though, or if you want to spot-check the accuracy of your tools, PowerShell does give you a simple way to check to see if a patch is installed. You can use this with `Invoke-Command`, which is great if you want to check it against a list of all your systems, for example.

```
Get-Hotfix -Id "KBXXXXXX" -ComputerName <servername>
```

Replace the `Xs` with the numbers in the actual Knowledge Base (KB), and the *`<servername>`* with the remote server's hostname. If you're running this command locally, you can omit the `-ComputerName` parameter altogether.

Checking running processes or services

Last but certainly not least are the administrative cmdlets to check for running processes and services.

Get-Process will return all the active processes by default. You can run it by itself, and you receive a table-formatted output with the active processes.

Get-Service is a little different in that it's totally normal to have services that are not running during the day. If you want to see only the running processes, you need to type a command similar to the following:

```
Get-Service | Where-Object {$_.Status -eq "Running"}
```

The $_.Status is a property of Get-Service. Using Where-Object in front of it will filter your output so that you only see running services.

What do you do if you want to check the services on multiple systems? You select multiple system names like this:

```
Get-Service -Name "WinRM" -ComputerName Server2022-DC,
   "Server2022-dc2", "Server2022-dc3" | Format-Table -Property
   MachineName, Status, Name, DisplayName -auto
```

This command will check for WinRM specifically on the three servers that I have included after the -ComputerName parameter.

PowerShell scripts

» Setting an execution policy for your PowerShell script

» Signing a PowerShell script with a Code Signing Certificate

» Creating a PowerShell advanced function with VS Code

Chapter 4

Creating Your Own Scripts and Advanced Functions

Tons of scripts are already purpose made and ready for download, but nothing is quite as satisfying as writing your own scripts and cmdlets. That feeling of accomplishment really can't be beat when the script you've been working on is put to use.

With Infrastructure as Code (IaC) gaining in popularity, being able to write custom scripts helps to make you a more marketable employee, not just from a system administration standpoint, but also from a DevOps standpoint. IaC allows you to script the deployment of a server so that you can respond to the need to scale much more quickly than if you had to manually build a server.

In this chapter, I show you how to create your own custom components of PowerShell. You discover how to create new shell extensions and your own PowerShell scripts, and you find out about creating your own cmdlets.

Creating a PowerShell Script

You've got this monotonous task you have to do every single day. This task regularly take you an hour to complete. You want to reclaim your hour. What do you do? You build a PowerShell script, of course!

PowerShell is probably one of my favorite scripting languages. The main reason is that when you have a grasp of the syntax and how to look up the properties of various cmdlets, you're really only limited by your imagination.

Before I move on to working with scripts and functions, let's get a few definitions out of the way:

- **Script:** A script is a series of commands that are run in order, dependent on whether you have conditional operations occurring within your script. Scripts often contain cmdlets, loops, and other elements that, when run together, accomplish some task. Scripts are the easiest method to automate repetitive work.
- **Advanced function:** An advanced function allows you to create and do the same things you can do with cmdlets, but without having to learn .NET Framework languages like C#, and without having to compile your code. You do have to follow the naming rules for PowerShell commands when creating a function — it should consist of the same verb/noun syntax as the usual PowerShell cmdlets.

Creating a simple script

When creating your first script, start with a small goal in mind and then build on it from there. If you try to do something really large and detailed, you may run into issues and get frustrated. Look at examples on the Internet to see how others have solved for the same issues, and experiment.

PowerShell scripts are always saved with a filetype of PS1. This tells your system that the file is a PowerShell script file and will suggest PowerShell or your favorite text editor to edit it or will open PowerShell to execute it.

If you ever find yourself in need of samples or you want to download pieces of code so that you're crafting code from scratch, I highly recommend PowerShell Gallery (`www.powershellgallery.com`). It's maintained by Microsoft and has modules, scripts, and even some examples of PowerShell DSC, which I talk about in

Chapter 5 of this minibook. Figure 4-1 shows you the PowerShell Gallery homepage. At the time of this writing, you can see that there are 4,260 unique scripts and modules available from PowerShell Gallery. What I really like is that you can see how many downloads of each resource have been made, and any questions or comments that have been made to the authors of the scripts or modules.

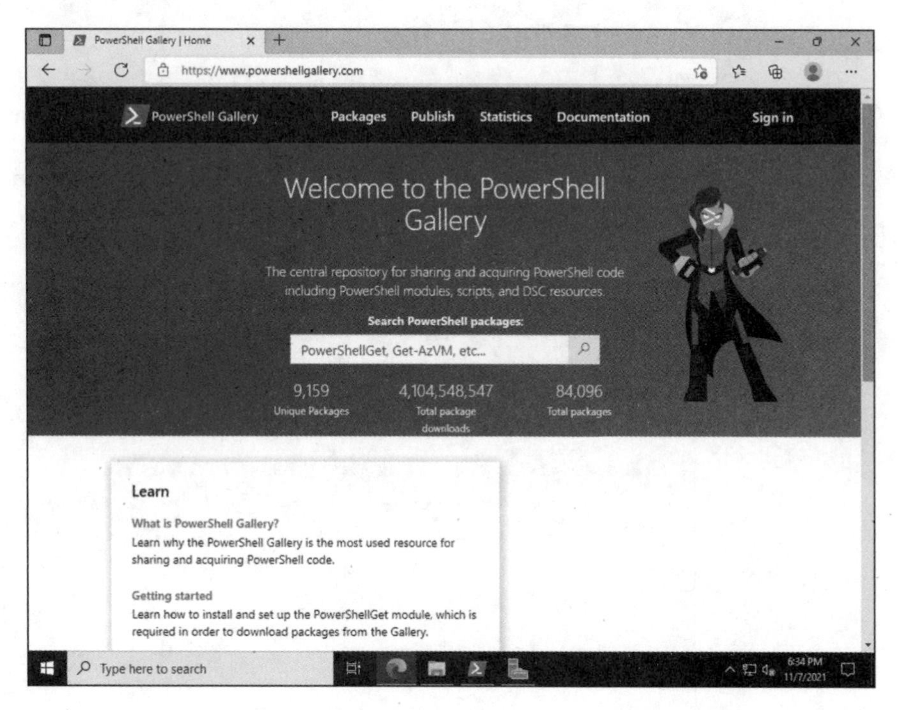

FIGURE 4-1: PowerShell Gallery is an excellent resource to download or see examples of various scripts or modules.

Before I get into an example of a script, I want to add that you should always use comments when you're writing a script. You can denote a comment in PowerShell by starting the line with a hash (#). Everything after the hash on that line is ignored and is treated like a comment. Comments are useful for you because they can form an outline of what you're trying to accomplish. They can serve as documentation for you later and can assist your co-workers in either using the script or helping you write it.

In Chapter 3 of this minibook, I show you an example of how to create a new user in Active Directory using PowerShell instead of using the graphical Active Directory Users and Computers. The real power of that little bit of code becomes apparent when it's put into a script that can take a CSV file (from HR perhaps) and import it into PowerShell, and then have PowerShell loop through each row and create each user.

In the following sections, I step through this script step by step. Then I show you the whole thing all together. Never fear! The script is on the GitHub repo created for this book, and you can download it from there, rather than having to retype it. The GitHub repository created for this book is located at `https://github.com/sara-perrott/Server2022PowerShell`.

Creating the CSV file

The CSV file is the most important piece of this whole exercise because it provides the input to the script. The column names are assigned in the first part of the script to variables so that they can be called when you get to the loop that processes each row in the CSV. Figure 4-2 shows you a sample of what the CSV should look like. After the CSV is complete, you can move on to the script.

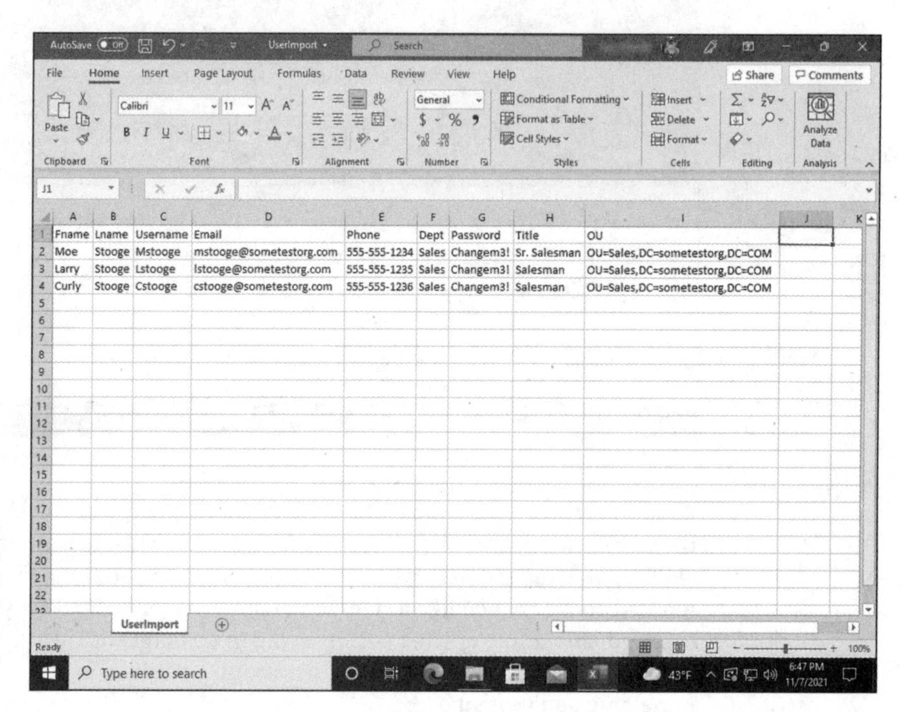

FIGURE 4-2: Using CSV files to import data sets for scripts is a simple way to deal with multiple inputs.

TIP

As a working security professional, I have to point out that storing passwords in plaintext CSV files is not a good thing to do. I suggest setting complex temporary passwords and destroying the CSV when it's complete. You can, of course, send the credentials to the supervisor of the new employee by whatever means is approved in your organization before destroying it. Don't leave spreadsheets hanging around with passwords in them, though. They're a treasure trove for hackers.

Creating the script

This script imports users from the CSV file you created earlier and creates an Active Directory account for each of them.

Assuming you aren't working on a domain controller, you need to import the Active Directory module so that you can work with the AD cmdlets. This process used to be manual, but now it's automatically imported when it's needed. Assuming that you have Remote Server Administration Tools (RSAT) installed, the module for Active Directory will load when you use a cmdlet that requires it.

The following line of PowerShell will import the CSV file when run and will store the contents in the `$ImportADUsers` variable.

```
$ImportADUsers = Import-Csv C:\PSTemp\UserImport.csv
```

The next line starts a `foreach` loop. This loop will go row by row through the CSV file. Each row is stored in the `$User` variable. Each of the column names is called by the `$User.columnname` section, and the value is stored in each of the variables that matches the same name. (Take note of the curly brace in between the `foreach` and the variable block.)

```
foreach ($User in $ImportADUsers)
{
   $FName = $User.FName
   $LName = $User.LName
                           $Username = $User.username
   $Email = $User.Email
   $Phone = $User.Phone
                                   $Dept = $User.Dept
                           $Password = $User.password
                                 $Title = $User.Title
   $OU = $User.OU
```

Next you want to check if the username you're creating already exists in Active Directory. The simplest way to accomplish this is with an `If...Else` statement. So, here's the `If` portion, which checks if the user account already exists, and if it does, prints the error to the screen:

```
if (Get-ADUser -Filter {SamAccountName -eq $Username})
{
Write-Warning "This user account already exists in Active
  Directory: $Username"}
```

Then you use the `Else` part to say that if the account was not already found let's create it. So the `New-ADUser` cmdlet is called with the various parameters that were captured in the CSV file.

```
else
   {
   New-ADUser `
            -SamAccountName $Username `
            -UserPrincipalName "$Username@sometestorg.com" `
            -Name "$FName $LName" `
            -GivenName $FName `
            -Surname $LName `
            -Enabled $True `
            -DisplayName "$LName, $FName" `
            -Path $OU `
            -OfficePhone $Phone `
            -EmailAddress $Email `
            -Title $Title `
            -Department $Dept `
            -AccountPassword (convertto-securestring $Password
  -AsPlainText -Force) -ChangePasswordAtLogon $True
}
```

Those are the separate pieces, so let's see the script in its entirety. ***Note:*** I've removed the comments from the example to make the print version more readable. The version on GitHub has all of the comments:

```
Import-Module ActiveDirectory
$ImportADUsers = Import-Csv C:\PSTemp\UserImport.csv
foreach ($User in $ImportADUsers)
{
 $FName = $User.FName
 $LName = $User.LName
 $Username = $User.username
 $Email = $User.Email
 $Phone = $User.Phone
 $Dept = $User.Dept
 $Password = $User.password
 $Title = $User.Title
 $OU = $User.OU
   if (Get-ADUser -Filter {SamAccountName -eq $Username})
```

```
    {
    Write-Warning "This user account already exists in Active
   Directory: $Username"
    }
    else
    {
    New-ADUser `
             -SamAccountName $Username `
             -UserPrincipalName "$Username@sometestorg.com" `
             -Name "$FName $LName" `
             -GivenName $FName `
             -Surname $LName `
             -Enabled $True `
             -DisplayName "$LName, $FName" `
             -Path $OU `
             -OfficePhone $Phone `
             -EmailAddress $Email `
             -Title $Title `
             -Department $Dept `
             -AccountPassword (convertto-securestring $Password
   -AsPlainText -Force) -ChangePasswordAtLogon $True
    }
}
```

Running the script

The first time you run the script, I highly recommend that you run it within Visual Studio Code. Visual Studio Code is helpful in troubleshooting issues with scripts because you can do debugging and execute one line of code at a time. Plus, it highlights issues for you, which can make them easier to find. The usual culprits are those darn braces on the loops.

To open and run in PowerShell ISE, follow these steps:

1. **Click Start, scroll down to Visual Studio Code, expand the folder and click Visual Studio Code.**
2. **Choose File ⇨ Open File.**
3. **Navigate to your script.**
4. **Select the script and click Open.**
5. **To run the script, choose Terminal ⇨ Run Active File.**

When you're sure that your script is working, there are two methods to run it:

- **Right-click the script and select Run with PowerShell.** I don't like this method because the PowerShell window pops up but closes right away when you're done, so you can't see if it encountered any errors.
- **Open a PowerShell window and run it by specifying the directory (Example 1) or running it from the same directory (Example 2):**
  - **Example 1:** `C:\PSTemp\UserImport.ps1`
  - **Example 2:** `.\UserImport.ps1`

Defining a Script Policy

Defining a script execution policy allows you to define what kind of scripts are allowed to run within your network. You can set the execution policy through Group Policy organization wide, or through the following PowerShell cmdlet. The execution policy is set to Restricted by default.

```
Set-ExecutionPolicy -ExecutionPolicy <policy>
```

Here are the policy types that can be used to set execution policy:

- **Restricted:** Prevents PowerShell scripts from running and will not load configuration files.
- **AllSigned:** For a script to run, it must be signed by a trusted certificate.
- **RemoteSigned:** Requires that any script that is downloaded from the Internet be signed by a trusted certificate. Scripts created locally do not have to be signed to run.
- **Unrestricted:** Allows you to run all scripts. You're prompted for permission before a script is run.
- **Bypass:** Similar to Unrestricted, but it doesn't prompt for permission to run.
- **Undefined:** Removes whatever execution policy is currently set, unless that execution policy is being set through Group Policy.

Signing a PowerShell Script

Depending on how your execution policy is set, you may be able to run scripts that you've created without any issue. If you're in a more secure environment, however, you may need to sign your script so that it will be trusted and allowed to run.

Check out Chapter 1 of this minibook for more on code signing. There, I walk you through the steps of requesting a code signing certificate and signing a PowerShell script.

Creating a PowerShell Advanced Function

For most system administrators, the idea of making your own tools in PowerShell can be a little intimidating. With PowerShell advanced functions, you can use much of what you've learned about PowerShell to create your own tool set that you can run just like you run PowerShell cmdlets. The biggest difference is that PowerShell cmdlets are written in .NET Framework languages like C#, and you must compile them to use them. Advanced functions are written using the PowerShell scripting language.

There are a few components that go into creating a PowerShell advanced function. I'll cover these components first, before I dig into creating your first advanced function.

- `[CmdletBinding()]`: This is what changes a function into an advanced function. It not only allows the function to operate like a cmdlet, but also allows you to use cmdlet features.
- `param`: This area is used to set the parameters that you want your advanced function to use.

You can see how these components are laid out in Figure 4-3.

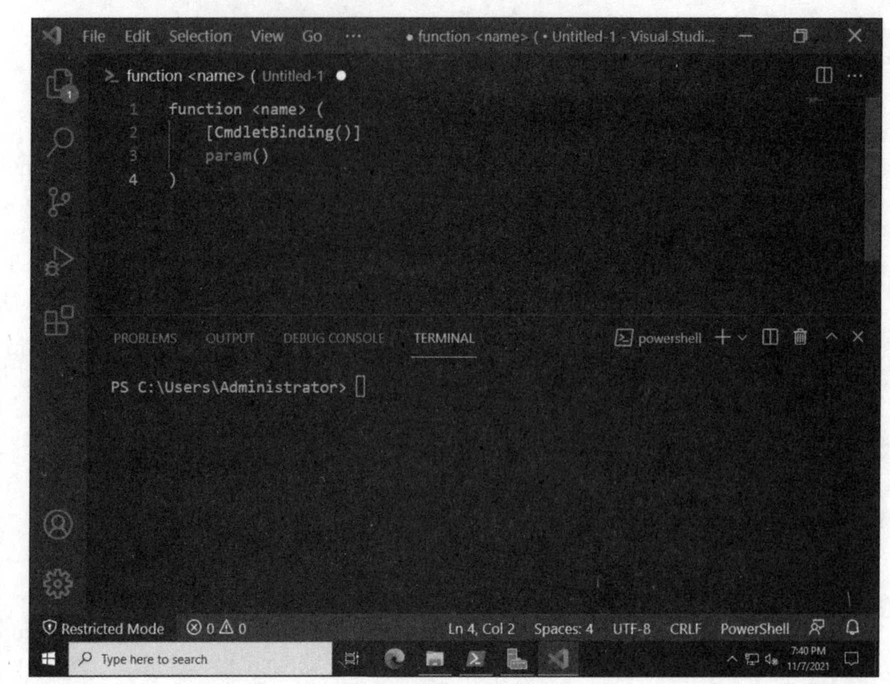

FIGURE 4-3: The basic anatomy of an advanced function includes `[CmdletBinding()]`, which allows the function to behave like a cmdlet and use cmdlet features.

Playing with parameters

Advanced functions give you a lot of granularity when it comes to working with parameters that you just don't have with basic functions. These are placed in the parameter block where you define the parameters for your function. Here are a few of my favorites:

- **Mandatory parameters:** When you specify a parameter as mandatory, the function will not be able to run if that parameter is not provided. In the following example, I've set the parameter to be mandatory, and I've indicated that the value for the parameter will come from the pipeline.

```
[Parameter(Mandatory,ValueFromPipeline)]
```

- **Parameter validation:** Parameter validation is very useful when you want to ensure that a parameter matches some form of expected input. This is done by typing **ValidateSet**, and then by specifying the strings you expect to see. If the parameter string does not match, then the function will not be able to run. See the following example:

```
[ValidateSet('String1','String2')]
```

Creating the advanced function

Now that you know the basic building blocks of advanced functions, I'll create an example of an advanced function. This advanced function will retrieve information about a system. I'll start with the individual components of the function, and then I'll show you the whole thing after it's done. The function is also available for download from this book's GitHub repository at `https://github.com/sara-perrott/Server2022PowerShell`.

First, to tell PowerShell that you want to create a function, you need to start the text in the file with function, followed by what you want to name your function. This absolutely has to follow the PowerShell syntax of verb–noun. Next up, you add `[CmdletBinding()]`, which tells PowerShell that this is an advanced function and should be treated similarly to a cmdlet. In the `Param` block, you define any parameters you want to use. In this case, you're defining one parameter, which is a variable named `computername`.

```
function Get-ReconData
{
    [CmdletBinding()]
   Param (
       [string[]] $computername
)
```

Next, I'll add some text in a `BEGIN` block. I like to use this to see which system it's on currently. You probably won't want to do this in production, especially if you have multiple systems that you're running this function against. But it's great for troubleshooting issues in your script.

```
   BEGIN {
      Write-Output "Gathering reconnaissance on $computername"
   }
```

Next up is where the magic happens. The `PROCESS` block is where I'm telling the function what I want it to do. In this example, I'm telling it to run the code block for every object that is passed to it through the `$computername` variable. Each object is assigned to the `$computer` variable.

Now you can make use of Windows Management Instrumentation (WMI) classes to get the information that you want. In this case, I'm using two WMI classes to query for the data that I want. I'm using the `Win32_OperatingSystem` class and the `Win32_ComputerSystem` class. You can see a listing of the properties that you can work with on the Microsoft documentation pages. `Win32_OperatingSystem` can be found at `https://docs.microsoft.com/en-us/windows/desktop/cimwin32prov/win32-operatingsystem` and Win32_ComputerSystem can be found at `https://docs.microsoft.com/en-us/windows/desktop/cimwin32prov/win32-computersystem`.

To tell PowerShell that I want to use the two WMI classes, I use the PowerShell cmdlet `Get-WmiObject` to assign the desired WMI class to a variable. I then chose a few of the properties that I felt were most useful for gathering some information about the system. I created names for them, and then mapped the name to the WMI variable I created earlier and the property that I'm interested in. Finally, I'm telling it to write the output of the function to the screen. In a production environment, if you were running this against multiple systems, you could export the data to a file.

```
    PROCESS {
       foreach ($computer in $computername) {
       $os = Get-WmiObject -class Win32_OperatingSystem
   -computerName $computer
       $comp = Get-WmiObject -class Win32_ComputerSystem
   -computerName $computer
       $prop = @{'ComputerName'=$computer;
                 'OSVersion'=$os.version;
                 'SPVersion'=$os.servicepackmajorversion;
                 'FreeMem'=$os.FreePhysicalMemory;
                 'OSType'=$os.OSType;
                 'Domain'=$comp.domain;
                 'Status'=$comp.Status}
       $sysinfo = New-Object -TypeName PSObject -Property $prop
       Write-Output $sysinfo}
    }
```

The last block is simply the `End{}` block. In this case, you don't need it to run anything after the function has ran, so it's left blank.

Here is the final function in all its glory!

```
function Get-ReconData
{
     [CmdletBinding()]
    Param (
        [string[]] $computername
    )
    BEGIN {
       Write-Output "Gathering reconnaissance on $computername"
    }
    PROCESS {
```

```
            foreach ($computer in $computername) {
            $os = Get-WmiObject -class Win32_OperatingSystem
    -computerName $computer
            $comp = Get-WmiObject -class Win32_ComputerSystem
    -computerName $computer
            $prop = @{'ComputerName'=$computer;
                    'OSVersion'=$os.version;
                    'SPVersion'=$os.servicepackmajorversion;
                    'FreeMem'=$os.FreePhysicalMemory;
                    'OSType'=$os.OSType;
                    'Domain'=$comp.domain;
                    'Status'=$comp.Status}
            $sysinfo = New-Object -TypeName PSObject -Property
    $prop
                    Write-Output $sysinfo}
    }
    END {}
}
```

Save your function as a .ps1 file just as you would a normal PowerShell script. Now let's try it and see what it looks like when it is run.

Using the advanced function

You can run functions from your code editor of choice or from PowerShell. They're all a little different as to how you should execute the code. In this section, I cover running the advanced function in VS Code, which you'll most likely do while testing, and running it in PowerShell, which is the more realistic production method.

Running an advanced function in VS Code

Now that you've written the advanced function, you want to try it out and see if it works. I'll use VS Code to test it. This is very common when wanting to validate that your function is working properly. I'll use the names and parameters from my previous example.

1. **Click the Start menu and scroll down to the Visual Studio Code folder.**
2. **Expand the folder and click Visual Studio Code to launch it.**
3. **Choose File ⇨ Open File.**
4. **Navigate to where your script is stored, select the file, and click Open.**
5. **In the Terminal, navigate to where you save your function.**

   In my case this is my PSTemp directory, so I typed **cd C:\PSTemp**.

6. **Type a period, followed by a space, and then** .\Get-ReconData.ps1.

   This tells it that you want to run the script from the current directory. It should look like this: . .\Get-ReconData.ps1.

7. **Type** Get-ReconData -computername <computername>.

The output you receive will contain all the information you asked the function to retrieve. In this case, server2022-svr3 is a Windows Server 2022 virtual machine, but you can see that I'm still able to get valuable information, shown in Figure 4-4.

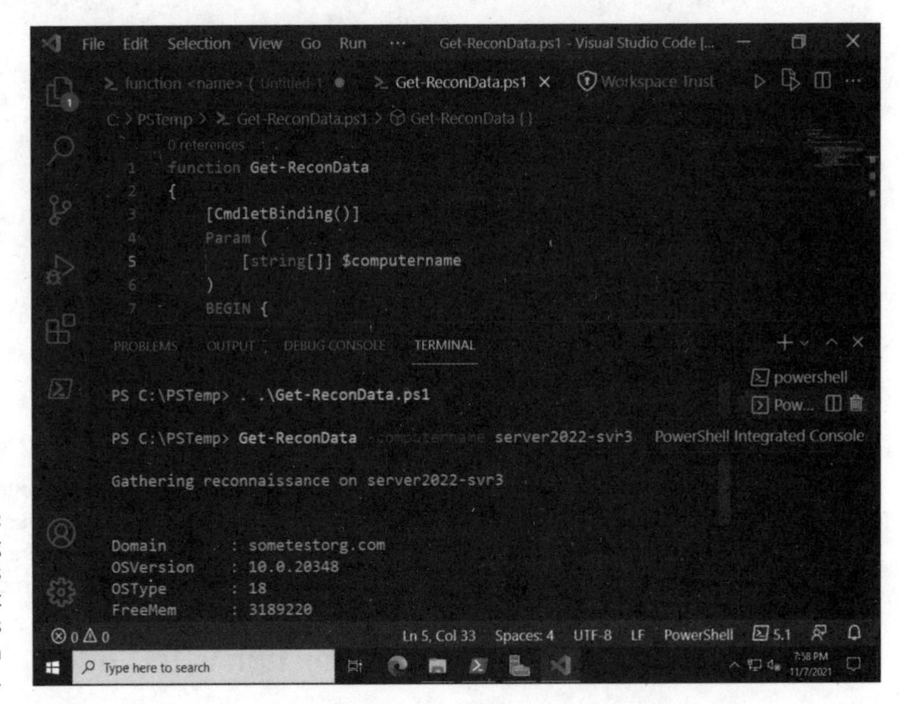

FIGURE 4-4: Testing your function is important to do, and it's simple from within VS Code.

Running an advanced function in PowerShell

In a production environment, it's far more likely that you'll choose to run the advanced function from PowerShell rather than a code editor. Here's how to do this:

1. **Right-click the Start menu and choose Windows PowerShell (Admin).**
2. **Navigate to the location where your function is saved.**

   In my case, it's in my PSTemp folder, so I typed **cd C:\PSTemp** to get to it.

3. **Type a period, followed by a space, and then** .\Get-ReconData.ps1.

   This tells it that you want to run the script from the current directory. It should look like this: . .\Get-ReconData.ps1

4. **Type** Get-ReconData -computername <computername>.

After the function has run, the information you requested is output to the screen, as shown in Figure 4-5.

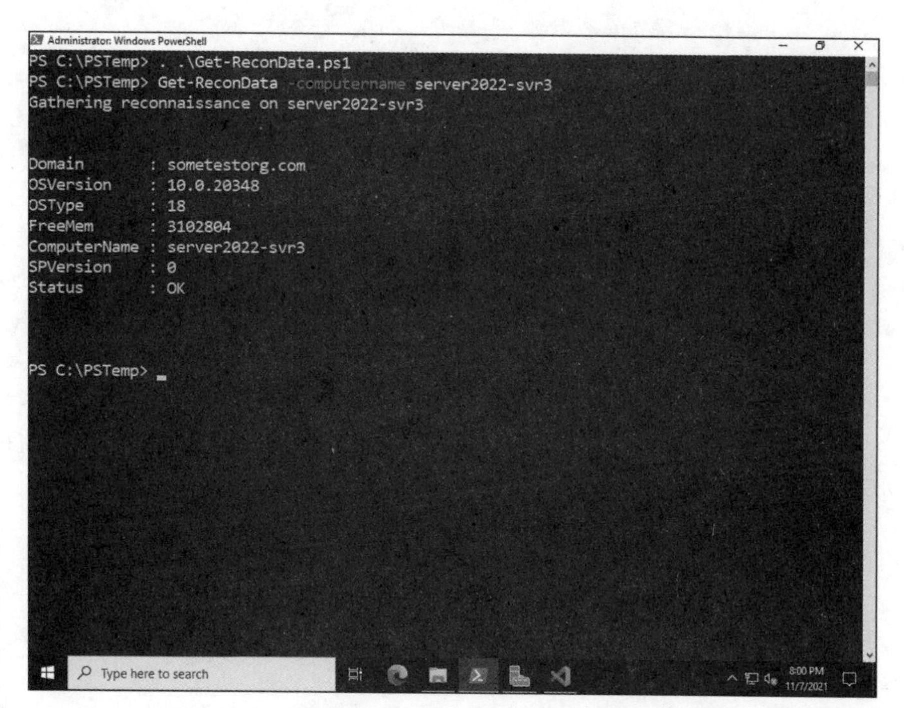

FIGURE 4-5: You can run your advanced function in your PowerShell console.

And that's all there is to it. If you've followed along you just created your first advanced function. From here, the sky's the limit. If you aren't sure what information you want to pull, I highly recommend that you check out the links for the various WMI classes. There is a wealth of information that you can pull.

» Making your first PowerShell Desired State Configuration script

» Using the PowerShell Desired State Configuration script

» Implementing PowerShell Desired State Configuration at scale

Chapter 5
PowerShell Desired State Configuration

PowerShell Desired State Configuration (DSC) enables system administrators to configure systems and keep them in compliance with set organizational baselines. It's sometimes referred to Configuration as Code.

You may be wondering when you would want to use something like this. In my own career, I've seen PowerShell DSC used to enforce security requirements like CIS Benchmarks or anti-malware software settings on servers. It allows system administrators to modernize server deployments with Infrastructure as Code (IaC) while also guaranteeing that build requirements are being met. If a user changes a security setting, PowerShell DSC can overwrite that setting back to the secure baseline that you've created. It truly is a powerful tool.

In this chapter, I introduce you to PowerShell DSC and tell you how to create a DSC script and how to apply it.

Getting an Overview of PowerShell Desired State Configuration

DSC was introduced in PowerShell version 4. It provided a simple way to specify what you wanted a system to be, instead of having to build a system one line of code at a time. I like to compare it to baking: Traditional PowerShell is like following the recipe; DSC is like saying, "I want this," and handing the picture of a cake to the baker.

DSC is exceptionally powerful. It can be used to install roles and features, copy files to specified locations, install software, and make changes to the Registry.

PowerShell DSC has three main components:

- **Configurations:** PowerShell scripts that are used to configure your resources according to your organization's requirements
- **Resources:** Code that keeps your system in compliance with the specified configuration
- **Local Configuration Manager (LCM):** Handles the interactions between the configurations and the resources

In the following sections, I cover each of these components in a little deeper detail and show you what they look like in an actual DSC script.

Configurations

Because PowerShell DSC files are saved as PS1 files, you may be wondering how PowerShell knows that the file is a PowerShell DSC file. That's a great question! It looks for the keyword *configuration*. This tells it that the file contains a DSC configuration. A sample in a DSC configuration may look something like this:

```
Configuration MyAwesomeWebsite {
```

TIP

You can name the configuration whatever you like. PowerShell doesn't care what it's called. I suggest giving it a name that makes sense to you so that you can look at it and know from the name what you're configuring with this particular configuration file.

A DSC script accomplishes its tasks with a slightly different feel from traditional PowerShell; it looks similar to an advanced function. With traditional PowerShell, you would write something along the lines of the following:

```
Install-WindowsFeature -Name "Web-Server"
```

This installs the Internet Information Services (IIS) web server on the Windows system that you run it on. You may also specify whether you want sub-features to be enabled and if you want the management tools for IIS to be installed.

With a DSC configuration, this is how you would accomplish the same thing as that line of PowerShell:

```
WindowsFeature WebServer {
            Ensure = "Present"
            Name = "Web-Server"
        }
```

Instead of telling it explicitly to install the `Web-Server` feature via PowerShell, you simply tell DSC that you want to make sure that `Web-Server` is present. If it is, then the script continues to execute and does not reinstall the `Web-Server`. If it is not installed, then the `Web-Server` feature will be installed.

Resources

Resources are the foundational pieces of PowerShell DSC. Resources make properties available that can contain PowerShell scripts and that can be used by LCM to implement the changes.

You may want to look up how to use a particular DSC resource. This is simple to do with the following command:

```
Get-DscResource Syntax <resource>
```

If, for instance, you run that command using `service` as the resource, you'll find some of the more common things that are used in relation to services, like the service name and the state of the service. Most administrators, for example, will want to check to see if a service is running. Maybe you want to verify that your antivirus service or patching service is running. The following example shows you how you can use a resource block to verify that a service is running on your local system, in this case IIS:

```
Configuration MyAwesomeWebsite
{
    Import-DSCResource -Name Service
    Node localhost
    {
        Service "W3SVC:Running"
```

```
        {
            Name = "W3SVC"
            State = "Running"
        }
    }
}
```

You may be wondering how I got that service name for IIS, the web server available in Microsoft Server 2022. You can go into the `Services.msc` panel and get your service names there. In the example of the web server, I scrolled down to World Wide Web Publishing Service and double-clicked it. The name that you need for DSC is listed as Service Name, shown in Figure 5-1 as W3SVC.

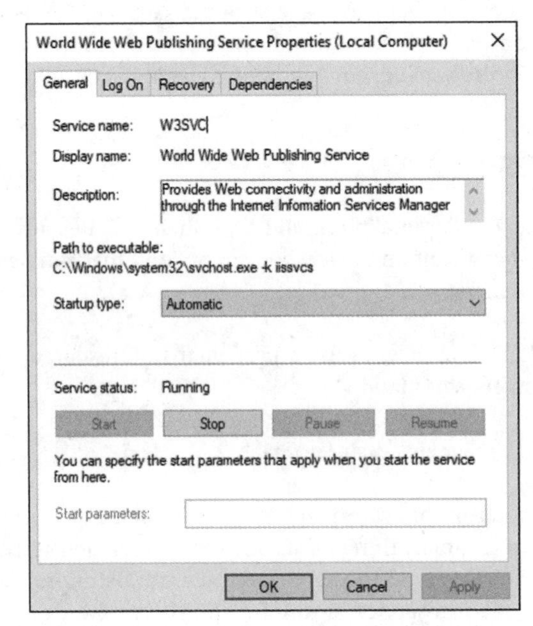

FIGURE 5-1: The service name is displayed within the services.msc panel for each service.

Of course, Service is not the only resource type available. The Import-DSCResource module contains many different resources that you can use in your DSC scripts. I have listed the more common resource types in Table 5-1.

If this table doesn't list a resource you need, you can download additional resources from the PowerShell Gallery (`www.powershellgallery.com`) and GitHub (`https://github.com/PowerShell/DscResources`).

TABLE 5-1 DSC Resources

Resource Name	Description
`file`	The `file` resource type can be used to copy files from source to destination and ensure that the files in the source and destination always match. It can use dates and hashes to compare the source and destination files. If they don't match, the source files are copied over the destination files.
`archive`	The `archive` resource type can unpack archive files like ZIP files. It can validate the integrity of the archive file against a `checksum`.
`environment`	The `environment` resource type can be used to work with environmental variables that you want to create, modify, or manage.
`group`	The `group` resource type can be used to work with groups. This includes the management of the groups and the users within the groups.
`log`	The `log` resource type is used for exactly what you may think: It writes to an event log. Specifically it writes to the Microsoft-Windows-Desired State Configuration, Analytic event log.
`package`	The `package` resource type is very useful when you want to ensure that something is installed. You can also use it to uninstall packages.
`registry`	The `registry` resource type allows you to work with the Registry, including the creation, modification, and deletion of Registry keys and their associated values.
`script`	The `script` resource type allows you to have a little fun and create your own script blocks. Your script may start with `Get`, `Set`, or `Test`.
`service`	The `service` resource type allows you to work with services that are on the system. You can ensure that they're present and that they're running or disabled.
`user`	The `user` resource type allows you to work with local user accounts on the system. You can create, modify, and delete users, as well as make changes to their accounts.
`WindowsFeature`	The `WindowsFeature` resource type is one of the more common types that you see if you're in a shop that is doing Infrastructure as Code. It can be used to work with both roles and features, and can install, uninstall, or modify both roles and features.
`WindowsProcess`	The `WindowsProcess` resource type allows you to work with windows processes. It allows you to start, stop, and configure them.

Local Configuration Manager

The LCM receives configurations that are sent to your systems and applies those configurations. It runs on every system that you're targeting with your DSC scripts.

LCM allows you to specify how you want to get configurations. You can use a push model or a pull model. You can also set how often each system reaches out to pull a configuration down.

If you want to see that the LCM settings are on a system on which you want to use DSC configuration scripts, you can run the following command:

```
Get-DscLocalConfigurationManager
```

This command displays the current settings for LCM. An example is shown in Figure 5-2.

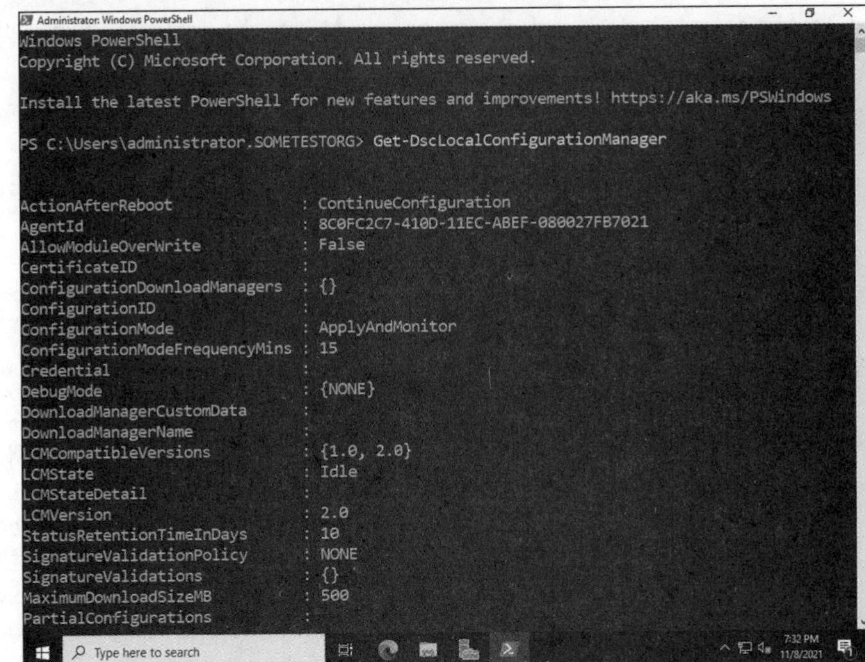

FIGURE 5-2: Checking the current settings for LCM is simple to do in PowerShell with the `Get-DscLocalConfigurationManager` command.

Creating a PowerShell Desired State Configuration Script

If you've read this chapter from the beginning, you've found out about the components of DSC at this point. Now how do you build something useful? In this section, you continue to build on the example of a web server. In your organization, you may need to be able to provision systems quickly to keep up with demand. In this example, you're going to use DSC to ensure that IIS is installed on the system and to copy files for the website over from a source that is specified to the new web server so that it can start serving out content. Normally, the source would not be on the same server; it would simply be available on the same network.

The script starts with the keyword configuration, which tells PowerShell this is a DSC file. The next step is to import `PSDesiredConfiguration`, which is needed to load the custom resources that you may need in your script. `Node` is used to specify the system that you want to run the DSC against. In this case, I'm running it on the computer that it resides on, so `Node` is set to `localhost` (my local computer). The `WindowsFeature` block is where I tell it that the `WebServer` feature must be present. This will check to see if Internet Information Services (IIS) is installed. If it is not installed, then DSC will install the feature. The last part of this script copies the website files from the source to the destination. This can be a great way to speed the provisioning of web servers if you need to scale quickly.

```
Configuration MyAwesomeWebsite {

    Import-DscResource -ModuleName PsDesiredStateConfiguration

    Node 'localhost' {
        WindowsFeature WebServer {
            Ensure = "Present"
            Name = "Web-Server"
        }
        File WebsiteGoodies {
            Ensure = 'Present'
            SourcePath = 'c:\PStemp\index.html'
            DestinationPath = 'c:\inetpub\wwwroot'
        }
    }
}
MyAwesomeWebsite
```

You'll want to write your script in your code editor of choice and save it to a PS1 file. I've made this simple DSC script available on the GitHub for this book. You can reach it at `https://github.com/sara-perrott/Server2022PowerShell`.

You may notice that this DSC configuration was written with just one node, `localhost`. In most production instances, you're going to want to target multiple hosts with your DSC configuration. This is a simple change to make. Instead of the single node being defined like this:

```
Node 'localhost' {
```

It gets defined like this to support multiple nodes:

```
Node @('localhost','Server2') {
```

Applying the PowerShell Desired State Configuration Script

After you have your script written, you'll want to actually apply it. The steps are fairly simple. First, run the PS1 file as you would any other PowerShell script. This will create a Managed Object Format (MOF) file. MOF files are what actually get used by PowerShell DSC to do its configuration work.

Compiling into MOF

Open your PowerShell window. You can reach this by right-clicking Start and then choosing Windows PowerShell Admin. Then follow these steps:

1. **Navigate to the directory where the new PS1 file is stored.**
2. **Run the following command:**

```
. .\AwesomeWebsite.ps1
```

You can see in Figure 5-3 that running the command generates the MOF file. The MOF file takes its name from the name supplied in the resource block. In this case, the name is `localhost`. If you have multiple systems, you'll have multiple MOF files, one for each system represented.

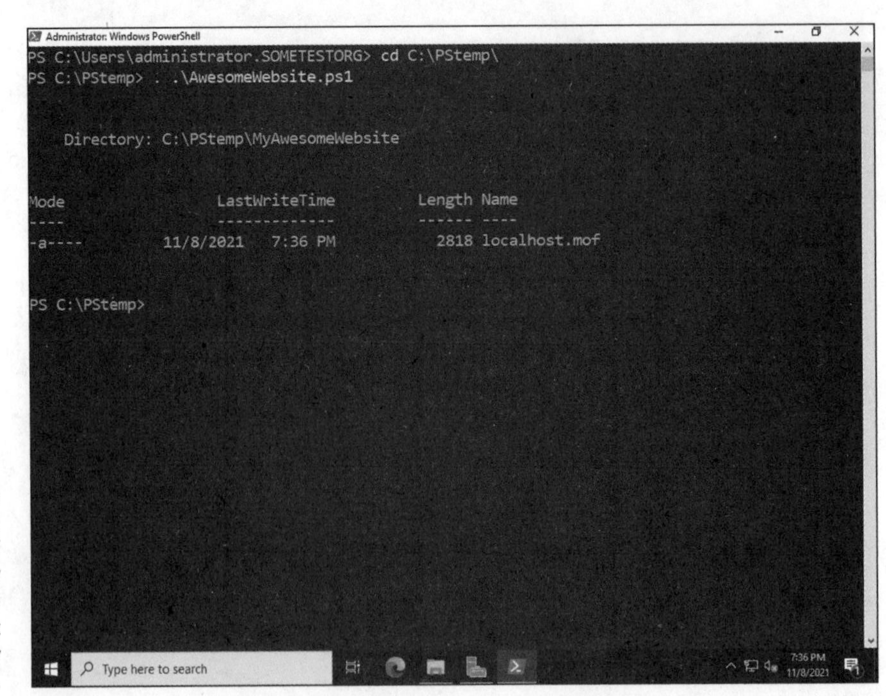

FIGURE 5-3: Compiling the PS1 into an MOF file is done by running the PowerShell script as you normally would.

If you browse to the directory where your script is located, you'll notice that you now have a folder with the name of your configuration on it. Inside of this folder is your MOF file.

Applying the new configuration

Now that you've created the MOF file, it's time to apply the configurations contained within it. You'll want to run the `Start-DscConfiguration` cmdlet, which will use the MOF file to run the configuration against the system. You run the cmdlet and specify the folder that the MOF file is in:

```
Start-DscConfiguration .\MyAwesomeWebsite
```

After this command is run, you get the state of the configuration, which should be Running, as shown in Figure 5-4.

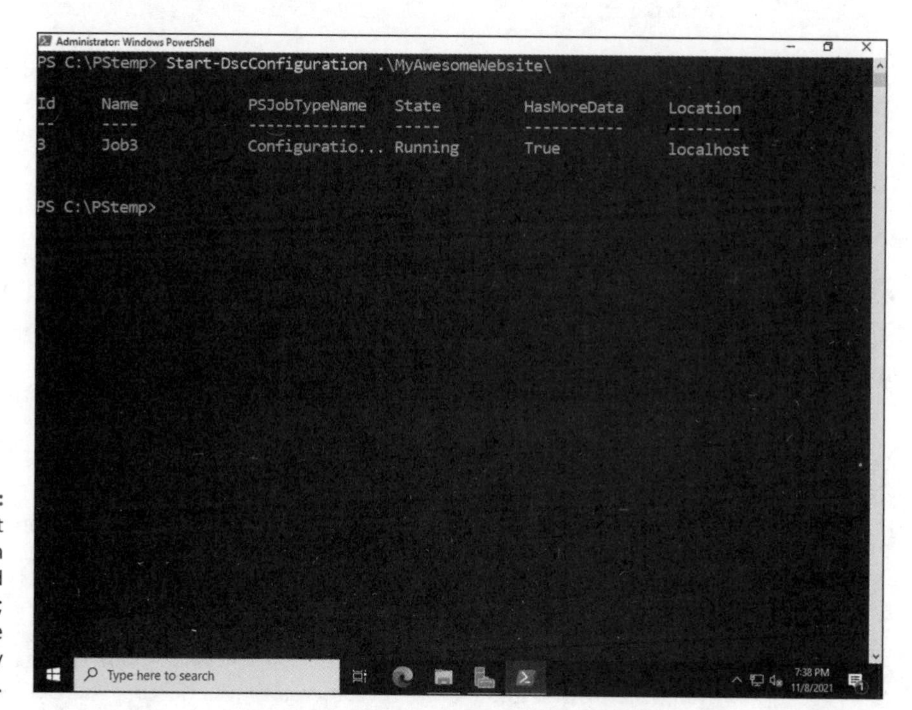

FIGURE 5-4: The DSC script has been compiled and has been run; you can see the State is currently Running.

Push and Pull: Using PowerShell Desired State Configuration at Scale

As you can see, DSC is a very powerful tool for configuring your systems. Chances are, you want to automate things a bit more. You don't want to log on to your server and run DSC manually. That's where you need to start looking at the two configuration modes: push mode and pull mode.

Push mode

In push mode, the configuration is pushed to the destination system. It's a one-way relationship. Push mode is done similar to the example in the "Applying the new configuration" section earlier in this chapter. You run the `Start-DscConfiguration` cmdlet, and you can use the `-ComputerName` parameter to target specific systems with the script.

TIP

You can easily automate this on a scripting server with scheduled tasks. By setting it up as a scheduled task, you can ensure that your configuration is pushed at regular intervals and that your systems are configured exactly the way you expect them to be configured.

When using push mode, you have a decision to make when it comes to the configuration mode that you want to use. You can see which one you're using currently by running the `Get-DscLocalConfigurationManager` command. There are three options:

- `ApplyOnly`: Applies the configuration once, but does nothing further.
- `ApplyandMonitor`: Applies the configuration and will write any discrepancies to the logs. This is the default value.
- `ApplyandAutoCorrect`: Applies the configuration, writes discrepancies to the logs, and then applies the current configuration again.

If you make a change to the configuration script, you need to regenerate the MOF. DSC does cache configurations, so if you have made changes to your script, you need to stop and restart the process that is hosting the DSC engine. You can do that with the following commands.

First, you need to get the process ID of the process that is hosting the DSC engine.

```
$ProcessID = Get-WmiObject msft_providers | Where-Object {$_.
   provider -like 'dsccore'} | Select-Object -ExpandProperty
   HostProcessIdentifier
```

Then you stop the process using the following command:

```
Get-Process -Id $ProcessID | Stop-Process
```

After you've stopped the process, simply run `Start-DscConfiguration` again, and it will use the newest version of the configuration script.

Pull mode

Pull mode is a little more complicated to get started. You need to set up a pull server, which will host the DSC service and contain all the configuration and resource scripts that the clients will pull.

The sequence for setting up pull mode is to create the configuration script, set up the pull server, and then set up DSC on the system that you want the pull to occur from.

Setting up the pull server

Before you can use the pull server, you need to configure it. First, you need to install a module named `xPSDesiredStateConfiguration`. This module available on PowerShell Gallery. With PowerShell open, you can run the command `Install-Module -Name xPSDesiredStateConfiguration`, which automatically downloads and installs the module for you. You may be prompted to accept a new provider named Nuget. Press Y if that occurs. You'll be asked if you're sure you want to install the module; press Y.

After the module is installed, you can run the script to set up the DSC pull server. I'm using the configuration script available from Microsoft's DSC pull page. You can copy the script from `https://docs.microsoft.com/en-us/powershell/dsc/pull-server/pullServer`.

Configuring DSC on the system to use the pull server

After the pull server is set up, the last step is to set up the system that will be pulling the configuration from the pull server. You can run the following to set up the client.

```
Configuration ConfigurationForPull
{
     LocalConfigurationManager
     {
       ConfigurationID =
   "registration_key_from_server_setup_script";
       RefreshMode = "PULL";
       DownloadManagerName = "WebDownloadManager";
       RebootNodeIfNeeded = $true;
       RefreshFrequencyMins = 30;
       ConfigurationModeFrequencyMins = 60;
       ConfigurationMode = "ApplyAndAutoCorrect";
       DownloadManagerCustomData = @{ServerUrl = "http://
   PullServer:8080/PSDSCPullServer/PSDSCPullServer.svc";
   AllowUnsecureConnection = "TRUE"}
     }
}
ConfigurationForPull -Output "."
```

Save the script and run it like you have before. And then use the `Start-DscConfiguration` cmdlet to apply the configuration.

7

Installing and Administering Hyper-V

Contents at a Glance

» **Knowing the difference between Type 1 and Type 2 hypervisors**

» **Installing and configuring Hyper-V on Windows Server 2022**

Chapter 1
What Is Hyper-V?

In the technology field, there is always the next hot thing that everybody starts talking about. Virtualization was one of those topics when it first made its appearance. Virtualization has enabled IT professionals to better use the resources that have been purchased and has led to the creation of cloud computing services.

In this chapter, you learn about Microsoft's virtualization product, which is called Hyper-V. I fill you in on the basics of virtualization, and then show you how to install and configure Hyper-V.

Introduction to Virtualization

When I first started in the technology field, every organization had physical servers. In most cases, they followed best practices and one server was dedicated to one application. This often led to wasted resources because the application didn't actually need all the central processing unit (CPU) and random access memory (RAM) it was given, so those resources would sit idle. At the same time, the organization was paying for power and cooling for a server that wasn't necessarily doing anything at the moment.

The amount of time it took to stand up a new server could be an issue for projects that were time-sensitive. With each physical server, you had to rack it, cable it, configure it, and install software on it. Provisioning new servers for large projects could take weeks to months, especially if multiple teams were involved.

Virtualization was a game changer. Instead of buying individual smaller servers to run single applications, an organization could purchase bigger, more powerful servers to run a hypervisor of some kind that would, in turn, run multiple virtual servers, referred to as *virtual machines* (VMs). By purchasing larger servers to run the smaller workloads, organizations were able to save on power and cooling costs. They were also able to reduce the amount of time needed to go to market, because the virtualization administrator was typically the one who would spin up the server operating system in a VM, set up the networking, and perform the basic configuration tasks like assigning IP addresses and other necessary steps.

Virtualization really streamlined the process for system administrators and organizations to be able to build servers quickly in response to the needs of other teams for projects or for expanding the existing capacity to support applications. It also simplified recovery efforts when configured properly because VMs on a failed host could be transferred to another host.

TECHNICAL STUFF

You sometimes hear hypervisors referred to as *hosts* and virtual machines referred to as *guests*. If you run into this terminology, don't let it confuse you. These terms are used across all types of virtualization technologies.

Type 1 and Type 2 Hypervisors

Before I dive into the difference between Type 1 and Type 2 hypervisors, I want to make sure that you understand what a hypervisor is. The *hypervisor* is essentially a process that allows you to create, run, and manage VMs.

The hypervisor is ultimately responsible for presenting resources to the VMs that are running on it, including CPU, RAM, networking, and storage.

TIP

Most of the hypervisors let you overprovision VMs, meaning that you can assign resources that are not necessarily available. This may work for you if your workloads are very small, but if there are spikes in the workloads, or if VMs take too many resources, then the hypervisor could become starved for resources, which could impact *all* your VMs that are running on that hypervisor. For this reason, I recommend not over-provisioning your VMs.

Type 1 hypervisors

Type 1 hypervisors are also referred to as *bare-metal hypervisors.* This is because the software for the hypervisor can run directly on the host system's hardware. Type 1 hypervisors provide the best performance and security of the hypervisors, but some of them are more complex that others to set up.

Here are some examples of the more common Type 1 hypervisors:

- Microsoft Hyper-V
- VMware ESXi
- Oracle VM Server
- KVM
- Citrix XenServer

Type 2 hypervisors

Type 2 hypervisors are referred to as *hosted hypervisors.* They require an operating system to be able to install and run. Type 2 hypervisors are usually easier to install and configure, but they're less secure and not as performant as Type 1 hypervisors because they don't have direct access to the host system's hardware.

Here are some examples of the more common Type 2 hypervisors:

- Oracle VirtualBox
- VMware Workstation
- VMware Fusion

Installing and Configuring Hyper-V

Windows Server 2022 offers Hyper-V, a Type 1 hypervisor. Hyper-V is a role that gets installed on a Windows Server 2022 operating system. Unlike with previous versions of Windows Server, there is no longer a stand-alone graphical user interface (GUI)–less Windows Server 2022 Hyper-V SKU. You'll need to install the role to work with Hyper-V. In this section, I show you how to install Hyper-V from the role. The lab systems that I'm using for this installation are joined to the domain `sometestorg.com`.

TIP

Windows 10 also has a version of Hyper-V available that you can install. It's a feature that can be enabled, and it will allow you to support virtual machines, virtual networking, and virtual storage. This is very helpful if you need to be able to run multiple operating systems in your normal day-to-day activities. This feature is only available if you're running Windows 10 Pro, Enterprise or Education editions. It is not available on Windows 10 Home edition. The Windows 10 version of Hyper-V does not support advanced functionality like live migration, Hyper-V Replica, or SR-IOV.

Installing Hyper-V

You need to make some basic configuration decisions during the installation of Hyper-V, but they can be changed after the installation, so if you change your mind or make a mistake, don't panic!

Follow these steps to install Hyper-V:

1. **From Server Manager, choose Manage ⇨ Add Roles and Features.**
2. **On the Before You Begin screen, click Next.**
3. **On the Select Installation Type screen, click Next.**
4. **On the Select Destination Server screen, click Next.**
5. **On the Select Server Roles screen, select Hyper-V.**
6. **Click Add Features in the dialog box that pops up, and then click Next.**
7. **On the Select Features screen, click Next.**
8. **On the first Hyper-V screen, click Next.**
9. **On the Create Virtual Switches screen, select the network adapter you want to use for the virtual switch.**

   As you can see in Figure 1-1, I have only one adapter to choose from right now, so I'll select it.

10. **Click Next.**
11. **On the Virtual Machine Migration screen, select the Allow This Server to Send and Receive Live Migrations of Virtual Machines on This Server check box and select the Use Credential Security Support Provider radio button (see Figure 1-2).**

TECHNICAL STUFF

   Live migrations enable you to move a virtual machine from one Hyper-V host server to another Hyper-V host server with no downtime. CredSSP is the simplest way to set up live migration, but it requires you to log into the server being migrated, so it isn't the best for automatically moving virtual machines.

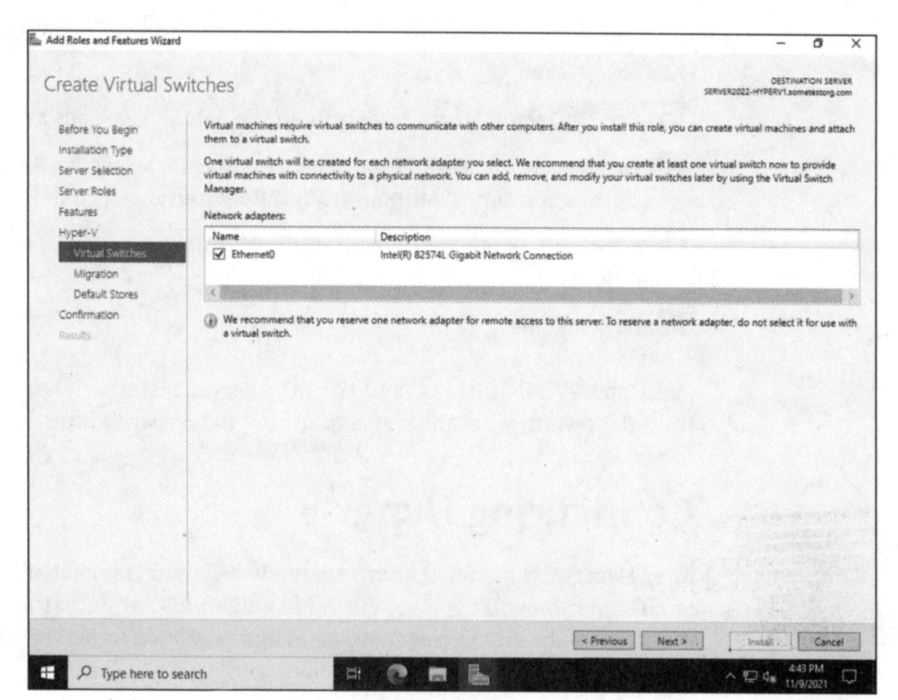

FIGURE 1-1: You must select a network adapter for the virtual switch to use. You can change this later.

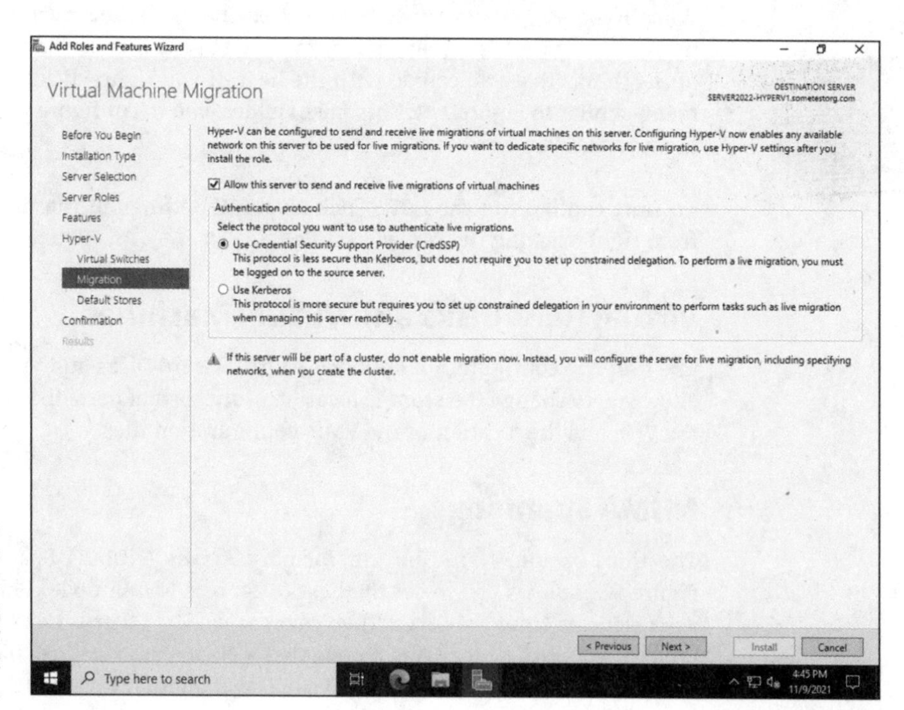

FIGURE 1-2: To enable live migration of VMs, select the check box to allow them and select CredSSP.

12. **Click Next.**
13. **On the Default Stores screen, keep the default locations and click Next.**
14. **On the Confirm Installation Selections screen, select the Restart the Destination Server Automatically If Required check box.**
15. **Click Yes on the dialog box that is confirming you want to allow the reboot.**
16. **Click Install.**

The Hyper-V role installs, and then the server restarts. When it comes back up from the restart, you can start configuring the Hyper-V host.

Configuring Hyper-V

After Hyper-V is installed, there are many different things that you can configure or change from what you set during installation. Getting to the Hyper-V console is similar to the other roles that you install on Windows Server 2022. From Server Manager, choose Tools ➪ Hyper-V Manager.

When Hyper-V Manager opens, you see the name of the server on which you just installed the role. Click that server, and you see the menus change to reflect some of the things that you can do with the host. If you right-click the host, you see a menu similar to Figure 1-3. This menu allows you to configure your Hyper-V host the way that you want to.

To start configuring the host, click Hyper-V Settings in the menu that you got from right-clicking the server's name.

Virtual Hard Disks and Virtual Machines

The first two configuration options — Virtual Hard Disks and Virtual Machines — allow you to change the storage location of the virtual hard disks that are used for the VMs and the location of the VM's configuration files.

NUMA Spanning

The third option, Non-Uniform Memory Access (NUMA) Spanning, shown in Figure 1-4, allows you to set the host to act as a NUMA node. This allows VMs to use resources from the server they're on as well as other servers that are configured to be NUMA nodes. This means that a virtual machine can have more CPU or RAM than what is on the one physical host, if another host which is also a NUMA node is sharing that resource. This has an impact on performance so I wouldn't recommend it unless you're using it in a lab or development environment. Avoid using this in production environments.

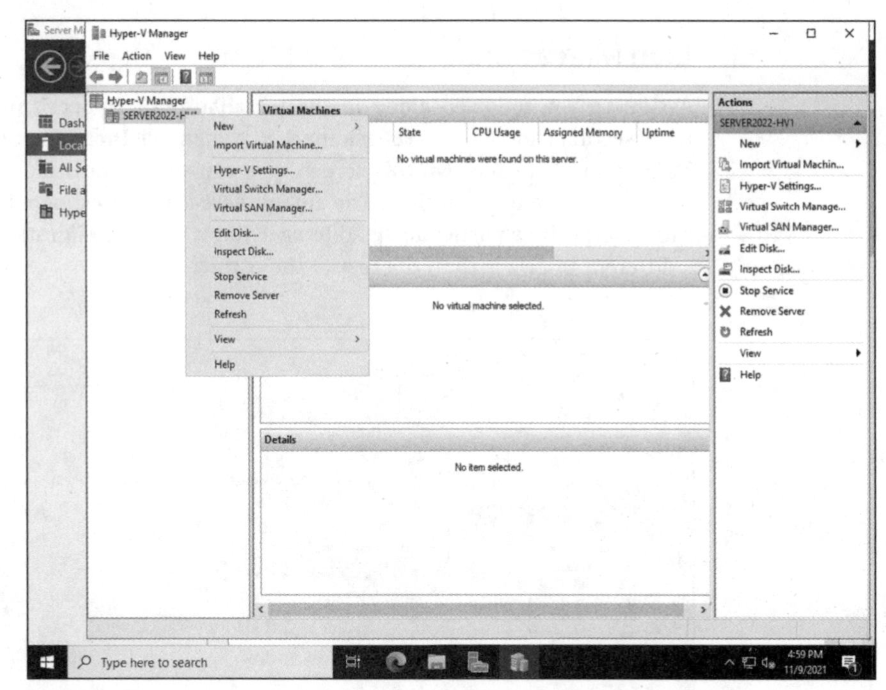

FIGURE 1-3: The menu for the host in Hyper-V Manager presents you with your configuration options for that host.

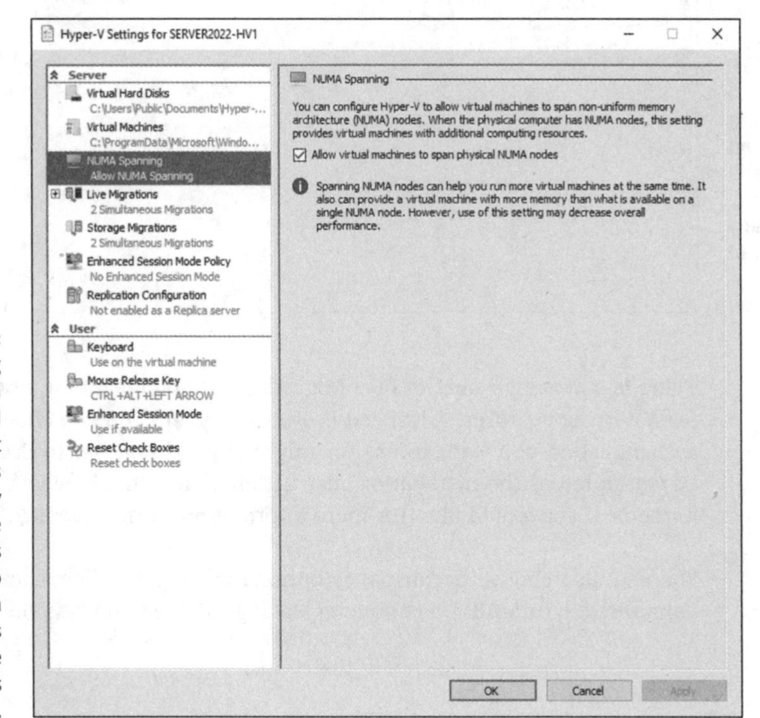

FIGURE 1-4: NUMA Spanning can be helpful in lab and development environments where you may not have the same resources that you do in production because it allows you to share resources across NUMA nodes.

Live Migrations

Assuming you followed along in the installation of Hyper-V, your Live Migrations section should have a check mark in the Enable Incoming and Outgoing Live Migrations check box. On this screen, you can specify how many live migrations can happen at any given time. The default here is two, as shown in Figure 1-5. You can also specify a particular IP address if you want Live Migration to happen over a different interface than the rest of the traffic.

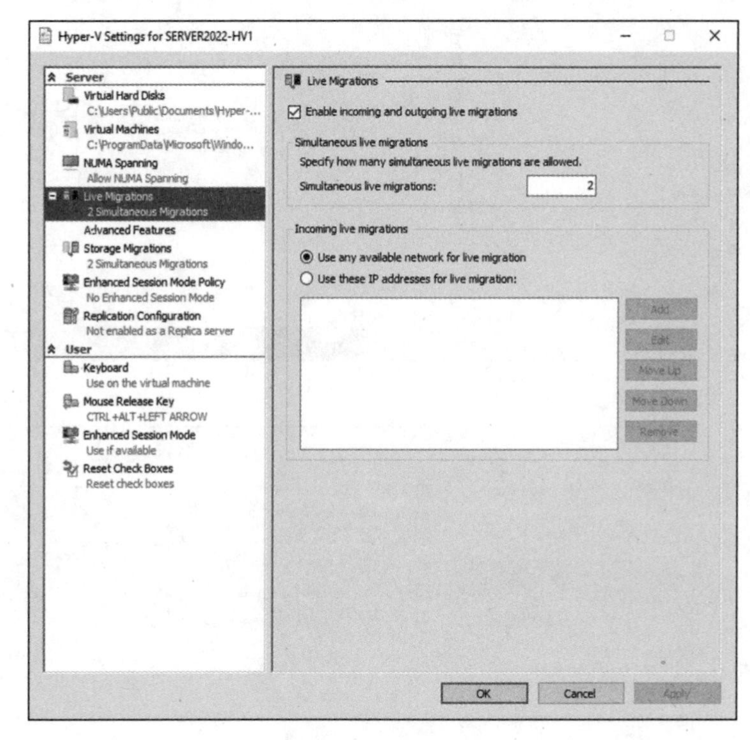

FIGURE 1-5: Live Migration allows virtual machines to move between hosts with no downtime, and you can specify how many migrations can happen at the same time.

There is a plus sign next to Live Migrations. If you click that, you get the option for Advanced Features. Advanced Features is where you can change what kind of authentication you want to use for migrations. This is set to CredSSP right now (if you followed the installation instructions), and this is where you can choose Kerberos if you would like (for more information, turn to Book 7, Chapter 5).

You can also choose performance options from here. Your choices are TCP/IP, Compression, or SMB. I recommend leaving this on Compression.

Storage Migrations

Storage Migrations allows you to move VM storage with no downtime to the virtual machine. It's very helpful when moving to a new storage array, or when getting ready to perform maintenance on a storage array because you can move the storage with the virtual machine still powered on. In this section, you can decide how many storage migrations you want to allow to happen at the same time. The default setting for this screen is two.

Enhanced Session Mode Policy

Enhanced Session Mode Policy allows your Hyper-V host to connect to your VMs over Remote Desktop Protocol (RDP). You may be wondering why you would want to allow that. When you use RDP to connect, you can pass local devices to your VMs like disk drives, flash drives, and other peripherals. You also gain a shared clipboard that allows you to copy and paste, and it improves support for viewing the VMs on a higher-resolution monitor. This setting is disabled by default on Windows Server 2022 so you need to enable it if you want to use this feature.

Replication Configuration

You can set up your Hyper-V host to act as a Hyper-V Replica. When a Hyper-V host is configured as a replica, VMs are copied to it from the primary Hyper-V servers. If the primary Hyper-V server ever experienced a major malfunction, the replica server can bring up the VMs that are kept in a powered-off state.

You can specify whether you want replication traffic to be sent plaintext or encrypted. I always recommend using encryption when it's available. And you can also select whether you want to allow replication from any server that can authenticate, or if you want to limit replication to specific servers. This screen is shown in Figure 1-6.

Keyboard

The Keyboard screen is one of the user settings. You can specify whether key combinations like Alt+Tab, for example, will apply to the physical computer the keyboard is attached to, the VM, or on the VM but only if the VM is full screen.

Mouse Release Key

If you haven't installed the VM drivers, you can set which key combination you want to use to release the mouse so that you can use it outside of the VM. Unless there is a good reason not to, I always recommend installing the VM drivers.

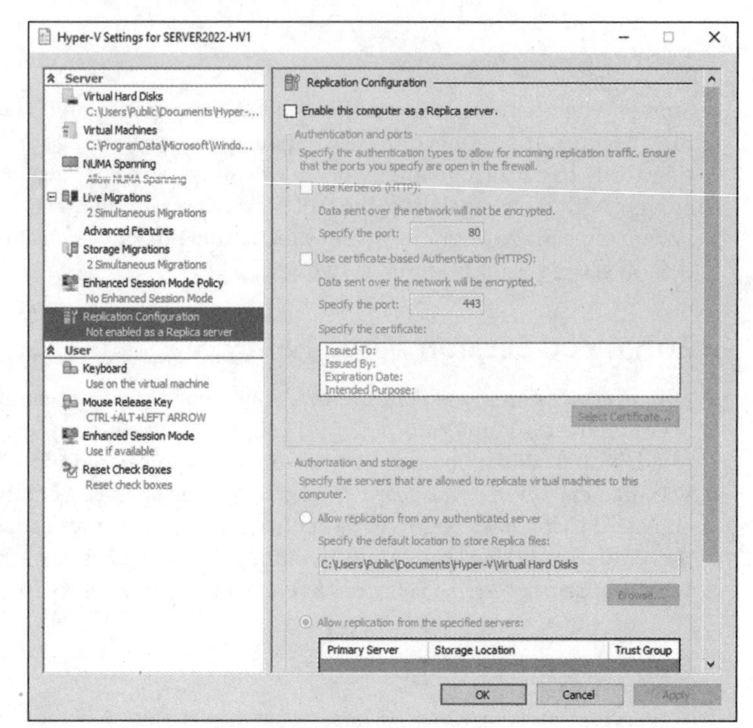

FIGURE 1-6: Configuring a Hyper-V Replica server is a great addition to your organization's disaster recovery strategy.

Enhanced Session Mode

Enhanced Session Mode is enabled for the user by default. It allows you to use a remote desktop connection to pass through drives, printers, and so on, and to use the shared clipboard.

Reset Check Boxes

All this setting does is reset check boxes that are used to hide pages or messages when they're checked. It doesn't reset anything else.

Virtual Switch Manager

When you right-click your Hyper-V host, you may notice an option for Virtual Switch Manager. This selection allows you to create virtual switches that your VMs can use to communicate on the network. There are three types of switches that you can use within Hyper-V:

- **External:** Allows you to connect to a physical network
- **Internal:** Allows the virtual machines to communicate with other virtual machines on the same switch and with the host

» **Private:** Only allows virtual machines to communicate with other virtual machines on the same switch

Having the right type of switch to support your use case is critical if you want your Hyper-V deployments to succeed. The screen is shown in Figure 1-7. You can learn more about configuring virtual networking with Hyper-V in Book 7, Chapter 3.

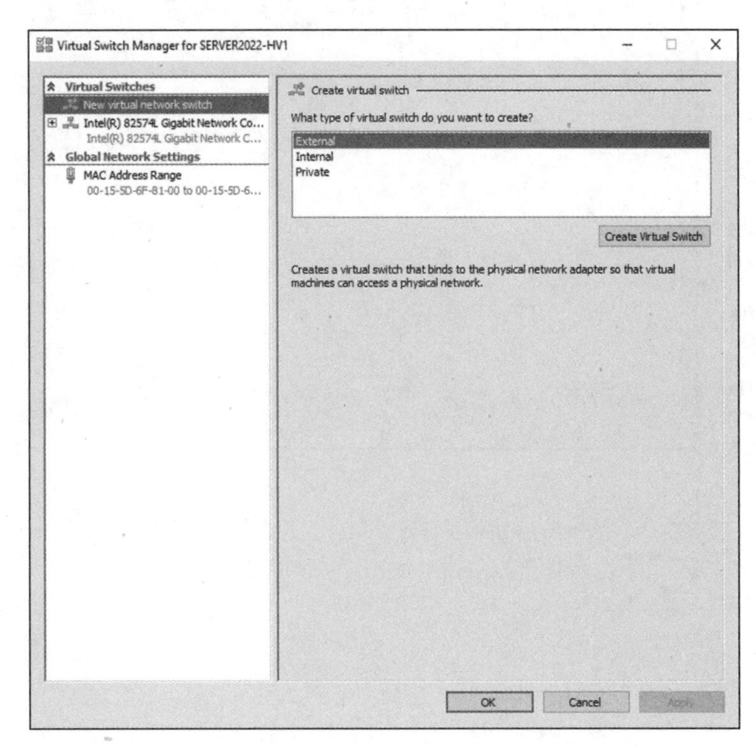

FIGURE 1-7: The Virtual Switch Manager allows you to configure virtual network switches for your Hyper-V environment.

Virtual SAN Manager

Also in the menu for your Hyper-V host is the Virtual SAN Manager. This allows you to connect your Hyper-V host to a Fibre Channel SAN. This is especially helpful for large organizations that have invested in Fibre Channel technology. You can see in Figure 1-8 that you can define the World Wide Node Name (WWNN) for the Fibre Channel port that is on the Hyper-V host.

REMEMBER

Fibre Channel SANs utilize special switching equipment to support high-speed, low-latency storage networks. Systems that use Fibre Channel need special storage network adapters installed, which are referred to as host bus adapters (HBAs).

Storage for Hyper-V servers is discussed in more detail in Book 7, Chapter 4.

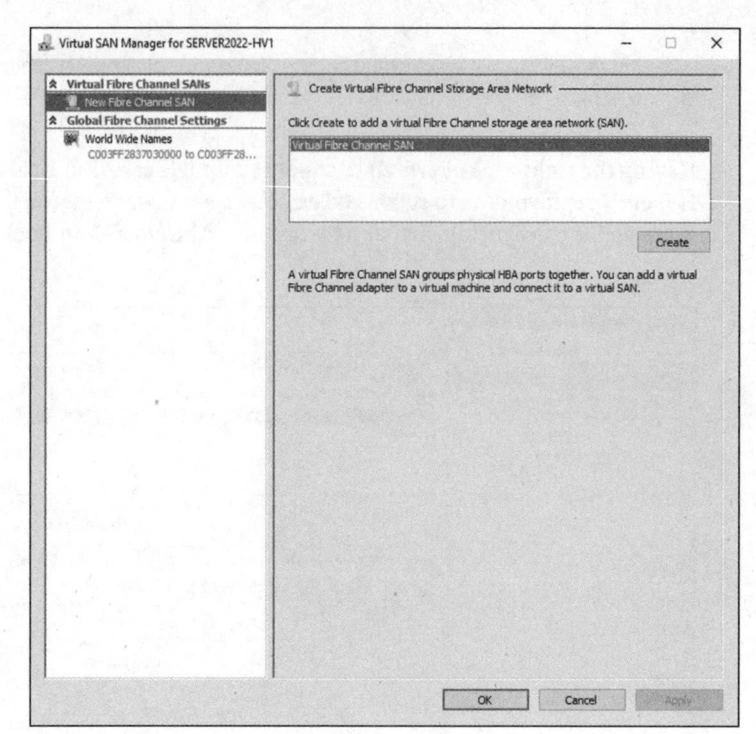

FIGURE 1-8: Hyper-V offers support for Fibre Channel SANs, which is a huge benefit to larger organizations that have invested in Fibre Channel technology.

» Configuring a virtual machine in Windows Server 2022 Hyper-V

» Safeguarding your virtual machines with shielded virtual machines

Chapter 2 Virtual Machines

In Chapter 1 of this minibook, I explain what virtualization is and show you how to install and configure Hyper-V. There's no point in having a host if you aren't going to have virtual machines (VMs). That's the fun part, after all! VMs on Hyper-V hosts and the Hyper-V hosts themselves can be centrally managed from the Hyper-V console. This makes day-to-day administration more efficient for system administrators because they can accomplish all their tasks in one console.

VMs, also referred to as *guests,* are the virtual servers that run on the host server. They're given resources by the host, and they're able to function much like a regular physical server would. You have a lot of configuration options when it comes to your VMs, and I explain those options in this chapter.

This chapter is all about VMs. Here, you find out how to create and configure Hyper-V VMs. You discover some of the choices you need to make when creating your VMs (some of them can't be changed later), and see why you may want to make those decisions for your VMs.

Creating a Virtual Machine

Creating a VM is the one of the most common activities that a system administrator will undertake. There are, of course, some very important decisions that you need to make in regard to your server, like which generation of VM you want it to be. You can't change the generation of a VM after it's created, so it's important that you choose the right one the first time.

Hyper-V supports multiple guest operating systems on both the Windows side and the Linux side. Supported Windows guests are Windows 7 with Service Pack 1, up to Windows 11, and Windows Server 2008 with Service Pack 2 all the way up to Windows Server 2022. Supported Linux guest VMs include Red Hat Enterprise Linux (RHEL) and CentOS, Debian, Ubuntu, Oracle, SUSE, and FreeBSD.

To get started, you need to start the New Virtual Machine Wizard. Follow these steps:

1. **From Server Manager, choose Tools ⇨ Hyper-V Manager.**
2. **Right-click your host server's name, and then select New ⇨ Virtual Machine (see Figure 2-1).**

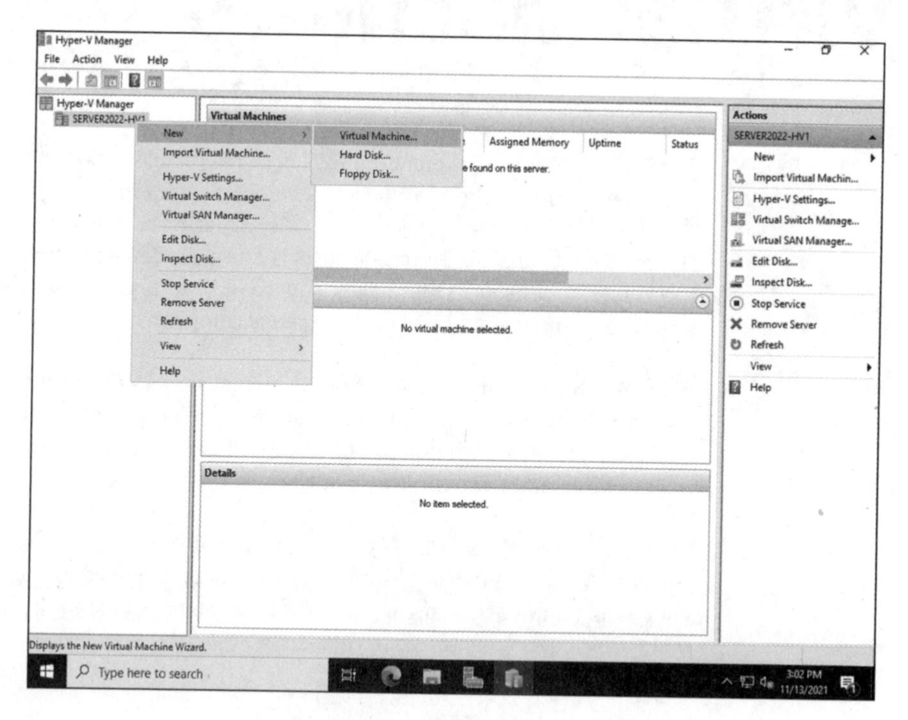

FIGURE 2-1: Launching the New Virtual Machine Wizard.

3. **On the Before You Begin screen, click Next.**
4. **On the Specify Name and Location screen, name your virtual machine.**

   I'll name mine VM1 (creative, I know).

5. **Select the Store the Virtual Machine in a Different Location check box if you want the VM's storage to be somewhere other than the default location.**

   I will leave that unchecked.

6. **Click Next.**

7. **On the Specify Generation screen, select Generation 2 and click Next.**

TECHNICAL STUFF

   Hyper-V supports Generation 1 and Generation 2 VMs. In most cases, you want Generation 2 VMs because they provide support for more advanced features. Generation 2 provides Unified Extensible Firmware Interface (UEFI) support, which is required if you need to use Secure Boot. Generation 1 is useful if you need to install a 32-bit operating system or if you need to support legacy hardware.

8. **On the Assign Memory screen, specify startup memory for the VM and, if you want, select the Use Dynamic Memory for This Virtual Machine check box.**

   For Windows Server 2022, I set the startup memory to 4096MB, and I do select the Use Dynamic Memory for This Virtual Machine check box.

TECHNICAL STUFF

   Startup memory is just that: memory used by the system at startup. This should not be confused with minimum random access memory (RAM), which you can set later on. Dynamic memory allows the host to control memory for the VM. As far as the VM knows, it has 4096MB of RAM, but when it's sitting idle, the host may let something else use the RAM if needed.

9. **Click Next.**

10. **On the Configure Networking screen, choose the virtual switch that you want to connect to.**

    In Figure 2-2, you can see that I've chosen the virtual switch that was created when I installed the Hyper-V role.

11. **Click Next.**

12. **On the Connect Virtual Hard Disk screen, you can choose to create a hard disk, attach to an existing hard disk, or choose to attach a hard disk later.**

    I'll create a hard disk, but I'll change the default 127GB size to 40GB, as shown in Figure 2-3. I'm creating a smaller disk because this is in my lab environment. You should create an appropriately sized disk based on what you're installing.

13. **Click Next.**

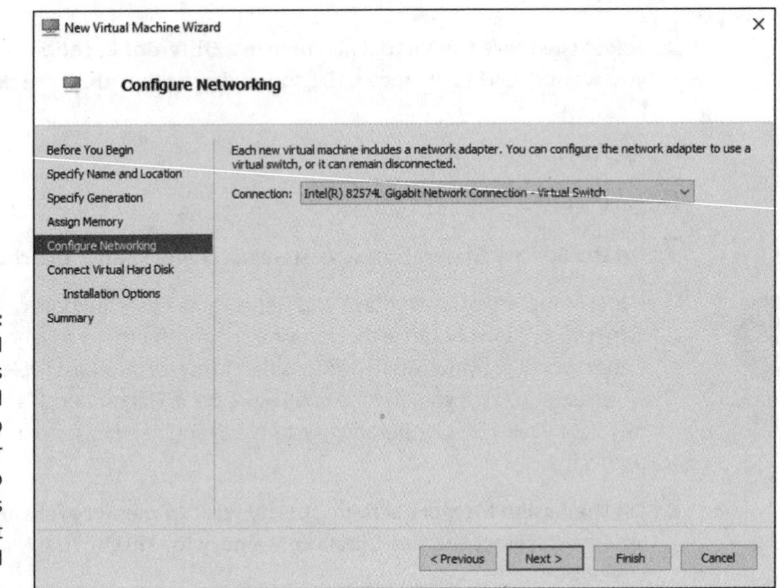

FIGURE 2-2: Selecting a virtual switch allows your virtual machine to connect over your network to other resources it may need, just like a traditional switch.

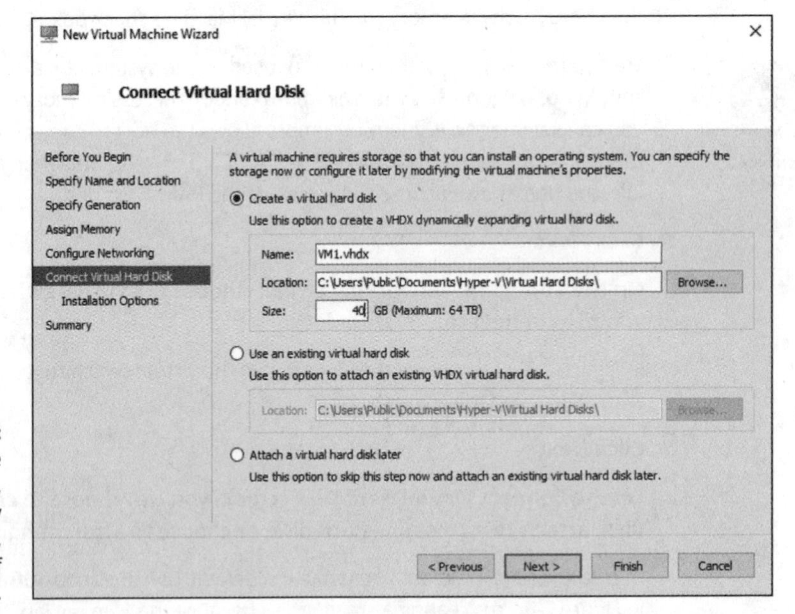

FIGURE 2-3: To create the virtual hard disk, you specify the name, size, and location of the disk.

14. **On the Installation Options screen, you can choose to install an operating system (OS) later, install from an ISO file, or install the OS from a network installation server like WDS.**

    For now, I'll leave Install an Operating System Later selected.

15. **Click Next.**

16. **On the Completing the New Virtual Machine Wizard screen, if everything looks correct, click Finish.**

Assuming there were no issues encountered during creation, you'll find yourself on a screen similar to Figure 2-4 with your newly create VM powered off and ready to be worked with.

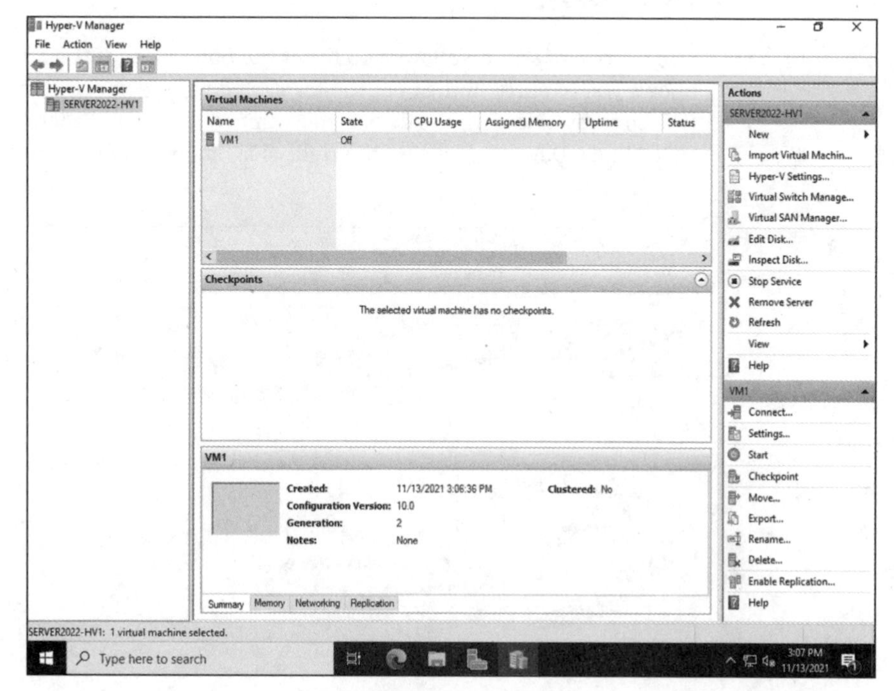

FIGURE 2-4: Your newly created virtual machine is listed in the Virtual Machines pane in Hyper-V Manager.

Configuring a Virtual Machine

When you create the VM, you have a limited set of initial configuration options. In most cases, you're going to want to customize the VM further. Similar to changing settings on a Hyper-V host, you can get into the configuration menu for a VM by right-clicking the VM and choosing Settings.

REMEMBER

The screenshots in this chapter are from a Generation 2 VM. Some of the options will be slightly different if you created a Generation 1 VM because the Generation 1 VM has legacy device support and does not support newer features.

Add Hardware

The Add Hardware section allows you to additional Small Computer System Interface (SCSI) controllers, network adapters, or Fibre Channel adapters. Simply select the device that you want, and click Add. The new device shows up in the menu on the left.

Firmware

The Firmware section allows you to set the boot order of the attached devices and allows you to entry in the boot file for each device. You can change the boot order if desired by selecting the device you want to move, and then by using the Move Up or Move Down buttons on the right side, shown in Figure 2-5.

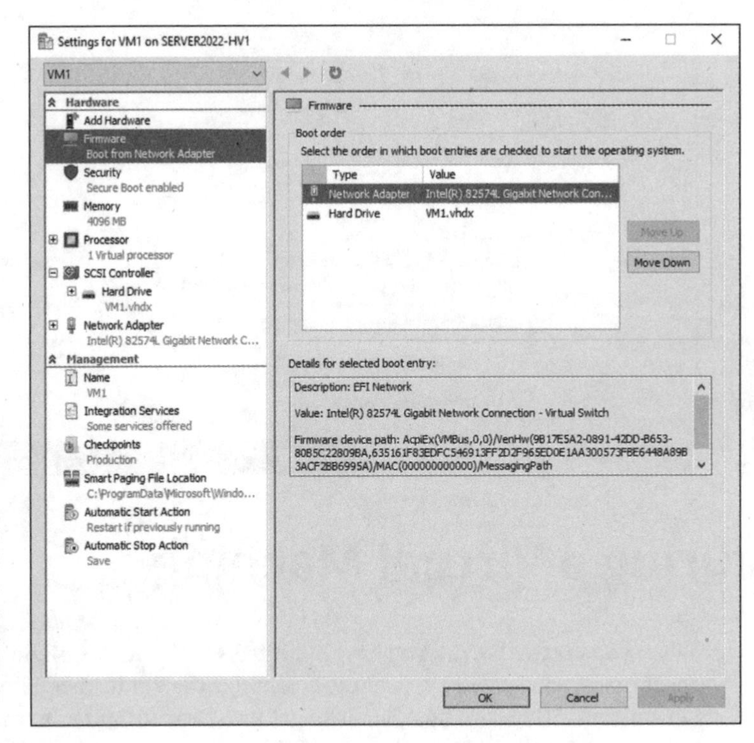

FIGURE 2-5: The Firmware screen allows you to change the boot order of the devices attached to your virtual machine.

Security

The Security section contains check boxes that allow you to provide further security to your VMs. Specifically, you can enable the following:

- **Secure Boot:** This feature was added with Windows Server 2012 R2 and is only supported on Generation 2 VMs because of its reliance of UEFI. This feature ensures that every driver and software component that is loaded during boot time is digitally signed and validated as being legitimate.
- **Trusted Platform Module:** The Trusted Platform Module (TPM) is a special chip that is used for cryptographic operations like the full disk encryption offered by BitLocker. The check box for the VM enables a virtual TPM (vTPM). As of Windows Server 2016, you can enable this even if the host system doesn't have a physical TPM. It does, however, require a Generation 2 VM. If you enable this setting, you can also enable encryption of system state and VM migration traffic. I highly recommend doing this.
- **Shielding:** This feature, which was introduced with Windows Server 2016, is only available on Generation 2 VMs. Shielded VMs take advantage of both Secure Boot and BitLocker to protect the integrity of their boot files and the privacy of their data. Shielding protects a VM from system administrators that are only supposed to administer the host. Those administrators will be able to power cycle the servers in question but will not be able to view the contents of the VM or change any of its settings.

Memory

The Memory section allows you to finetune how your VM will be assigned its memory. When you created the VM, you were only asked how much startup RAM you wanted to assign the machine. The value that you specified there becomes the amount that the host will present to the VM.

If you enable dynamic memory, you can set the minimum and maximum amount of RAM that you want the host to be able to assign. If the VM absolutely needs 512MB to run, you'll want to ensure that the Minimum RAM is set to 512MB. You can set the maximum RAM if you need to make sure that the VM is not allowed to go over a certain amount of RAM. Make sure that you're giving the VM what it needs. If it doesn't have enough memory assigned to it, the performance will suffer or the system may crash.

WARNING

Some applications do not play well with dynamic memory. Your vendor will let you know if this is the case. If it isn't able to take advantage of dynamic memory, then the amount you specified as the startup RAM is the amount of RAM the system will have.

The memory buffer is used to tell Hyper-V how much memory it should set aside for a VM, when you're using dynamic memory. Think of it as the Hyper-V equivalent of a rainy-day fund. When the VM needs more memory, the buffer is used in between the time that the VM makes it known that it needs the RAM and when the RAM is actually granted. So the VM gets the memory it needs, even though it hasn't technically been allotted yet. By default, this value is set to 20%.

The last setting on this screen is Memory Weight. This is where you can tell the host how important this VM is. If, for example, a mission-critical application is running on it, and you need to make sure that it's never starved for resources, you would set the Memory Weight to High. By default, this setting is on Medium, which you can see in Figure 2-6.

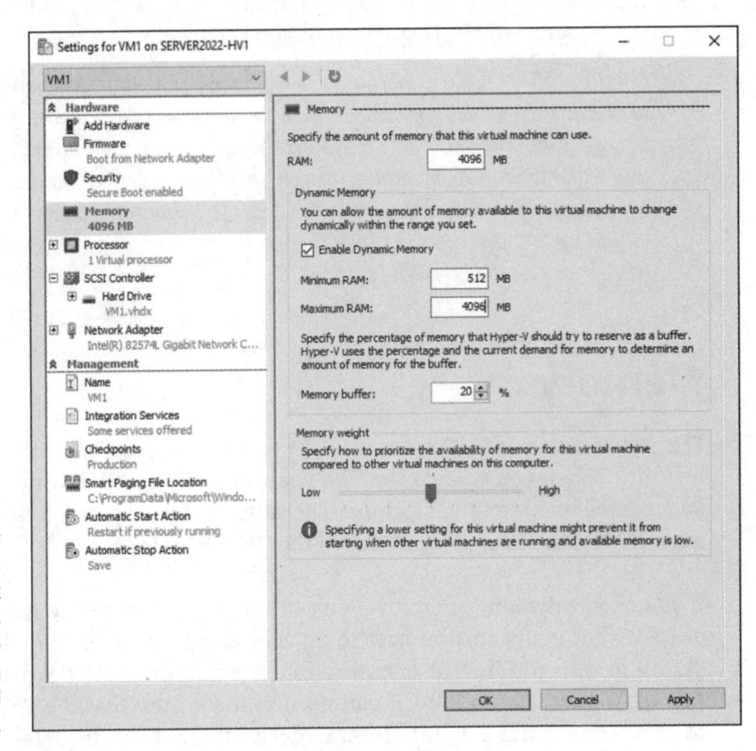

FIGURE 2-6: The Memory screen allows you to adjust how memory is handled by the host for each individual virtual machine.

Processor

The Processor section allows you to set the number of virtual processors that are assigned to the VM. If you click the plus (+) sign next to Processor, you get additional screens for Compatibility and NUMA. If you check the check box on the Compatibility screen, the processor features presented to the VM will be limited, but this will allow you to move to a system with a different processor version. The NUMA configuration screen allows you to set things specifically for NUMA node use. This can improve performance on VMs that have more than one processor assigned to them, if they're sharing CPU resources made available by a NUMA node.

SCSI Controller

The SCSI Controller section allows you to add additional hard drives, DVD drives, or shared drives. When you select the device that you want to add, all you need to do is click the Add button and the new device will appear in the menu. With each device, you're taken to a page that will allow you to configure the new device that you've added.

If you click the plus (+) sign next to SCSI Controller, you see all the SCSI-based devices that are currently connected to your VM. If you select one of the devices (the hard drive, for instance), you can see its location, you're presented with a few options to work with the virtual hard disk file, and you're also presented with the option to remove the hard disk file completely (as shown in Figure 2-7).

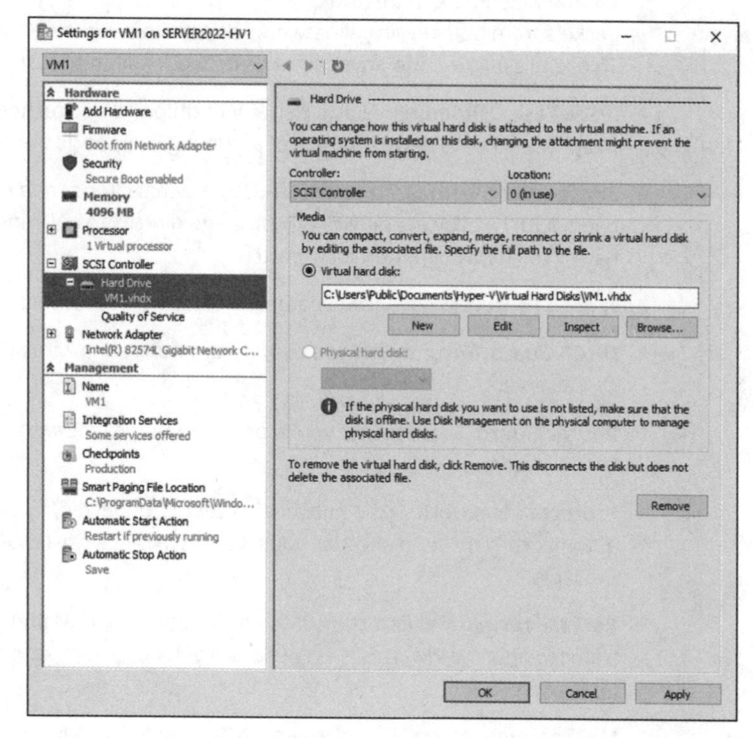

FIGURE 2-7: Each SCSI device has a configuration page that allows you to configure the device or remove it.

Below your hard drive in SCSI Controller, you see an option for Quality of Service. If the VM is hosting an application that has specific Input/Output Operations Per Second (IOPS) needs, you can specify minimum and maximum IOPS on this screen. The vendor will usually specify minimum IOPS if its application is really dependent on low-latency storage. The storage you're using must be capable of meeting the IOPS requirements; making the setting change here only tells Hyper-V to potentially give greater importance to storage traffic for this drive.

Network Adapter

The network adapter section allows you to do several things. You can change the virtual switch that the network adapter is attached to, you can enable vLAN identification to support vLAN tagging, and you can enable and configure bandwidth management. If you click the plus sign (+) next to Network Adapter, you have two options:

A virtual LAN (vLAN) is a logical network made up of endpoints that are on different physical LANs. This allows devices on the same vLAN to communicate as if they were on the same physical wire.

- **Hardware Acceleration:** Hardware Acceleration includes the following:
  - **Virtual Machine Queue (VMQ):** Improves performance by delivering packets from outside virtual networks directly to a virtualized system. To use this feature, the physical host's network adapter must support it.
  - **IPSec Task Offloading:** Allows Hyper-V to offload IPSec-related tasks to the physical host's network adapter assuming it supports the feature.
  - **Single-Root I/O Virtualization (SR-IOV):** Lowers latency and improves bandwidth by allowing network traffic to go directly to a VM instead of having to go through the virtual switch.
- **Advanced Features:** Advanced Features includes the following:
  - **DHCP Guard:** Ensures that you're only communicating with authorized DHCP servers.
  - **Router Guard:** Ensures that you're only communicating with authorized DHCP routers.
  - **Protected Network:** If the network is ever disconnected, the VM can automatically move to another node if that VM is in a cluster and is highly available.
  - **Port Mirroring:** Makes a copy of the traffic going to a VM and sends the traffic to another VM. This is very useful if you need to be able to do traffic inspection.
  - **NIC Teaming:** Allows you to create a NIC team for the VM. The operating system of your VM needs to support NIC teaming for this to work.

NIC teaming allows you to logically join multiple network adapters, so that they behave as if they were one big network adapter. NIC teaming offers some fault tolerance, assuming the connections in the team are going to separate infrastructure switches. NIC teaming can also be used to load-balance traffic coming to the server.

- **Device Naming:** Used to ensure that the VM knows the name of the Hyper-V network. This is helpful when a system has multiple network adapters, especially if the system is multi-homed.

Name

The Name section allows you to change the name of the VM and add any notes that you want to have visible when someone clicks on the VM within Hyper-V Manager.

Integration Services

The Integration Services section allows you to expose services on the host Hyper-V server to the VM you're configuring. This includes things like operating system shutdown, time synchronization, data exchange, heartbeat, backup, and guest services.

Checkpoints

Checkpoints are a point-in-time backup of your VM. They're very useful before a change is made within the operating system that could potentially be destructive. They should not be used in place of backups.

By default, checkpoints are enabled and will be created as production checkpoints. You have the choice between production and standard checkpoints.

- **Production checkpoint:** Uses whatever backup technology is available within the VM's operating system to create checkpoints that can protect data, but that are not aware of running applications.
- **Standard checkpoint:** Captures information on running applications.

You can set your VMs to create automatic checkpoints, and you can also set the save location for all your checkpoints.

Smart Paging File Location

This screen allows you to set the location for the smart paging file for the VM. Smart paging is only used at boot time and can be used to ensure that a VM will be able to boot, even if the host can't provide the minimum startup RAM that the VM requires.

Automatic Start Action

The Automatic Start Action section allows you to set what you want the VM to do when the host is started. You have a choice between doing nothing, automatically starting if it was running when the service stopped, and always starting this VM automatically.

Automatic Stop Action

The Automatic Stop Action section determines what the VM will do if the Hyper-V host is shut down. The options are to save the VM state, turn off the VM, or shut down the guest operating system.

Shielded Virtual Machines

In the "Configuring a Virtual Machine" section, earlier in this chapter, I show you the check box that allows you to turn a VM into a shielded VM. Shielded VMs require more than that simple check box to function properly. You must set up the appropriate server infrastructure to support shielded VMs.

In Windows Server 2016, the Host Guardian Service was introduced. The Host Guardian Service protects shielded VMs by ensuring that the host they're on is a trusted Hyper-V host (guarded host). Essentially, a shielded VM may only run on a Hyper-V host that is able to pass its health attestation to the system or systems running the Host Guardian Service. There are two types of attestation that a guarded host can use:

- **TPM-trusted attestation:** Verifies the guarded host's identity using the guarded host's TPM chip. This method of attestation provides the best security but is more complicated to set up.
- **Host key attestation:** Verifies the guarded host's identity with a key. It's easier to set up than the TPM-based type and is intended to be used on systems that don't have a TPM chip.

For VMs to support being shielded VMs, they must be Generation 2 VMs and have a virtual TPM. Shielded VMs are encrypted by BitLocker.

TIP

You can convert existing VMs into shielded VMs if you want. They just need to be running at least Windows Server 2012.

» Understanding how to create a virtual switch

» Digging deeper with Hyper-V networking

Chapter 3
Virtual Networking

One of the most foundational topics when you're learning about virtual machines (VMs) and Hyper-V is virtual networking. Virtual networking can be a difficult shift for those who are used to physical network equipment. Virtual networking not only requires changing the way that things are done, but also opens the door for so many more possibilities.

For example, in the past, when you got a new physical server, you needed someone to configure a switch port for your server, and then someone had to run the cables. With Hyper-V virtual switches, that need is taken out of the equation — you can simply assign a VM to a virtual switch, and Hyper-V handles a lot of the configuration for you. You can get into the weeds if you want to, of course, but you don't have to.

The virtual switch in Hyper-V allows VMs to communicate with each other, with the host system they're on, and with systems that are on other networks.

In this chapter, I introduce you to virtual networking in Hyper-V, fill you in on the types of virtual switches, and explain how to create switches and do some more advanced network configurations.

Identifying the Types of Virtual Switches

Hyper-V gives you three different types of virtual switches to work with: external, internal, and private. The type of virtual switch you create will depend entirely on what kinds of communication you want to allow to occur. In the following sections, I introduce you to each type of switch and show you where they would be the best fit.

TECHNICAL STUFF

Hyper-V hosts can have multiple virtual switches. A virtual machine can have one virtual switch per network adapter. So if you have a virtual machine that needs access to two virtual switches, you need two separate virtual network adapters.

External

When you first configure Hyper-V, you're asked to assign a network adapter for a virtual switch. This is an external switch. External switches allow you to reach out to other networks, including the Internet. An external switch is close to the traditional network switch, given the way that it's used.

The biggest differentiator between external virtual switches and internal or private virtual switches is that external virtual switches are the only ones that are assigned to a physical network adapter. This is what allows them to access your physical network.

Internal

The internal virtual switch allows VMs to talk to one another (as long as they're on the same host) and to their Hyper-V host. VMs may not communicate with VMs on other Hyper-V hosts or anywhere else on the network. Internal virtual switches are not assigned a network adapter on the physical host; they're assigned a vNIC, which facilitates communication between the VMs and the Hyper-V host.

Internal virtual switches are most useful for traffic related to monitoring by the Hyper-V host, including *heartbeat traffic,* which is used to confirm that a system is up by sending packets back and forth. If the host doesn't receive a packet back from the VM guest, it can restart the VM.

Private

The private virtual switch allows you create a truly isolated network segment. It allows VMs on the same Hyper-V to communicate with one another. It does not allow the VMs to communicate with the Hyper-V host. Private virtual switches are not assigned a network adapter.

Private virtual switches are very useful when you need to isolate a network environment, as you do in a development, test, or lab network.

Creating a Virtual Switch

Now that you have an understanding of the types of switches available to you in Hyper-V, you're probably ready to actually create a virtual switch and start having some fun. You can create a virtual switch through Hyper-V Manager or through PowerShell. I walk you through both options in this section.

Hyper-V Manager

Creating a virtual switch in Hyper-V Manager is how many system administrators start their journey into Hyper-V networking. Hyper-V Manager is an easy-to-use interface, and creating the virtual switch takes just a few minutes. For this example, I'll create an external virtual switch. The creation of an internal or private virtual switch is identical, except you don't select a network adapter with either the internal or private virtual switch.

Follow these steps:

1. **From Server Manager, choose Tools ➪ Hyper-V Manager.**
2. **Right-click the name of your Hyper-V server and choose Virtual Switch Manager.**
3. **With External selected, click the Create Virtual Switch button.**

   The Virtual Switch Properties screen appears. Here, you can customize the settings of the virtual switch.
4. **Give your virtual switch a name.**
5. **Ensure the radio button for External Network is selected, and from the drop-down list, select the network adapter the switch will use, as shown in Figure 3-1.**
6. **Click Apply, and then click OK.**

Virtual Networking

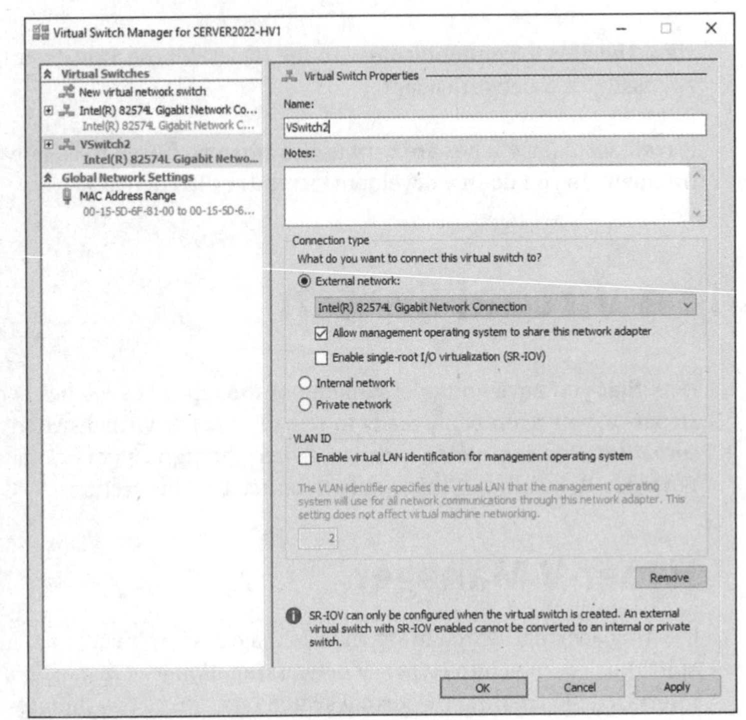

FIGURE 3-1: Under Connection Type, you need to select External Network and select a network adapter to create an external virtual switch.

PowerShell

Creating a virtual switch in PowerShell can be just as simple as it is through the graphical user interface (GUI) when you understand which parameters PowerShell needs to create the type of switch you want.

External

To create an external virtual switch in PowerShell, you need to find out what the name of the network adapter is that you want to use. To do this, you can run the following command:

```
Get-NetAdapter
```

TECHNICAL STUFF

You can only have one virtual switch to one network adapter. You'll get an error if you try to bind more than one virtual switch to the same adapter.

In my case, my network adapter is named Ethernet1, because Ethernet0 is being used by the default switch. So, to create the external virtual switch, the command looks like this:

```
New-VMSwitch -Name PSExternalSwitch -NetAdapterName Ethernet1
   -AllowManagementOS $true
```

This command tells PowerShell that I want to create a new virtual switch, that I want to name it PSExternalSwitch, that I want it to be an external switch (because I'm specifying a network adapter), and that I'll allow the management operating system (OS) to use it. (In Hyper-V Manager, there was a check box that was checked by default to allow this same thing.)

Start to finish, Figure 3-2 shows you what creating an external virtual switch looks like.

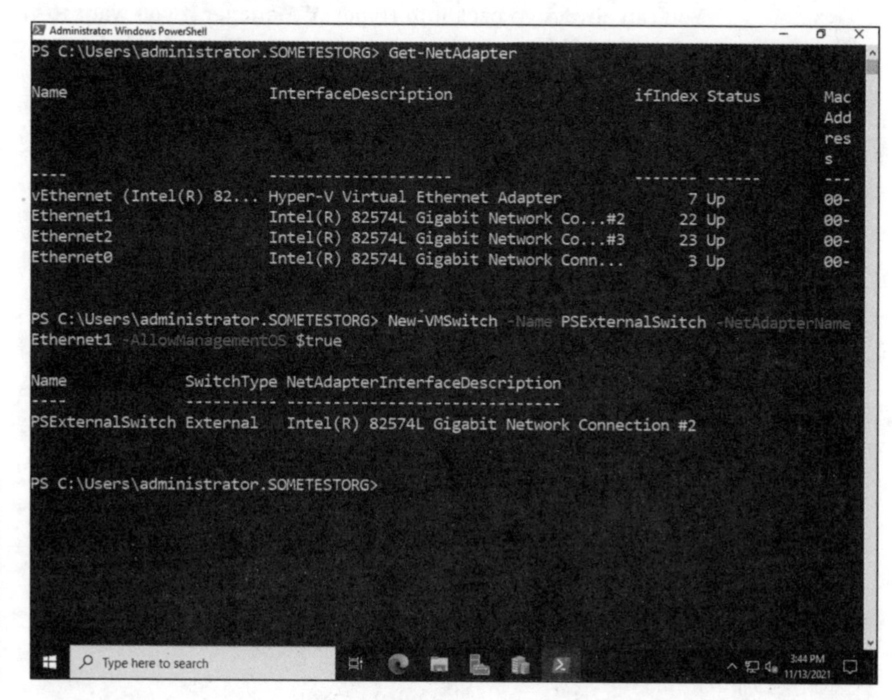

FIGURE 3-2: Creating an external virtual switch requires finding the name of the network adapter you want to use.

Internal

Creating an internal virtual switch does not require any knowledge of the network adapters on the system because it doesn't use them. Instead, you specify what you want your switch to be named and what type of switch you want it to be.

The code to create an internal virtual switch looks like this:

```
New-VMSwitch -Name PSInternalSwitch -SwitchType Internal
```

Private

Private virtual switches are very similar in creation to internal virtual switches. The only difference is that the -SwitchType parameter is used to set the virtual switch as a private virtual switch:

```
New-VMSwitch -Name PSPrivateSwitch -SwitchType Private
```

You can always go back into Hyper-V Manager if you want to view your switches in a graphical format. Figure 3-3 shows the Virtual Switch Manager with the three virtual switches that were created in this section.

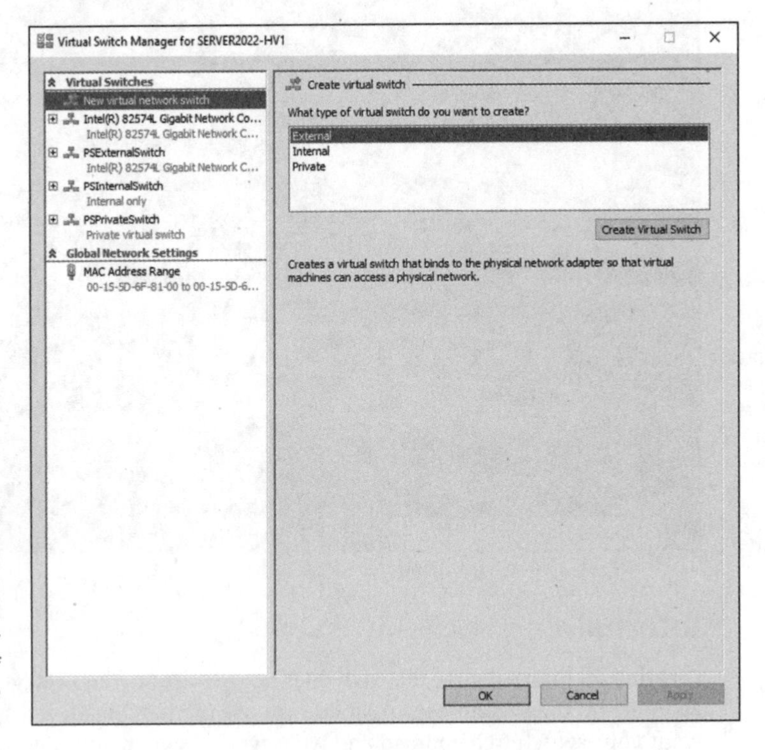

FIGURE 3-3: All your virtual switches will show up in the Virtual Switch Manager regardless of where they're created.

Getting into Advanced Hyper-V Networking

Networking in Hyper-V offers a wide array of features that make it usable and scalable in an enterprise situation. In this section, I cover a few of the more common features that come in handy in enterprise environments.

Virtual local area network tagging

Virtual local area networks (VLANs) are used to isolate network traffic. You may want to do this for security reasons, or you may do it to improve network performance by decreasing the size of the broadcast domain.

A *broadcast domain* is essentially the area where a broadcast is being forwarded. If you have a large broadcast domain, network performance can be negatively impacted because all the systems in that broadcast domain will receive any broadcast traffic and potentially respond. It's better to have a smaller broadcast domain so that it doesn't become saturated with a large amount of broadcast traffic.

VLAN tagging is used to identify packets as they travel through trunks (which are used for communications between switches). You don't have to give up the ability to do VLAN tagging when you make the move to a virtualized environment. You do need to ensure that both the network adapters and the physical network that switch the host is connected to support VLAN tagging. Assuming they do, you turn on VLAN tagging on the virtual external switch, and then on the individual VMs. In the following sections, I cover the steps involved in enabling VLAN tagging.

Enabling VLAN tagging on the switch

The first step to using VLAN tagging is to enable it on the virtual switch. This will add the VLAN tag to traffic that is going through the virtual switch for the Hyper-V host (management OS). Follow these steps:

1. **From Hyper-V Manager, right-click the name of the Hyper-V server and choose Virtual Switch Manager.**
2. **Select the external virtual switch you want to configure, and select the Enable Virtual LAN Identification for Management Operating System check box.**

3. **Set the VLAN ID number to whatever you need to use (see Figure 3-4).**
4. **Click Apply, and then click OK.**

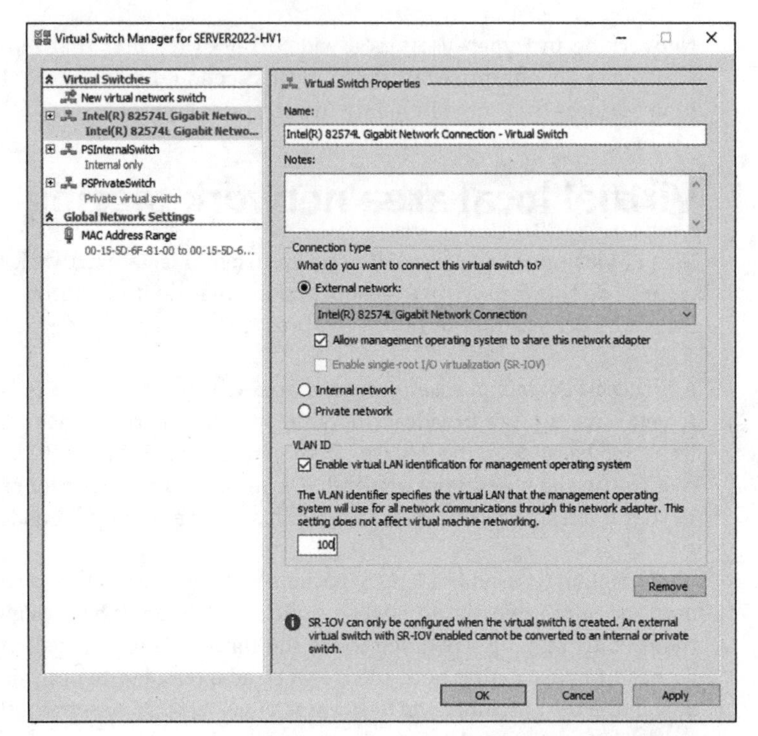

FIGURE 3-4: To enable VLAN tagging on the virtual switch, you need to select the check box and set which ID number you want the switch to use.

Enabling VLAN tagging on the virtual machine

Now that VLAN tagging is enabled on the switch, you're ready to enable it for a VM. Follow these steps:

1. **From Hyper-V Manager, right-click the VM and choose Settings.**
2. **Select the network adapter that is connected to the external switch where you just enabled VLAN tagging.**
3. **Select the Enable Virtual LAN Identification check box and set the VLAN ID number, as shown in Figure 3-5.**

   The VLAN ID number should match what is set on the switch.

4. **Click Apply, and then click OK.**

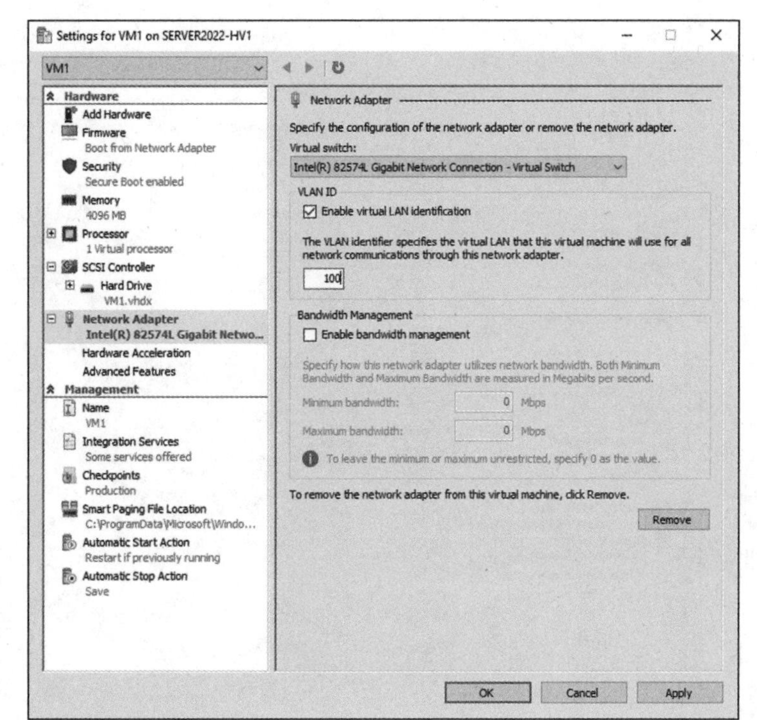

FIGURE 3-5: To enable VLAN tagging on the VM, select the check box and set the VLAN ID.

Bandwidth management

For each network adapter that is connected to a VM, you can set up bandwidth management. Bandwidth management allows you to specify a minimum or maximum amount of bandwidth in Megabits per second (Mbps) that the network adapter is allowed to use. Setting either bandwidth setting to 0 means that it's unrestricted. In my example in Figure 3-6, you can see that I've set the minimum to unrestricted but the maximum to 1024 Mbps.

To make changes to the bandwidth, follow these steps:

1. **From Hyper-V Manager, right-click the VM and choose Settings.**
2. **Select Network Adapter.**
3. **Select the Enable Bandwidth Management check box and make the changes to Minimum and Maximum that you want.**
4. **Click Apply, and then click OK.**

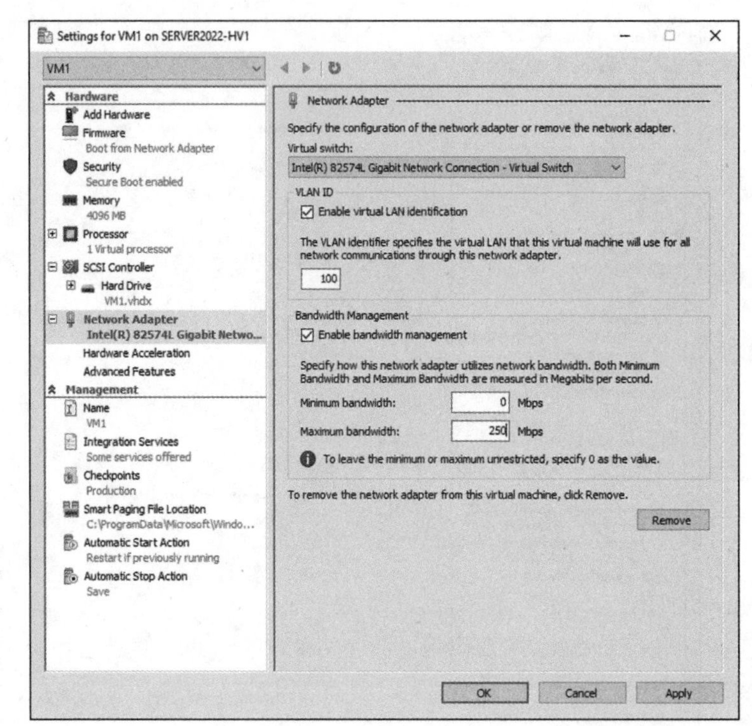

FIGURE 3-6: Setting a maximum amount of bandwidth for a VM can prevent the VM from saturating a network adapter.

Network interface card teaming

Network interface card (NIC) teaming can help to reduce bandwidth constraints by allowing you to pair two network adapters so that they act like one network adapter. There are some prerequisites that you need to meet before you can configure your VM to use NIC teaming:

- The Hyper-V host must have at least two network adapters to support NIC teaming.
- If the network adapters on the Hyper-V host are on separate physical switches, they must be in the same subnet.
- You need to create a separate external virtual switch and attach it to each network adapter in the NIC teaming pair.
- The VM that you want to use NIC teaming with has to be connected to both virtual switches.

In the following sections, I walk you through the steps needed to configure NIC teaming in Hyper-V.

Creating the two virtual switches

First, you need to create the two virtual switches and assign them to network adapters. Follow these steps:

1. **From Hyper-V Manager, right-click the Hyper-V host and select Virtual Switch Manager.**
2. **Select New Virtual Network Switch and click Create Virtual Switch.**
3. **Name the switch, ensure that it's set to External Network, and choose one of the unused network adapters, as shown in Figure 3-7.**
4. **Click Apply.**
5. **Repeat steps 2 through 4 for the second virtual switch.**
6. **Click OK.**

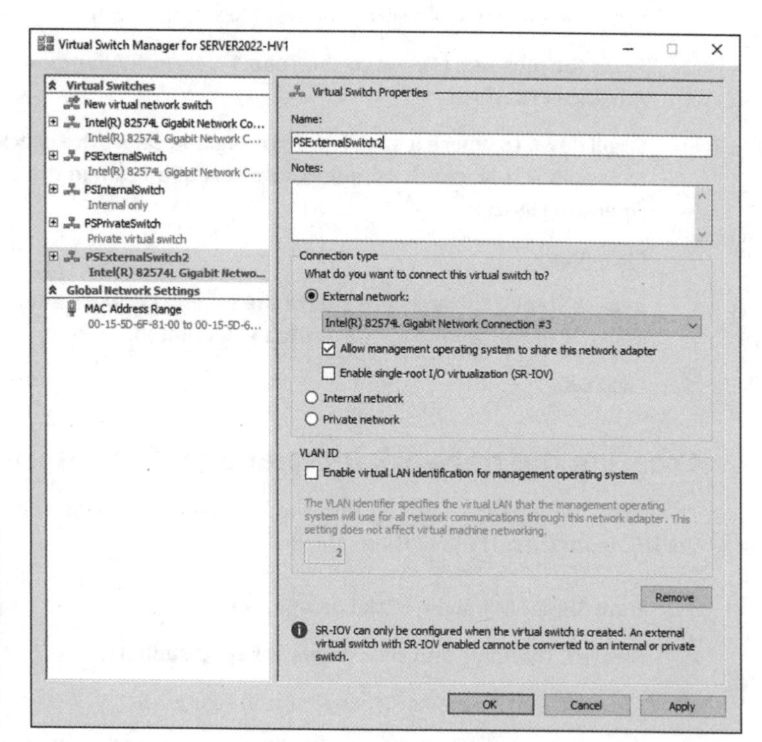

FIGURE 3-7: To support NIC teaming, you must create two virtual switches connected to two separate network adapters.

Virtual Networking

Creating the network adapters on the virtual machine and enabling network interface card teaming

Now that the virtual switches are created, they have to be attached to the VM that you want to create the team on. To do this, you add two network adapters to the VM. Each of these network adapters will be assigned to one of the switches you created previously. Follow these steps:

1. **Right-click the virtual machine and select Settings.**
2. **Click Add Hardware and choose Network Adapter.**
3. **Click Add.**
4. **On the configuration page for the network adapter, choose one of the virtual switches you just created.**

   In my case, that virtual switch is named PSExternalSwitch.
5. **Press the plus sign (+)next to the Team1 Network Adapter and select Advanced Features.**
6. **Scroll down to where it says NIC Teaming and select the check box next to Enable This Network Adapter to Be Part of a Team in the Guest Operating System.**
7. **Click Apply.**
8. **Repeat Steps 2 through 7 to create the second network adapter and assign it to the second virtual switch you created.**
9. **Click OK.**

Creating the network interface card team

The final step involves logging into the VM and completing the configuration of the NIC team there. Follow these steps:

1. **From Server Manager, click Local Server.**
2. **Find NIC Teaming, and click where it says Disabled.**

   You get the NIC Teaming screen similar to Figure 3-8.
3. **Select Ethernet 2 and Ethernet 3 (yours may be named differently).**

   You can hold down the Ctrl key to select them both.

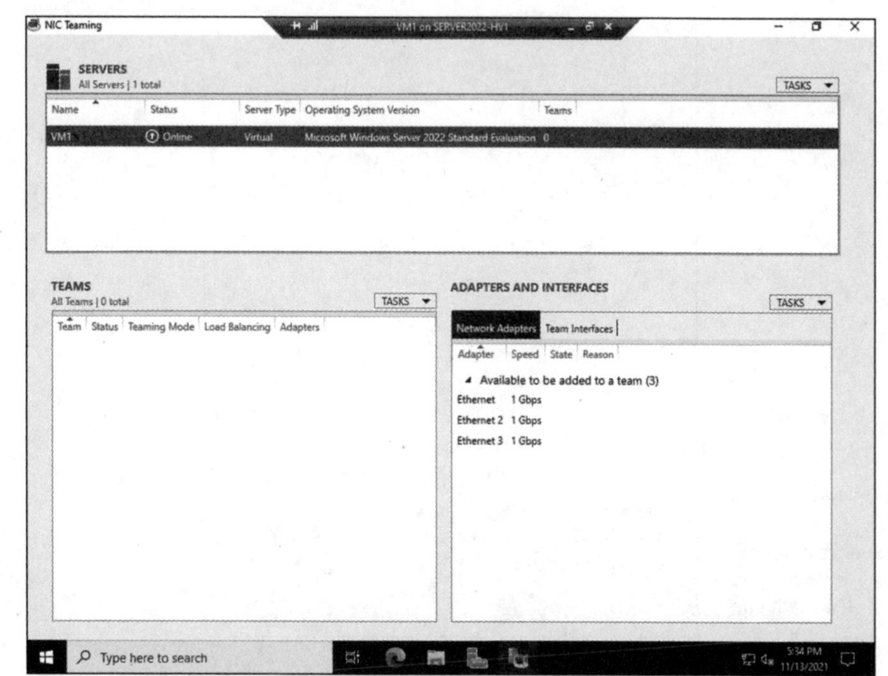

FIGURE 3-8: The NIC Teaming screen is where you can set up the network adapters for teaming inside of the operating system.

4. **With the two network adapters highlighted, click Tasks and then click Add to New Team.**
5. **Name your team.**
6. **Click Additional Properties and make your selections.**

   I've set Teaming Mode to Switch Independent, Load Balancing Mode to Address Hash, and Standby Adapter to None, as shown in Figure 3-9.

   Here's what these settings mean:

   - **Teaming Mode:** With Teaming Mode, you have two options. You can choose Switch Independent or Switch Dependent. With Switch Independent, the switches are not aware of the NIC team's existence and make no determination on how traffic should be distributed. Switch Dependent implies that the switch will determine how to distribute traffic.
   - **Load Balancing Mode:** There are three load balancing modes — Address Hash, Hyper-V Port, and Dynamic.

     Address Hash creates a hash that is based on the addressing of the packet. The hash is then assigned to one of the adapters.

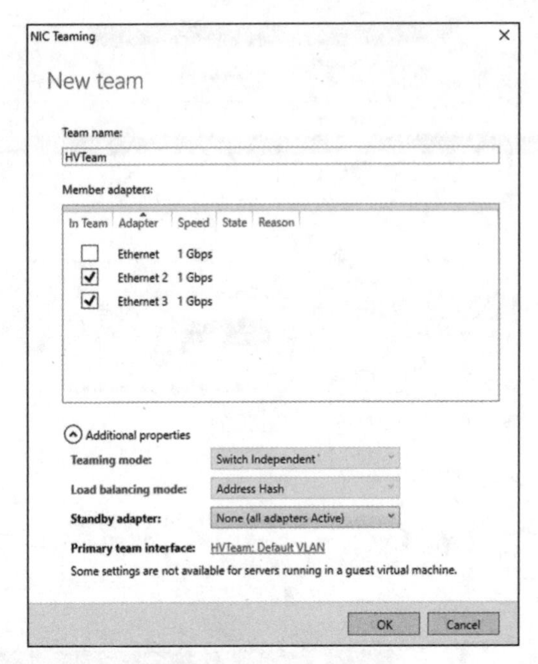

FIGURE 3-9: Configuring the team can be simple, or it can be a bit more advanced by expanding the Additional Properties.

In Hyper-V Port, VMs get independent MAC addresses. This can't be used on a NIC team inside of a VM. It's recommended to use address hashing instead of Hyper-V port for that use case.

If you select Dynamic, outbound traffic is based on a hash made from the TCP port and IP address. Inbound loads are treated the same as Hyper-V Port style. Dynamic has the best performance of the three modes.

7. **Click OK.**

It will take a moment for the teamed NICs to come up, but when they do, they'll both show Active and you'll begin to get traffic statistics for them (as shown in Figure 3-10).

Looking at single-root I/O virtualization

Single-root I/O virtualization (SR-IOV) can reduce latency for virtual machines by bypassing the software layers of Hyper-V networking and allowing more direct access to the host's network card. To use SR-IOV, you must have a network adapter that supports it, and the system's motherboard and firmware must support it as well.

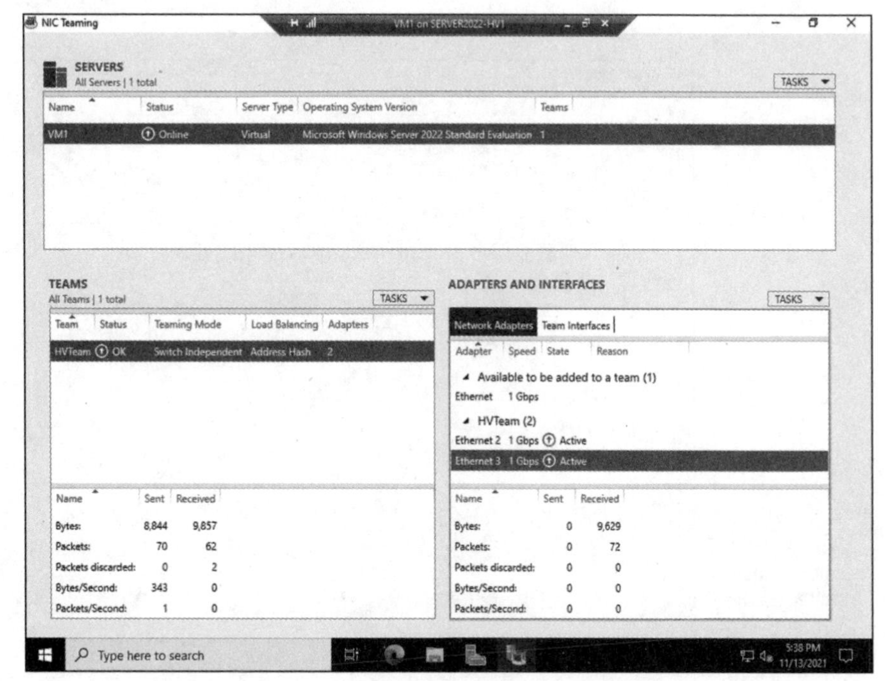

FIGURE 3-10: The configured NIC Team gives you the health status and statistics on how much traffic is passing through.

To begin using SR-IOV, it needs to be enabled for the virtual switch, and then for the virtual network adapter of the VM that you want configure to use SR-IOV. You can't enable SR-IOV after a switch has been created — you need to create a new virtual switch. Follow these steps:

1. **From Server Manager, choose Tools ➪ Hyper-V Manager.**
2. **On the right side of the screen, click Virtual Switch Manager.**
3. **Click New Virtual Network Switch, and click External.**
4. **Click the Create Virtual Switch button.**
5. **Name your switch and then select the Enable Single-Root I/O Virtualization (SR-IOV) check box, as shown in Figure 3-11.**
6. **Click OK.**
7. **When you're warned about your pending change affecting network connectivity, click Yes.**

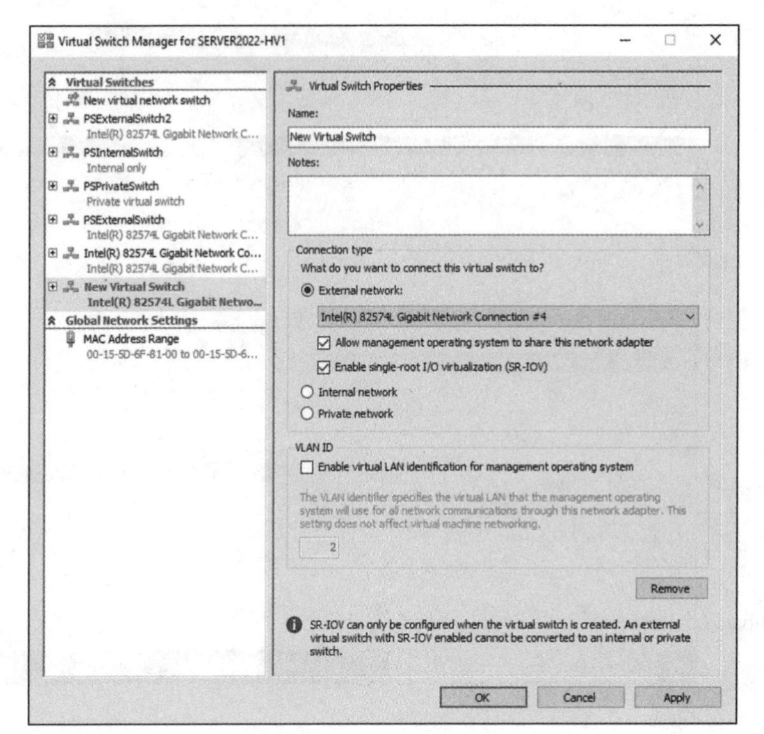

FIGURE 3-11: You must enable SR-IOV on the virtual switch before you can enable it on a virtual machine.

Now SR-IOV is enabled on the new switch, so you can assign the switch to a VM and check the SR-IOV option for the VM as well. Follow these steps:

1. **From Hyper-V Manager, right-click the virtual machine you want to configure, and click Settings.**
2. **Add the new switch you just created by clicking Add Hardware, Network Adapter and clicking Add.**
3. **Select the new switch from the drop-down list and click Apply so that the change is saved.**
4. **Click the plus sign next to Network Adapter, and click Hardware Acceleration.**
5. **Select the Enable SR-IOV check box (shown in Figure 3-12), and then click OK.**

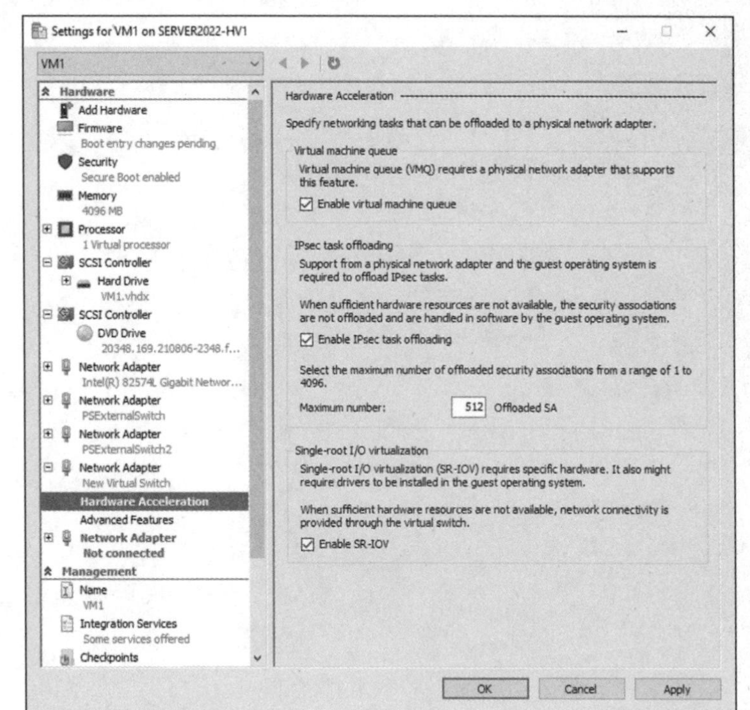

FIGURE 3-12: The option to enable SR-IOV is in the Hardware Acceleration section of the Network Adapter.

» Identifying types of disks

» Adding storage to the host

» Adding storage to the virtual machine

Chapter 4
Virtual Storage

After you build your virtual machine (VM) and your networking is set up, the next thing most system administrators will look at is storage for the application that's going to be installed on the VM. There are multiple reasons for this:

- You don't want to install applications on the operating system (OS) drive.
- The application may have very strict size and/or performance requirements for its storage.

This chapter discusses how to add storage to both the Hyper-V host and the VMs that reside on the Hyper-V host. Here, you learn about the different types of disks and the disk formats as well.

Understanding Virtual Disk Formats

There are two disk formats to be aware of when you start talking about virtual disks: VHD and VHDX. Each format has pros and cons that you should consider before you decide which one you want to use:

- **VHD:** VHD was the original disk format. It's supported by Generation 1 VMs, and by Azure Infrastructure as a Service (IaaS) offerings. A VHD can be a maximum of 2TB in size. To change the size of the disk, you have to power down the VM it's attached to. In most cases, you'll want to use a VHDX format unless you need to do the following:

- Use the drive with a Generation 1 VM (legacy hardware support). Generation 2 VMs don't support VHD.
- Move the VM to Azure.

» **VHDX:** VHDX is the newer disk format available. It's supported by both Generation 1 and Generation 2 VMs and can reach a maximum of 64TB in size. You can change the storage size of a VHDX file while a VM is powered on, which is a huge win for a mission-critical system running on a VM.

Considering Types of Disks

There are four different types of disks used in Hyper-V: fixed, dynamic, differencing, and pass-through. Each has a use case that makes the most sense. When you try to create a disk, you may not see all these options in the menu to choose from.

VHDs in Hyper-V are created as a file that lives in a location that you determine. They may be located locally on your Hyper-V host, or they may be located on a shared volume on network storage.

Fixed

A fixed disk is similar to your standard traditional hard drive. If you provision a 60GB hard drive, then the file for your VM will be a full 60GB. This disk choice offers the best performance, but it can lead to a lot of wasted space because it can't be shrunk down after it's created. Fixed disks are also referred to as *thick provisioning.*

Dynamic

A dynamic disk can save on space. You may have an application that says it requires 200GB for storage. If you provision a dynamic disk, the OS and the application will see a 200GB disk. The actual disk file will be much smaller — the size of the actual data. For instance, if you're only using 30GB of a 200GB dynamic disk, then the size of the disk file will be 30GB. The performance is not as good as a fixed disk, and you have to be careful to not overprovision the storage available to the VMs on the Hyper-V host. Dynamic disks are also referred to as *thin provisioning.*

Differencing

Differencing disks are different from your traditional VHD/VHDX disks. A differencing disk is a dynamic disk that is tied to a *parent disk*. The disk that is tied to the parent disk is known as a *child disk*. Child disks store changes that are made to the parent disks. This makes them excellent for troubleshooting but can make it easy to overprovision storage because they're using dynamic disks.

Pass-through

Pass-through disks are not VHDs. These are physical drives that are attached to the Hyper-V host and then passed through to the VM. Using pass-through disks also allows you to take advantage of physical network storage that is connected to a Hyper-V host. This enables you to maximize the benefit you receive from an expensive physical storage array in your virtual environment. Plus, it can simplify backups and restores, because the data is stored in its native NTFS format.

Adding Storage to the Host

Adding storage to a Hyper-V host is a common administrative task. The Hyper-V host is typically storing the VM configuration files and the VM's disk files on local storage, unless you have it configured to save them on network storage of some kind.

Adding the storage to the Hyper-V host is similar to adding storage to any Windows 2022 Server. Let's create two 50GB drives and add them to the host. Then I'll show you how to change the default save location for the virtual hard disks and the VMs.

Adding the drives

Before you can add the drives to the server for use, they must be installed. After they've been installed, you can go into Server Manager and initialize them.

Follow these steps:

1. **From Server Manager, click File and Storage Services.**
2. **Click Disks.**

3. **Right-click the first disk that says Offline, and click Bring Online, as shown in Figure 4-1.**

   You get a dialog box warning about data loss if the disk is in use elsewhere.

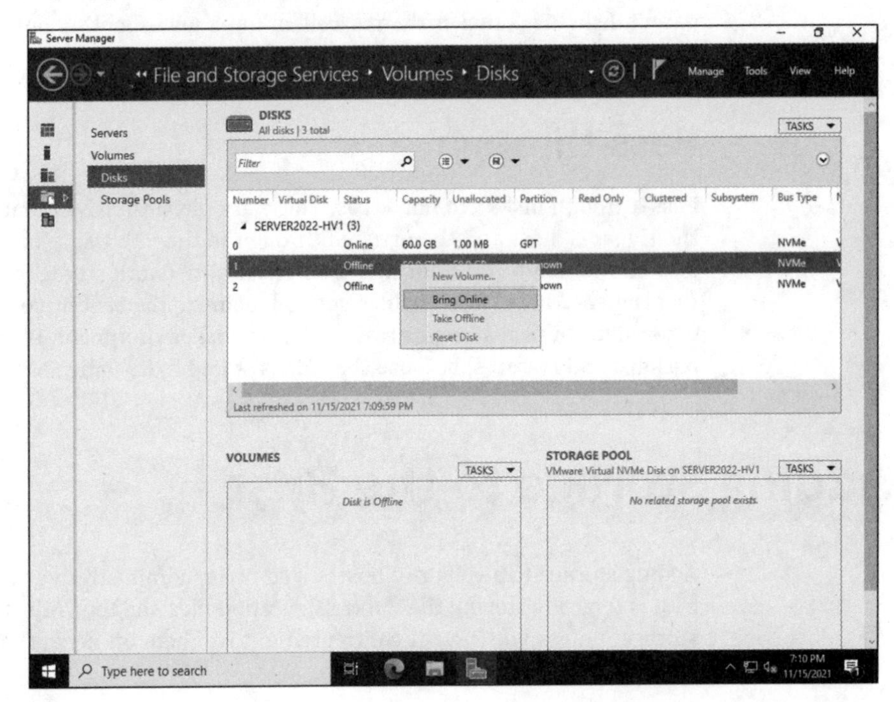

FIGURE 4-1: After the disks are installed, they must be brought online and initialized.

4. **Click Yes.**
5. **Right-click the disk again and choose Initialize.**
6. **On the Initialize Disk warning box that appears, click Yes.**
7. **Right-click the disk again and choose New Volume.**
8. **On the Before You Begin screen, click Next.**
9. **Select the disk you brought online and initialized and click Next.**
10. **On the Specify the Size of the Volume screen, click Next.**
11. **On the Assign to a Drive Letter or Folder screen, click Next.**
12. **On the Select File System Settings screen, click Next.**
13. **On the Confirm Selections screen, click Create.**
14. **After the disk has been created, click Close.**

TIP

If the disk disappears from the list after you've created the volume, simply refresh by clicking the Tasks button and then clicking Refresh.

15. **Repeat Steps 3 through 14 for the second offline disk.**

At this point, your screen should look similar to Figure 4-2, with all three disks showing online and partitioned.

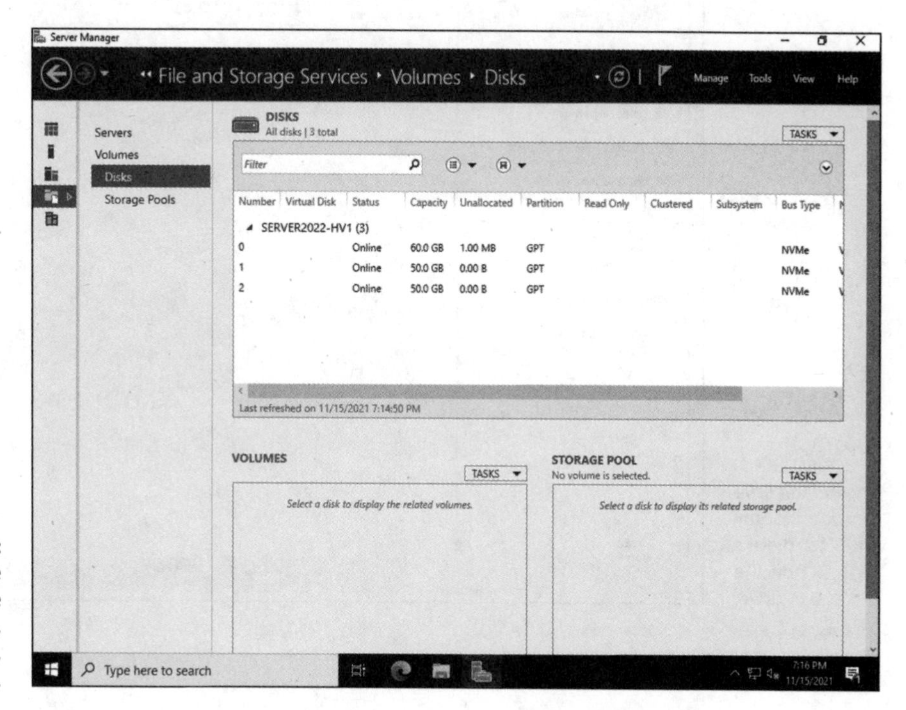

FIGURE 4-2: With the storage added to the server, initialized, and formatted, it's ready for use.

Changing the default save locations of virtual disk files

After you have the new storage added to the Hyper-V host, you can use it for all kinds of things. One of the most common reasons to add storage is to create save locations for the VMs and for the virtual disk files.

Follow these steps to change the default save locations for both VMs and virtual disks:

1. **From Hyper-V Manager, right-click the Hyper-V host and choose Hyper-V Settings.**

   Virtual Hard Disks is selected by default, so let's change that one first.

2. **Click Browse and select one of the new drives you added earlier.**

3. **Click Select Folder.**

   The new location will appear in the box, as shown in Figure 4-3.

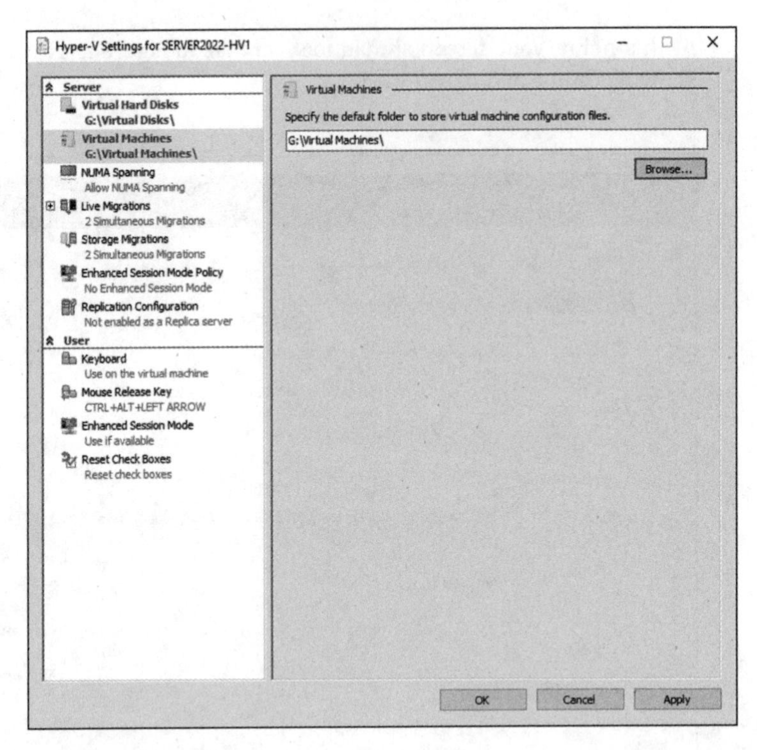

FIGURE 4-3: The default save location for my virtual hard disks is now the `G:\` drive.

4. **Click Apply.**
5. **Click Virtual Machines, underneath Virtual Hard Disks.**
6. **Select Browse and choose the other drive that was added earlier.**
7. **Click Select Folder.**
8. **Click Apply, and then click OK.**

This does not change VMs that were previously created. They would need to be moved to the new locations via storage migration (see Chapter 5 of this minibook).

Adding Storage to the Virtual Machine

Adding storage to a VM is an everyday event for a system administrator. Sometimes you need to expand a virtual drive; other times, you need to add additional virtual drives.

Adding a new virtual drive

With a Generation 1 VM, you're given a choice between a VHD- or VHDX-formatted drive. If you have a Generation 2 VM, you don't have a choice — it can only be a VHDX-formatted drive — so you're asked what type of drive you want to create rather than what format you want to use.

Because the wizards are a little different for the two generations of VMs, I've included the instructions for each.

Adding a disk to a Generation 1 virtual machine

With the Generation 1 VM, you need to make a decision as to what kind of disk you want to create. If you're going to move the VM to Azure, you need to select the VHD format; otherwise, you can typically choose the VHDX format. Follow these steps:

1. **Right-click the VM and click Settings.**

   You have an IDE controller and a SCSI controller in the configuration menu on the left. You can use either one. I'm going to use the IDE controller because my other hard drive is using it (this is for tidiness — you can mix and match controllers).

2. **Select the IDE controller and choose Hard Drive and then click Add (see Figure 4-4).**

   In my case, it's labeled IDE Controller 0. The virtual hard disk button will be selected.

3. **Click New.**
4. **On the Before You Begin screen, click Next.**
5. **On the Choose Disk Format screen, choose the format you want and click Next.**

   I'll select VHDX.

6. **On the Choose Disk Type screen, select Dynamically Expanding, and then click Next.**

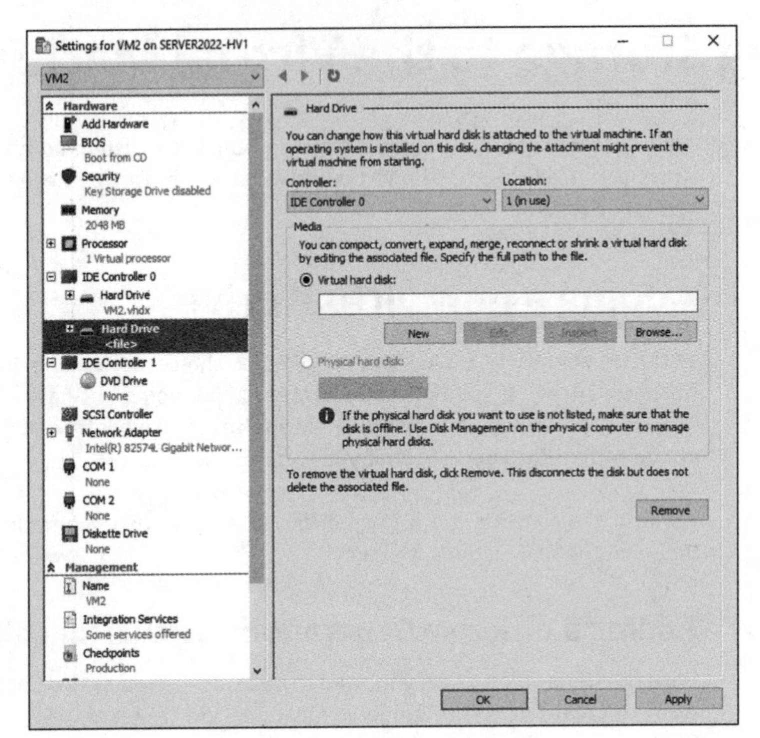

FIGURE 4-4: Creating a disk in a Generation 1 VM gives you the option of an IDE controller or a SCSI controller.

7. **On the Specify Name and Location screen, you can name the disk whatever you would like.**
8. **If you changed the default location in the earlier section of the chapter, that will show here. Change it if needed, and then click Next.**
9. **On the Configure Disk screen, select Create a New Blank Virtual Hard Disk and change the size to whatever you want; then click Next.**

   On the last page of the wizard, you're given a summary of your selections.
10. **If everything looks correct, click Finish.**
11. **On the VM Settings page, click OK.**

Adding a disk to a Generation 2 virtual machine

The Generation 2 VM does not have the IDE controller as an option, so you'll select the SCSI controller when you create this hard drive (unless you have a Fibre Channel Adapter). Follow these steps:

1. **Right-click the VM and click Settings.**
2. **Select the SCSI controller, select Hard Drive, and click Add.**

   Virtual Hard Disk will be selected.
3. **Click New.**
4. **On the Before You Begin screen, click Next.**
5. **On the Choose Disk Type screen, select Dynamically Expanding, and click Next.**
6. **On the Specify Name and Location screen, name your disk and either accept the default location for the disk or choose another location; then click Next.**
7. **On the Configure Disk screen, select Create a New Blank Virtual Hard Disk, set the desired size, and click Next.**
8. **On the final summary page, verify that the settings are correct, and then click Finish.**
9. **Back on the VM Settings screen, click OK.**

Expanding a disk drive

If you want to expand a VHDX file, you can do it while the VM is running. This is one of the great things about VHDX-formatted disks. If you're using a VHD-formatted disk, you need to power down the VM to expand it. Follow these steps:

1. **Right-click the VM and click Settings.**
2. **Select the hard drive you want to expand and click Edit, as shown in Figure 4-5.**
3. **On the Locate Virtual Hard Disk screen, click Next.**
4. **On the Choose Action screen, select Expand, and then click Next.**
5. **On the Expand Virtual Hard Disk screen, set the desired size, and then click Next.**
6. **On the summary page, verify that the settings look correct, and then click Finish.**
7. **Back on the VM Settings screen, click OK.**

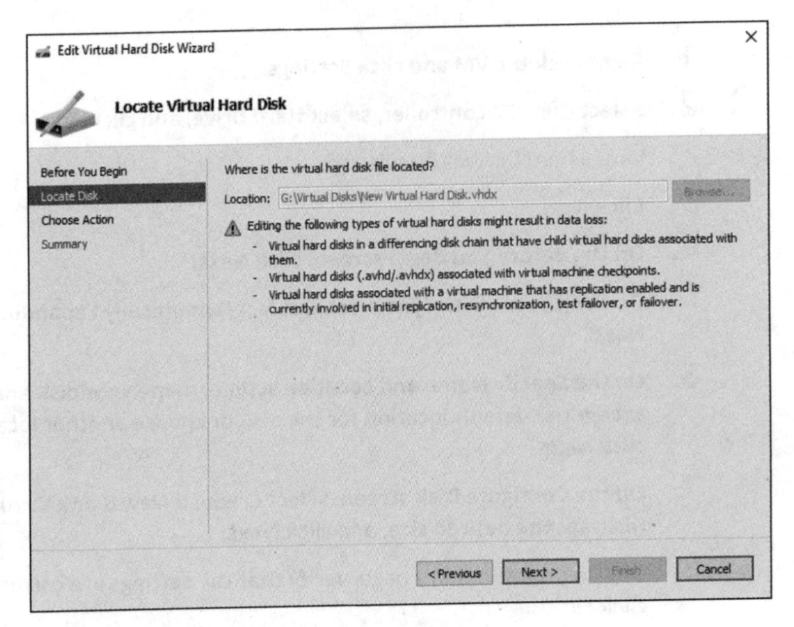

FIGURE 4-5: Expanding a disk is done by selecting Edit when the hard drive is selected.

Adding a pass-through disk

For the VM to see the pass-through disk, it must be offline on the host. If it isn't offline, it won't show up as an available disk. You can verify that there is an offline disk available for you by checking the Disks area of File and Storage Services in Server Manager, as shown in Figure 4-6.

After you've verified that you do have an available offline disk, open Hyper-V Manager and follow these steps:

1. **Right-click the VM you want to add the disk to and click Settings.**
2. **Select SCSI controller and then Hard Drive, and then click Add.**

   Virtual Hard Disk is selected by default.
3. **Select Physical Hard Disk, as shown in Figure 4-7.**

   If there is no hard drive listed, ensure that the disk shows offline on the Hyper-V host.
4. **Click Apply, and then click OK.**

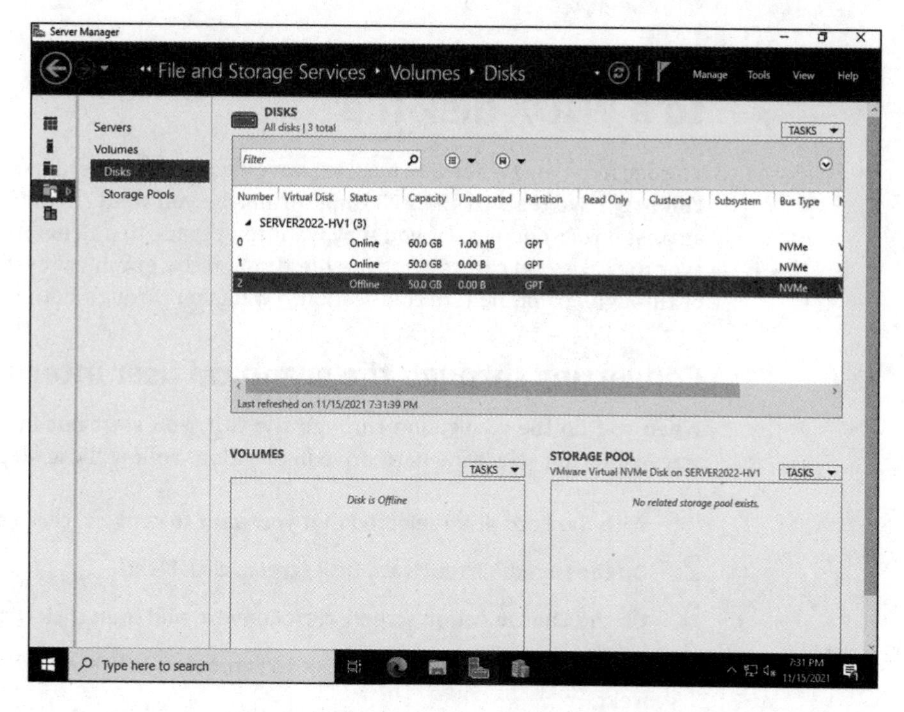

FIGURE 4-6: You need to verify that you have an available disk that is showing as offline. You can do this from Server Manager.

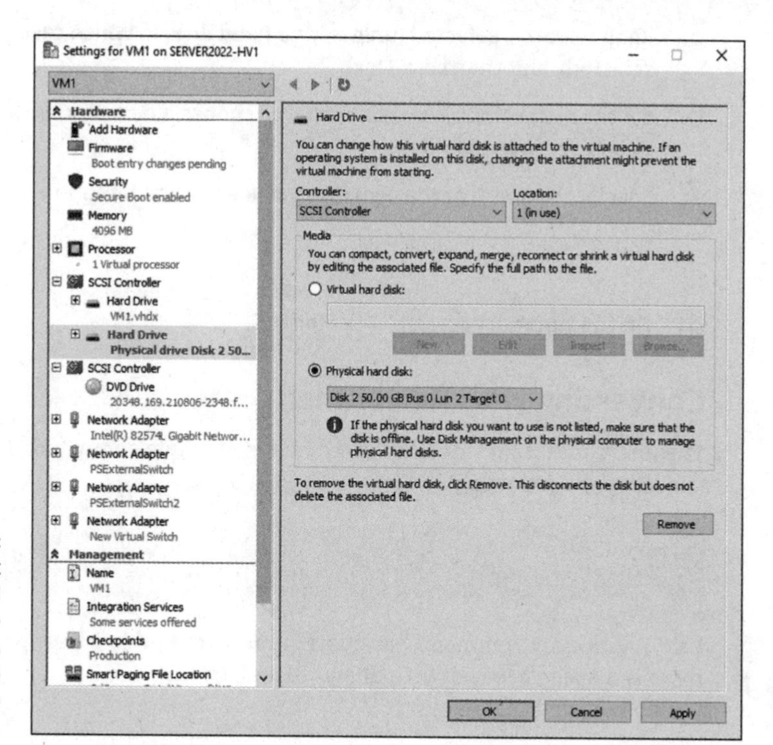

FIGURE 4-7: When you select physical hard disk, the offline disk that was added to the Hyper-V host will show as available.

Converting a VHD disk file to a VHDX disk file

The day may come when you need to convert your VHD disk file to a VHDX disk file. You may have reached the 2TB limit, or maybe you want the improved resiliency against power outages. Or you may want to upgrade to a Generation 2 VM. Whatever the case, you can convert the file through the graphical user interface (GUI) or through PowerShell. In this section, I walk you through both methods.

Converting through the graphical user interface

When you do the conversion through the GUI, you start out in the VM Settings screen and you select the hard drive in question. Follow these steps:

1. **With the hard drive selected that you want to convert, click Edit.**
2. **On the Locate Virtual Hard Disk screen, click Next.**
3. **On the Choose Action screen, click Convert, and then click Next.**
4. **On the Convert Virtual Hard Disk screen, select VHDX, and then click Next.**
5. **Keep Dynamic selected (unless it's a fixed disk, in which case keep Fixed selected), and then click Next.**
6. **On the next screen, you don't need to choose a new storage location unless you want to do so.**
7. **Specify the name of the converted disk and click Next.**

   You can leave it the same — you just need to specify `.vhdx` as the file extension.
8. **On the summary screen, click Finish.**

Converting with PowerShell

The disk conversion through PowerShell is a simple command. The command is similar to the following:

```
Convert-VHD -Path "G:\Virtual Disks\VHD_PS.vhd" -DestinationPath
  "G:\Virtual Disks\VHD_PS.vhdx"
```

TIP

You'll notice my command has quotes around the path. I have done this because there is a space between Virtual and Disks. If your path has no spaces in it, then you do not need the quotes.

In Figure 4-8, you can see that I have the VHD file in my `G:` drive. I run the preceding command, and then I have the VHDX file. That's all there is to it.

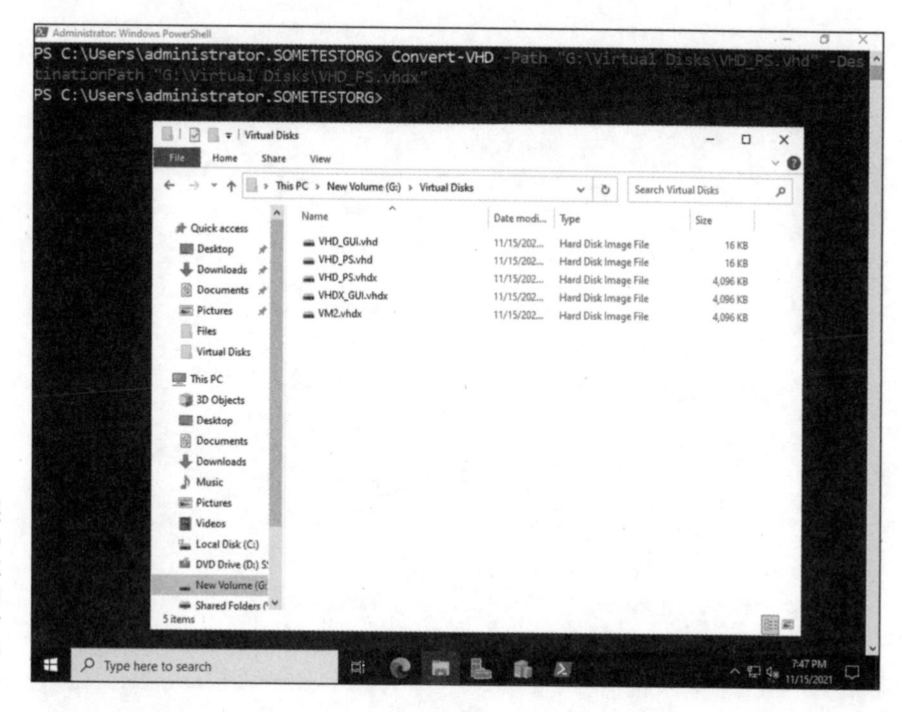

FIGURE 4-8: Converting a VHD file to a VHDX file in PowerShell involves far fewer steps than it does through the GUI.

Attaching the converted drive to the virtual machine

The converted disk drive does not get attached automatically to the VM. You have to specify that you want the VM to use the VHDX file after the conversion is complete. This process is similar to creating a new hard drive, except instead of clicking New or Edit, you click Browse and select the VHDX file. After you select it and click OK, the converted VHDX file will then be attached to the VM.

IN THIS CHAPTER

» Understanding Hyper-V replica

» Performing live migration

» Performing storage migration

» Configuring Failover Clustering

Chapter 5
High Availability in Hyper-V

No discussion about Hyper-V would be complete without a discussion on how to make it highly available. The biggest win for most organizations is the ability to reduce downtime when the underlying physical hardware fails.

In this chapter, you find out how to make your Hyper-V deployments highly available. These technologies can be used to prevent or minimize disruption due to hardware failure, or maintenance on the underlying hosts.

Hyper-V Replica

Hyper-V Replica is a great solution for disaster recovery situations. It allows you to replicate a live virtual machine (VM) to an offline Hyper-V Replica. In the event of an issue that takes the active host down, the VM can be powered on at the replica server.

Hyper-V Replicas may be in the same physical datacenter, or they may be located in geographically distant datacenters.

To use Hyper-V Replica, you must first set up the Hyper-V hosts. Then you can configure the VMs that you want to replicate.

Setting up Hyper-V Replica on the Hyper-V hosts

Before you can use Hyper-V Replica, you need to enable the replica server and configure where you want to allow replication to happen from. This can be done over an unencrypted Kerberos connection or a certificate-based (HTTPS) connection. For production environments, it's recommended to use a certificate-based connection because the replication traffic will be encrypted. For a test or development environment where encryption would introduce overhead, the Kerberos connection would work well. For this example, let's set up an unencrypted replica.

On the replica server, follow these steps:

1. **Open Hyper-V Manager, and right-click the Hyper-V host.**
2. **Select Hyper-V Settings.**
3. **Select Replication Configuration and select the Enable This Computer as a Replica Server check box.**
4. **Select the Use Kerberos (HTTP) check box.**
5. **In the Authorization and Storage section, select Allow Replication from any Authenticated Server.**
6. **Set a storage location for the replicated VM hard disks if you want something other than the default.**
7. **Click Apply, and then click OK.**

   Your screen should look similar to Figure 5-1.

Now that the Hyper-V host is set up as a replica, you'll want to verify that Windows Firewall will allow replication traffic in. There are two rules that you may need to enable:

- **Hyper-V Replica HTTP Listener (TCP-In):** This rule needs to be enabled if you're using the Kerberos-based connection.
- **Hyper-V Replica HTTPS Listener (TCP-In):** This rule needs to be enabled if you're using the certificate-based connection.

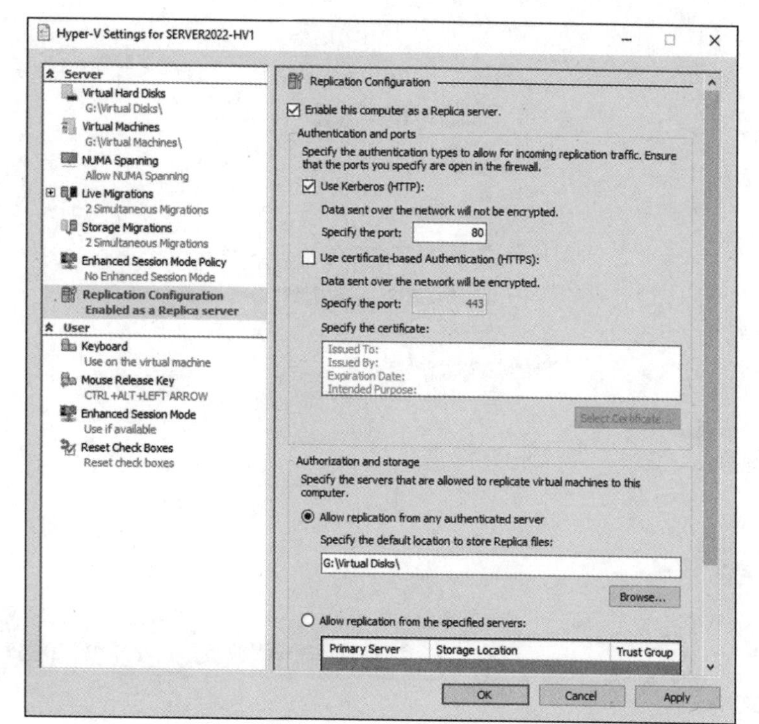

FIGURE 5-1: You must set up the replica server first, before you can replicate VMs.

To enable the rule in the Windows Firewall, follow these steps:

1. **Click Start, and then click the gear icon to access the Settings menu.**
2. **Click Network & Internet.**
3. **On the Status page, scroll down and select Windows Firewall.**
4. **Click Allow an App through Firewall.**
5. **Scroll down to Hyper-V Replica and enable the applicable rules.**

   I've enabled the Hyper-V Replica HTTP rule for the Domain profile, as shown in Figure 5-2.
6. **Click OK.**

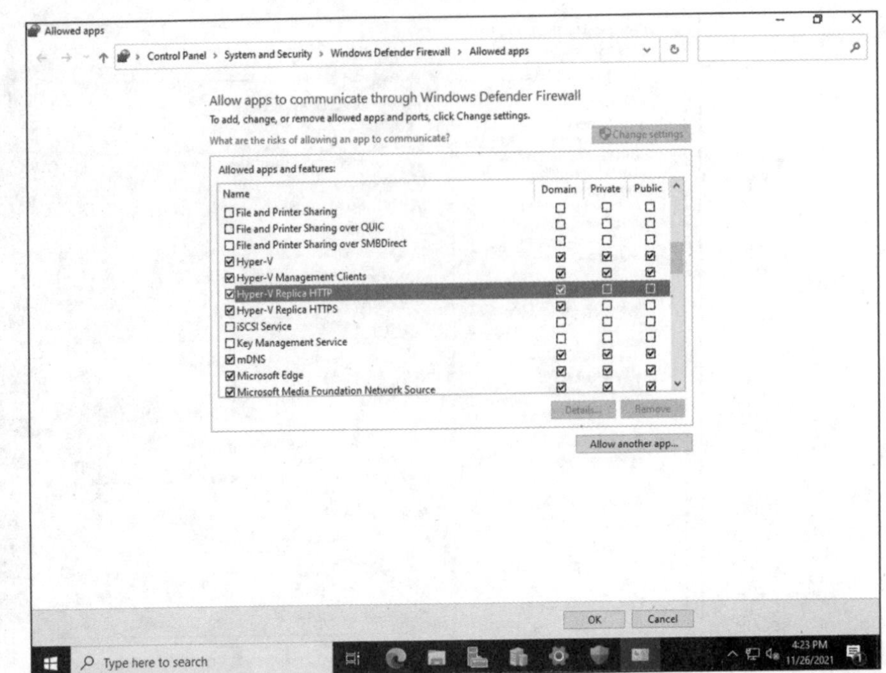

FIGURE 5-2: Hyper-V Replica communications need to be allowed through the firewall for replication to occur.

Setting up replication on the virtual machines

After the Hyper-V Replica host is all set, it's time to configure the VM to replicate to it. Follow these steps:

1. **From Hyper-V Manager, right-click the VM and select Enable Replication.**
2. **On the Before You Begin screen, click Next.**
3. **On the Specify Replica Server screen, enter the name of the server you configured in the prior section, and click Next.**
4. **On the Specify Connection Parameters screen, verify that the information presented is correct.**

   It should look similar to Figure 5-3 if you've been following along.
5. **Click Next.**
6. **On the Choose Replication VHDs screen, select which virtual disks you want to replicate and click Next.**

   This will usually be all of them, but you can uncheck one of the drives if you don't want it to replicate.

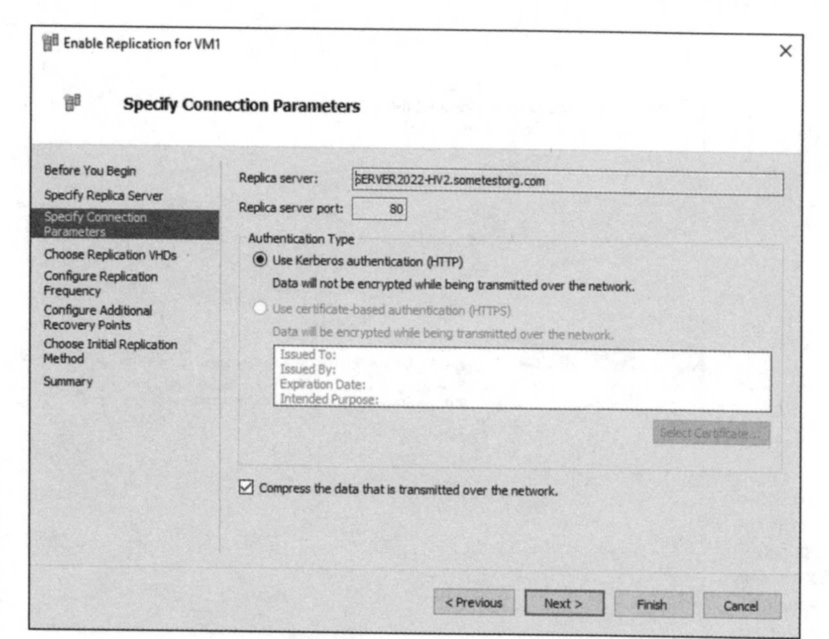

FIGURE 5-3: Make sure that the connection parameters match what you set on the Hyper-V Replica.

7. **On the Configure Replication Frequency screen, set how much time can be between replication cycles and click Next.**

   You can choose among 30 seconds, 5 minutes (default), or 15 minutes. I'll leave it on 5 minutes.

8. **On the Configure Additional Recovery Points screen, leave this set on Maintain Only the Latest Recovery Point and click Next.**

TECHNICAL STUFF

   If you're concerned about data corruption with the data on the drive, you can set the replication to save additional recovery points, every hour. If your system is ever hit by ransomware or some other malicious activity or entity, this will give you the ability to restore to a good recovery point, rather than the potentially damaged data that was replicated over.

9. **On the Choose Initial Replication Method screen, choose Send Initial Copy over the Network and Start Replication Immediately, as shown in Figure 5-4.**

   You can start immediately, or you can schedule replication. If it's a large system, you can export the initial copy and then import on the replica.

10. **On the summary page, confirm that your settings look correct and then click Finish.**

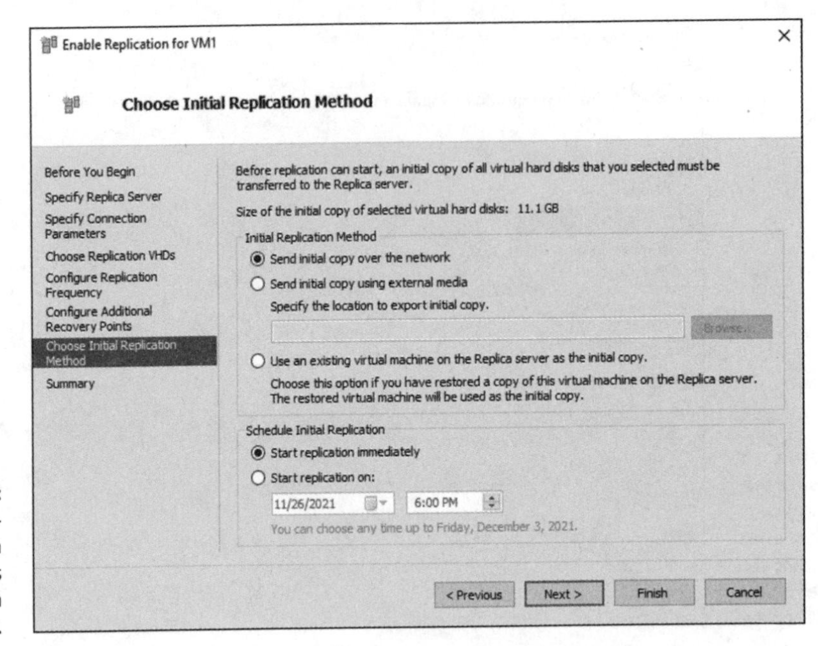

FIGURE 5-4: Setting the initial replication parameters gets the replication started.

At this point, replication is enabled, and the VM is being copied over to the replica server. You should get a message similar to Figure 5-5 if your replication was successful. This tells you that you need to connect the network adapters for the VM.

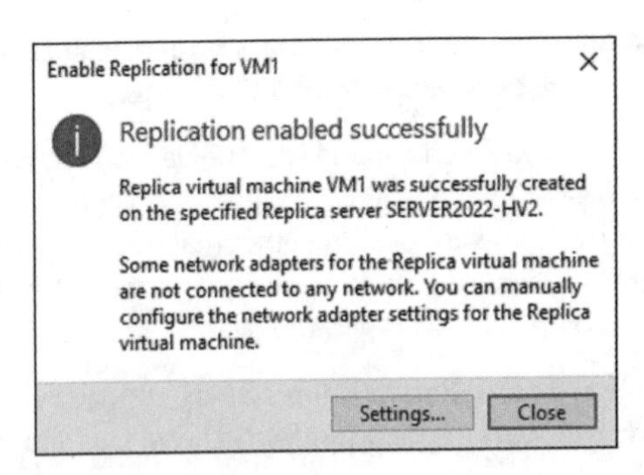

FIGURE 5-5: This message indicates that the initial replication was successful and that you need to connect to the other Hyper-V host and connect the virtual network adapters to a virtual switch.

Live Migration

Hyper-V Replica is great for disaster recovery purposes, but sometimes you just need to move VMs to another Hyper-V host so you can do maintenance. You may also have a host that's starting to have resource constraints so you may want to move VMs for that reason.

Live migration is a great fix for both of these scenarios because it allows you to move a VM with near zero downtime. When I've tested this in a lab environment, I've found when pinging a server that I lose a single ping at the most. Pretty impressive technology.

There are two ways that you can set up live migration. CredSSP is the simplest, but it's also the least secure. Kerberos is the most secure, but setting up constrained delegation makes it more complex to implement.

Setting up live migration

Live migration must be set up on any Hyper-V host that will be using it. The setup itself is pretty simple, and it can be used right away with CredSSP or after some additional configuration with Kerberos. (The additional configuration is covered in the "Kerberos" section.)

Follow these steps to configure your first Hyper-V host to support live migration. Be sure to do this on all your hosts or it won't work!

1. **From Hyper-V Manager, right-click the Hyper-V host and choose Hyper-V Settings.**
2. **Click Live Migrations and select the check box next to Enable Incoming and Outgoing Live Migrations.**

   Simultaneous live migrations defaults to 2. Unless you have a reason to change it, it can be left as is.
3. **Under Incoming Live Migrations, leave Use Any Available Network for Live Migration selected and click OK.**

CredSSP

CredSSP is the simplest method of authentication to set up when supporting live migration. It requires you to sign into the server that you want to move. If you move the server and then want to move it back after a maintenance, for example, you need to sign in to the VM before you can move it back. If you aren't signed in when you try to move the VM, you get an error that will indicate "No credentials are available in the security package."

Because CredSSP requires no further configuration, enabling live migration is all you need to do.

Kerberos

Using Kerberos to support live migration is preferred in enterprise environments because you don't have to sign in to a server before you move it. If your organization wants to automate things, this is the way to go.

The only downside to using Kerberos for authentication in live migrations is that it does take a little more time to set it up properly than CredSSP does. You must configure what is referred to as *constrained delegation* to allow live migration to work with Kerberos. This is essentially giving the systems permissions to work with one another directly, and it's done in Active Directory.

To set up constrained delegation for Kerberos, you need a system that has the Active Directory Users and Computers RSAT installed. Follow these steps:

1. **From Server Manager, choose Tools⇨Active Directory Users and Computers.**
2. **Double-click the Computers folder.**
3. **Right-click the Hyper-V host that is going to be the source server, and choose Properties.**
4. **Click the Delegation tab and select Trust This Computer for Delegation to Specified Services Only and select Use Any Authentication Protocol.**
5. **Click Add, and then click Users or Computers.**
6. **Type the name of the destination server and then click OK.**
7. **Under Available Services, select the following (hold down Ctrl to select them both):**
   - **cifs:** This allows you to move the VM storage.
   - **Microsoft Virtual System Migration Service:** This allows you to move VMs.
8. **Click OK.**

   Your screen should look similar to Figure 5-6.
9. **Click OK again to save the changes.**
10. **Repeat steps 3 through 9 for any other Hyper-V hosts that will participate in live migration.**

After this is complete, you can kick off a live migration without having to log in to the VM.

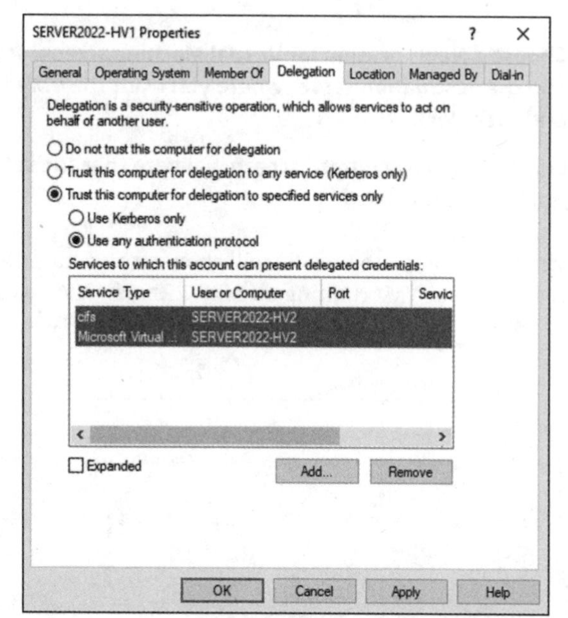

FIGURE 5-6: You must grant permissions on the source system for the destination system to enable Kerberos authentication for live migrations.

Kicking off a live migration

Now that you have Kerberos configured and live migration configured, I'm sure you want to test it to verify that it's working properly. Follow these steps:

1. **On the source Hyper-V host, open Hyper-V Manager.**
2. **Right-click the VM you want to migrate and choose Move.**
3. **On the Before You Begin screen, click Next.**
4. **Select Move the Virtual Machine, and click Next.**
5. **On the Specify Destination Computer screen, type the name of the destination Hyper-V host, and click Next.**
6. **On the Choose Move Options screen, keep the default Move the Virtual Machine's Data to a Single Location and click Next.**

TECHNICAL STUFF

You have a few options when moving your VM:

- **Move Your Virtual Machine's Data to a Single Location:** Moves the VM and its hard disks to the same location on the destination host.
- **Move the Virtual Machine's Data by Selecting Where to Move the Items:** Allows you to select different locations for the VM and for its disks.
- **Move Only the Virtual Machine:** Moves the VM but not its hard disks.

7. **On the Choose a New Location for Virtual Machine screen, set the location on the destination server where you want the VM's files to be stored, and click Next.**
8. **On the Completing Move Wizard screen, validate that the settings look correct, and then click Finish.**

When the migration is complete, the VM will show up on your other Hyper-V host. In Figure 5-7, you can see VM1 running on SERVER2022-HV2, which is my second Hyper-V host. I did the live migration from SERVER2022-HV1.

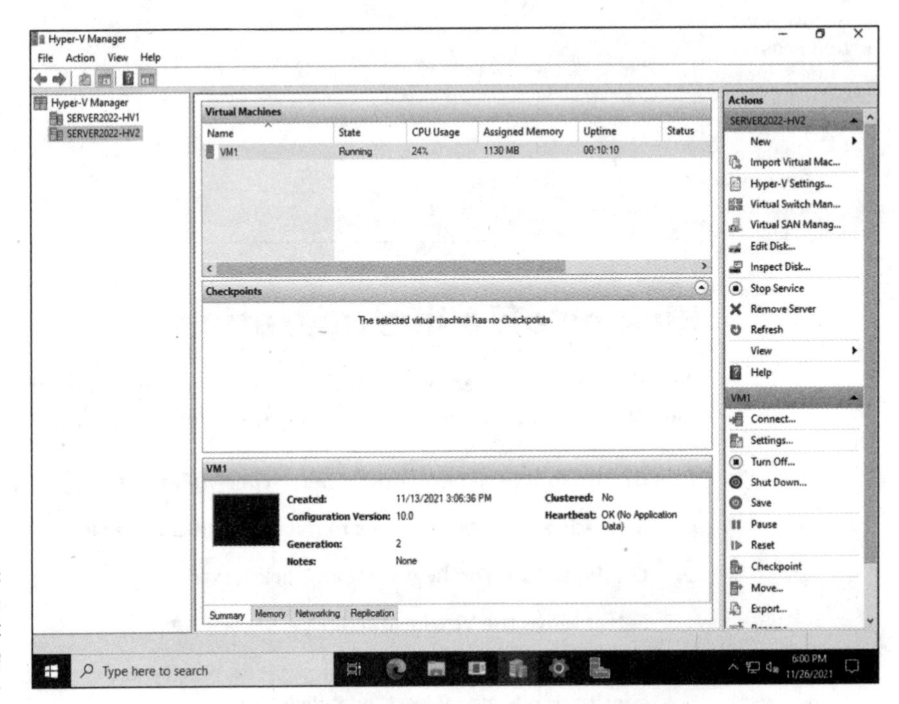

FIGURE 5-7: Live migration does not impact the uptime of the machine.

Storage Migration

Moving VMs is very useful, but sometimes you just need to move their storage. You may have purchased a lightning-fast storage area network (SAN), or you may have just added a new drive to support the virtual disks of your Hyper-V VMs. No matter the use case, storage migration can do that for you.

For this example, I have the virtual disk for a VM in my `E:` drive, but I want it to live in my `F:` drive. Here's how you can move storage around:

1. **On the source Hyper-V host, open Hyper-V Manager.**
2. **Right-click the VM you want to migrate and choose Move.**
3. **On the Before You Begin screen, click Next.**
4. **Select Move the Virtual Machine's Storage, and click Next.**
5. **Accept the default Move All the Virtual Machine's Data to a Single Location, and click Next.**
6. **On the Choose a New Location for Virtual Machine screen, enter the path that you want the storage to be moved to and click Next.**
7. **On the Completing Move Wizard screen, verify that everything looks correct and click Finish.**

Failover Clustering

Failover Clustering is not new to Windows Server 2022. It provides a way for system administrators to ensure that their systems are highly available. Applications or services talk to a cluster address. In the case of Hyper-V, the cluster address is the Hyper-V Replica Broker, and it handles communications for the clustered Hyper-V hosts.

Installing Failover Clustering

Failover Clustering is a feature that is available for installation in Windows Server 2022. Installing it is similar to installing other features. Follow these steps:

1. **From Server Manager, choose Manage⇨Add Roles and Features.**
2. **On the Before You Begin screen, click Next.**
3. **On the Select Installation Type screen, click Next.**
4. **On the Select Destination Server screen, click Next.**
5. **On the Select Server Roles screen, click Next.**
6. **On the Select Features screen, select Failover Clustering.**
7. **Click Add Features, and then click Next.**
8. **On the Confirm Installation Selections screen, click Install.**
9. **When you get the message that installation succeeded, click Close.**

Repeat these steps on the other members of the failover cluster before continuing on to the next step.

Configuring Failover Clustering

After the Failover Clustering feature is installed, you can actually create the cluster. First, you'll run the validation, which will let you know if there are issues that will prevent the cluster from provisioning properly; then you'll build the cluster.

1. **From Server Manager, choose Tools⇨Failover Cluster Manager.**
2. **Click Validate Configuration.**
3. **On the Before You Begin screen, click Next.**
4. **Enter the names of the servers in the failover cluster, and then click Next.**
5. **Accept the default Run All Tests, and click Next.**
6. **On the Confirmation screen, click Next.**

   The test will begin running, and you'll see something similar to Figure 5-8.

TIP

   Warnings won't necessarily prevent you from creating your failover cluster, but you should address the warning items in the report before relying on your failover cluster for a production deployment.

7. **When the test finishes, select the check box next to Create the Cluster Now Using the Validated Nodes, and then click Finish.**

   The Create Cluster Wizard launches.

8. **On the Before You Begin screen, click Next.**

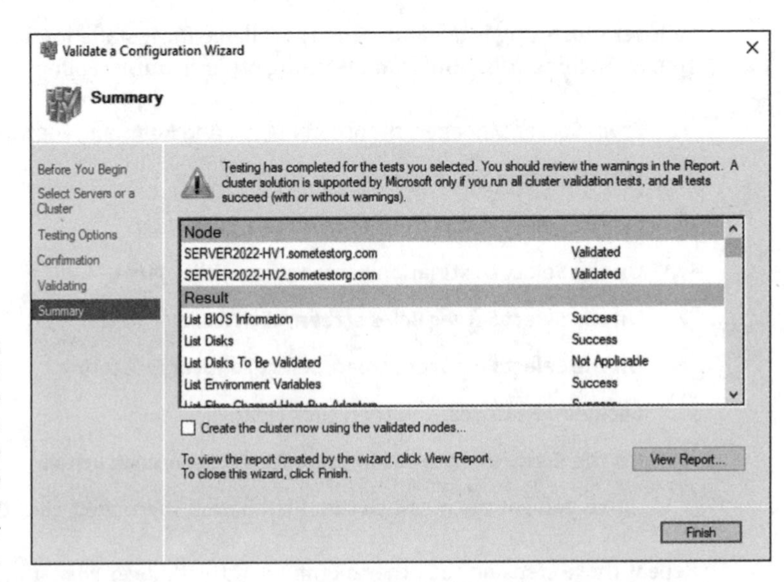

FIGURE 5-8: It's always a good idea to run the cluster validation tools to ensure there are no issues with creating your cluster.

9. **On the Access Point for Administering the cluster screen, choose a cluster name, and then click Next.**
10. **On the confirmation screen, click Next.**
11. **When the summary page launches, click Finish.**
12. **Click Configure Role.**
13. **On the Before You Begin screen, click Next.**
14. **On the Select Role screen, select Hyper-V Replica Broker and click Next.**
15. **On the Client Access Point screen, enter a name for the cluster role and click Next.**
16. **On the confirmation screen, click Next.**
17. **On the Summary page, click Finish.**

Your screen should look similar to Figure 5-9 at this point if you've been following along.

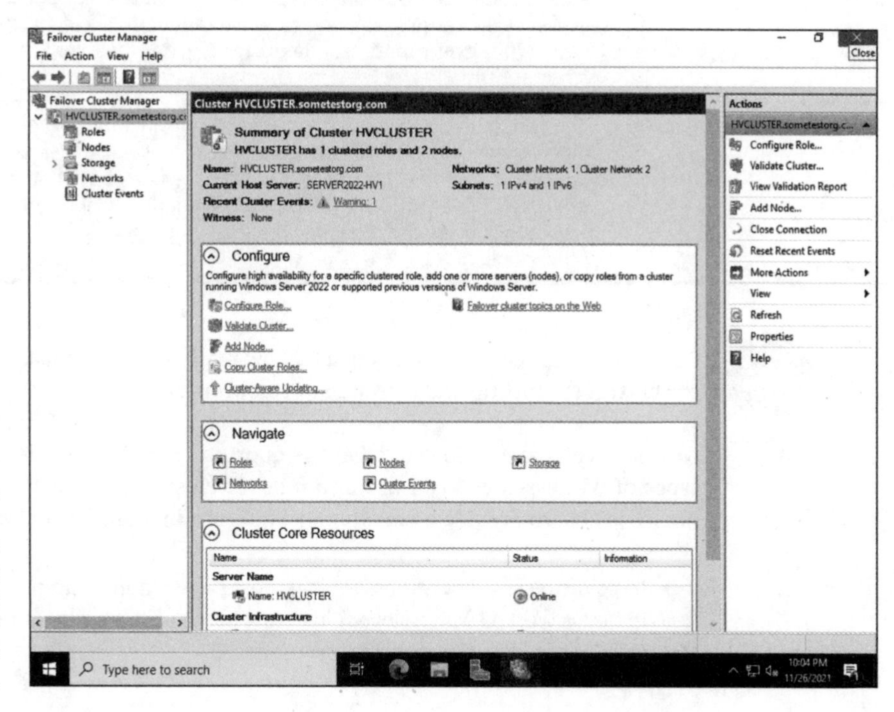

FIGURE 5-9: The failover cluster has been created and has been assigned the Hyper-V Replica Broker role.

Configuring a witness for your failover cluster

Before I jump into witness types, let's take a step back and review the different quorum types in Windows Server Failover Clustering (see Table 5-1). Remember each node is treated as one vote.

TABLE 5-1 Quorum Types

Type	Description
Node Majority	This type of quorum is used when there is an odd number of nodes. The nodes are divided into smaller subsets and whichever subset has the most nodes has the most "votes." For instance, five nodes divide into a subset of three and a subset of two; the subset of three will be active because three votes is more than two.
Node & Disk Majority	This type of quorum works really well when there are an even number of nodes and a clustered disk (disk witness). Each of the nodes gets a vote, and the clustered disk gets a vote, too. The disk witness adds an additional vote to even-numbered subsets so that you know which subset should be active. As an example, say you have four nodes, in two subsets of two nodes each. Whichever subset has the cluster disk has the third vote and is the active subset.
Node & File Share Majority	This type of quorum is similar to Node & Disk Majority, except instead of a disk witness, you get a file share witness. It provides the same service as the disk witness but is usually placed in a datacenter where both failover clusters can reach. For example, say you have four nodes, in two subsets of two nodes each. Whichever subset has the file share witness has the third vote and is the active subset.

TECHNICAL STUFF

Quorum in this sense is determined by how many systems must be operational for the cluster to start up and continue to run properly.

Now that you have an idea of what the quorum types are, let me fill you in on the types of witnesses you can use in a failover cluster. Each of these types of witnesses is able to provide a vote to make quorum in a failover cluster:

» **Disk witness:** A disk witness is what most system administrators are used to if they've worked with Failover Clustering in the past. It's a small drive that can failover between nodes in a failover cluster.

- **File share witness:** A file share witness is essentially a file server that has a share configured that stores clustering information. The use case with file share witness was to have the file server with the share in a different datacenter from the other nodes in the failover cluster. If one datacenter went down, the file share witness was able to cast a deciding vote to the existing datacenter.
- **Cloud Witness:** Introduced in Windows Server 2016, the Cloud Witness allows you to create blob storage in Azure (hence, the reason why *cloud* is in the name) and use it to store clustering information. Cloud Witness was introduced as a cost-effective means of getting similar functionality to a file share witness, but without the cost of an additional file server in another datacenter.

A file share witness is relatively simple to set up, so that's what I use for the example in this chapter. Let's add the file share witness now. I've created a folder on a server that isn't in the failover cluster, and I've created a share on that folder with the cluster name that I used earlier. You'll find it created a DNS object when you went through the wizard to create the cluster. Give it read and change permissions. Now go to one of the systems that is in the failover cluster and follow these steps:

1. **In the Failover Cluster Manager, select the cluster to which you want to add the file share witness.**
2. **In the Actions menu on the right, click More Actions and then click Configure Cluster Quorum Settings.**
3. **On the Before You Begin screen, click Next.**
4. **On the Select Quorum Configuration Option, choose Select the quorum witness and then click Next.**
5. **On the Select Quorum Witness screen, select Configure a File Share Witness and then click Next.**

   Figure 5-10 shows some of your quorum witness options.
6. **Enter the path of the file share that you've configured and then click Next.**
7. **On the Confirmation screen, click Next.**
8. **Click Finish.**

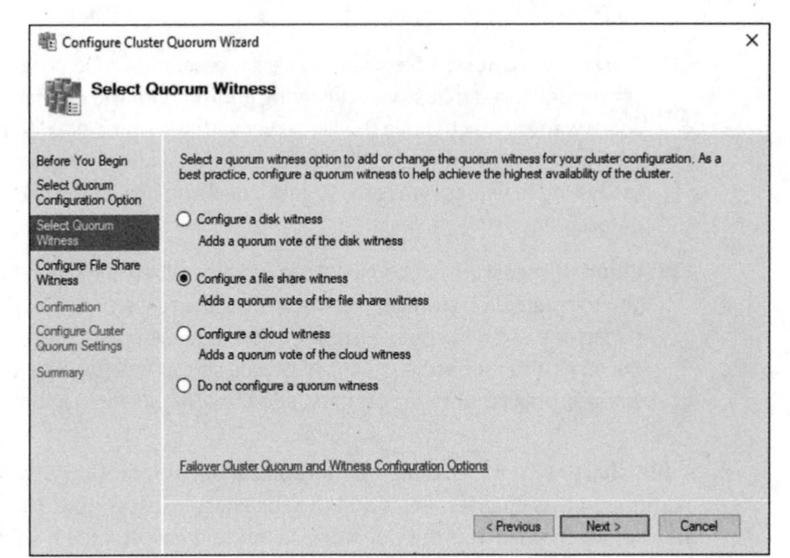

FIGURE 5-10: You have several options for configuring a quorum witness.

8

Installing, Configuring, and Using Containers

Contents at a Glance

Windows Server 2022

» Examining common use cases for containers

» Deciding what type of containers you want to use

» Implementing containers at scale

Chapter 1
Introduction to Containers in Windows Server 2022

Virtualization drastically changed the way that IT operated in organizations of all sizes, but containers have had a large impact as well. You may be wondering why someone would want to use containers. They're just virtual machines (VMs), right? Well, not exactly. The technologies may seem similar, but containers and VMs are not the same. VMs are presenting virtual hardware to the user. Containers don't expose the hardware or the operating system; they're meant to run applications in isolation.

In this chapter, I fill you in on containers — what they are, why you would use them, and how the Windows Server implementation will work for you.

Understanding Containers

VMs can be thought of as Infrastructure as a Service (IaaS). Although VMs do present virtual hardware to system administrators, the administrators of virtual servers don't have to be concerned about the underlying hardware. They can focus on the operating system and applications that they're responsible for.

Containers take this idea and refine it to where each container is responsible for running an application. The application is baked into the image so the containers can be stood up and torn down constantly. This is great for Platform as a Service (PaaS) scenarios where developers just want to test their code and not worry about getting servers provisioned to test against. Developers don't generally care about hardware or operating systems; they just want to know that their code works in the way they expect it to.

The main idea behind containers is that the application inside of each container has all the resources that it requires to function within the same container. This means that you can drop the container on any container host, and all the application's requirements will still be met because those requirements (.NET, for example) move with the application inside the container.

Knowing what a container looks like

You may be wondering what containers look like. Let me use the example of containers in Windows specifically. At a high level, the architecture looks something like Figure 1-1.

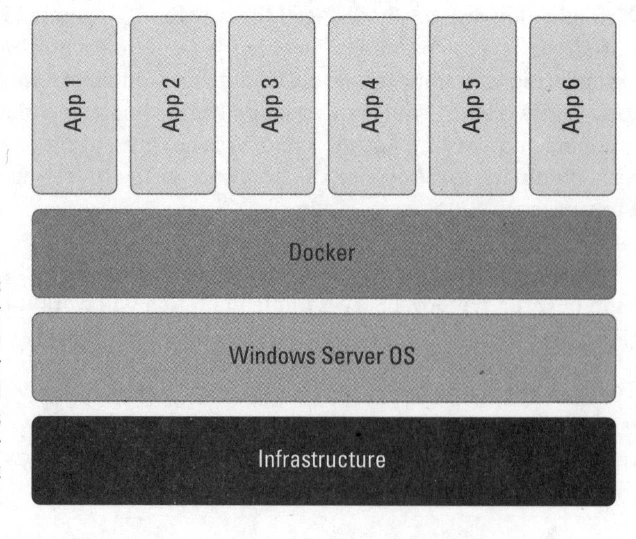

FIGURE 1-1: Container architecture on Windows Server involves several layers and utilizes the Docker Engine to work with containers.

In a Windows Server operating system, after you enable the containers feature, you install the Docker Engine. The Docker Engine is responsible for packaging and deploying the containers. Microsoft partnered with Docker for the first time with Windows Server 2016 to support running containers on a Windows operating system.

Defining important container terms

As with most newer technologies, there are new terms that you need to understand to be on the same page as other system administrators who work with containers. Here are the most important terms:

- **Container host:** The container host is the system that is configured with the Windows Container feature. It can be a physical host or, through the joys of nested virtualization, a virtual host. All the containers on the container host share the host's resources.
- **Container image:** When you create a container image, you create a deployable image that contains the changes you made to the original image, which were stored in the sandbox. The container image does not contain the operating system (OS); instead, when you deploy custom container images, they're a layer of customization that is added on top of the container OS image.
- **Sandbox:** The sandbox saves changes as they're made to the container image. This can include modifications to the file system and Registry, and any new applications you might install. Changes saved in the sandbox can be saved as container images so they can be reused.
- **Container OS image:** Not to be confused with the container image, the container OS image can't be modified. It is the first layer in the container sandwich and provides the operating system that the container will use.
- **Container repository:** Container images along with any dependencies they may have are stored in a container repository so that they can be reused. They can be stored in a local repository, or if you plan on using the image across multiple container hosts, you can create private or public repositories on Docker Hub. Repositories may also be referred to as registries; Docker Hub, for instance, is often referred to as a container registry.

Seeing how containers run on Windows

You may have noticed that I've referenced Docker multiple times. That's because containers use the Docker Engine to run on Windows Server! Containers were first introduced in Windows Server 2016, but the technology and, of course, Docker itself have been around a lot longer than that.

Docker is the engine that is responsible for packaging and delivering container images. Those container images can be based on Windows or Linux operating systems and can run in your datacenter an Windows Server 2022.

You can find more information on Docker in Chapter 2 of this minibook.

Considering Use Cases for Containers

There are several different use cases for containers. The use cases generally depend on what interests you most. System administrators, for example, will have very different use cases from developers. Thankfully, containers can support the majority of the use cases.

Developers

Because containers have everything your application depends on to run, you can move those containers anywhere (so long as it's running Windows Server 2016 or newer). This means that you can start your development work from just about anywhere and finish by moving your container image into production. Plus, Docker Hub has more than 180,000 applications already packaged and available, which can greatly speed up your development work!

System administrators

I don't know many system administrators who enjoy building environments over and over again — pulling out the dreaded checklist to ensure that you don't forget to install this one thing or configure that one feature.

Rejoice, system administrators! With containers, you have an instant packaged environment for all your teams from the developers to your production support teams. No more dreaded checklist! When you need to make a change or update the containers, you simply update the container image that the containers are built from, and — *voilà!* — the next time they build their container from the container image, they're all up to date. Something breaks because of an update or a change? Roll them back to the previous container image.

Deciding What Type of Containers You Want to Use

Windows Server allows you to use two different methodologies to manage your containers:

- **Windows Server containers:** Windows Server containers are simple and are good for workloads where security and isolation are not as important as simply getting them up and going.
- **Hyper-V containers:** Hyper-V containers provide a higher degree of security and isolation and are the perfect choice for when you need to more strongly enforce these on your containers.

You aren't stuck with your decision. You can create a container as a Hyper-V container, for instance, and change it to a Windows container later on.

I cover the installation of containers in Chapter 3 of this minibook.

Windows Server containers

Windows Server containers provide a simple method of application isolation that leverages process and namespace isolation. Because this type of container shares the kernel of the container host that it's on, and by extension all the other containers on that host, you don't get true isolation. If you write code that you think could be destructive, this is not the ideal environment for that. Windows Server containers should be used for trusted or non-impactful code that does not require a strong security component.

REMEMBER

Because the kernel is shared, it's important to remember that the host and containers must all be on the same kernel version. This may cause issues for developers — you do lose some flexibility for testing because of this requirement.

Hyper-V containers

Hyper-V containers provide a true degree of isolation. Each of the containers is run in its own VM and does not share the kernel with either the container host or the other containers on that same host. These containers are perfect for workloads that require a higher degree of security, or when running untrusted code, because damage is contained in the individual container. If the untrusted code causes damage, you can simply spin up another Hyper-V container and try again after the issue is fixed.

Because the kernel isn't shared between the container host and the containers that reside on it, you can have different kernel versions throughout all the containers.

Managing Containers at Scale

Even if you're convinced of the benefit of containers, you may be wondering how difficult it will be if containers catch on at your workplace. How are you going to manage all those containers?!

Dockerfile allows you to automate the creation of your container images. Think of Dockerfile like a script for your containers; Dockerfile specifies which features to install and configures the container.

Microsoft maintains a GitHub repository with tons of examples of Dockerfiles to get you started. You can check it out at `https://github.com/MicrosoftDocs/Virtualization-Documentation/tree/master/windows-container-samples`. What I love most about this repository is that there are so many use cases represented, and not all are with Microsoft products. There are Dockerfiles for Apache, MongoDB, MySQL, and nginx, as shown in Figure 1-2.

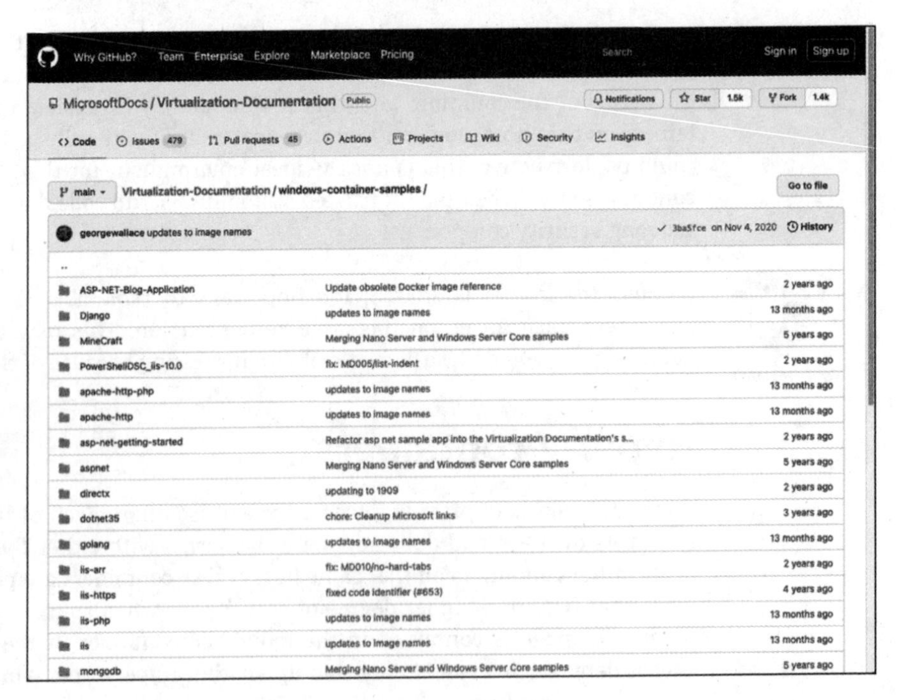

FIGURE 1-2: The GitHub repository where Microsoft stores examples of Dockerfiles you can use with Windows Server container implementations.

» Getting acquainted with Docker Hub

Chapter 2 Docker and Docker Hub

Before I introduce you to containers in Windows Server, you need a good understanding of the technology that Windows containers are based off of. That would be Docker!

In this chapter, I introduce you to Docker and Docker Hub and explain how you can use them in conjunction with Windows Server to begin your container journey.

Introduction to Docker

Docker is an open-source platform that assists you in packaging and deploying applications. You can run multiple containers on a container host, and because they share the container host's kernel, they use fewer resources than virtual machines (VMs) because you don't need the overhead of a hypervisor to manage them.

Docker architecture

Docker is architected to use a client–server model. The Docker client talks to the Docker server component, which is called a *daemon.* Your Docker client can be on the same server as the Docker daemon, or you can run the Docker client from your workstation.

The Docker server

The Docker server is the brains of the operation. It manages much of what goes on in Docker, including the various objects that are created, and communications with the Docker application programming interface (API). The server component is referred to as a daemon.

The Docker client

The Docker client is where you perform most of your work with containers. Whenever you run a Docker command, you're running it from the Docker client.

The Docker registry

Docker images are stored in the Docker registry.

TECHNICAL STUFF

You may also hear this referred to as a *repository. Registry* is the official word in Docker documentation, but many developers are used to calling this type of construct a *repository.* Both words work — be aware that you may see them used interchangeably.

Docker objects

Docker objects is a term used to refer to a multitude of different components, like images, containers, and services.

Basic Docker commands

Docker commands always start with `docker` and include keywords that determine the action that you want to take. Table 2-1 lists some of the more common commands that you should remember.

TABLE 2-1 Common Docker Commands

Command	Description
`docker pull`	Pulls a container image from whichever registry you have configured to store your container images
`docker push`	Pushes your container image to whichever registry you have configured to store your container images
`docker run`	Pulls the container image if it is not available already and then creates the new container from the container image
`docker images`	Lists all the container images that are stored locally on the container host
`docker login`	Used to log in to a registry; not required for public registries, but required to access private registries
`docker stop` Book	Stops the running container that was named
`docker ps`	Lists all the containers that are running at that time

Introduction to Docker Hub

Docker Hub is a public registry owned by Docker that is available for storing container images in individual repositories. Businesses can use Docker Hub to create their own private repositories to store proprietary container images in as well. Many of the images that are available are from large open-source projects, but there are also plenty of container images from organizations that are not open source. For example, Microsoft has a public repository that has about 68 container images at the time of this writing.

You may be asking, "How do I get to Docker Hub? It sounds pretty cool." You can access Docker Hub by going to `https://hub.docker.com`.

Finding public images

Public images are the easiest ones to find. You don't need an account to search for public images, nor do you need an account to do a `docker pull` on one.

To find an image that you're interested in, you can simply type your query into the search box at the top. For example, if you want to search for Windows Server images, just type **Windows Server** and press Enter. In the search results, you'll find Windows Server Core. If you click it, you can find out more about the container, as shown in Figure 2-1.

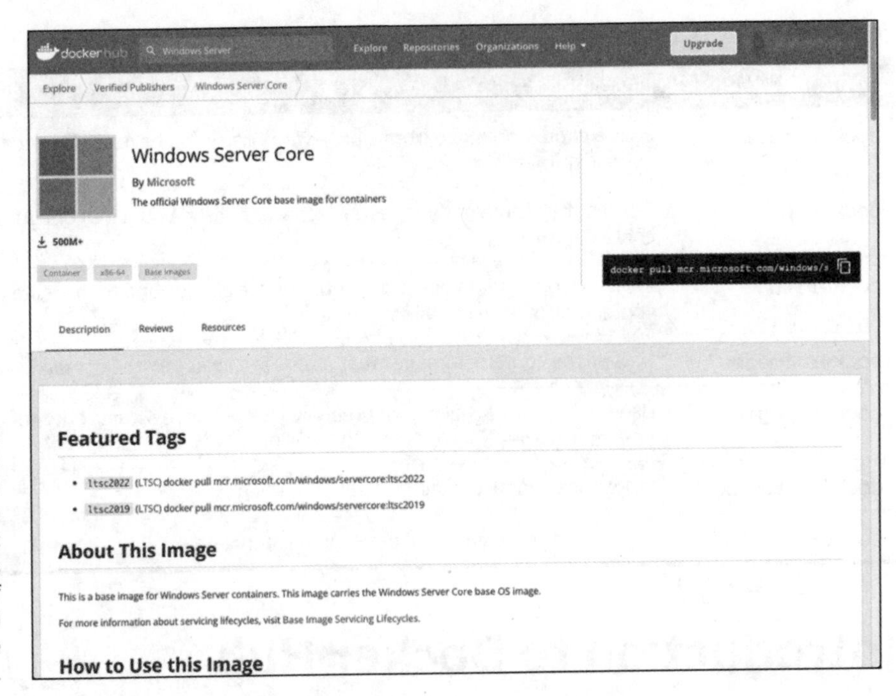

FIGURE 2-1: The search box on Docker Hub makes it simple to find public container images from hundreds of organizations and find out more about them.

If only one container image matches your query, you're taken to a page that is dedicated to that container image. If you type the name of an organization, or your search returns multiple results, you're presented with search results. If I had searched for Microsoft, for example, I could have gotten any container image that has to do with Microsoft. Official Microsoft container images can be filtered on by selecting Verified Publisher from the filters on the left side of the screen, as shown in Figure 2-2.

One of the really great things about Docker Hub is that you can click a container image to learn more about it. The page that you click into is the same one you get if you search for a product and there is only one result. You're presented with a description of the container image, which includes available tags and commands needed to use the container image. These commands are often used to accept licensing agreements. The Microsoft SQL Server container image, for example, tells you to run this command to start an MS SQL server instance running SQL Express:

```
docker run -e 'ACCEPT_EULA=Y' -e 'SA_PASSWORD=yourStrong(!)
  Password' -e 'MSSQL_PID=Express' -p 1433:1433 -d mcr.
  microsoft.com/mssql/server:2019-latest
```

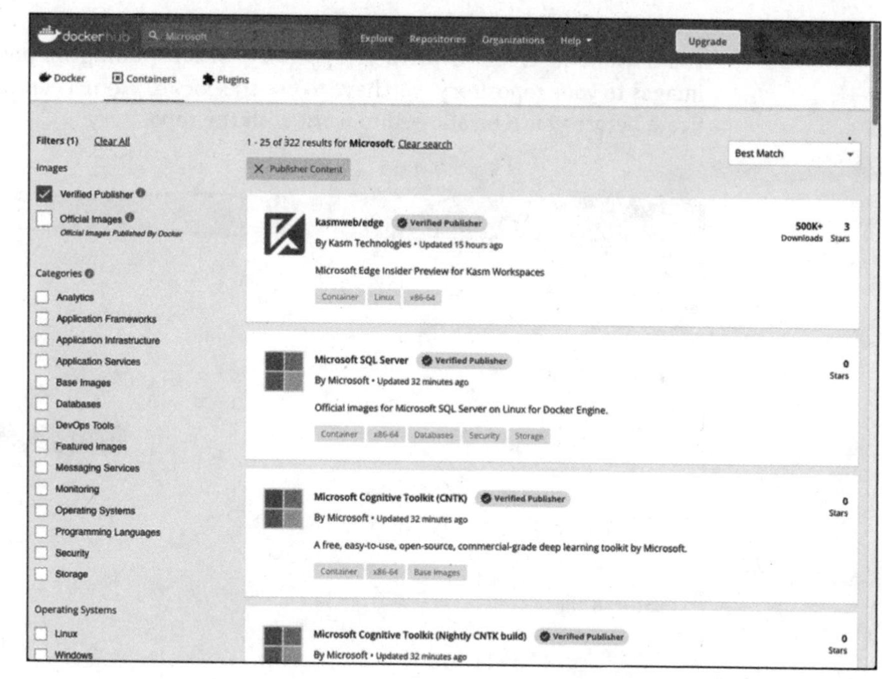

FIGURE 2-2: You can filter for just verified publishers in Docker Hub, which ensures that you're getting an official container image.

The information on the container image will also cover software requirements and available environment variables, along with a full listing of tags. Tags allow you to choose different versions of a container image. If you don't specify a tag, then by default you get the container image with the "latest" tag.

You're also given the command to pull an image if you're interested in it. For example, to pull this MS SQL container image into Docker, you would run the following:

```
docker pull mcr.microsoft.com/mssql/server
```

One last thing that I find to be really helpful is that you can see how many times a container image has been pulled. This information is useful if you aren't familiar with the organization that supplied the container image. Underneath the name next to a logo of a down arrow is a number that tells you how many times it has been pulled. Microsoft SQL Server, at the time of this writing, had been pulled more than 50 million times, as shown in Figure 2-3.

Creating a private repository

Public repositories make acquiring container images convenient, but if you're working on container images and you don't want them to be publicly available, you'll want to create a private repository. When pulling or pushing container images to your repository, you have to use the `docker login` command to authenticate before you'll be allowed to work with the repository.

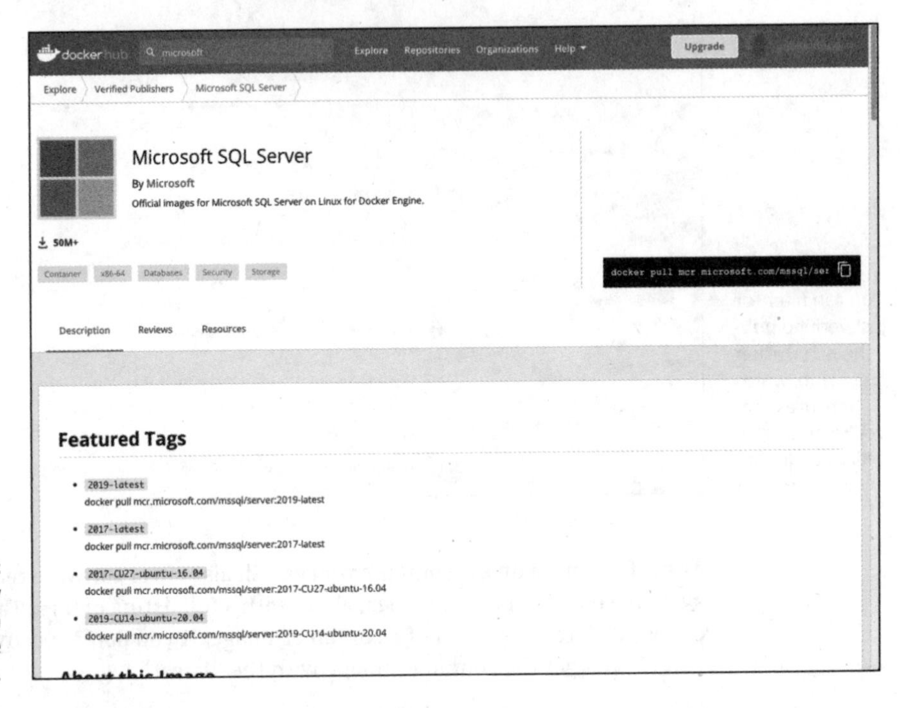

FIGURE 2-3: You can see how many times a container image has been pulled.

By default, you get one free private repository in Docker Hub. If you need more private repositories than that, you can upgrade to a paid plan. At the time of this writing, you could pay $5 a month for unlimited private repositories with the Pro plan.

Creating an account

Creating an account on Docker Hub is simple and free. From the home page, click the Sign Up link in the upper-right corner. Choose a Docker ID, enter your email address and password, accept Docker's terms, check the box on the CAPTCHA, and then click Sign Up, as shown in Figure 2-4.

You'll get an email to verify your email address. Click the link in the email to activate your account.

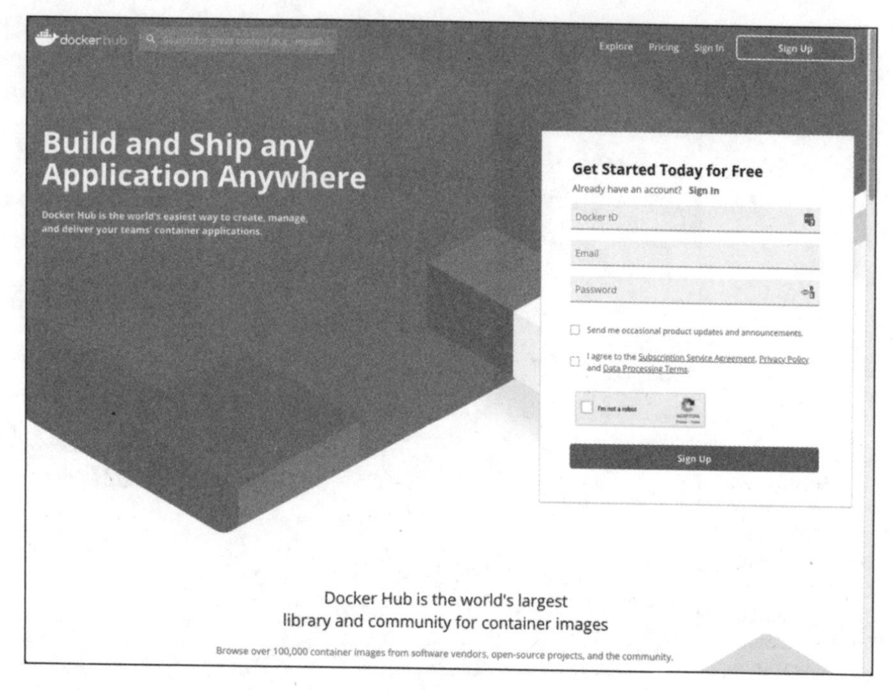

FIGURE 2-4: You can get started with Docker Hub with very minimal information — all you need is a Docker ID, an email address, and a password.

Creating your private repository

When you log in to Docker Hub after creating your account, you're asked whether you want to create a repository or create an organization.

1. **Click Create a Repository.**
2. **Enter a name for your repository and a description.**
3. **Change visibility to Private.**
4. **Click Create.**

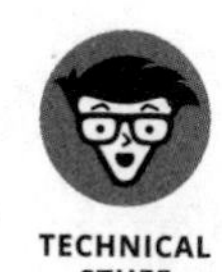

You can choose to link your repository to your GitHub or Bitbucket accounts to do automated container image builds. This menu is located in the repository creation menu, though you can go back in later and set it if you need to.

After your repository is created, it will be blank, but it will give you a sample of the command you would need to run to push things to your repository, as shown in Figure 2-5. In the image, I've blurred my username, but you can see the repository name of server2022 and the push command that I would use.

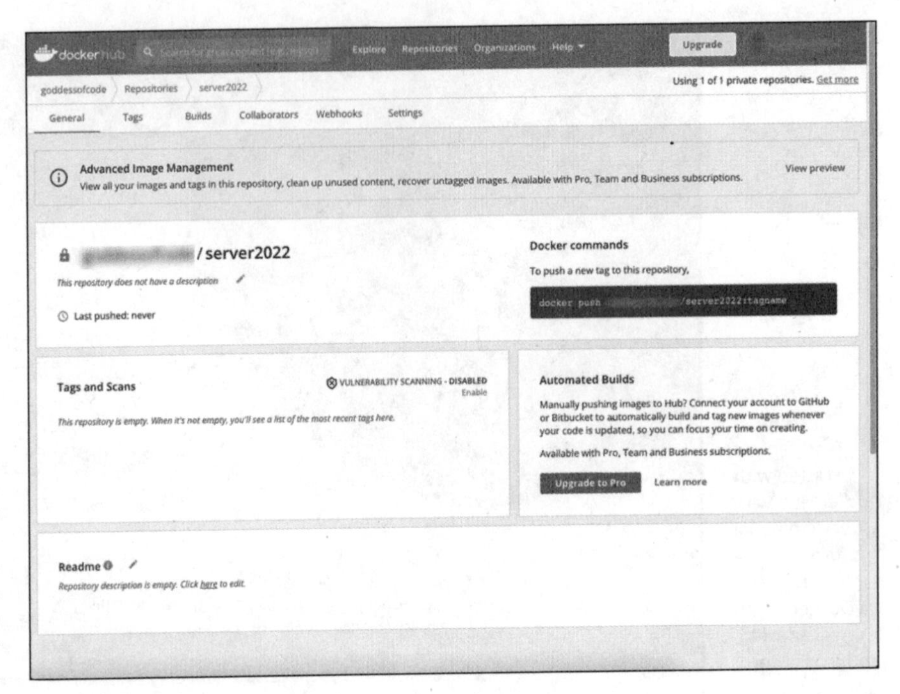

FIGURE 2-5: Your repository is empty when you first create it, but it contains the command to push container images to it.

Using a private repository

To use your private repository, you first have to log in to Docker; then you can push and pull container images as much as you want. To log in, enter the following command:

```
docker login
```

I'm going to pull the standard Nano Server image from Microsoft's repository. I can do that with the following command:

```
docker pull mcr.microsoft.com/windows/nanoserver:ltsc2022
```

I'll add the command that will let me push the container image to my repository. You would normally do this after you made changes to the image; in this case, I'm going to do it to demonstrate how you would do it.

```
docker push <dockerid>/<reponame>:ltsc2022
```

The command uses my Docker ID, followed by the name of my repository, and then the tag I used for my container image. In this case, I created a tag with a value of `ltsc2022`. You can see the command line part in Figure 2-6.

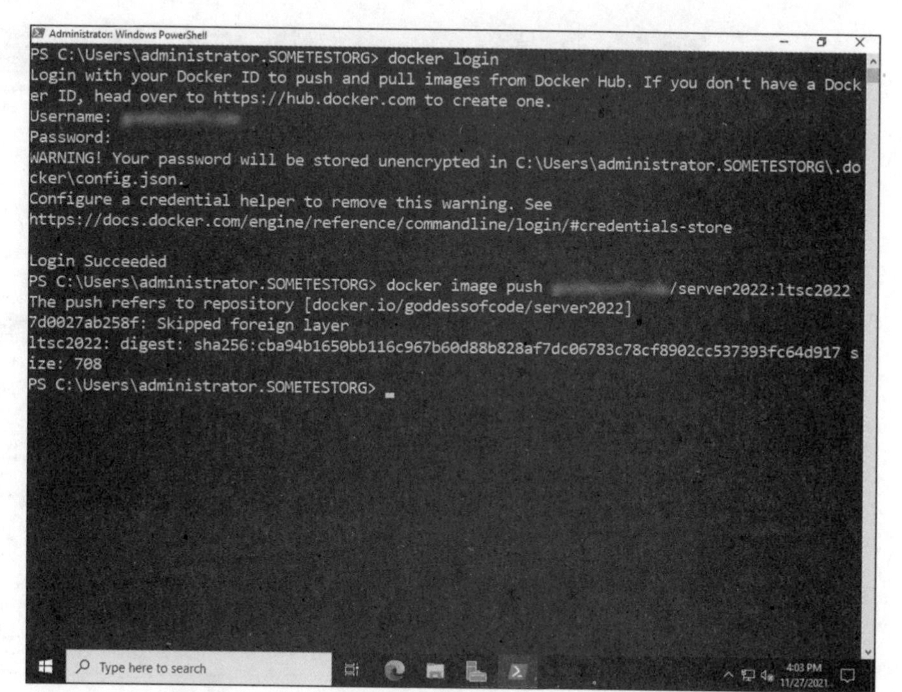

FIGURE 2-6: You can use the Docker commands to push images to your private repositories after you log in with the `docker login` command.

After the container image has been pushed, it will show up in your repository in Docker Hub. All your tags that are pushed to Docker Hub show up in your portal. You can't alter the container images from inside of Docker Hub; in fact, the only thing you can do is delete them. To modify your container images, you need to pull them, make your changes, and then push them again. Figure 2-7 shows you what Docker Hub looks like after I pushed my tagged container image.

To pull the container image down to modify it, I would issue a very similar command to what I used to push the tagged image:

```
docker pull <mydockerid>/server2022:ltsc2022
```

After I make the changes that I need to make (like updating the container image), I can push it back up to my private repository where it's accessible to any system from which I can log in to my Docker repository.

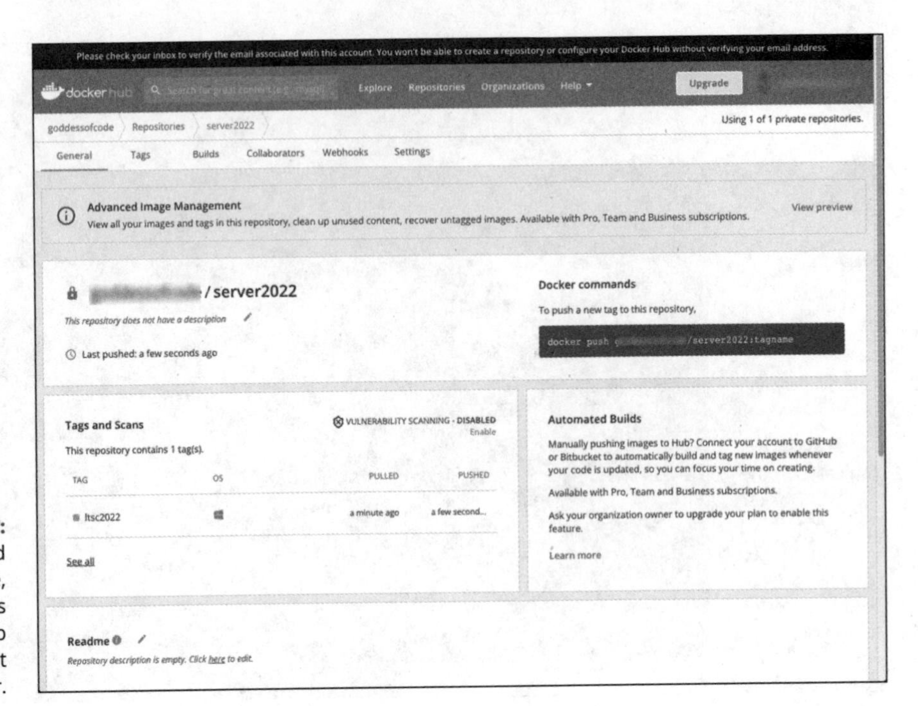

FIGURE 2-7: My tagged container image, `ltsc2022`, shows up in Docker Hub after I push it from my server.

IN THIS CHAPTER

- » Implementing Windows Server containers
- » Implementing Hyper-V containers
- » Installing Docker on Windows Server 2022
- » Testing your containers after Docker has been installed

Chapter 3
Installing Containers on Windows Server 2022

Containers are a game-changing technology — especially for teams that have developers who need dynamic environments to work from. A developer can launch a container that supports the needs of her application within minutes, and many of the container images are purpose build with the various programming frameworks called out in the title of the container image.

Windows Server 2022 supports two variations on containers:

- » **Windows container:** The Windows container is the traditional container model. It's fast, lightweight, and easy to use. The downside is that it shares the kernel with the host operating system (OS).

» **Hyper-V container:** If you have a workload that requires different versions of the kernel, or highly secure workloads that can't share a kernel, the Hyper-V container is the better choice. The Hyper-V container has a higher performance hit on the host server, but because it runs each virtual machine (VM) in its own container, you can have containers that have different versions of the kernel, and you have true isolation because the container is not sharing the kernel of the host OS with the host and other containers.

The best thing about this conversation is that you don't need to decide on one type or the other type. Containers can go from being Windows containers to Hyper-V containers.

In this chapter, I show you how to install Windows containers and Hyper-V containers, as well as how to install the Docker pieces that are needed to make everything work.

Installing Windows Containers

Installing Windows containers is simple. You just enable the feature, and then install Docker. This section covers installing the feature. After you've completed the steps in this section, continue with the "Installing Docker" section.

1. **From Server Manager, choose Manage⇨Add Roles and Features.**
2. **On the Before You Begin screen, click Next.**
3. **On the Select Installation Type screen, click Next.**
4. **On the Select Destination Server screen, click Next.**
5. **On the Select Server Roles screen, click Next.**
6. **On the Select Features screen, select Containers (shown in Figure 3-1), and then click Next.**
7. **On the Confirm Installation Selections screen, click Install.**
8. **Click Close and restart the server.**

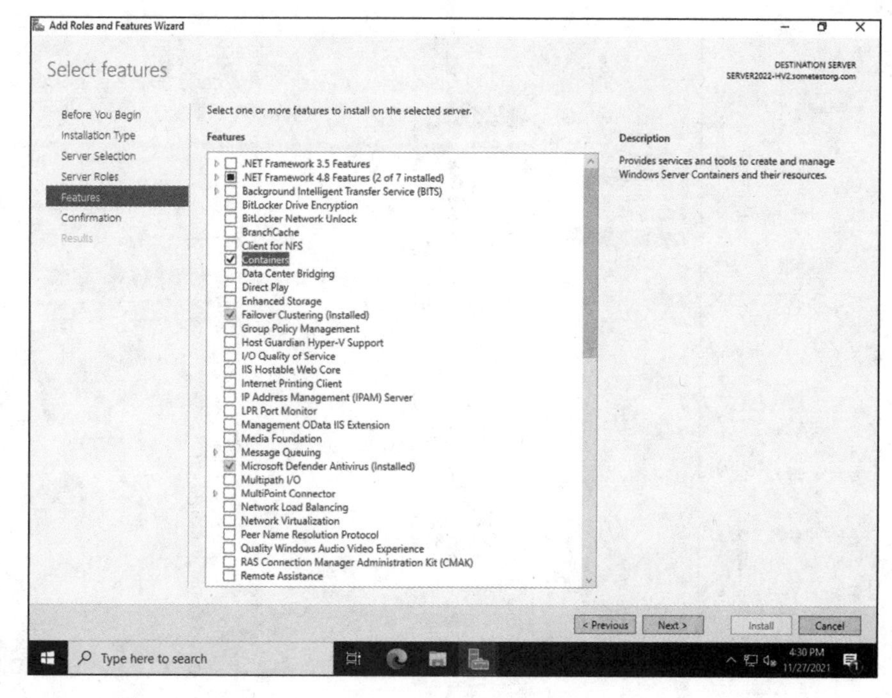

FIGURE 3-1: To use Windows Containers, you only need to install the Containers feature and then install Docker.

Installing Hyper-V Containers

To install Hyper-V containers, you also have to install the Hyper-V role. You can install them both at the same time. Follow these steps:

1. **From Server Manager, choose Manage⇨Add Roles and Features.**
2. **On the Before You Begin screen, click Next.**
3. **On the Select Installation Type screen, click Next.**
4. **On the Select Destination Server screen, click Next.**
5. **On the Select Server Roles screen, select Hyper-V, click Add Features, and then click Next.**
6. **On the Select Features screen, select Containers and then click Next.**
7. **On the Hyper-V screen, click Next.**
8. **On the Create Virtual Switches screen, select your network adapter, and click Next (see Figure 3-2).**
9. **On the Virtual Machine Migration screen, click Next.**
10. **On the Default Stores screen, click Next.**

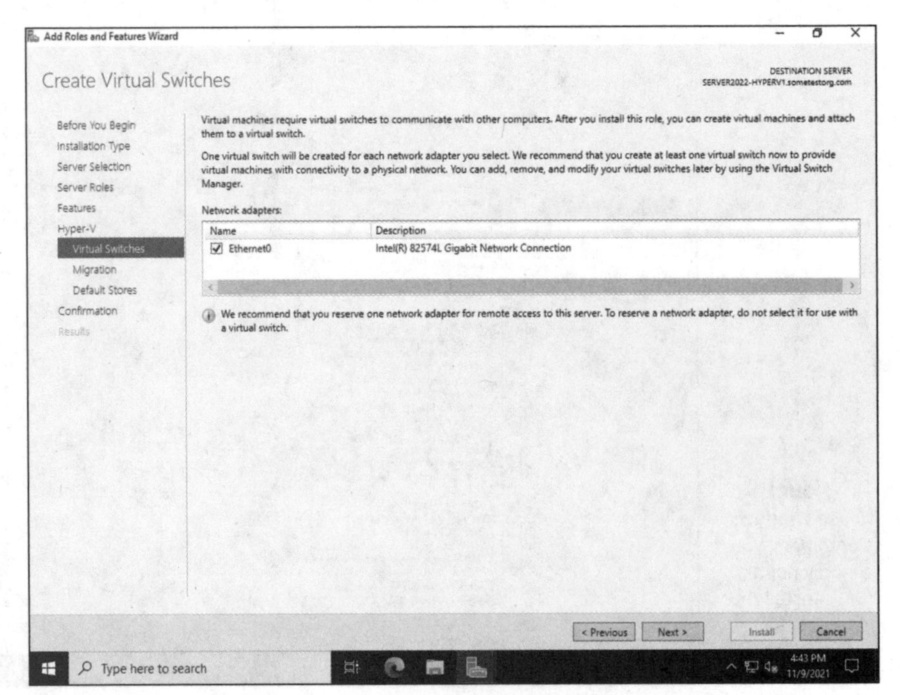

FIGURE 3-2: When setting up Hyper-V, you can provision the virtual switch at the same time.

11. **On the Confirm Installation Selections screen, click Install.**
12. **After the installation is complete, click Close and then restart the server.**

Installing Docker

At this point, you've at least got the containers feature installed. You may have even installed the Hyper-V role and the containers feature at the same time. Now you need to install the Docker Engine. This is the piece that really ties all the other pieces together.

You'll need to open PowerShell to run these commands, as well as the commands in the "Testing Your Container Installation" section. To open PowerShell, right-click on Start and select Windows PowerShell (Admin).

After you've opened PowerShell, your first step is to install the Microsoft Package Provider for Docker. This is done with the following command:

```
Install-Module -Name DockerMsftProvider -Repository PSGallery
   -Force
```

Now you can install the latest version of Docker with the following command:

```
Install-Package -Name docker -ProviderName DockerMsftProvider
```

After Docker is installed, you need one more restart. You can do this through the graphical user interface (GUI), or you can just type the following into PowerShell:

```
Restart-Computer -Force
```

These commands are shown in Figure 3-3. If everything went well, you get no output. The PowerShell prompt will simply return, and you can run the next command.

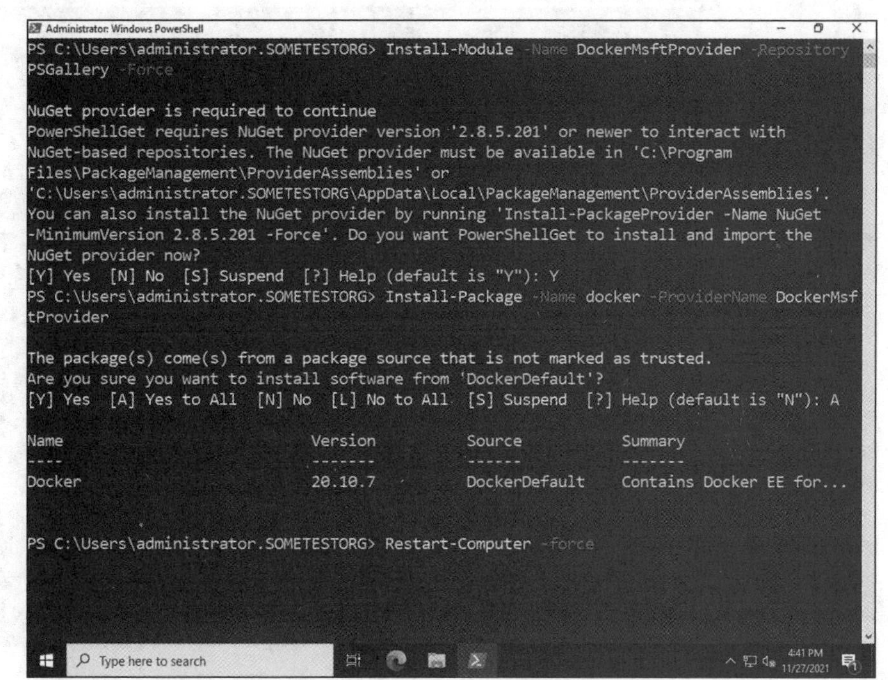

FIGURE 3-3: It takes two PowerShell cmdlets to install Docker on Windows Server 2022, three if you count the restart cmdlet at the end.

Testing Your Container Installation

After your server is configured and Docker is installed, you'll want to test to ensure that your container installation is working properly.

Windows container

There is a simple way to test that your Windows container installation is installed properly: Download and run a container image. One of my favorites is a sample image because it prints out a "Hello world"–style message and then exits.

To run this test, you use the `docker run` command (shown in Figure 3-4). Because the container image is not downloaded yet, it will download the container image first, and then run it.

```
docker run hello-world
```

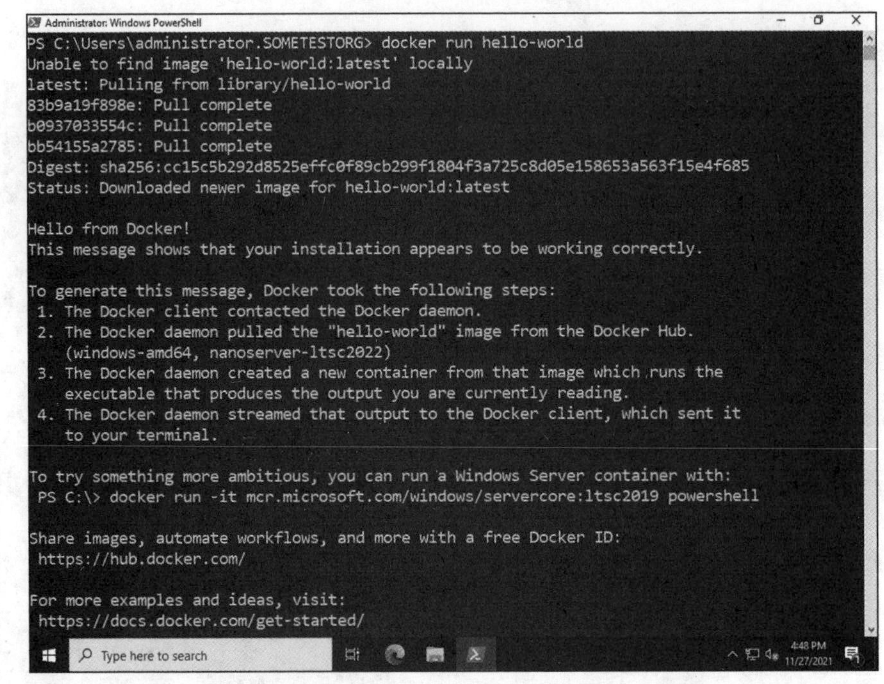

FIGURE 3-4: The simple hello world container can be used to quickly test your Docker installation using the `docker run` command.

If you want to stage the image so you can play with it later, you can use the `docker pull` command instead of docker run, and it will only download the container image. Here is the command to download a container (in this case, I'm downloading Windows Server Nano):

```
docker pull mcr.microsoft.com/windows/nanoserver:ltsc2022
```

Note that the download may take a few minutes because it's pulling down a copy of Nano Server. You can watch the progress on the screen. See Figure 3-5 for the output from running the command. You can see when I use Docker images that I've downloaded the new container.

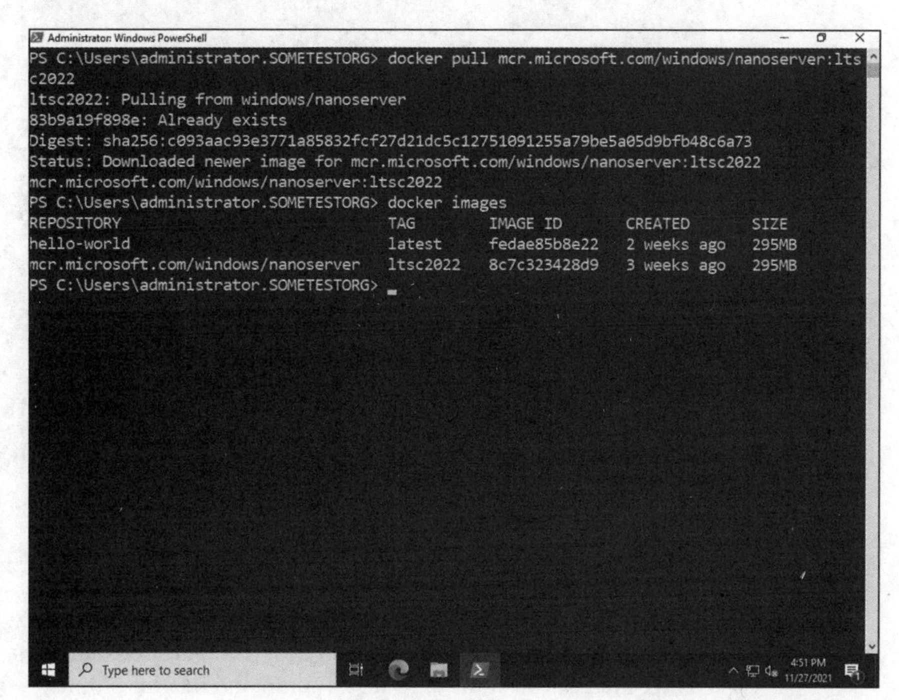

FIGURE 3-5: Downloading the sample container from Docker Hub is simple using the `docker pull` command.

REMEMBER

The container images must use the same kernel version as the container host. If you try to run the container with a kernel version that doesn't match the container host's kernel version, you'll get an error similar to the screenshot in Figure 3-6. Notice the line in the error that ends in "The container operating system does not match the host operating system." I'm getting this error because I tried to run Nano Server 1909, instead of ltsc2022. Nano Server 1909 is older than the kernel in use by the container host.

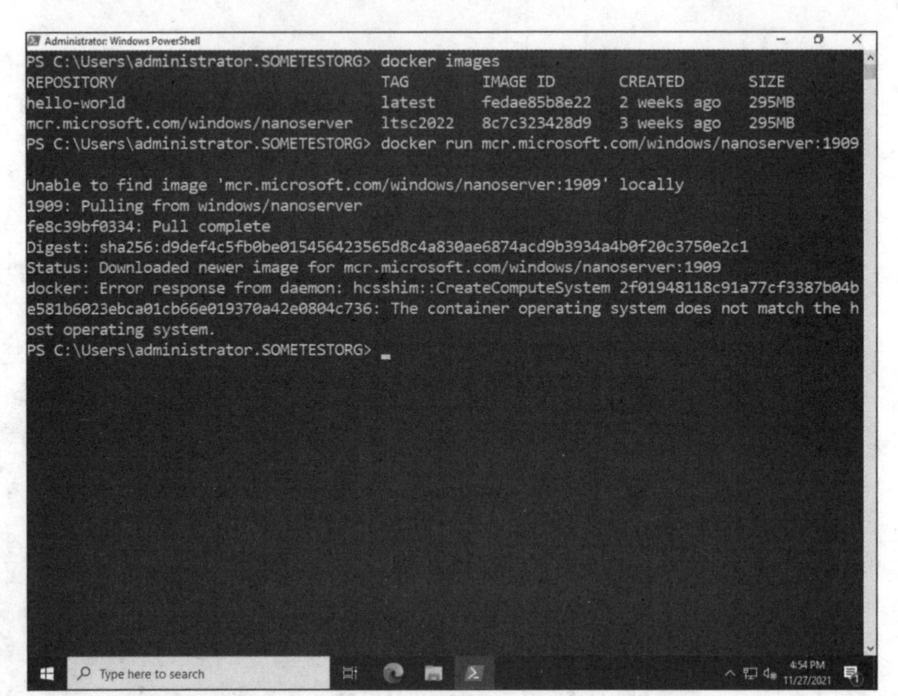

FIGURE 3-6: You get an error if you try to run a container image that does not match the container host's OS.

Hyper-V container

Testing the Hyper-V container is similar to testing a Windows container, but because the kernel isn't shared, you have far more freedom as far as which container images you want to run. The command itself is similar — you just need to include `--isolation=hyperv` to tell it that you want it to launch the container as a Hyper-V container rather than a Windows container.

```
docker run --isolation=hyperv mcr.microsoft.com/windows/
  nanoserver:1909
```

As you can see in Figure 3-7, the container image (which I download earlier, in the "Windows container" section) was able to run even though it had an older kernel version because I was using Hyper-V isolation on it.

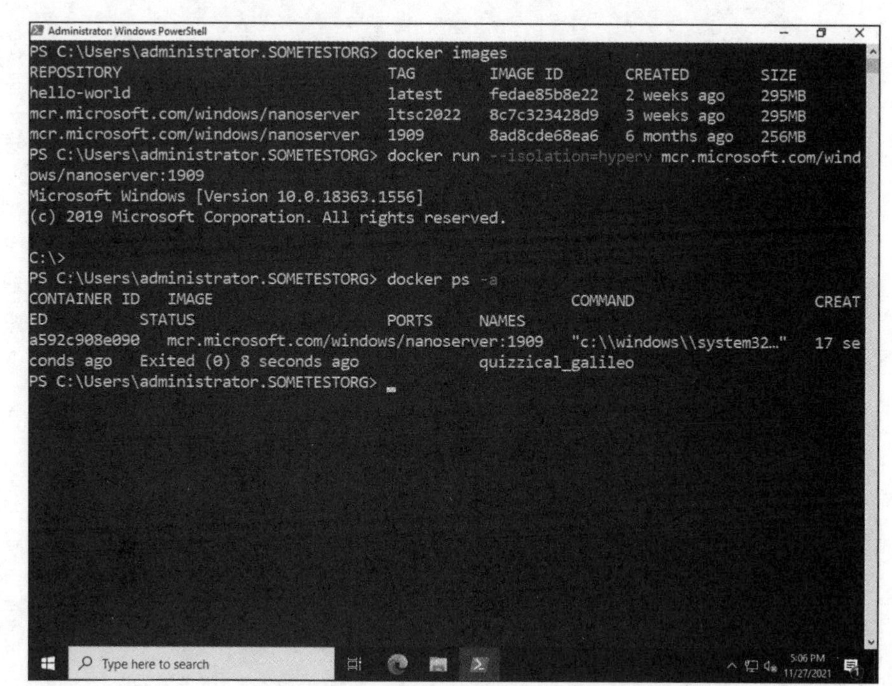

FIGURE 3-7: The container image was used to create a container named quizzical_galileo because I didn't specify a name.

» Applying custom metadata to containers and other objects

» Configuring containers

» Configuring the Docker daemon with daemon.json

Chapter 4 Configuring Docker and Containers on Windows Server 2022

Installing containers is the easy part, and you can certainly have some fun while doing it. In an enterprise environment, however, you'll be asked to configure the containers so that they can do useful things and provide value to the business.

In this chapter, you find out how to configure and customize your containers so that your organization can really reap the benefits of containerization.

Working with Dockerfile

A *dockerfile* allows you to specify how a container should be built. It identifies the container image that you'll use and contains any commands that you may want to run. The changes that you make to the container image are saved as a layer on top of the original container image.

When you create a dockerfile on Windows, you need to ensure that it does not have a file extension. You can do this by saving it as `"dockerfile"` with the quotes and change Save as Type to All Files, as shown in Figure 4-1.

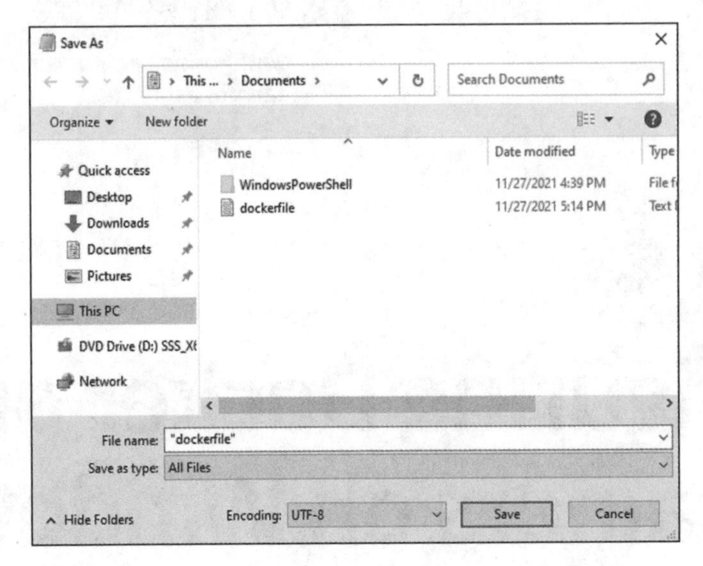

FIGURE 4-1: Creating the dockerfile with no extension in Notepad.

Inside the dockerfile, you'll add the instructions that will be used to build your container. These instructions are shown in Table 4-1.

TABLE 4-1 Docker Instructions

Name	Description
`FROM`	Used to specify which container image your image will be built from.
`LABEL`	Labels are used to organize your objects within Docker. Labels are stored in key-value pairs as strings.
`RUN`	Used to run commands within the operating system (OS) like PowerShell, the Command Prompt, or any other executable that can be called.
`CMD`	Used to set a default command that will be run any time the container is created and deployed.
`COPY`	Copies files or folders from the specified source to the specified destination. The files or folders must be in a location with the dockerfile or relative to it; otherwise, it won't be able to copy.
`ADD`	Similar to the `COPY` instruction, except it can copy from sources on remote hosts with URLs.
`WORKDIR`	Allows you to set your working directory, which may be needed to things with the `RUN` and `CMD` instructions.

You may be ready to build your first dockerfile. I'll walk you through creating a simple dockerfile that will pull the Windows Server Core Image and install Internet Information Services (IIS). Then you'll create a simple "Hello World"-style of page that will be placed in the default website.

TIP

The hash symbol (#) is used to make comments in the dockerfile, which can make it easier to revisit later and make changes if needed.

Open Notepad and type the following, or copy the dockerfile from the GitHub repo for this book, available at `https://github.com/sara-perrott/Server2022PowerShell`.

```
# Sets the base container image to Windows Server 2022 Core
FROM mcr.microsoft.com/windows/servercore:ltsc2022

# Metadata indicating an image maintainer.
LABEL "version" = "1.0" "Description" = "Core with IIS"

# Use dism.exe to install IIS
RUN dism.exe /online /enable-feature /all /
   featurename:iis-webserver /NoRestart

# Creates the Hello World file and adds our message to it.
RUN echo "Hello World - I was created from a Dockerfile!" >
   c:\inetpub\wwwroot\index.html
```

When you're ready to create your container from the dockerfile, open PowerShell and navigate to the directory that your dockerfile is saved in. In my case, it's saved in a folder named `dockerfiles`. When you're in the correct folder, type the following command:

```
docker build -t coreiis .
```

This command creates a container image named `coreiis`, built from the dockerfile you created. Note that the dot (`.`) tells it that the dockerfile is in the current directory. You can also put a whole path into the command instead. In Figure 4-2, you can see the commands from the dockerfile as they kick off.

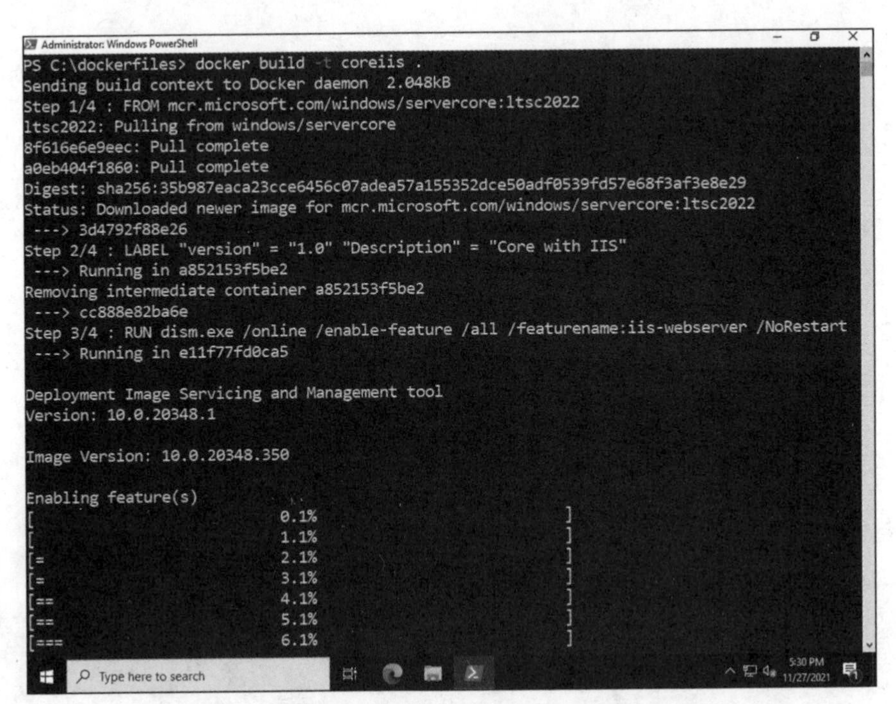

FIGURE 4-2: The docker build command allows you to build a container from the instructions in a dockerfile.

Applying Custom Metadata to Containers and Other Objects

When you hear the word *metadata* in relation to containers, you should automatically think "labels." You can use labels to organize different kinds of objects in Docker — everything from images and containers to networks and volumes.

A *label* is a key-value pair. The key of a label is on the left side of the key-value pair and is typically the thing being used to group your objects. For instance, if you've made some special container images for a particular project (and this happens often), you may create a key of `Project`. The value, which goes to the right side of the key-value pair, would in this case be the name of the project. Let's call the project "NewApp." So, the full key-value pair that you would attach to these special container images would be `Project = NewApp`. Container images can use more than one label so you add a description label with the project stakeholder's name or any other relevant information. The sky's really the limit on this one.

TIP

If your container image's parent (specified by `FROM` in the dockerfile) has labels, your container image will inherit the labels that are set on the parent. If you set a newer value for a label that already exists, your newer value will overwrite the existing value.

Labels are created in the dockerfile that you use for your image. In this section, I walk you through how to set labels and how to view labels after they're set.

Creating labels

Creating labels within the dockerfile is pretty simple. Here is an example based on the scenario I mentioned in the introduction to this section.

Single label

For the example of a single label, the entry in the dockerfile would look something like this:

```
LABEL "Project" = "NewApp"
```

Multiple labels

If you wanted to create multiple labels, you can follow the same format as the single label — you just use a space to separate labels:

```
LABEL "Project" = "NewApp" "Author" = "John Doe" "version" =
  "1.0"
```

If you find that your labels are getting too long and are wrapping around the screen, you can increase the readability with the forward slash. This is the same example, just edited to look more readable with the slashes.

```
LABEL "Project" = "NewApp" \
"Author" = "John Doe" \
"version" = "1.0"
```

Viewing labels

If you ever want to view what labels are set on a container image, you can type the following:

```
docker inspect <containername>
```

This command lists any labels that are associated with the container image. In Figure 4-3, you can view the labels attached to the container image that you created in the dockerfile section of this chapter.

```
PS C:\dockerfiles> docker inspect coreiis
[
    {
        "Id": "sha256:d4d4564beaa4c25a1c5d8258d18de345cd0cd65ff5d4adf458a33769ce6b99c5",
        "RepoTags": [
            "coreiis:latest"
        ],
        "RepoDigests": [],
        "Parent": "sha256:b0fc7f9c9ace0b4a1da27097e041784af168a900c58c8f32c1ada25161d5766e"
,
        "Comment": "",
        "Created": "2021-11-28T01:29:09.2675748Z",
        "Container": "2398b261378fe02e14d7c02a579f7abfb13c8d2f1cd710d3dd45e5ca567e8e32",
        "ContainerConfig": {
            "Hostname": "",
            "Domainname": "",
            "User": "",
            "AttachStdin": false,
            "AttachStdout": false,
            "AttachStderr": false,
            "Tty": false,
            "OpenStdin": false,
            "StdinOnce": false,
            "Env": null,
            "Cmd": [
                "cmd /S /C echo \"Hello World - I was created from a Dockerfile!\" > c:\\in
etpub\\wwwroot\\index.html"
            ],
            "Image": "sha256:b0fc7f9c9ace0b4a1da27097e041784af168a900c58c8f32c1ada25161d576
```

FIGURE 4-3: The docker `inspect` command allows you to view the metadata of a container image, including the labels that are currently assigned.

Configuring Containers

At this point, you've played with a dockerfile and deployed a container image. There are some other forms of configuration that you may be interested in as well. I cover a few of these topics in this section.

Starting containers automatically

For containers that are running important services, you may want to make sure that they stay up and running. When a container exits, for instance, you can use restart policies to ensure that the container restarts automatically.

To use the restart policy, you need to manually start the container, and that container has to be up for at least 10 seconds. At that point in time, Docker is able to monitor it. You can still manually stop a container, and Docker will not use the restart policy if you've stopped a container.

You can tell Docker that you want to use a restart policy with the `--restart` flag. The command looks something like this:

```
docker run -dit --restart <policy> coreiis
```

You're telling docker to set a restart policy for the coreiis containers, this is the container image you created previously if you've been following along. You've also specified that it will run in detached mode (-d), keep a connection open even when it is not attached (-i), and create a remote terminal session (-t). There are four different kinds of restart policies, shown in Table 4-2.

TABLE 4-2 **Restart Policy Flags**

Flag Name	Description
no	Enforces the default behavior in Docker, which is to not automatically restart the container.
on-failure	Great for error handling. If the container exits due to an issue, it will be restarted automatically.
unless-stopped	Unless you manually stop the container, it will automatically restart if it exits for any reason.
always	Use carefully. The container will automatically restart, even if it was stopped manually by an administrator.

Limiting a container's resources

This is possibly one of the most important configurations to understand and know how to implement. When you're running multiple containers on a container host, you may need to limit what resources the container can use. This prevents the container from taking all the resources and potentially starving the other containers or the container host of resources. In the following sections, I show you how to limit both the CPU and memory.

Memory

When you create a container, you can specify the amount of memory that the container is allowed to use with the --memory flag. For instance, to restrict our coreiis containers to 512MB of ram, the command would look like this:

```
docker run -it --memory 512m coreiis
```

CPU

CPU can be limited in much the same way that memory can. The flag is --cpus. If you wanted to ensure that your container could only use 1 CPU and 512 MB of RAM, the command would look like this:

```
docker run -it --cpus 1 --memory=512m coreiis
```

Configuring the Docker Daemon with daemon.json

The last topic that I want to cover is the configuration file for the Docker daemon, which is the server component of Docker. Modifying this file allows you a great deal of customization, but it's worth noting that at the time of this writing, the `daemon.json` file for the Docker daemon on Windows Server operating systems does not support all the configuration options available. The configuration options available in the `daemon.json` file are shown in Table 4-3. Note that you don't need to add all these parameters to the `daemon.json` file. You only need to add the parameters that you want to make changes to. The `daemon.json` file is located by default at `C:\ProgramData\docker\config\daemon.json`.

TABLE 4-3 Allowed Configuration Options in the Windows Docker daemon.json File

Parameter	Description
`authorization-plugins`	Allows you to specify the name or location of an additional authorization source by name or by the location of its specification file.
`dns`	Allows you to specify specific DNS servers for all the containers.
`dns-opts`	Allows you to set options to use with your DNS servers.
`dns-search`	Sets the DNS search domain for all the containers.
`exec-opts`	Lets you specify options for runtime execution.
`storage-driver`	Allows you to select which storage driver you would like to use. Windows Server container hosts support either windowsfilter for Windows containers or lcow for Linux containers running on your Windows host.
`storage-opts`	Allows you to set options specific to the storage driver that you've chosen to use. Windowsfilter has one configuration item that can be set with `storage-opts` and that is related to size. Lcow (Linux containers on Windows) has quite a few more options available.
`labels`	Replaces the labels from the daemon with a new set of labels.
`log-driver`	The default driver for logs from the containers.
`mtu`	Allows you to specify the maximum transmission unit (MTU) that the container network will use.
`pidfile`	Lets you set the path for the daemon's pidfile. ***Note:*** The pidfile contains the process identification (pid) number of the daemon. This allows other programs to locate the process id.

Parameter	Description
`data-root`	Tells Docker where you want to store containers and their images. If this is not used, the default location of `c:\ProgramData\docker` is used.
`cluster-store`	Points the container to the system and port it can use to get a certificate to support transport layer security (TLS) communication in the cluster. With the cluster-store-opt flag, you can tell it where the `.pem` files are located in the file system.
`cluster-advertise`	Specifies the IP address and the port number that the daemon should use to advertise itself to a cluster.
`debug`	Will swap the daemon into debug mode if it's set to true. You should only do this if you're actively troubleshooting an issue.
`hosts`	Lets you set the IP address and port number of systems for the Docker daemon to connect to.
`log-level`	Allows you to change the logging level of the daemon. Valid values are debug, `info`, `warn`, `error`, and `fatal`.
`tlsverify`	Uses TLS to verify the remote connection.
`tlscacert`	Tells the daemon which certificate authority it can trust certificates from.
`tlscert`	Specifies the location of the of TLS certificate file.
`tlskey`	Specifies the location of the TLS key file.
`group`	Allows you to change the group used for connections. The default group is docker.
`default-ulimits`	Sets the default ulimit for each container. If a ulimit is not specified when a container is built, it will use the setting specified by this parameter. ulimit sets the number of open files allowed at any one time. The default is 1,024 open files in a Docker container.
`bridge`	Attaches your containers to a network bridge.
`fixed-cidr`	Sets the IPv4 subnet for your containers.
`raw-logs`	Ensures that logs have full timestamps and no American National Standards Institute (ANSI) coloring.
`registry-mirrors`	Lets you select Docker registry mirrors that you want to use.
`insecure-registries`	Allows you to communicate with registries that are not considered to be secure. This is typically because they're only listening on HTTP and don't support TLS, or because they're using an untrusted certificate.

A *daemon* is a process that runs continuously in the background, whose sole purpose is to fulfill requests from services. Users do not interact directly with a daemon.

Now that you have an idea of which options are available to you, let's create a simple `daemon.json` file.

The first thing you'll need to specify is which port you want Docker to accept incoming communications on. In this example, you don't care where the connections come from, only the port that they're coming in on. That line looks like this:

```
"hosts": ["tcp://0.0.0.0:2375"]
```

Next, let's change the storage location for your containers and your images so that it isn't on the system drive. You need a second slash in the file location, so don't forget it!

```
"data-root": "D:\\DockerStuff"
```

When these items are put together into the json file, they look like this:

```
{
"hosts": ["tcp://0.0.0.0:2375"]
"data-root": "D:\\DockerStuff"
}
```

That's all there is to creating a `daemon.json` file to configure your Docker daemon. There are a lot of options — the best way to learn is to start playing around, so I encourage you to build out a lab environment and start practicing!

» Pushing images to Docker Hub

» Pulling images from Docker Hub

» Image versioning

Chapter 5
Managing Container Images

As you make changes to your container images, you'll get to the point where you're happy with where the current container image is and you want to save the changes so that people can simply launch your new container image instead of having to run the image from a dockerfile and configure and tweak the image from there.

In this chapter, I show you how to save changes that you've made to your container images. I also explain more about pulling and pushing images to repositories.

Making Changes to Images and Saving the Changes You Make

When you launch containers, you can interact with them on a similar level to a regular system from the command line. You can create users and groups, and you can run PowerShell to install roles and features. After you've made these changes, you may decide you don't need them anymore. The great thing about containers is that when you don't need them anymore, you can exit out of the container and launch a new one with a fresh container image. But what if you like the changes that you made, and you want to create a new container image that contains those changes? This is where the `docker commit` command comes in.

The `docker commit` command creates a new container image that contains all the changes you made from the base container image. When you use the command, you specify the name of the running container that you want to save, and the name of the new container image that you want to create.

In the following sections, I walk you through the steps of creating a Docker container, connecting to the container, making some changes, and then saving the changes.

Create the container

To be able to modify the container, you need to actually create it first. So let's create a container. I'm going to assume that you have Docker installed and have the PowerShell window open.

1. **At the PowerShell prompt, run the command** `docker images`**.**
2. **Locate the container image you want to use.**

   In my case, I'll use the `mcr.microsoft.com/windows/servercore:ltsc2022` image.
3. **Type the following command to create the container and a session to it:**

   ```
   docker run -dit mcr.microsoft.com/windows/servercore:
     ltsc2022
   ```
4. **Press Enter.**
5. **When the PowerShell prompt returns, type** docker ps **and press Enter.**
6. **Note the name that was assigned to the container, found in the last row.**

   In my case, the name of the container is `crazy_chaplygin`.

You can see the container creation steps in Figure 5-1. I tell it to run in detached, interactive mode. Then I use the command `docker ps` to list the running containers. You can see the container that was just created, as well as the name it was given, its uptime, and its container ID.

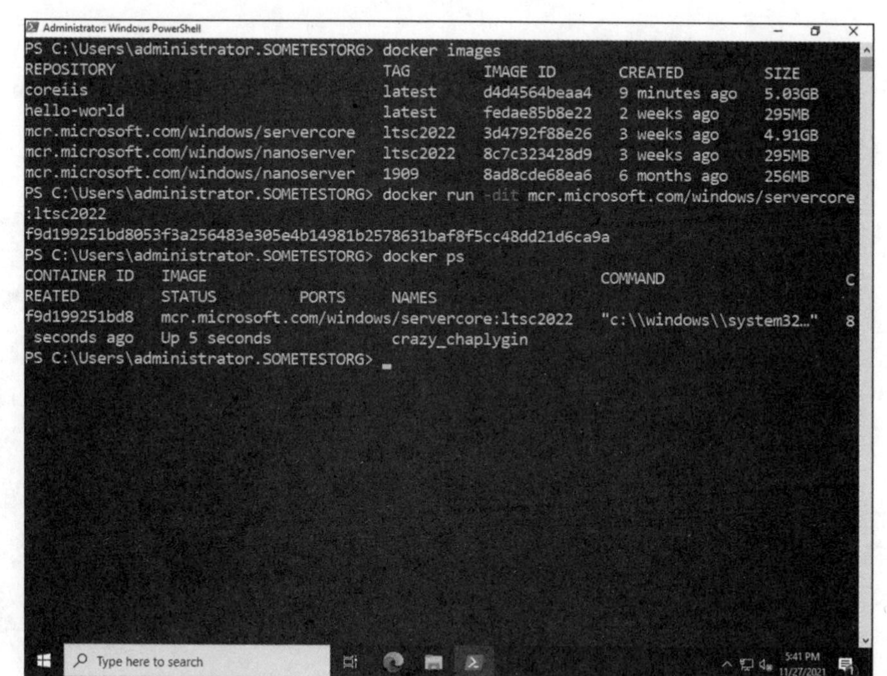

FIGURE 5-1: Creating the container is as simple as identifying the container image and issuing the command to create the container from that image.

Connecting to the container and making changes

Now that you've created your container, you're ready to connect to it to interact with it. You use the `docker exec` command in this case to connect to the `cmd.exe` on the container that you just created. You need the name of the container that you took note of in the previous section. The command should look like this:

```
docker exec -it <containername> powershell.exe
```

In my case, I'll type the following command:

```
docker exec -it crazy_chaplygin powershell.exe
```

After you press Enter, you're connected to the container. You can validate this by running the hostname command. You get a string of letters and numbers that are randomly generated.

On this container that I've run, I want to install Internet Information Services (IIS). So, from inside the container, I'll run the following PowerShell command:

```
Install-WindowsFeature Web-Server
```

After the web server is installed, I'll type **exit** and then press Enter, which will drop me back out to the container host. The commands I've run in the container are shown in Figure 5-2.

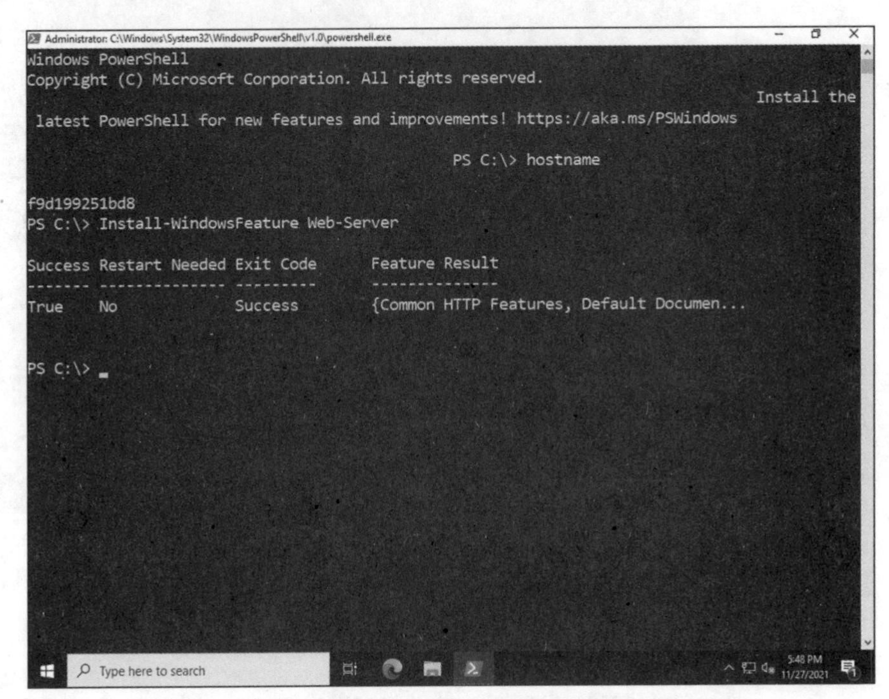

FIGURE 5-2: After you connect to a container, you can make changes to it like adding/removing directories, installing roles and features, and so on.

Saving the changes to the container

Now that you've installed IIS, the next step is to save the container as a container image so that you don't have to repeat these steps the next time you need to launch a Server Core with IIS container.

First, you have to stop the container because you can't do a commit against a running container in Windows. The command to stop the container is similar to the following:

```
docker stop <containername>
```

After the container is stopped, you can do the commit. The command follows the format of:

```
docker commit <containername> <reponame>:<imagename>
```

You can see these commands start to finish in Figure 5-3.

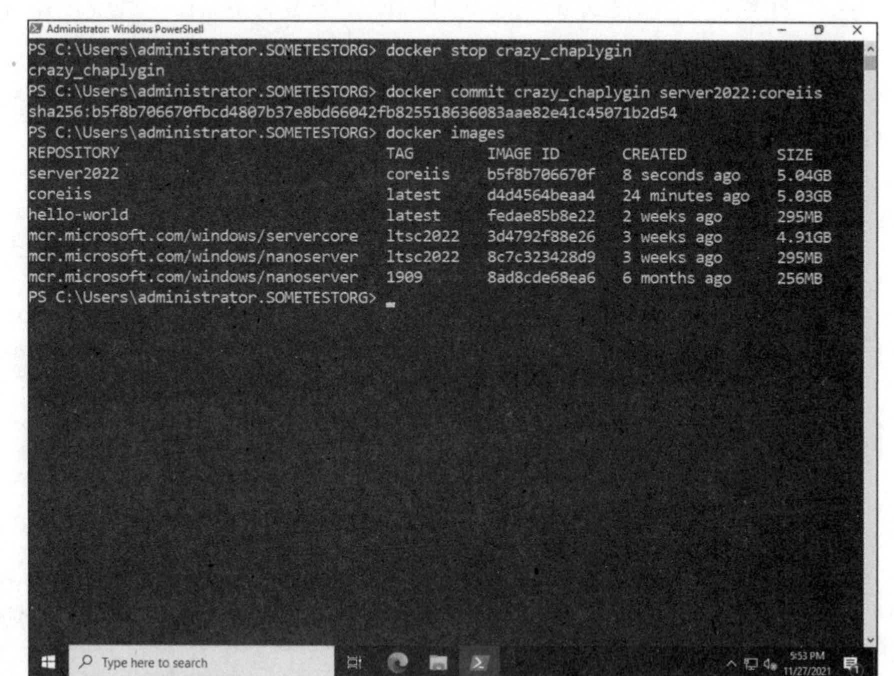

FIGURE 5-3: After you stop the container, you can commit the changes you have made to it.

Pushing Images to Docker Hub

When you issue a `docker push` command, it's assumed that you're targeting Docker Hub unless you specify a different registry. If you ever need to log in to a different registry, you specify the registry address in the `docker login` command. It would look like this:

```
docker login some.registry.com:8000
```

You're prompted for your username and password, and then you'll be connected.

If you ever need to rename your images to make them suitable to upload to a repository, you can use the `docker image tag` command. For example, I downloaded a Server Nano image from the Microsoft repository. Say I want to upload that image to my own repository. My container image may be `server2022:ltsc2022`. When I log in to Docker Hub, I need to specify my repository name, which in this case is a combination of my Docker ID and *<reponame>* and then the image name. The command ends up looking like this:

```
docker image tag mcr.microsoft.com/windows/nanoserver:ltsc2022
  <dockerid>/server2022:ltsc2022
```

You may need to do this if you created a local repository and now want to push to a repository on Docker Hub. For example, when I committed the container image that I worked on earlier, I saved it as `server2022:coreiis`. To be able to then push it to Docker Hub, I run the following command:

```
docker image tag server2022:coreiis <dockerid>/
  server2022:coreiis
```

After this is complete, I can push the container image with a simple `docker push` command. See "Pushing to a repository," later in this section, to see how to use `docker push` to push the image to a private or public repository.

Understanding private versus public repositories

On Docker Hub, you're given the option of creating a private repository or a public repository. For both types, you need a valid login to push images, but when you pull images there are differences between the two.

Public repositories don't require a login to pull an image. You can simply enter the `docker pull` command with the proper repo and image name and your container image will be downloaded. Private repositories require a valid login before you can view the contents of the repository or pull an image. Private repos are great for businesses who want to store their container images in a repo but don't want those images to be available to the public. After you log in with the `docker login` command, you can do a `docker pull` to download the image. Assuming that you have permissions to pull the container image, it will be downloaded for you.

Pushing to a repository

Pushing to a repository requires a login for a user who is authorized to upload to the repository; this is true of both private and public repositories. You log in, and then you do the `docker push` command.

To log in to the repo, type the following:

```
docker login
```

Fill in your username and your password. After you're logged in successfully, you can push the container image to your repository.

```
docker push <dockerid>/reponame:imagename
```

The image in Figure 5-4 shows you what the process looks like from start to finish.

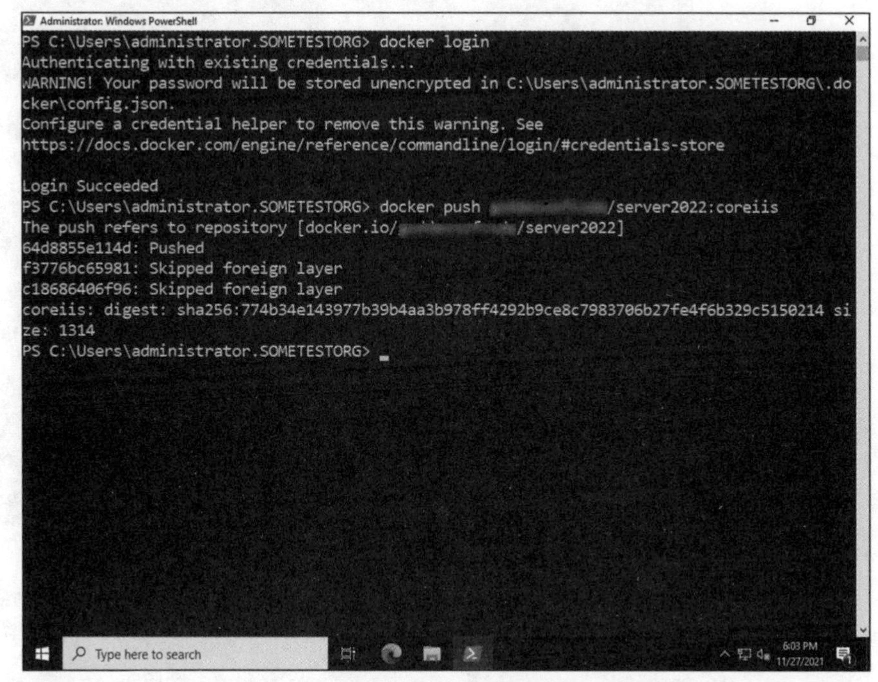

FIGURE 5-4: You can use the `docker push` command to push to your private repository after you're logged in.

Pulling Images from Docker Hub

Repositories are a great way to make container images available. You may be making container images available to people in your organization, or you may be taking advantage of repositories set up by other companies like Microsoft. Having a central location for container images makes it far simpler to maintain your images and ensure that your customers are always getting the most up-to-date image from you.

Pulling from a public repository

Pulling from a public repository is one of the simplest things to do. One of the things I love about Docker Hub, especially for people just getting started with container technology, is that you can click an image, and you're presented with the `docker pull` command to get the container image, as shown in Figure 5-5.

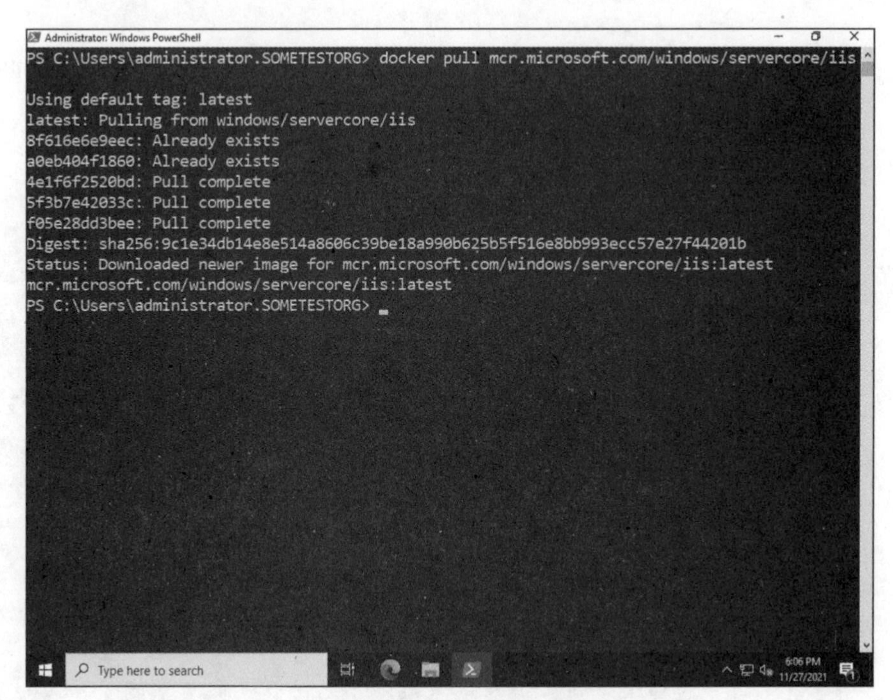

FIGURE 5-5: Docker Hub makes it easy to get started pulling container images because it gives you the pull command in the container image's page.

For instance, the command to pull down a Windows Server Core container with IIS is available to be copied. It looks like this:

```
docker pull mcr.microsoft.com/windows/servercore/iis
```

You can copy and paste this directly into the PowerShell window of your Docker container host, and it will pull the container image for you.

Pulling from a private repository

A private repository uses the same commands as a public repository does. The only difference is that you must log in first with the `docker login` command.

Private repo on Docker Hub

When you don't specify a registry, it's assumed that you want to connect to Docker Hub. So you log in with the docker login command; then you do your pull just as you would do with a public repository.

Private repo with a different registry

Because the docker login command assumes that you want to connect to Docker Hub, you need to specify which registry service you want to connect to. You can do that with the docker login command as well, with a slight modification, as follows:

```
docker login some.registry.com:8000
```

After that, you can issue the pull command as you normally would and it will be able to successfully pull the image from your repository on the other registry service.

Handling Image Versioning

As you create new versions of your container images, you may want a way to track the version of the container that you're on. One of the most common ways to do this is to utilize tags to track new container image versions. For instance, you may have the first version of your container image in production, and then you need to update a few settings on it. You save the new version of the container image with docker commit, and you want to make sure that people who perform a docker pull get the latest version. This is when you apply a tag to the container image. The syntax of the command is this:

```
docker tag <image_id> image/tag
```

Here's how you can tag an image. First, get the image ID number. To get the image ID, you run the docker images command.

```
docker images
```

Take note of the image ID. In this case, I'm going to use the container image that has an image ID of b5f8b706670f, which is my coreiis container image. Now tag the image as a v1.0 image:

```
docker tag b5f8b706670f coreiis:v1.0
```

Now when you run docker images again, you can see the new tag. This section is shown in Figure 5-6. Using tags for version numbers, build dates, and so on is one of the most common use cases for this feature.

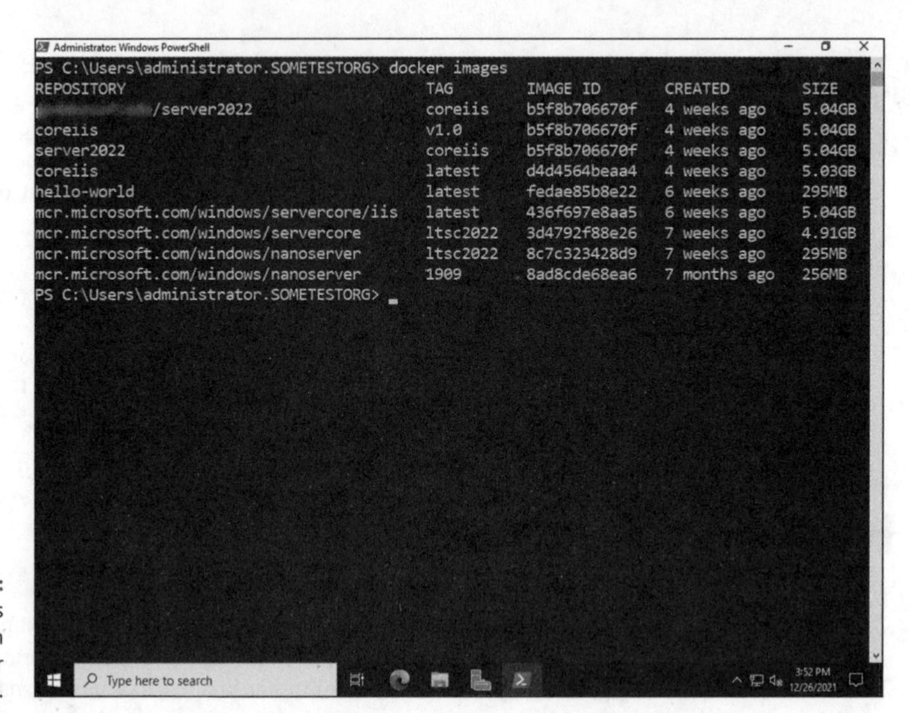

FIGURE 5-6: You can use tags to track version numbers on your container images.

IN THIS CHAPTER

» Looking at types of network connections

» Viewing your network adapters and virtual switches

» Configuring a NAT network connection

» Configuring a transparent network connection

» Configuring an overlay network connection

» Configuring an l2bridge network connection

» Configuring an l2tunnel network connection

» Connecting to a network

Chapter 6

Container Networking

Foundational to most server concepts is networking, and containers are no exception to that. Containers are configured by default to use a Network Address Translation (NAT) connection, but you can change that to customize your container environment.

In this chapter, you learn the ins and outs of container networking, what types of network connections you can make, why you would use them, and how to configure them.

Considering the Different Types of Network Connections

With Windows container networking, there are five different modes that you can use:

- **NAT:** NAT is the default network mode that is configured for your containers the first time that Docker is run on your system. If a different network is not specified when a container is run, it will be assigned to the NAT network and assigned an IP address from the NAT's internal IP address range, which is 172.16.0.0/16. The NAT connection is useful when you want a container to be able to communicate with the container host, and potentially the other containers on the host. This is typically used by developers or for home labs.
- **Overlay:** Overlay networks allow containers to communicate with other containers on other container hosts when the Docker Engine has been configured to run in Swarm mode. Your container host must be at least Windows Server 2016 to use this type of network. Overlay networks are required if you're using Docker Swarm in a multi-node architecture.
- **Transparent:** Transparent networks give you a more traditional network connection. Containers are given a connection to an external virtual switch, which allows them to connect and communicate directly over a physical network. Transparent networks are good for developers (because they mirror traditional networks) or for small-scale deployments.
- **L2bridge:** L2bridge networks are an interesting concept. The containers exist within the same subnet as their container hosts, but they're also connected to the physical network via an external switch. This type of network is typically used for Kubernetes and for Microsoft Software Defined Networking (SDN).
- **L2tunnel:** L2tunnel networks are a special case. They work similarly to l2bridge networks, but they're designed with the Microsoft Cloud Stack in mind and can be used in conjunction with SDN policies. This type of network can only be used in Azure.

Viewing Your Network Adapters and Virtual Switches

When you first log on to your container host, you may want to validate that the virtual switches you expect to have configured are there and that the container networks you need are present as well.

By running the `docker network ls` command, you can list all the networks that you've defined for your containers. If you haven't configured any, you'll see the default NAT connection that is created when the Docker Engine is run for the first time, shown in Figure 6-1.

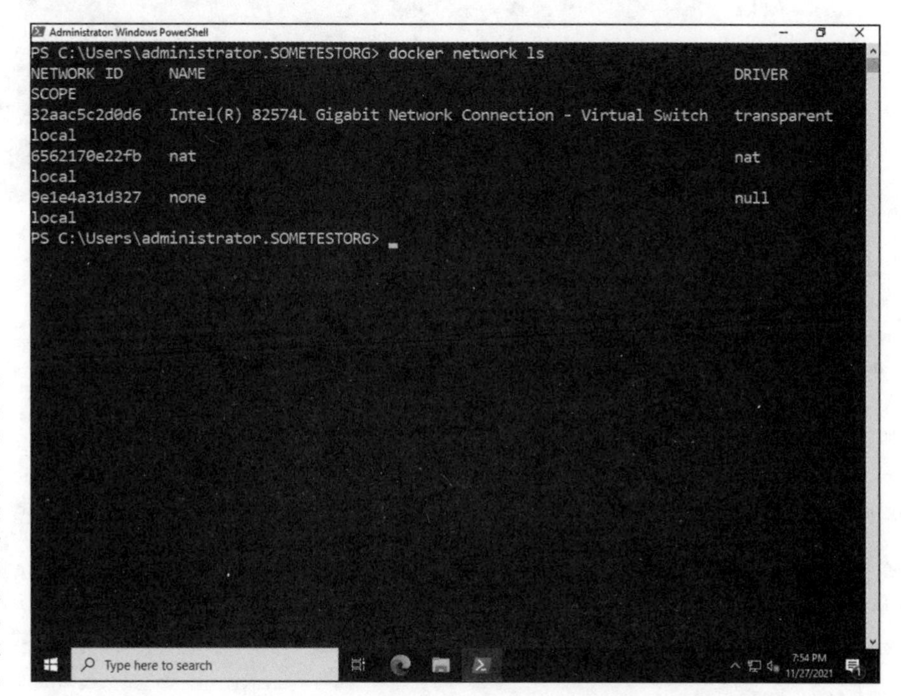

FIGURE 6-1: When you run the Docker Engine for the first time, a NAT network connection is created; by default, all containers are connected to the NAT network connection.

If you want to see the virtual switches that are on the machine currently, you can run the command `Get-VMSwitch`. If this is a fresh installation, you'll only have one virtual switch. It's an internal virtual switch, and it will be named NAT, as shown in Figure 6-2. In my case, I also have the default external virtual switch from Hyper-V.

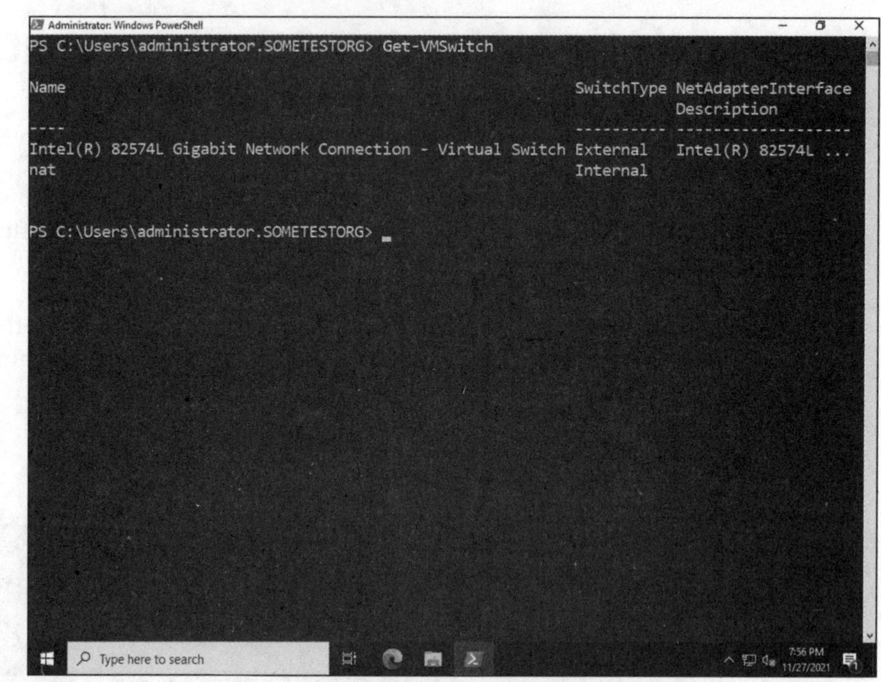

FIGURE 6-2: You can check to see which virtual switches are defined on your container host already by running the `Get-VMSwitch` command.

Configuring a Network Address Translation Network Connection

The network address translation (NAT) network is the simplest to use because one is created by default. But you're limited to the IP range that Microsoft assigns to it, which is 172.16.0.0/16. That is a very large subnet, and you may want to define a smaller subnet. You can do that when you configure your own NAT. Assuming that you have a subnet to use, the following command creates the NAT connection:

```
docker network create -d nat --subnet=<subnet with cidr>
  --gateway=<network_gwaddress> <nat_network_name>
```

When you run this command, your NAT connection is created, and you can view it using the `docker network ls` command, shown in Figure 6-3.

In my example, you can see the NAT network called MyNat was created. It was listed after I ran the `docker network ls` command.

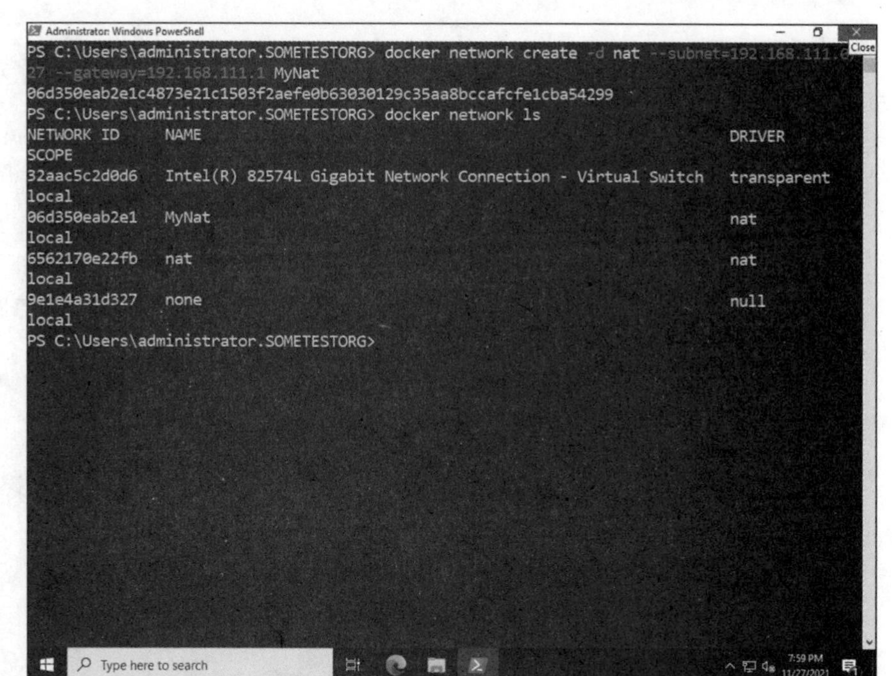

FIGURE 6-3: You can create a NAT network for container hosts with the `docker network create` command.

Configuring a Transparent Network Connection

The transparent network connection is similar to the traditional network connection in that your containers can interact directly with the outside world. The command to create a transparent network is simpler than the one used to create a NAT network, because you don't need to define a subnet or default gateway. Here's the command:

```
docker network create -d transparent <network_name>
```

This creates the transparent network and corresponding external virtual switch if you didn't have one already. The virtual switch is given a random name of letters and numbers (see Figure 6-4).

You can also choose to bind to a specific network interface if you want to control which network adapter your container traffic is able to go through.

```
docker network create -d transparent -o com.docker.network.
  windowsshim.interface="net_adapter_name" <network_name>
```

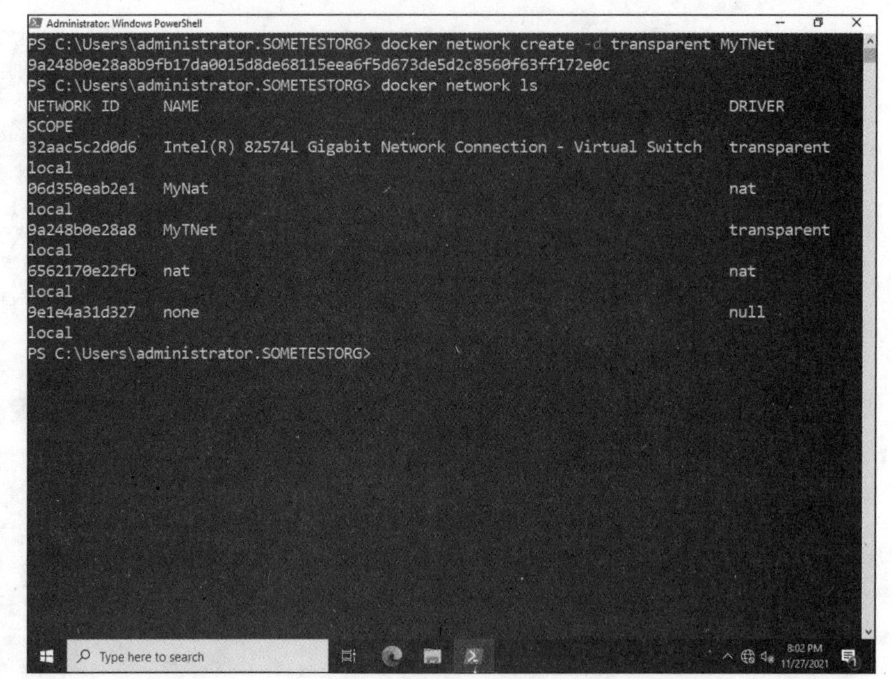

FIGURE 6-4: Creating a transparent network connection automatically creates the corresponding external virtual switch to support the desired communication to outside physical networks.

Configuring an Overlay Network Connection

Overlay networks are used to support container-to-container communication across container hosts when using Docker Swarm. This book does not cover Docker Swarm; just know that it allows system administrators to manage whole clusters of Docker nodes.

WARNING

At the time of this writing, it should be noted that encrypted overlay networks are not supported on Windows container nodes. If you put a Windows container node on an encrypted overlay network, you won't get any errors — but you won't get any communication either.

```
docker network create -d overlay <network_name>
```

To support overlay communications, you need to open the following firewall ports on the container hosts to allow the nodes to talk to one another:

- **TCP 2377:** Cluster management traffic
- **TCP/UDP 7946:** Inter-node communication
- **UDP 4789:** Overlay network communication

Configuring an l2bridge Network Connection

The l2bridge network creates containers within the same subnet as the container host but also allows them to communicate directly to the outside physical network via an external virtual switch. The command is very similar to the NAT command that you used before.

```
docker network create -d l2bridge --subnet=192.168.1.0/24
  --gateway= 192.168.1.1 BridgeNet
```

After you've run the command, you can verify that the network connection was created properly by running `docker network ls` and `Get-VMSwitch`. These commands show you your new network, as well as your new virtual switch, as shown in Figure 6-5.

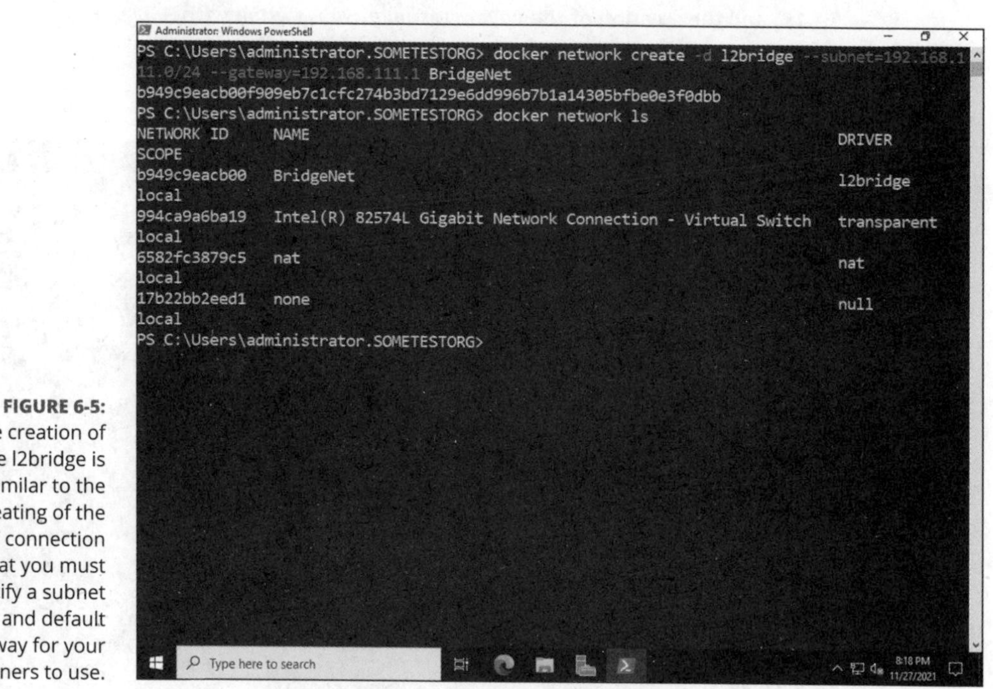

FIGURE 6-5: The creation of the l2bridge is very similar to the creating of the NAT connection in that you must specify a subnet and default gateway for your containers to use.

Configuring an l2tunnel Network Connection

The l2tunnel connection is designed specifically for the Microsoft Cloud Stack and Microsoft Azure. It's similar to l2bridge, but there is one big difference: All of the communication between two containers is sent to the physical Hyper-V host. This is the case regardless of whether there are containers in other subnets or on different virtual container hosts.

Configuring the l2tunnel connection is very similar to creating the l2bridge connection:

```
docker network create -d l2tunnel --subnet=192.168.1.0/24
   --gateway= 192.168.1.1 TunnelNet
```

As usual, you can validate the creation of the new network with `docker network ls`, and the creation of the new external virtual switch with `Get-VMSwitch`, shown in Figure 6-6.

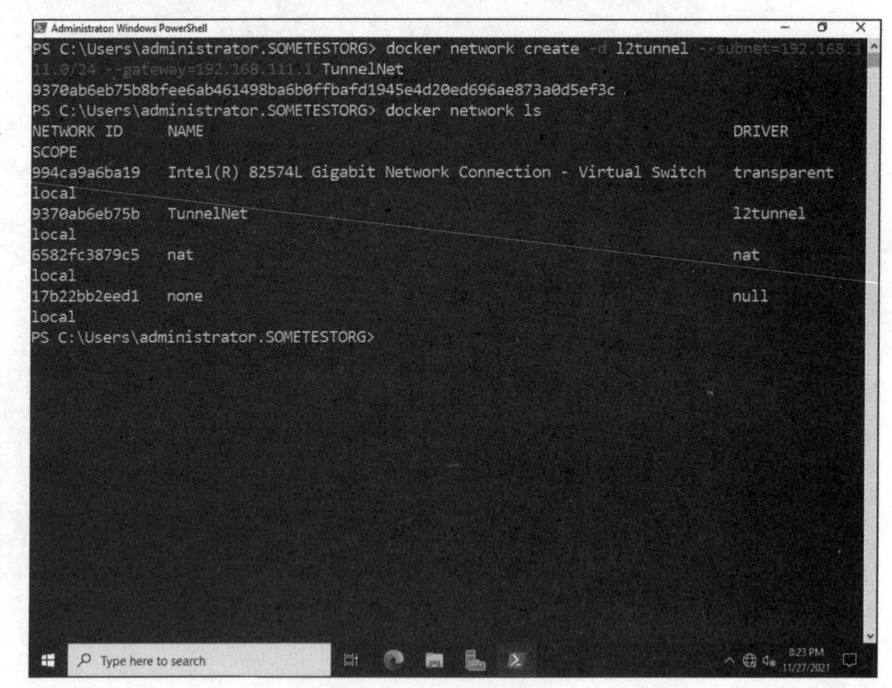

FIGURE 6-6: The `docker network create` command can be used to create the l2tunnel network and is similar in structure to the l2bridge command.

Removing a Network Connection

You may run into a scenario where you need to remove a configured network connection. For instance, NAT, l2bridge, and l2tunnel can't share the same adapter, and you'll get an error message if you try to add them.

Removing the old connection is simple. Just run the following command:

```
docker network rm <network name>
```

For instance, in the case of l2bridge, which I named BridgeNet, I would run the following command:

```
docker network rm BridgeNet
```

After that, I can create the new network connections without any issue.

Connecting to a Network

You can specify which network you want your container to connect to at runtime with the following command:

```
docker run -dit --network=<network_name> <image_name>
```

This command runs the container in detached mode, which allows it to run in the background, and will connect the container to the network specified in the *<network_name>* field. The container will be launched from whichever container image was specified by the *<image_name>* field (see Figure 6-7).

TIP

For some types of network connections, you may have to enable Media Access Control (MAC) spoofing because all the containers may be sharing the host's MAC address. MAC spoofing allows the virtual machine to change the outgoing packets to a different MAC address than their own. In this case, the virtual machine assigns the MAC address of the container host. The modes that typically require this are transparent and l2bridge (if the container host is a virtual machine). To enable MAC address spoofing on the container host, run the following command:

```
Get-VMNetworkAdapter -VMName <container_host_name> |
   Set-VMNetworkAdapter -MacAddressSpoofing On
```

FIGURE 6-7: You can specify your desired network connection at runtime with the --network flag.

Container networking isn't too complicated when you get down to it. You create the network connection and then you specify which connection you want it to use when you run the container. Pretty simple right?

» Creating a volume inside of a container

» Working with persistent volumes

Chapter 7
Container Storage

The last chapter of this book covers another foundational topic of computing: storage. Creating containers is fine, but sometimes you need to add storage to your individual container images. Maybe you need to create a new volume inside the container, or maybe you just want to add external storage so that the container has a defined area to copy files from when it's launching from a dockerfile.

This chapter is all about container storage. I cover what kinds of storage are available and why you should (or shouldn't) use a type of storage with containers.

Getting Acquainted with Container Storage

Storage in Windows containers is referred to as *layer storage* because the changes being made to a container are a layer on top of the base container image. Layers are stored by default within the `image` and `windowsfilter` directories located under `C:\ProgramData\docker`.

TIP

You can change the storage location by putting a data-root entry into the Docker Engine's configuration file, `daemon.json`, and add the path where you want the Docker Engine to store layers. As a best practice, this should be somewhere other than the system volume.

WARNING

Don't make changes to the layers directly through the file system. You'll most likely break something. Instead, use the Docker commands (like `docker images`, `docker rmi`, and `docker pull`) to manage your layers.

Creating a Volume Inside of a Container

The whole idea of a container is to be a self-contained and easily deployable object that contains all the dependencies to run a particular application. If your application has a dependency that requires its own volume, you can create the volume at run time. This allows you to support the needs of the application and can be used to add additional storage space if needed (which is great if you need to save data from the container to work with later). The volume points to a location on the container host, so the volume doesn't technically live inside the container.

To create a volume inside the container, run the following command.

```
docker run -it -v <volumepath> <imagename>
```

This command creates a symbolic link that makes the container think it has a volume named Data under its `C:` drive, when in fact the volume lives on the container host. You can still interact with the directory just as you normally would. For instance, if I run the following command:

```
docker run -it -v C:\Data coreiis
```

Docker will create a container for me based on my coreiis container image and will create a symbolic link (symlink) to the physical location on the host within the container. This is visible by running the `dir` command inside the container when it launches, as shown in Figure 7-1.

The physical location on the host is within Docker's directory. By default, volumes are stored in `C:\ProgramData\docker\volumes\volID\_data`. As you can see in Figure 7-2, I created the file inside the container while in the symlink `C:\Data` location, and it appeared on the container host under the `_data` folder.

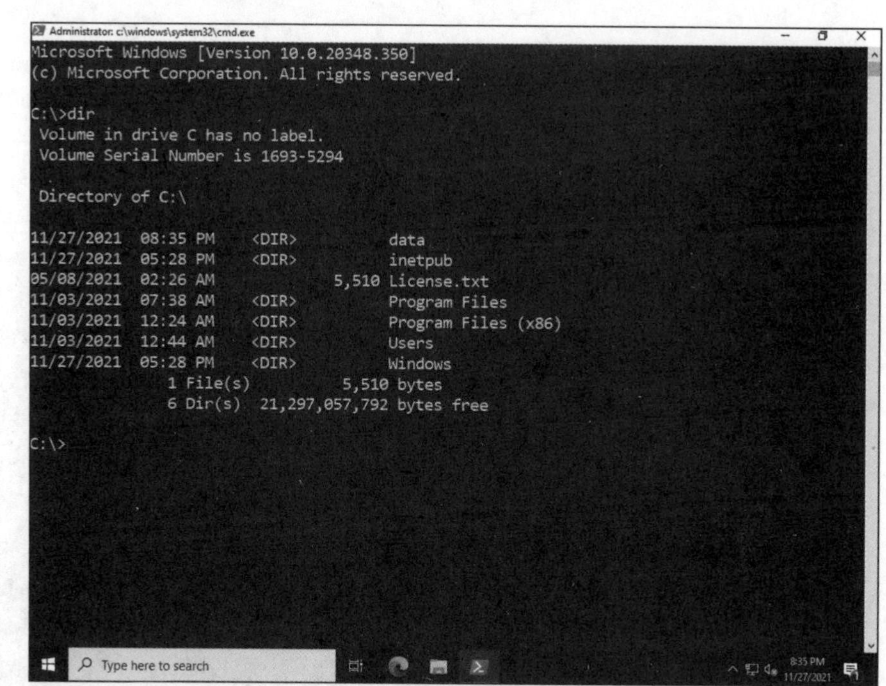

FIGURE 7-1: A symbolic link maps the volume that you created on the container to the physical location on the container host.

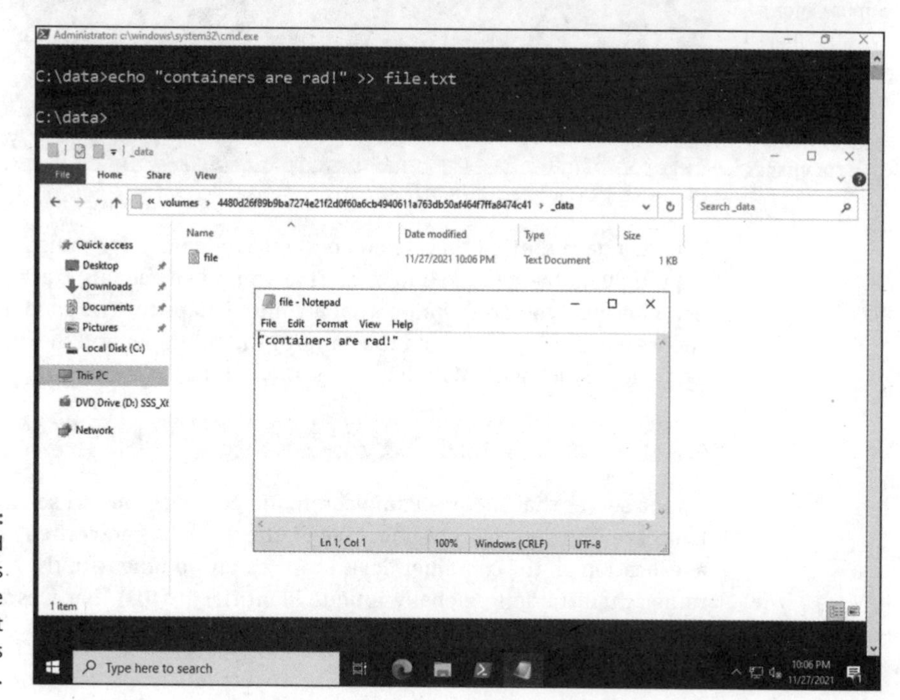

FIGURE 7-2: Volumes created for containers reside on the container host within Docker's directories.

If you need to see which volumes are on the container host, you can always run `docker volume ls`. To get more detailed information on the volume (like its storage location), you can run `docker volume inspect` *<volumeid>*, as shown in Figure 7-3.

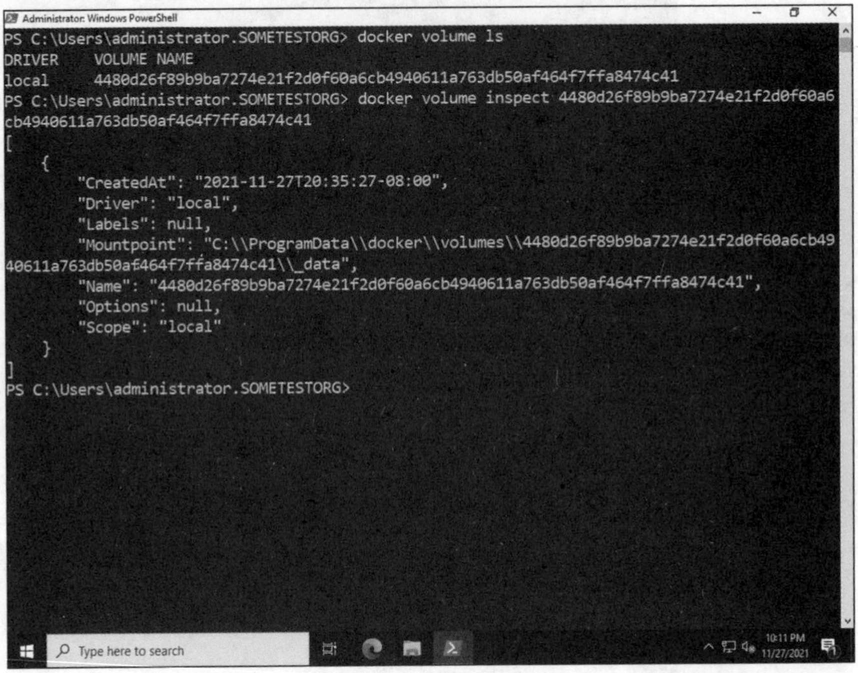

```
Administrator: Windows PowerShell
PS C:\Users\administrator.SOMETESTORG> docker volume ls
DRIVER    VOLUME NAME
local     4480d26f89b9ba7274e21f2d0f60a6cb4940611a763db50af464f7ffa8474c41
PS C:\Users\administrator.SOMETESTORG> docker volume inspect 4480d26f89b9ba7274e21f2d0f60a6
cb4940611a763db50af464f7ffa8474c41
[
    {
        "CreatedAt": "2021-11-27T20:35:27-08:00",
        "Driver": "local",
        "Labels": null,
        "Mountpoint": "C:\\ProgramData\\docker\\volumes\\4480d26f89b9ba7274e21f2d0f60a6cb49
40611a763db50af464f7ffa8474c41\\_data",
        "Name": "4480d26f89b9ba7274e21f2d0f60a6cb4940611a763db50af464f7ffa8474c41",
        "Options": null,
        "Scope": "local"
    }
]
PS C:\Users\administrator.SOMETESTORG>
```

FIGURE 7-3: You can get more information on container volumes with the `docker volume ls` and `docker volume inspect` commands.

You may have noticed that if you don't specify names for things, they tend to get long, ugly, auto-generated names. That can make keeping track of volumes difficult because you don't know what a volume maps to. The good news is, you can choose a name. I recommend the name of the container and the drive letter. In my example, I stuck with MyVolume to demonstrate:

```
docker run -it -v MyVolume:c:\data coreiis
```

On the server that the preceding command created, you can see that the symlink looks the same on the container (see Figure 7-4), but now when I check the storage location on the container host, I can see my volume with the name I specified, rather than the long globally unique identifier (GUID) that was assigned to the other volumes.

```
Microsoft Windows [Version 10.0.20348.350]
(c) Microsoft Corporation. All rights reserved.

C:\>dir
 Volume in drive C has no label.
 Volume Serial Number is 1693-5294

 Directory of C:\

11/27/2021  10:12 PM    <DIR>          data
11/27/2021  05:28 PM    <DIR>          inetpub
05/08/2021  02:26 AM             5,510 License.txt
11/03/2021  07:38 AM    <DIR>          Program Files
11/03/2021  12:24 AM    <DIR>          Program Files (x86)
11/03/2021  12:44 AM    <DIR>          Users
11/27/2021  05:28 PM    <DIR>          Windows
               1 File(s)          5,510 bytes
               6 Dir(s)  21,297,074,176 bytes free

C:\>exit
PS C:\Users\administrator.SOMETESTORG> docker volume ls
DRIVER    VOLUME NAME
local     4480d26f89b9ba7274e21f2d0f60a6cb4940611a763db50af464f7ffa8474c41
local     myvolume
PS C:\Users\administrator.SOMETESTORG> _
```

FIGURE 7-4: By using a custom name on my volume, I can more easily know which volume is the one that I just created.

Working with Persistent Volumes

Volumes are the preferred method of making data available to containers. Using persistent volumes can enable multiple containers to potentially share the same volume, and it ensures that the data persists (it's in the name!) as containers are created or destroyed.

Looking at volume types

There are two types of volumes that you can work with when creating volumes for containers. Volumes can be configured as a bind mounts or as named volumes.

Bind mounts

Bind mounts are great when storage needs the best performance possible. The storage on the host is mounted onto the container. The downside to bind mounts is that they don't have as much functionality as volumes do.

TIP

Docker recommends using named volumes over bind mounts in any new development work that you're doing.

Note that in the following examples, `C:\ContainerData` is the actual location on the host that you're binding the storage to, and `C:\Data` is where the symlink will be on the container.

By default, bind mounts are read/write enabled so you don't need to do anything other than specify the binding. Here's an example:

```
docker run -v c:\ContainerData:c:\Data <imagename>
```

If you only want the containers to have read access, you must specify that as part of the binding command:

```
docker run -v c:\ContainerData:c:\Data:RO <imagename>
```

Figure 7-5 shows an example where I have run both commands. The first command created a read-write binding, and the second created a read-only binding.

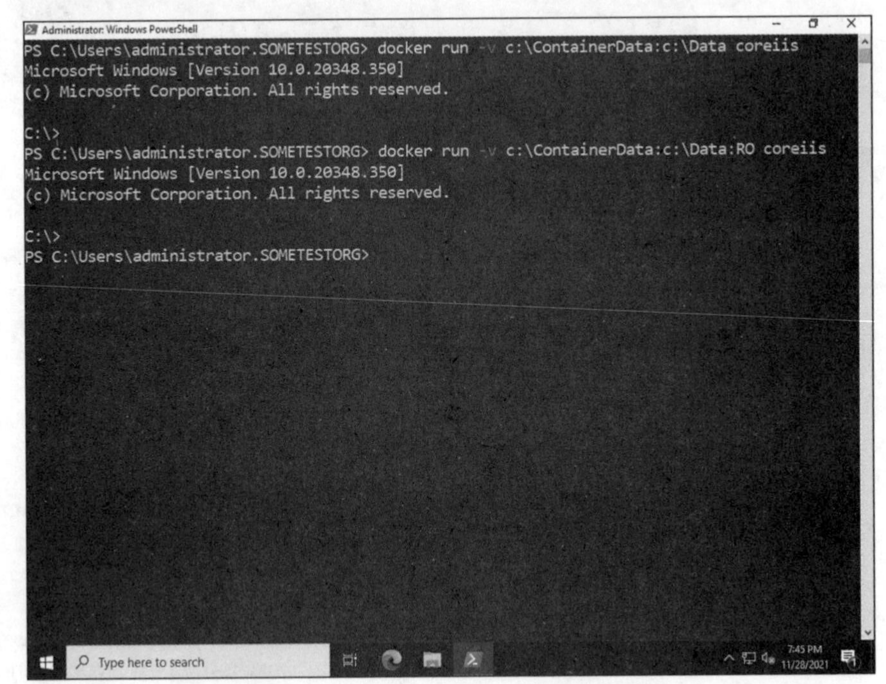

FIGURE 7-5: I've created a bind mount at run time for two containers; the first is a read-write binding, and the second is a read-only binding.

Named volumes

Named volumes are the preferred method for storing data outside of containers. Bind mounts are very much dependent on the container host, but named volumes are managed completely by Docker and can be shared across multiple containers.

You can create a named volume without having to start a container at the same time. This is done with the following command:

```
docker volume create volume1
```

When you run a container, you can specify a volume that already exists by name, or if the volume does not exist it will be created. Previously, you created a volume named Volume1, so let's look at the command to create a container and tell it to use the volume.

```
docker run -d -v volume1:C:\Data coreiis
```

An example of this method is shown in Figure 7-6 along with a listing of the volumes so that you can see that it used the volume that was created earlier.

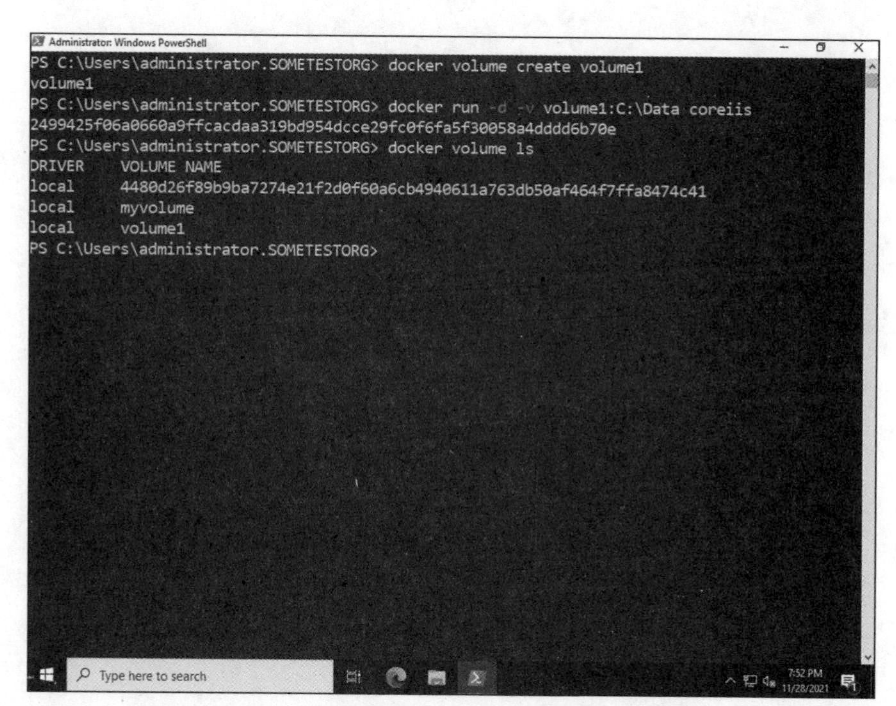

FIGURE 7-6: You can create a container and attach it to a previously created named volume.

Removing volumes

The day will inevitably come when you need to do some cleanup on your container host. One of the cleanup tasks may be to remove volumes that are no longer in use.

You may want to issue the `docker volume ls` command to see all your volumes first; then you can issue the `rm` command using the volume name.

TIP

If the volume names are the long GUID format, I recommend tweaking your PowerShell properties so that you can Ctrl+Shift+C and Ctrl+Shift+V within the windows to copy and paste. You can make this change by right-clicking the PowerShell window and clicking Properties. In the Properties dialog box, select the Use Ctrl+Shift+C/V as Copy/Paste check box.

The command to remove the volume has the following format:

```
docker volume rm <volume_name>
```

So, to remove Volume1 from earlier, I would type the following:

```
docker volume rm volume1
```

Index

Symbols and Numerics

A

B

C

D

E

F

G

H

I

K

L

M

N

O

P

Q

R

S

T

U

V

W

X

Z

About the Author

Sara Perrott is an information security professional with a systems and network engineering background. She shares her passion for all things information technology by teaching classes related to Windows Server, Amazon Web Services (AWS), and networking and virtualization, as well as other classes when needed at a local community college. She enjoys speaking at public events and presented most recently at the RSA conference in 2019. Sara also enjoys technical editing and technical proofreading and has had the pleasure to work on a few projects doing this type of work.

When Sara is not working or writing, she enjoys spending time with her husband playing Destiny 2 and Elder Scrolls Online, and doing various arts and crafts. She also loves playing with her two pugs. Sara has a website, `www.saraperrott.com`, where you can see some of the things she has been up to. You can also follow her on Twitter (@PerrottSara) and Facebook (@PerrottSara).

Dedication

This book is dedicated to my husband, Chris Perrott. I couldn't have written it without your support, assistance, and encouragement. Thank you for always supporting me and for always encouraging me to be the best version of me.

Author's Acknowledgments

This book would not have been possible without a fantastic team of people supporting me every step of the way.

I would like to thank my agent, Carole Jelen, who helped me get my foot in the door with Wiley.

I would like to thank Steve Hayes at Wiley for reaching out to me to do the revision of this book. It really is a labor of love!

I would like to send my sincere gratitude to my amazing technical editor, Robert Shimonski. With your help, this book went from great to amazing!

Huge thanks and shout-out to my editor, Elizabeth Kuball, for your patience and knowledge, and for helping me keep on track with all the deadlines. I couldn't have done this without you!

Thanks to everyone at Wiley who helped this book become a reality from the drafts to the layouts and then the final printed product.

Publisher's Acknowledgments

Executive Editor: Steve Hayes
Project Editor: Elizabeth Kuball
Copy Editor: Elizabeth Kuball
Technical Editor: Robert Shimonski

Production Editor: Saikarthick Kumarasamy
Cover Image: © Gorodenkoff/Shutterstock